Socio-Environmental
Vulnerability Assessment
for Sustainable Management

Socio-Environmental Vulnerability Assessment for Sustainable Management

Editors

Szymon Szewrański
Jan K. Kazak

MDPI • Basel • Beijing • Wuhan • Barcelona • Belgrade • Manchester • Tokyo • Cluj • Tianjin

Editors
Szymon Szewrański
Wrocław University of
Environmental and Life Sciences
Poland

Jan K. Kazak
Wrocław University of
Environmental and Life Sciences
Poland

Editorial Office
MDPI
St. Alban-Anlage 66
4052 Basel, Switzerland

This is a reprint of articles from the Special Issue published online in the open access journal *Sustainability* (ISSN 2071-1050) (available at: https://www.mdpi.com/journal/sustainability/special_issues/Socio-environmental_Vulnerability_Assessment_Sustainable_Management).

For citation purposes, cite each article independently as indicated on the article page online and as indicated below:

LastName, A.A.; LastName, B.B.; LastName, C.C. Article Title. *Journal Name* **Year**, *Article Number*, Page Range.

ISBN 978-3-03943-651-4 (Hbk)
ISBN 978-3-03943-652-1 (PDF)

Contents

About the Editors

Szymon Szewrański Full Professor—Director of the Institute of Spatial Management, Wrocław University of Environmental and Life Sciences (Wrocław, Poland). He graduated in environmental engineering. Experienced in research on GIS, decision support systems, sustainable development, environmental and spatial management, environmental law and policy, and environmental impact assessments. He has practical experience in business intelligence and GIS combined systems, such as Tableau and ArcGIS/QGIS. He completed postgraduate study in spatial planning, urban designing, geographical information systems, and environmental law, as well as an Executive MBA. He is a certified tutor and also graduated in pedagogical studies.

Jan K. Kazak Associate Professor at the Institute of Spatial Management at the Wrocław University of Environmental and Life Sciences (Wrocław, Poland). He holds a PhD in environmental sciences and a DSc in engineering and technology sciences. His research interests cover, among other things, sustainable urban and regional management, urban adaptation to climate change, energy performance and policymaking process, decision support systems, and environmental impact assessments. He broadened his scientific and technical experience during internships in the Netherlands, Spain, Scotland (UK), and Australia. Since 2018, he has served as a Management Committee member in the European Network COST Action 17133: implementing nature-based solutions for creating a resourceful circular city. Currently, he is the leader of the leading research group, Sustainable Cities and Regions (https://upwr.edu.pl/en/research/leading-research-group/sustainable-cities-and-regions-scr), where he coordinates the work of an interdisciplinary team of urban planners, landscape architects, geographers, economists, real estate professionals, and IT experts to solve key problems in the functioning of socio-environmental systems.

Preface to "Socio-Environmental Vulnerability Assessment for Sustainable Management"

In the face of the global challenges encountered by human civilization in the 21st century, it is necessary to develop new study areas in the context of social and environmental systems. The ongoing parallel processes of urbanization, the aging of society, population growth, and increasing consumption interplay with threats caused by climate change, extreme weather events, and the depletion of resources. As climate-related hazards impact both human and environmental elements, there is a need to explore, analyze, and understand the vulnerability of socio-environmental systems. Therefore, adaptation to climate change as well as sustainable development require a knowledge-based approach and intelligent solutions for the integrated assessment of the state of the environment and society. The new complex approach incorporates studies on environmental impact assessment, human impact assessment, and adaptation to climate change; energy poverty and climate justice; the aging of society, environmental threats, and sustainability risk assessment; resilience assessment and mapping, supported by geospatial and artificial intelligence tools. The common framework should be attractive for a wide spectrum of specialists in the domains of environmental engineering, urban planning, geography, public policy, and other disciplines and cross-disciplinary fields.

Szymon Szewrański, Jan K. Kazak
Editors

Editorial

Socio-Environmental Vulnerability Assessment for Sustainable Management

Szymon Szewrański * and Jan K. Kazak

Institute of Spatial Management, Wrocław University of Environmental and Life Sciences, ul. Grunwaldzka 55, 50-357 Wrocław, Poland; jan.kazak@upwr.edu.pl
* Correspondence: szymon.szewranski@upwr.edu.pl

Received: 15 September 2020; Accepted: 18 September 2020; Published: 24 September 2020

Research on complex socio-environmental systems (also known as socio-ecological systems) has a long tradition in scientific considerations. Their theoretical basis was defined already in the 1970s [1]. In the following decades, the concept of a holistic consideration of complex interactions between social, economic, and environmental systems became permanently embedded in the paradigm of sustainable development. In the 21st century, in the face of global challenges faced by human civilization, it was necessary to develop new study areas in the context of research on social and environmental systems. Such concepts as resilience, integrated assessment of ecosystem services, socio-ecological system frameworks, coupled human and natural systems, and vulnerability frameworks appeared [2].

The conceptual framework that considers the vulnerability of complex human–environment systems proposed by Turner et al. (2003) is a rapidly developing research perspective [3]. In the context of the latest research, it has become crucial to seek answers to the question of who or what is most vulnerable to global environmental changes and where this vulnerability is the most crucial in terms of the geo-spatial point of view. Research conducted all over the world indicates that vulnerability to change is not a simple function of exposure to hazards, but also depends on the sensitivity and resilience of complex systems at a particular place and time [4].

Ongoing parallel processes of urbanization, aging of society, population growth, and increasing consumption interplay with threats caused by climate change, extreme weather events, and depletion of resources. As climate-related hazards impact both human and environmental elements, there is a need to explore, analyze, and understand the vulnerability of socio-environmental systems. Therefore, adaptation to climate change as well as sustainable development required a knowledge-based approach and intelligent solutions for integrated assessment of the state of the environment and society. With the purpose of illustrating the dynamics of research on the vulnerability of socio-environmental systems, we have conducted a query of the term "socio-environmental vulnerability" in the Web of Science and Scopus databases. Since 2006, we have observed a gradual increase in interest in the problem of socio-environmental vulnerability. In recent years, on average, six to eight works indexed in databases have appeared annually (Figure 1). Finally, we identified 53 articles in Web of Science and 57 in the Scopus database.

According to the classification of research areas in Web of Science, the largest number of works had been assigned to Environmental Sciences (and Studies) (30%), Geography (25%) and Public, Environmental, and Occupational Health (19%). A total of 11% of the papers were in the field of Green and Sustainable Science and Technology and 8% in Education Research. Some articles (4%) had been tagged as Biodiversity Conservation, Economics, Law, Social Sciences, Interdisciplinary, and Urban Studies. Regarding the Scopus classification, Social Sciences covered 58% of the papers. In Environmental Science, it was 47%, while in Medicine, it was about 25%. Some papers had been assigned to Agricultural and Biological Sciences (14%), Earth and Planetary Sciences, or Energy (11%). In the field of Business, Management, and Accounting, there were 9% of the publications. Decision

Sciences covered 5%, and Arts and Humanities covered 4%. Among all records, 5% were recognized as Engineering.

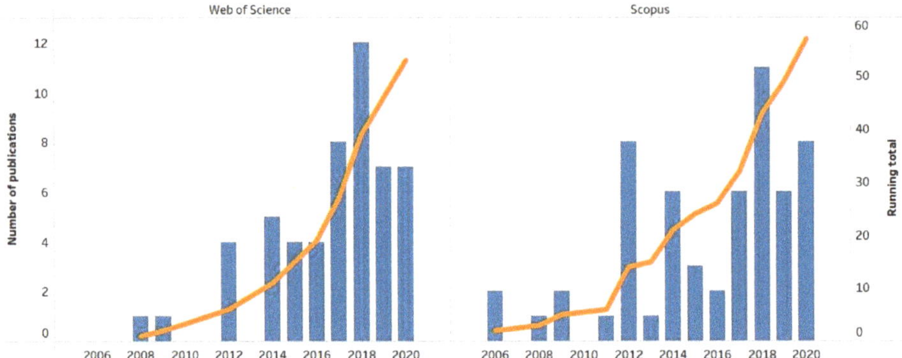

Figure 1. The annual and the running total numbers of scientific publications tagged literally with "socio-environmental vulnerability" in Web of Science and Scopus databases.

The interdisciplinary research conducted in this domain addresses such issues as evaluation of the quality of life in urban and suburban environments [5,6], issues related to public health protection [7,8], environmental injustice [9], engineering and infrastructure safety [10–12], energy security [13], income and environmental risk [14,15], hydrological and climate change risks [16–18], and mapping techniques [19–22]. All of them had been incorporated into socio-environmental vulnerability assessments, which present a broad perspective of this domain.

This Special Issue also explores cross-disciplinary approaches, methodologies, and applications of socio-environmental vulnerability assessments that can be incorporated into sustainable management. The volume collects 20 different points of view, which cover environmental protection and development, urban planning, geography, public policymaking, participation processes, and other cross-disciplinary fields (Figure 2).

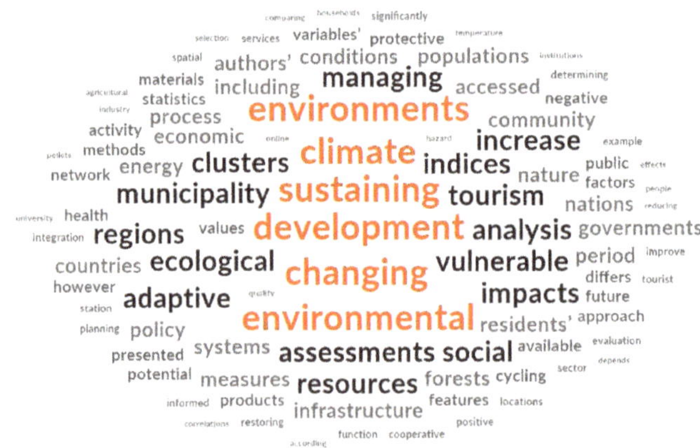

Figure 2. Word cloud generated from 20 publications collected in the Special Issue on "Socio-Environmental Vulnerability Assessment for Sustainable Management" in *Sustainability*.

The articles collected in this volume come from all over the world and seek answers to multidimensional questions.

- What is the current state of the world's environmental and social systems in local, regional, and national terms [23–25]?
- How can the resilience of environmental and social systems to changing climate or hydrological threats be assessed? Multidimensional and multi-factorial issues require new approaches and analytical tools. Hierarchical methods, clustering, and ranks have been successfully tested by the authors whose work is included in this volume [26–29].
- How does one implement sustainable development in practice? How can the principles of social participation and partnership support modernization processes [30,31]? Is it possible to formulate a progressive environmental and development policy [32]?
- What is the future of social–environmental systems? How will demographic change, particularly in an aging society, affect social and environmental resilience [33]? How should we supply ourselves with energy [34–36]? How should we shape our transport systems [37]? How can technical infrastructure and spatial management support the development of tourism in environmentally valuable areas [38]? Can we leverage our efforts by applying nature-based solutions [39,40]? How can open data and artificial intelligence support us [41,42]?

These and other questions will be answered in this Special Issue. We hope that dissemination of this broad spectrum to the scientific community will be helpful and may possibly open new horizons for future research.

Acknowledgments: The work has been created as a result of scientific activity conducted within the Leading Research Group: Sustainable Cities and Regions at the Wrocław University of Environmental and Life Sciences.

Conflicts of Interest: The authors declare no conflict of interest.

References

1. Holling, C.S. Resilience and Stability of Ecological Systems. *Annu. Rev. Ecol. Syst.* **1973**, *4*, 1–23. [CrossRef]
2. Pulver, S.; Ulibarri, N.; Sobocinski, K.L.; Alexander, S.M.; Johnson, M.L.; McCord, P.F.; Dell'angelo, J. Frontiers in socio-environmental research: Components, connections, scale, and context. *Ecol. Soc.* **2018**, *23*. [CrossRef]
3. Turner, B.L.; Kasperson, R.E.; Matsone, P.A.; McCarthy, J.J.; Corell, R.W.; Christensene, L.; Eckley, N.; Kasperson, J.X.; Luers, A.; Martello, M.L.; et al. A framework for vulnerability analysis in sustainability science. *Proc. Natl. Acad. Sci. USA* **2003**, *100*, 8074–8079. [CrossRef] [PubMed]
4. Turner, B.L.; Esler, K.J.; Bridgewater, P.; Tewksbury, J.; Sitas, J.N.; Abrahams, B.; Chapin, F.S.; Chowdhury, R.R.; Christie, P.; Diaz, S.; et al. Socio-Environmental Systems (SES) Research: What have we learned and how can we use this information in future research programs. *Curr. Opin. Environ. Sustain.* **2016**, *19*, 160–168. [CrossRef]
5. Roggero, M.A.; Ziglio, L.; Miranda, M. Socio-environmental vulnerability, analysis of health situation and indicators: Implications for the quality of life in the city of Sao Paulo. *Confin. Fr. Geogr. Fr. Geogr.* **2018**, *36*, 17. [CrossRef]
6. Kran, F.; Ferreira, F.P.M. Life quality in Palmas-To: An analysis through housing and urban environmental indicators. *Ambient. Soc.* **2006**, *9*, 123–141. [CrossRef]
7. Londe, L.R.; Marchezini, V.; da Conceição, R.S.; Bortoletto, K.C.; Silva, A.E.P.; dos Santos, E.V.; Reani, R.T. Impacts of socio-environmental disasters on public health: Studies of scenarios in the Brazilian states of Santa Catarina in 2008 and Pernambuco in 2010. *Rev. Bras. Estud. Popul.* **2015**, *32*, 537–562. [CrossRef]
8. Houghton, A.; Prudent, N.; Scott, J.E.; Wade, R.; Luber, G. Climate change-related vulnerabilities and local environmental public health tracking through GEMSS: A web-based visualization tool. *Appl. Geogr.* **2012**, *33*, 36–44. [CrossRef]
9. de Mendonca, M.B.; da-Silva-Rosa, T.; Monteiro, T.G.; de Souza Matos, R. Improving Disaster Risk Reduction and Resilience Cultures Through Environmental Education: A Case Study in Rio de Janeiro State, Brazil. In *Climate Change and Health*; Springer: Cham, Switzerland, 2016; pp. 279–295.

10. Silva, R.T. Integration of hydraulic infrastructure in metropolitan São Paulo. Prospects of change in a context of growing vulnerability. *Geogr. Helv.* **2011**, *66*, 92–99. [CrossRef]

11. Zanetti, V.B.; Junior, W.C.S.; De Freitas, D.M. A climate change vulnerability index and case study in a Brazilian Coastal City. *Sustainability* **2016**, *8*, 811. [CrossRef]

12. Goncalves, K.D.; Siqueira, A.S.P.; de Castro, H.A.; Hacon, S.D.S.; Gonçalves, K.S.; Siqueira, A.S.P.; de Castro, H.A.; Hacon, S.D.S. Indicator of socio-environmental vulnerability in the Western Amazon. The case of the city of Porto Velho, State of Rondonia, Brazil. *Cienc. Saude Coletiva* **2014**, *19*, 3809–3817. [CrossRef]

13. Westin, F.F.; dos Santos, M.A.; Martins, I.D. Hydropower expansion and analysis of the use of strategic and integrated environmental assessment tools in Brazil. *Renew. Sustain. Energy Rev.* **2014**, *37*, 750–761. [CrossRef]

14. Castillo-Oropeza, O.A.; Delgado-Hernández, E.; García-Morales, Á. Gentrification and disaster in La Condesa Zone. *Bitacora Urbano Territ.* **2018**, *28*, 35–43. [CrossRef]

15. Szewrański, S.; Świąder, M.; Kazak, J.K.; Tokarczyk-Dorociak, K.; van Hoof, J. Socio-Environmental Vulnerability Mapping for Environmental and Flood Resilience Assessment: The Case of Ageing and Poverty in the City of Wrocław, Poland. *Integr. Environ. Assess. Manag.* **2018**, *14*, 592–597. [CrossRef] [PubMed]

16. Barros, M.V.F.; Mendes, C.; de Castro, P.H.M. Social-environmental vulnerability to floods in urban areas of Londrina city-PR. *Confin. Fr. Geogr. Fr. Geogr.* **2015**, *24*, 21. [CrossRef]

17. Castro, B.; Filho, W.L.; Caetano, F.J.P.; Azeiteiro, U.M. Climate Change and Integrated Coastal Management: Risk Perception and Vulnerability in the Luanda Municipality (Angola). In *Climate Change Impacts and Adaptation Strategies for Coastal Communities*; Springer: Cham, Switzerland, 2018; pp. 409–426.

18. Claeys, C.; Giuliano, J.; Tepongning Megnifo, H.; Fissier, L.; Rouadjia, A.; Lizée, C.; Geneys, C.; Marçot, N. Interdisciplinary analysis of socio-environmental vulnerabilities: Urbanized cliffs on the Mediterranean coastline. *Nat. Sci. Soc.* **2017**, *25*, 241–254. [CrossRef]

19. Johnson, F.A.; Hutton, C.W.; Clarke, M.J. The Socio-Environmental Vulnerability Assessment Approach to Mapping Vulnerability to Climate. In *Solutions to Climate Change Challenges in the Built Environment*; GeoData Institute, School of Geography, University of Southampton Highfield: Southampton, UK, 2012; pp. 283–301.

20. Mavromatidi, A.; Briche, E.; Claeys, C. Mapping and analyzing socio-environmental vulnerability to coastal hazards induced by climate change: An application to coastal Mediterranean cities in France. *Cities* **2018**, *72*, 189–200. [CrossRef]

21. Norman, L.M.; Villarreal, M.L.; Lara-Valencia, F.; Yuan, Y.; Nie, W.; Wilson, S.; Amaya, G.; Sleeter, R. Mapping socio-environmentally vulnerable populations access and exposure to ecosystem services at the U.S.–Mexico borderlands. *Appl. Geogr.* **2012**, *34*, 413–424. [CrossRef]

22. Ho, H.C.; Lau, K.K.-L.; Yu, R.; Wang, D.; Woo, J.; Kwok, T.C.Y.; Ng, E. Spatial variability of geriatric depression risk in a high-density city: A data-driven socio-environmental vulnerability mapping approach. *Int. J. Environ. Res. Public Health* **2017**, *14*, 994. [CrossRef]

23. Ponomarenko, T.; Nevskaya, M.; Marinina, O. An assessment of the applicability of sustainability measurement tools to resource-based economies of the commonwealth of independent states. *Sustainability* **2020**, *12*, 5582. [CrossRef]

24. Nyairo, R.; Machimura, T.; Matsui, T. A combined analysis of sociological and farm management factors affecting household livelihood vulnerability to climate change in rural burundi. *Sustainability* **2020**, *12*, 4296. [CrossRef]

25. Derlukiewicz, N.; Mempel-Śniezyk, A.; Mankowska, D.; Dyjakon, A.; Minta, S.; Pilawka, T. How do clusters foster sustainable development? An analysis of EU policies. *Sustainability* **2020**, *12*, 1297. [CrossRef]

26. Chaudhary, S.; Wang, Y.; Dixit, A.M.; Khanal, N.R.; Xu, P.; Yan, K.; Liu, Q.; Lu, Y.; Li, M. Eco-environmental risk evaluation for land use planning in areas of potential farmland abandonment in the high mountains of Nepal Himalayas. *Sustainability* **2019**, *11*, 6931. [CrossRef]

27. Dumieński, G.; Mruklik, A.; Tiukało, A.; Bedryj, M. The Comparative analysis of the adaptability level of municipalities in the nysa klodzka sub-basin to flood hazard. *Sustainability* **2020**, *12*, 3003. [CrossRef]

28. Li, R.; Han, R.; Yu, Q.; Qi, S.; Guo, L. Spatial heterogeneous of ecological vulnerability in arid and semi-arid area: A case of the Ningxia Hui autonomous region, China. *Sustainability* **2020**, *12*, 4401. [CrossRef]

29. Long, J.W.; Steel, E.A. Shifting perspectives in assessing socio-environmental vulnerability. *Sustainability* **2020**, *12*, 2625. [CrossRef]

30. Furmankiewicz, M.; Campbell, A. From single-use community facilities support to integrated sustainable development: The aims of inter-municipal cooperation in Poland, 1990–2018. *Sustainability* **2019**, *11*, 5890. [CrossRef]

31. Hsu, C.H.; Lin, H.H.; Jhang, S. Sustainable tourism development in protected areas of rivers and water sources: A case study of Jiuqu Stream in China. *Sustainability* **2020**, *12*, 5262. [CrossRef]

32. Akbar, I.; Yang, Z.; Han, F.; Kanat, G. The influence of negative political environment on sustainable tourism: A study of Aksu-Jabagly world heritage site, Kazakhstan. *Sustainability* **2020**, *12*, 143. [CrossRef]

33. Sobczak, E.; Bartniczak, B.; Raszkowski, A. Aging society and the selected aspects of environmental threats: Evidence from Poland. *Sustainability* **2020**, *12*, 4648. [CrossRef]

34. Besser, A.; Kazak, J.K.; Świader, M.; Szewrański, S. A customized decision support system for renewable energy application by housing association. *Sustainability* **2019**, *11*, 4377. [CrossRef]

35. Greinert, A.; Mrówczyńska, M.; Szefner, W. The use of waste biomass from thewood industry and municipal sources for energy production. *Sustainability* **2019**, *11*, 3083. [CrossRef]

36. Adynkiewicz-Piragas, M.; Miszuk, B. Risk analysis related to impact of climate change on water resources and hydropower production in the Lusatian Neisse River Basin. *Sustainability* **2020**, *12*, 5060. [CrossRef]

37. Jaszczak, A.; Morawiak, A.; Zukowska, J. Cycling as a sustainable transport alternative in polish cittaslow towns. *Sustainability* **2020**, *12*, 5049. [CrossRef]

38. Kulczyk-Dynowska, A.; Stacherzak, A. Selected elements of technical infrastructure in municipalities territorially connected with national parks. *Sustainability* **2020**, *12*, 4015. [CrossRef]

39. Pardela, Ł.; Kowalczyk, T.; Bogacz, A.; Kasowska, D. Sustainable green roof ecosystems: 100 years of functioning on fortifications—A case study. *Sustainability* **2020**, *12*, 4721. [CrossRef]

40. Peczkowski, G.; Szawernoga, K.; Kowalczyk, T.; Orzepowski, W.; Pokładek, R. Runoff and water quality in the aspect of environmental impact assessment of experimental area of green roofs in Lower Silesia. *Sustainability* **2020**, *12*, 4793. [CrossRef]

41. Czernecki, B.; Glogowski, A.; Nowosad, J. Climate: An R package to access free in-situ meteorological and hydrological datasets for environmental assessment. *Sustainability* **2020**, *12*, 394. [CrossRef]

42. Mrówczyńska, M.; Sztubecka, M.; Skiba, M.; Bazan-Krzywoszańska, A.; Bejga, P. The use of artificial intelligence as a tool supporting sustainable development local policy. *Sustainability* **2019**, *11*, 4199. [CrossRef]

Article

An Assessment of the Applicability of Sustainability Measurement Tools to Resource-Based Economies of the Commonwealth of Independent States

Tatyana Ponomarenko *, Marina Nevskaya * and Oksana Marinina *

Faculty of Economics, Saint Petersburg Mining University, 199106 Saint Petersburg, Russia
* Correspondence: stv@spmi.ru (T.P.); nevskaya_ma04@spmi.ru (M.N.); Marinina_OA@pers.spmi.ru (O.M.); Tel.: +7-911-295-9667 (O.M.)

Received: 31 May 2020; Accepted: 8 July 2020; Published: 10 July 2020

Abstract: The concept of sustainable development (SD) is aimed at ensuring public well-being for the present and future generations. Hundreds of methods have been proposed for assessing and comparing the sustainable development of countries and analyzing their contribution to the future of the world. When applied to resource-based economies (RBEs), assessment tools do not take into account the value and impact of mineral resources on SD indicators. The purpose of the study is to reveal the limitations of applying some tools by taking into consideration the specific features of RBEs. Research methods include a correlation analysis between gross national income (GNI) per capita and aggregated indices (the Sustainable Society Index (SSI), the Human Development Index (HDI), and the Environmental Performance Index (EPI)), a comparative analysis of these indices and mining companies' performance indicators. Object Eurasian RBEs were selected, but Norway was analyzed separately from the sample. The results of the study show that correlations between GNI per capita and SD indicators are heterogeneous. There is no statistically significant correlation between GNI per capita and SSI, a strong correlation with HDI, and a weak correlation with EPI. The EPI and SSI structures do not reflect the specific features of RBEs.

Keywords: sustainable development; assessment; indicators; resource-based economy; mineral resources

---

1. Introduction

Currently, the global community considers laying the groundwork for ensuring universal human well-being based on the principles of sustainability to be one of the highest priorities. Despite the fact that a lot of studies have been conducted over a long time [1–6], the concept of sustainable development (SD) has not yet been shaped into a scientific theory. SD is perceived as a fundamental principle or model for managing economic systems at macro, meso, and micro levels, as public administration ideology, or a new philosophy at the corporate level.

In this regard, the problem arises of sustainability assessment in terms of economic, environmental, and social indicators at the macro level (that of individual countries). An objective way of solving this problem is to carry out quantitative assessment based on a set of indicators or a combination of indicators in an integrated form. There are some well-known indicators which are used at the macro level to monitor economic, environmental, and social parameters, to develop strategies and programs for economic and social development, to identify indicators signaling problem areas, and to compare countries using different parameters.

The relevance of measuring sustainability in resource-based economies (RBEs) stems from the fact that their number and share in global GDP is growing. For example, at the end of the 20th century, 58 countries accounted for 18% of global GDP, whereas in 2011, 81 countries accounted for 26% of the global economy [7]. RBEs are characterized by: a) a high share of income from mining operations in their GDP and export structure, b) being largely dependent on tax revenues from the mining sector, c) the depletion of mineral resources, d) huge amounts of waste accumulated in the mining sector, e) significant anthropogenic impacts on the environment, and f) massive investments in environmental protection measures. These characteristics give rise to the following research questions:

Is there a correlation between the economic indicator (The gross national income (GNI) per capita) and a number of integrated sustainability indicators for RBEs?

What is the impact of the Russian mining industry on the natural and social environment?

How are the specific features of RBEs (such as the depletion of resources, huge amounts of mining waste, anthropogenic impact, and environmental costs) reflected in the most popular sustainability assessment methods?

The concept of sustainable development is still debatable and has dozens of definitions, which makes it impossible to develop a single conceptual framework that could be used in further research [6]. Many researchers and institutions are trying to expand the famous definition given by the Brundtland Commission by adding to it the specifics of either particular countries or activities [8]. For example, the two definitions given by the Brundtland Commission are as follows: the Sustainable Society Foundation (SSF) has extended the definition of Brundtland with a third sentence, as follows: "A sustainable society is a society that (a) meets the needs of the present generation; (b) does not compromise the ability of future generations to meet their own needs; and (c) in which each human being has the opportunity to develop itself in freedom within a well-balanced society and in harmony with its surroundings".[9]. As the components of sustainable development are complex, interconnected, interdependent, ambiguous, and difficult to assess, sustainable development can be called a global goal that is difficult to reach and the progress towards which is very slow. In 2002, [10] it was concluded that global economic, environmental, and social problems had not been solved. This was the beginning of a new wave of interest in this topic demonstrated by both developed and developing countries.

So far, three main approaches to building a theory of sustainable development have been developed: the ecosystem approach, the anthropocentric approach, and the triune approach. An analysis of these approaches has shown that while none of them is universally recognized, the triune approach is becoming predominant in SD research. This is a result of the activities of international organizations that are involved in monitoring and forecasting change in both the global economy and countries with different levels of economic development.

In its modern interpretation, the ecosystem approach is a viewpoint which is based on the priority of the environmental component and argues that sustainable development at both global and national levels means that the economy, population, and people's needs should grow within the resource and environmental limits of the planet or an individual country. This approach is based on the principles of environmental economics and the concept of strong sustainability [1,11–13].

An alternative to the ecosystem concept is the anthropocentric approach, which implies that economic growth is fostered by the development of human capital and technological advances, which weaken the impact of limited natural resources on the economy. This concept is based on the principles of neoclassical economics and the concept of weak sustainability. The main idea is that economic growth is independent of its impact on the environment [14]. Research and development (R&D) become an integral part of sustainable development strategies [15].

The concept of weak sustainability is based on the requirements specified by people as to the quality of the environment in order to satisfy their needs. However, this approach does not challenge the need for the coordinated development of the economy and environment, the rational use of resources and natural resource restoration, biodiversity conservation, and environmental safety [16]. At the same time, there is a point of view which states that there is no such thing as weak stability, since the current economic system can be characterized as unstable and weak stability will not result in any form of sustainable future [4]. The main arguments against the concept of weak sustainability are that innovative solutions are not always available [17], and that the use of environmentally friendly technologies often requires much more energy and resources if the whole production process is analyzed [4].

The anthropocentric approach to sustainable development laid the foundation for the concept of a green economy, which also has global and national (sectoral, territorial) components. On a global scale, UNEP experts define a green economy as being economic activities that improve human well-being and promote social justice while significantly reducing environmental risks and preventing environmental degradation [18]. On a national scale, the concept of a green economy is mainly applicable to developed European countries that are rich in innovations, investment, and human resources but lack natural resources, especially mineral ones, which they have to import.

In resource-based countries (those rich in natural resources but lacking innovations), including Russia, the concept of a green economy and its derivatives (green energy, green mining) are also considered to be components of sustainable development. However, they are mainly regarded as tools for improving economic performance and upgrading production facilities in terms of their environmental characteristics. As the primary industries of the Russian economy, including the mining industry, lag behind in upgrading their facilities and making them greener [19], mineral resources can be greatly underestimated. Moreover, there is a danger that environmentally unfriendly facilities will be moved to countries with a high mineral resource potential.

One of the main directions in the development of green economy ideas is the concept of a circular economy, which states that it is of bigger priority to rearrange material flows than improve production facilities [20]. Considering the fact that this concept rests on the ideas of resource cycles and industrial metabolism, we believe that it can be considered to be a compromise between the concepts of strong and weak sustainability. The economic transformation of the "take, make, waste" principle into the "take, make, reuse" principle corresponds to weak sustainability. Compliance with the principles of resource efficiency and zero waste corresponds to strong sustainability [21].

SD relying on circular economy ideas has in its foundation the following five principles (5 Rs):

- reducing energy and resource consumption,
- replacing non-renewable resources with renewable ones,
- recovering necessary components from waste that has been recycled,
- recycling waste,
- reusing products.

From an economic point of view, recycling can increase the material value through efficient resource processing, providing new opportunities for innovative companies, and having a positive impact on society and the environment [22]. The circular economy implies that there is enhanced control over the supply of natural resources which relies on renewable resources in order to protect and enhance natural capital. It also means that it is necessary to optimize consumption processes through reusing and recycling products. Moreover, it is aimed at identifying the sources of negative impact from production processes on the environment and preventing negative consequences in order to improve the efficiency of economic and environmental systems [21].

The ecosystem and anthropocentric approaches define the place and role that people and society play in sustainable development in different ways. In the strong sustainability model, society is seen as a source of needs that should be reasonably limited, while the weak sustainability model defines it as a source of human resources that need to be transformed into human capital for innovative development. The inclusion of people as a particular subject in the framework of sustainable development is embedded in the triune approach.

In the context of this approach, there are three main goals of sustainable development: environmental integrity, eco-efficiency, and environmental justice. The economic component of the triune approach is based on the Hicks-Lindahl concept of the maximum income stream, which can be produced provided that the total capital used for its production is maintained [23]. In this concept, it is not taken into account which kind of capital (human, natural, or material) needs to be maintained or increased. Thus, the two previously discussed concepts of strong and weak sustainability are taken into account.

The social component of the triune approach implies that cultural and social systems should be preserved, with material wealth justly distributed. The social component of SD is understood as development aimed at improving well-being, achieving greater social justice, and improving the quality of human and social capital to make it meet the principles of reproducibility, balance, and involvement [24].

The environmental component implies that biological systems should be preserved and their stability over time should be maintained. It can be said that the triune approach combines the concepts of strong and weak sustainability, complementing them with a social element (Figure 1).

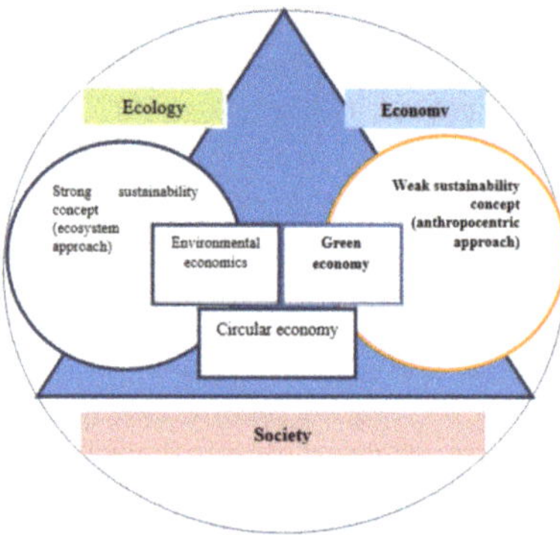

Figure 1. Relationships between sustainable development concepts within the triune concept. Source: compiled by the authors.

The triangle is a three-pronged approach with equivalent environmental, economic, and social goals for SD.

Circles are the two main concepts of sustainable development (strong sustainability and weak sustainability).

Rectangles are the main economic tools and principles that are characteristic of the concepts of strong sustainability (tools and principles of environmental Economics), weak sustainability (tools and principles of "green" economy), and the triune approach (tools and principles of circular economy).

It should be emphasized that the anthropocentric approach to SD is more characteristic of developed countries with high levels of development and income. Such countries are characterized by a big share of human and social capital (up to 80%), a small share of produced capital (less than 20%), and a very small share of natural capital (up to 5%), (Figure 2).

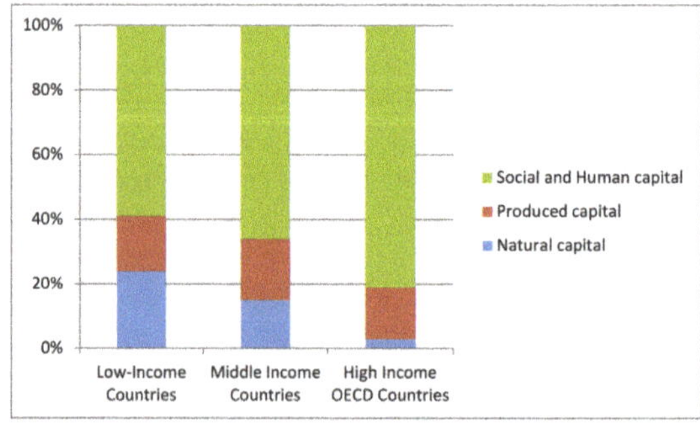

Figure 2. Capital structure by different levels of economic development [25].

The ecosystem approach based on the principles of natural resource restoration and environmental damage mitigation can be seen in resource-based countries. It should be noted that these countries are very different in terms of their levels of economic and institutional development. So far, no clear differentiation criteria have been developed that take into account the country's industry specialization which determines its place in the global economy and the specific conditions for its sustainable development.

Certain conflicts between the ecosystem and anthropocentric approaches have been identified in terms of how important natural resources, including mineral ones, are considered to be for countries' social and economic development. This is due to the fact that the role that mineral resources play at a macroeconomic level varies a lot. In some developed countries, the share of the mining industry in the country's GDP can amount to 10% (for example, in Sweden), whereas in resource-based countries, it can exceed 50% [26]. Resource-based countries have different quantities of mineral reserves. They vary in how developed their mining and mineral processing industries are and how big their impact on the environment is. Also, they differ in terms of multiplicative effects witnessed in related industries. As a result, they differ in their economic, environmental, and social indicators and are at different levels of sustainable development.

Among resource-based countries, the leading ones are developing countries with economic development levels that are both above the average and low. Most of them are former colonies and former Soviet countries, including Russia. In the case of these countries, which traditionally specialize in producing mineral and energy resources, a sustainability assessment is impossible without taking into account the specific impact of the natural resources sector on social and economic development and how this impact can be measured. However, as the issue of how to define sustainable development has not yet been solved, no framework for measuring it at the macro level has been created.

Studies by S. Kuznets in assessing national product and income as indicators of change marked the start of using macroeconomic indicators in assessing countries' levels of economic development. In the 1970s, the first attempts were made to switch from national product and national income to more complex indicators which would make it possible to take into account the influence of non-market forces on economic development at the macro level. It should be noted that S. Kuznets himself drew attention to the fact that it is possible to expand the range of assessment tools to include those that can take into account non-market activities and emphasized that assessments depend on the social values of time and place, the nature of family, industrial, and governmental organization, and the problems to which the figures are designed to apply [27]. Thus, in his approach, S. Kuznets emphasized the connection between social well-being and stakeholders' interests.

Sustainability measurement tools include about 500 indices. It should be noted that they hardly take into account the specialization of the economy or company. However, resource-based countries have several features which determine how a sustainable development concept will be designed and implemented:

- the exploitation of natural resources and the associated depletion of non-renewable mineral resources;
- the accumulation of mining waste and the human impact which reduces the efficiency of ecosystem services;
- the need to bear additional environmental costs (including both investment and operational costs).

In their studies, a number of authors have analyzed the most common and relevant indices [28–36]. However, there have been few attempts to systematize them. There is no single classification of indicators, and neither is there a universal method for assessing sustainability in an objective way. Such a situation, on the one hand, makes it more difficult to choose assessment methods and indicators. On the other hand, it indicates that these methods are constantly being improved (some indicators and indices become replaced with others, the number of assessment indicators is changing, and more countries are being included) but, unfortunately, not in keeping with real-life social, economic, and environmental processes.

All the assessment methods that have been developed so far can be classified on the basis of several characteristics:

- How the method is developed: systems of indices and indicators; aggregated indicators.
- What parameters are assessed: environmental; economic; social; social and environmental; environmental and economic.
- How information is accumulated: through open-access international databases (secondary); through the SNA and using statistical data (secondary); polls (secondary).
- How different indicators are combined: using ranking systems; using weighting factors.
- What methods are used: calculation; expert-based evaluation.
- What countries are covered: all countries; OECD countries; countries by continents; countries grouped by other parameters.
- What issues are covered [37]: environmental rankings; rankings reflecting how individual countries impact the environment on a global scale; rankings including both environmental and economic parameters; social development rankings; rankings based on sustainable development indicators; rankings reflecting progress in the green economy; ratings reflecting the quality of life and including the environmental component; other rankings including the environmental component.
- According to Hezri and Dovers [38] (p. 87), the main approaches to developing sustainability indicators are as follows: (1) extended national accounts, (2) bio-physical accounts, (3) weighted indices, (4) eco-efficiency and dematerialization approaches, and (5) indicator sets.

Among the systems of indicators (indices), the most common are World Development Indicators [39], the OECD system of indicators [40], and the system of indicators by the UN Commission on Sustainable Development (UN CSD). Their advantage is that they are focused on the SDGs. However, they rely on a lot of input data and cannot be used to compare individual countries. This is why it is difficult to apply them to sustainability assessment in real life.

The most common aggregated SD indicators include the Adjusted Net Savings (ANS) [41], the Genuine Progress Indicator (GPI), the Ecological Footprint, the Environmental Performance Index (EPI), the Legatum Prosperity Index, the Human Development Index (HDI), the Living Planet Index (LPI), the Sustainable Society Index (SSI), and others.

Sustainability assessment indicators should meet the following requirements:

Indicators should not be in conflict with the requirements of the SDGs and the indicator system developed by the UN CSD.

They should be universal to be applied at different levels (that of individual countries, industries, or companies) and they should adequately reflect the situation and processes.

Indicators should be obtained or calculated using relevant and reliable sources of information (statistical ones) that are publicly available.

Indicators should take into account the economic, environmental, and social aspects of SD and how they are interconnected.

2. Materials and Methods

The statistical correlation between the key economic indicator that characterizes the level of economic development of a country (GNI per capita) and indices that characterize the three aspects of SD was revealed as follows.

2.1. The Object of the Study

As an object of the study, Commonwealth of Independent States (CIS) resource-based countries were selected. The main criteria for studying RBEs were proposed by the IMF, the Natural Resource Governance Institute (NRGI) [42], and several researchers. Studies devoted to resource-based countries arose due to the fact that there was an ambiguous attitude to natural resources as, on the one hand, an essential asset for improving the economy and fostering economic growth, and, on the other hand, an asset resulting in the so-called resource curse [43,44]. In 2005, the latter concept was discussed by the UN and it was shown that an important role in this issue is played by institutions and how developed they are. Australia, Canada, and Scandinavian countries were used as examples. This was when a resource-based economy was given its definition which states that in such an economy, natural resources account for more than 10% of the country's GDP and 40% of exports. The IMF suggests that as a benchmark, the share of natural resources in total exports should be at least 25% [45]. An important characteristic of RBEs is that they are highly sensitive to external changes due to the volatility of prices for different types of raw materials [46]. In this study, two criteria (GDP and export shares) were selected as the main ones [47].

As objects, we select countries that correspond to two conditions at the same time:

(1) CIS countries with transition economies:

Source: https://www.imf.org/external/np/exr/ib/2000/110300.htm.

(2) countries with economies that can be characterized as resource-based.

Thus, the object of the study is the following six countries: Russia, Kazakhstan, Kyrgyzstan, Uzbekistan, Turkmenistan, and Azerbaijan.

2.2. Selection of Assessment Indicators

Among the indices that characterize SD at the macro level, SSI, HDI, and EPI were selected. The choice was made using Kostanza's classification by the method of measuring progress in SD [29] using each group of indicators.

As an integral indicator covering economic, social, and environmental aspects, the Sustainable Society Index (SSI), which has been calculated by the Sustainable Society Foundation (SSF) since 2006 once every two years, was selected. This index connects economic performance with social and environmental well-being. It has scores ranging from zero (minimum SD) to ten (maximum SD) and takes into account 21 indicators in three areas: human well-being, environmental well-being, and economic well-being [48]. The methodology for calculating the index varies depending on the set of indicators and their relative shares [49]. The Sustainable Society Index (SSI) aims to be a comprehensive and quantitative method to measure and monitor the health of coupled human-environmental systems at a national level worldwide. The SSI framework departs from a purely protectionist approach that would aim to maintain natural systems with minimal human impact [9].

As a social indicator, the Human Development Index (HDI) was selected, which is used to compare standards of living in different countries.

HDI is calculated as follows [50]:

$$HDI = \sqrt[3]{LEI \cdot EI \cdot II} \tag{1}$$

where LEI is the life expectancy index, EI is the education index, and II is the income index.

The index has a 30-year-long history; since 1990, HDI indicators for different countries have been published annually in the Human Development Reports by the United Nations Development Programme (UNDP).

As an environmental indicator, the Environmental Performance Index (EPI) was selected, which is used as a tool for quantitative assessment and comparative analysis to analyze the environmental situation and whether countries' environmental policies are effective. The purpose of the index is to reduce environmental impact and, as a result, the negative impact on human health, and to foster the vitality of environmental systems and the sustainable management of natural resources.

The EPI system ranks countries based on their performance in several categories, which are combined into two groups: ecosystem vitality and environmental health [51]. EPI rankings are published every two years and are calculated using the methodology developed by the Yale Center for Environmental Law and Policy together with a group of independent international experts. According to the developers, the higher the score, the more the country cares about the environment.

Based on the 2018 methodology, the index measures the country's performance using 24 indicators across 10 issue categories, which reflect various environmental aspects including those of ecosystem vitality [52]. They include protecting biodiversity, climate, and public health, the scope of economic activity and its impact on the environment, as well as the effectiveness of environmental policies [51]. The calculation method, the number of indicators, and the number of countries vary in different years. For example, the number of indicators was 22 in 2012, 19 in 2014, and 24 in 2018. The weighting factors for different groups and categories can also change. Following the adoption of the Global SDGs in 2015, the EPI indicators have been aligned with the goals.

The indicators described above were selected due to several reasons [51]:

- the most important factor for health is clean air. A study conducted in 2016 by the Institute for Health, Metrics and Evaluation showed that 2/3 of all diseases and deaths are connected with air quality;
- the quantity and quality of biomass (both in the sea and on land) are of great importance, which corresponds to the SDGs;
- many countries have improved air quality by reducing CO_2, NO, methane, and black carbon emissions, which is in line with the 2015 Paris Climate Agreement;
- over 20 years of research, experience has been accumulated which shows that when developing indices, two opposite patterns should be taken into account: environmental health, which improves with economic growth and development, and ecosystem vitality, which worsens with industrialization and the expansion of economic activity.

In view of this, countries with a growing and developing resource-based economy which depends on the mining industry can obviously rank very low.

2.3. Relevance Score

An analysis of the correlation between the selected assessment indices and GNI per capita in RBEs was performed. The per capita indicator was selected due to the need to ensure that income indicators for different RBEs can be compared. For the purpose of comparability, GDP values were adjusted and presented as normalized values. Normalization was performed for the entire set of countries in the sample. Then, we get a range of values from 0 to 1 on normalized GDP scales. The indexes characterize the country's place in the global rankings and estimate normalized scoring. The method of regression and correlation analysis makes it possible to test the hypothesis of the study which states that there should be a rather strong correlation between GNI per capita, human development, environmental indicators, and social sustainability since the country's income provides financial support for social and environmental programs. If this correlation is positive, there is no evidence of a "resource curse" in the macroeconomic sense, mineral resources in RBEs have a positive impact on the development of the economy, the environment, and society. If this correlation is negative, then according to the institutional interpretation of the "resource curse" [53], RBEs have an undeveloped institutional environment.

The method of morphological analysis was used to analyze the SD indices (SSI, HDI, EPI, Adjusted net savings (ANS), and GPI) and reveal how their structures reflect the specific features of resource-based countries. The initial data were obtained from official reports by the UN and the WB [54–59].

Using Russia as an example, changes in the indicators showing waste accumulation, investment and operational costs associated with environmental protection, and investment patterns in the mining sector were analyzed.

3. Results

The initial data were obtained from official reports by the UN and the WB (Table 1).

To find correct correlations between the indices and GNI per capita values, the latter were normalized and the indices were found.

Table 1. Gross national income (GNI) per capita in Commonwealth of Independent States (CIS) resource-based economies (RBEs), USD (thousands).

Country	2010	2011	2012	2013	2014	2015	2016	2017	2018
Kyrgyzstan	0.78	0.87	0.83	0.9	0.99	1.21	1.25	1.17	1.1
Uzbekistan	0.91	1.1	1.28	1.51	1.72	1.88	2.09	2.16	2.22
Turkmenistan	2.84	3.42	3.79	4.8	5.41	6.88	8.02	7.38	6.67
Azerbaijan	3.83	4.84	5.33	5.29	6.22	7.35	7.6	6.56	4.76
Kazakhstan	6.16	6.92	7.58	8.26	9.78	11.55	11.85	11.39	8.81
Russia	9.66	9.34	9.9	10.65	12.7	13.85	13.22	11.72	9.72

Normalization was performed for all the six countries in the sample for each year. The model for calculating GNI indices has the following form (I_{GNI}):

$$I_{GNI} = \frac{(GNI_f - GNI_{min})}{(GNI_{max} - GNI_{min})} \tag{2}$$

where GNI_f, GNI_{min}, GNI_{max} are factual, minimum, and maximum values of GNI per capita (thousand \$/person).

The results of calculating GNI per capita are shown in Table 2.

Table 2. Normalized GNI per capita in CIS RBEs.

Country	2010	2011	2012	2013	2014	2015	2016	2017	2018
Kyrgyzstan	0.000	0.000	0.000	0.000	0.000	0.000	0.000	0.000	0.000
Uzbekistan	0.015	0.027	0.050	0.063	0.062	0.053	0.070	0.094	0.130
Turkmenistan	0.232	0.301	0.326	0.400	0.377	0.449	0.566	0.589	0.646
Azerbaijan	0.343	0.469	0.496	0.450	0.447	0.486	0.530	0.511	0.425
Kazakhstan	0.606	0.714	0.744	0.755	0.751	0.818	0.886	0.969	0.894
Russia	1.000	1.000	1.000	1.000	1.000	1.000	1.000	1.000	1.000

HDI and SSI in CIS RBEs are shown in Tables 3 and 4.

Table 3. HDI in CIS RBEs.

Country	2010	2011	2012	2013	2014	2015	2016	2017	2018
Azerbaijan	0.732	0.731	0.736	0.741	0.746	0.749	0.749	0.752	0.754
Kazakhstan	0.764	0.772	0.782	0.791	0.798	0.809	0.808	0.813	0.817
Kyrgyzstan	0.636	0.639	0.649	0.658	0.663	0.666	0.669	0.671	0.674
Russia	0.78	0.789	0.797	0.803	0.807	0.813	0.817	0.822	0.824
Turkmenistan	0.673	0.68	0.686	0.691	0.696	0.701	0.706	0.708	0.71
Uzbekistan	0.665	0.672	0.681	0.688	0.693	0.696	0.701	0.707	0.71

Table 4. SSI components in CIS RBEs.

Country	2010			2012			2014			2016		
	Human Well-Being Index	Environmental Well-Being Index	Economic Well-Being Index	Human Well-Being Index	Environmental Well-Being Index	Economic Well-Being Index	Human Well-Being Index	Environmental Well-Being Index	Economic Well-Being Index	Human Well-Being Index	Environmental Well-Being Index	Economic Well-Being Index
Azerbaijan	6.9	4.9	5.6	6.9	5.5	5.7	7.1	5.1	5.9	7.3	3.9	5.7
Kazakhstan	7.4	2.1	4.2	7.4	2.8	4.2	7.5	2.5	4.6	7.6	2.7	5.3
Kyrgyzstan	6.9	4.9	3.1	7	6.2	3.5	7.1	4.9	3.2	7	4.9	2.2
Russia	6.9	2.3	5.2	6.8	2.5	5.1	6.8	2.3	5.4	6.9	2.5	5.5
Turkmenistan	5.7	1.5	4.6	5.6	1.7	4.6	5.8	1.8	4.9	5.8	1.7	4.9
Uzbekistan	6.4	5.0	3.7	6.2	5.3	3.8	6.2	5.1	3.9	6.6	5.1	4

The average SSI values were calculated as the arithmetic mean due to the absence of this value in the initial data (Table 5).

Table 5. Average SSI in RBEs.

Country	Average SSI Values			
	2010	2012	2014	2016
Azerbaijan	5.78	6.03	6.03	5.63
Kazakhstan	4.58	4.80	4.87	5.20
Kyrgyzstan	5.00	5.57	5.07	4.70
Russia	4.78	4.80	4.83	4.97
Turkmenistan	3.92	3.97	4.17	4.13
Uzbekistan	5.04	5.10	5.07	5.23

EPI in CIS RBEs are shown in Table 6.

Table 6. EPI in CIS RBEs.

Country	EPI				Average SSI			
	2010	2012	2014	2016	2010	2012	2014	2016
Azerbaijan	59.1	43.11	55.47	83.78	5.78	6.03	6.03	5.63
Kazakhstan	57.3	32.94	51.07	73.29	4.58	4.80	4.87	5.20
Kyrgyzstan	59.7	46.33	40.63	73.13	5.00	5.57	5.07	4.70
Russia	61.2	45.43	53.45	83.52	4.78	4.80	4.83	4.97
Turkmenistan	38.4	31.75	–	70.24	3.92	3.97	4.17	4.13
Uzbekistan	42.3	32.24	43.23	63.67	5.04	5.10	5.07	5.23

Figures 3–5 show the correlations between 1) GNI per capita and HDI, 2) GNI per capita and SSI, and 3) GNI per capita and EPI.

The sample of RBEs is characterized by a strong correlation between GNI per capita and HDI, which is almost linear. This may be due to the fact that GNI per capita is included in the HDI calculation model. It is very possible that for these countries the contribution of GNI per capita to the HDI model is more important than that of other components (such as the education index and the life expectancy index). This is why a lower GNI value correlates with a lower HDI value.

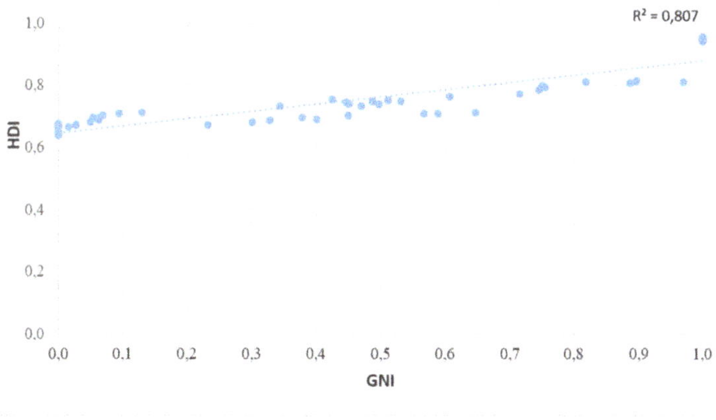

Figure 3. The correlation between GNI and HDI in CIS RBEs.

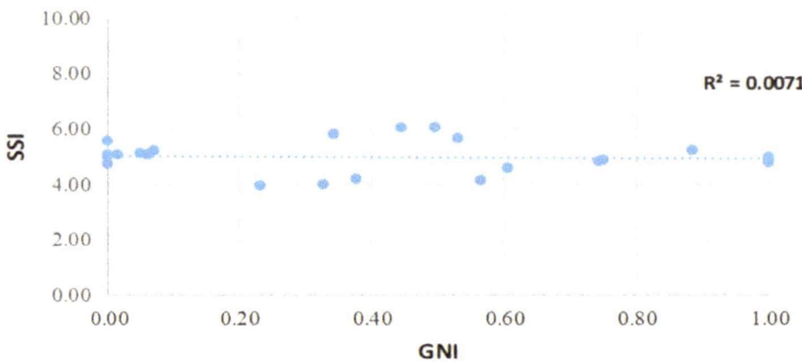

Figure 4. The correlation between GNI and SSI in CIS RBEs.

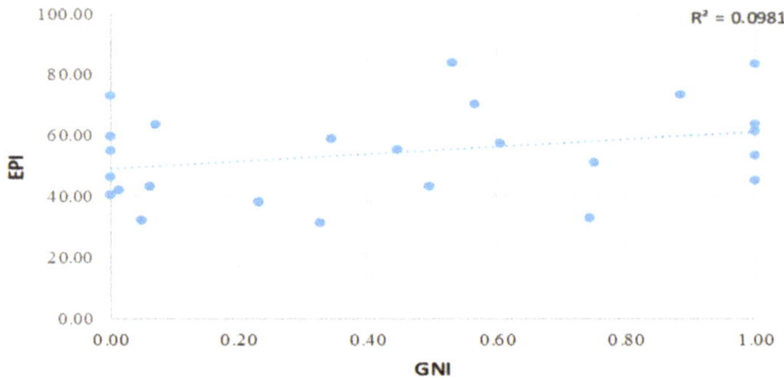

Figure 5. The correlation between GNI and EPI in CIS RBEs.

The results of finding the strength of the correlation between GNI per capita and SSI and that between GNI per capita and the SSI components are heterogeneous. For the group of countries, there is almost no correlation with SSI. SSI includes 21 indicators that can change their values in different directions, which leads to a complete lack of any correlation. It should be noted that SSI values for previous years are recalculated as the methodology changes, which ensures that the results are comparable.

The values of the correlation ratios between GNI per capita and EPI show a high variability with a low degree of correlation. EPI is based on 24 indicators characterizing completely different aspects of environmental health and ecosystem vitality. Such a number of indicators and the fact that changes are frequently made to the methodology result in a rather weak correlation.

Based on the statistical data for Russia, an analysis of the environmental impact of the mining sector was carried out, which showed that it strongly influences all elements of the natural environment including the atmosphere, biosphere, hydrosphere, and lithosphere. In terms of impact complexity, this sector ranks first among all other sectors of the economy. However, individual elements of the environment are impacted in different ways. According to the report titled Basic Environmental Protection Indicators which was issued in 2019 by the Russian Federal State Statistics Service (Rosstat) [60], the mining sector accounted for 10.8% of the total amount of greenhouse gas emissions (whereas the share of the energy sector was 78.8%). The volume of effluent did not exceed 6% of the total amount. At the same time, the mining sector is a leader in air pollutant emissions (with a share of more than 28%) and the generation of waste. A bar chart reflecting the situation with waste generation in Russia is shown in Figure 6.

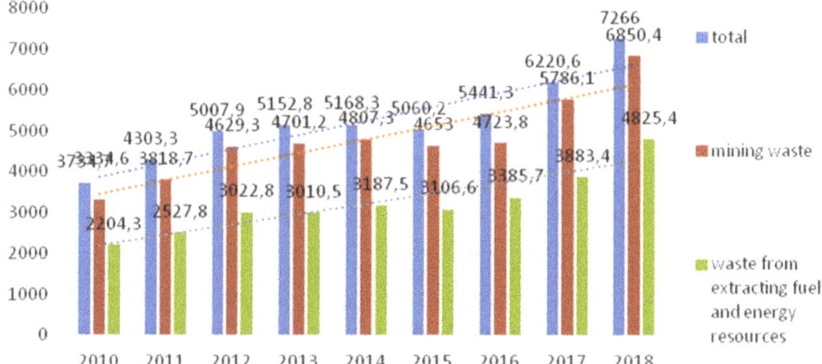

Figure 6. Waste generation in Russia, mln tons (compiled by the authors according to Rosstat data on environmental issues [60]).

More than 90% of waste generated in Russia is accounted for by the mining sector, of which about 70% is waste associated with the extraction of fuel and energy resources. More than half of the waste produced in the mining sector is recycled (Figure 7), and the rest goes to spoil tips or tailings dams, which can act as secondary sources of pollution.

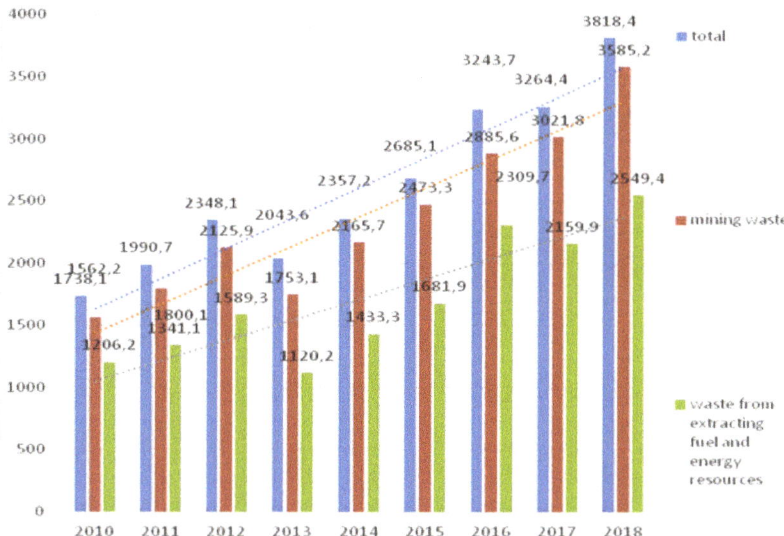

Figure 7. Waste recycling in the Russian Federation, mln tons (compiled by the authors according to Rosstat data on environmental issues [60]).

The complex nature of the environmental impact results in a high share of investment and operational costs associated with environmental protection in the mining sector. In 2017, out of the total amount of investment in environmental protection, which was 154.0 billion rubles, the mining sector accounted for more than 30%. In 2018, this share slightly decreased and amounted to 23.1% of 157.6 billion rubles [60]. Over the last 8 to 10 years, there has been an increase in both investment and operating costs associated with environmental protection in Russia (Figures 8–11).

Compared to 2012, there has been a 33% growth in operational costs associated with environmental protection. The main items are effluent disposal and treatment, waste management, air quality protection, and climate change prevention. Over a ten-year period, the amount of investment in fixed assets associated with environmental protection almost doubled, and the main expenses were aimed at protecting the atmosphere and water resources. The share of investment in fixed assets in the GDP of the Russian Federation in 2017 and 2018 was 0.16% and 0.15%, respectively, and the share of operational costs in GDP did not exceed 0.3%, which is clearly not enough if we compare it with that demonstrated by developed countries. For example, the share of environmental control costs is approximately 1.47% of GDP in the United States, 1.5% of GDP in Germany, and 1.25% of GDP in Japan [61].

More than 50% of investment was aimed at protecting the atmosphere; a little more than 20% of investment was allocated for protecting water resources; a little more than 10% of investment was invested in waste management; less than 10% of investment was invested in land conservation. We believe that this can be explained by the fact that Russian legislation concerning mineral resource production considers mining waste to be a potential source of raw materials, and the activities associated with its use are qualified as activities related to the use of mineral resources.

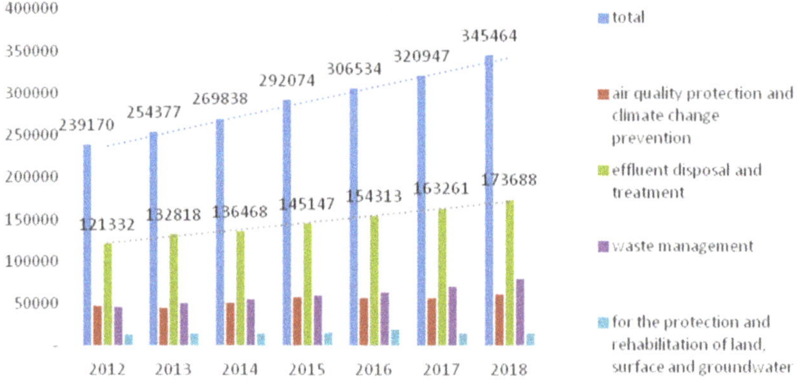

Figure 8. Operational costs associated with environmental protection in Russia, mln RUB (compiled by the authors according to Rosstat data on environmental issues [60]).

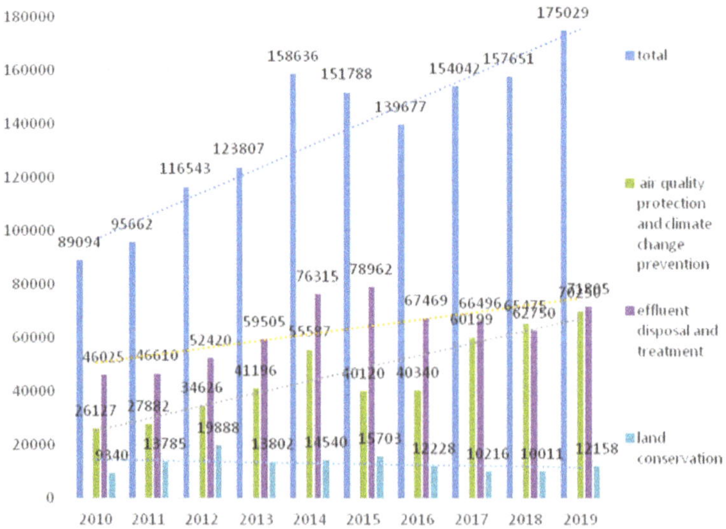

Figure 9. Investments in fixed assets associated with environmental protection, mln RUB (compiled by the authors according to Rosstat data on environmental issues [60]).

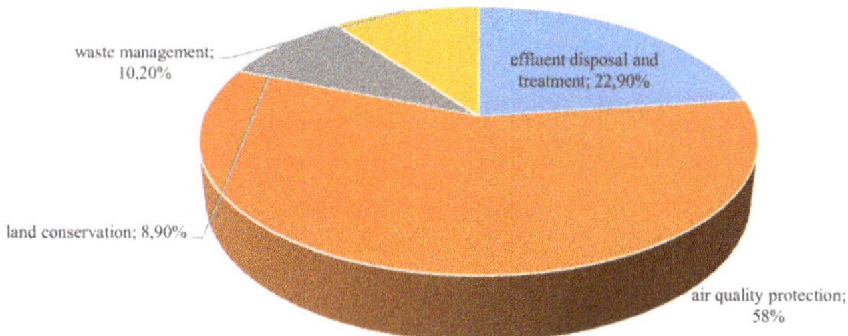

Figure 10. The structure of investment costs associated with environmental protection in the Russian mining sector in 2018 (compiled by the authors according to Rosstat data on environmental issues [60]).

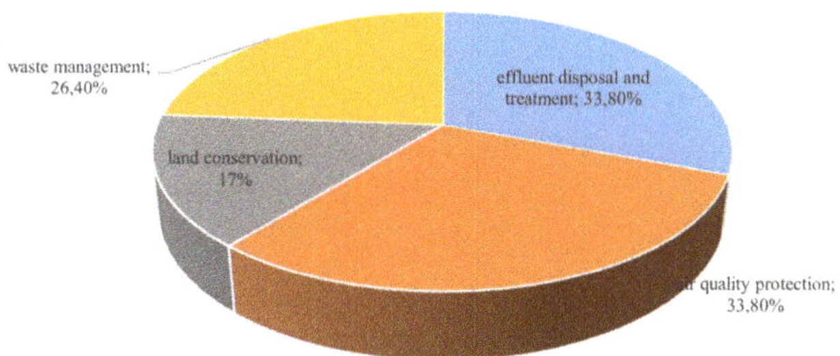

Figure 11. The structure of operational costs associated with environmental protection in the Russian mining sector in 2018 (compiled by the authors according to Rosstat data on environmental issues [60]).

4. Discussion

Results obtained by other researchers [30] who tested correlations between the SSI components for 151 countries over the years 2006, 2008, 2010, and 2012 prove a strong positive correlation between the three indices and a strong negative correlation between the human well-being index and the environmental well-being index.

An analysis of the SSI and EPI structures showed that they contain some indicators reflecting such aspects as agriculture, biodiversity, climate, water resources and air quality, energy use, water, and energy conservation. Issue-specific indicators reflect negative impact on the environment by categories: heavy metals, air quality and pollution, climate, and energy [61–63] In issue-specific indicators, greenhouse gases are separately analyzed as a negative factor affecting the climate. Russia, Kazakhstan, and Turkmenistan have been characterized by low scores over a long time [64,65] Russia takes 4th place in the world [66] after the USA, China, and India) in terms of greenhouse gas emissions, most of which are accounted for by the fuel and energy sector. Within the Eurasian Economic Union (EAEU), Russia and Kazakhstan account for 97% of all pollutant emissions [48].

EPI and SSI do not contain indicators reflecting the situation with mineral resources as a component of natural capital or indicators for depletion assessment. Only the methodologies for calculating adjusted net savings (ANS) and the Genuine Progress Indicator (GPI) reflect the cost of mineral resource depletion. However, they do not take into account any indicators characterizing the impact of the natural resources sector in a resource-based country on ecosystems.

None of the indices discussed above reflect the amount of mining waste and its impact on the environment. It is possible that this impact (from both primary and secondary sources of pollution) is included in the indicators reflecting air and water quality but it is not considered as an individual component.

RBEs make the biggest contribution to environmental impact, as a result of which environmental costs are increasing. This has an impact on social well-being, but environmental costs (both investment and operational ones) are not reflected in SD indicators. This means that conclusions cannot be made about the contribution of the mining sector to either pollution or environmental protection.

As has been shown, there is no single universally accepted theoretical and methodological approach to the analysis and assessment of sustainable development. All the indicators which have been developed have their drawbacks.

First, when using systems of issue-specific indicators, the problem arises of selecting indicators that adequately reflect the processes of social, economic, and environmental development, and can be used for their quantitative assessment. Using quantitative indicators based on the SNA can be problematic due to the fact that there are differences between countries' statistical systems and they may present not enough data. Some of the indicators proposed by the UN CSD are mainly focused on assessing the social component of SD, which does not make it possible to analyze the correlation between economic development, its sources, and results in the social and environmental spheres.

Second, when using aggregated indices for comparing or ranking countries, the problem arises of how they or their indicators can be compared. With the development of calculation methods, the number of indicators and the calculation models are changing, which makes it almost impossible to use such indices to reveal trends.

Third, developing qualitative indicators (either issue-specific or aggregated ones) is associated with the use of information that can only be obtained through surveys, which reduces the reliability of the results. Qualitative indicators make it possible to evaluate many social factors, but at the same time, the risk of subjectivity increases (for example, when choosing and substantiating weighting factors that characterize the contribution of various indicators to the total result and can vary). Another problem is how qualitative indicators can be transformed into quantitative ones (for example, in the case of GPI).

Fourth, none of the methods discussed takes into account the specific features of RBEs and their impact on SD, and neither do they take into account the national characteristics of natural resource consumption or the country's place in the international division of labor. RBEs are characterized by the depletion of mineral resources, the accumulation of mining waste, human impact on the environment, and significant investments in environmental protection. This puts resource-based countries at a disadvantage in comparison with countries importing natural capital.

Fifth, aggregated indicators cannot serve as objective criteria for assessing SD in RBEs since they poorly reflect their specific features, including the accumulation of mining waste and human impact on the environment. Also, they do not reflect the value of mineral resources for the economy and the social development of present and future generations, and neither do they cover an aspect such as resource depletion.

Sixth, of all the considered methods for assessing SD, only ANS and GPI contain information on the depletion of mineral resources.

5. Conclusions

It has been found that there is no statistically significant correlation between GNI per capita and SSI for the group of countries discussed. Also, there is a moderate negative correlation between GNI per capita and environmental sustainability, a fairly strong positive correlation with economic sustainability, and a weak correlation with social sustainability. The sample is characterized by a strong correlation between GNI per capita and HDI, which is almost linear. The values of the correlation ratios between GNI per capita and EPI show a high variability with a low degree of correlation. EPI is based on 24 indicators characterizing completely different aspects of environmental health and ecosystem vitality. Such a number of indicators and the fact that changes are frequently made to the methodology result in a rather weak correlation.

An analysis of the SSI and EPI structures showed that they contain some indicators reflecting such aspects as agriculture, biodiversity, climate, water resources and air quality, energy use, water, and energy conservation. Issue-specific indicators reflect negative impact on the environment by categories: heavy metals, air quality and pollution, climate, and energy. In issue-specific indicators, greenhouse gases are separately analyzed as a negative factor affecting the climate.

EPI and SSI do not contain indicators reflecting the situation with mineral resources as a component of natural capital or indicators for depletion assessment. Only the methodologies for calculating adjusted net savings (ANS) and the Genuine Progress Indicator (GPI) reflect the cost of mineral resource depletion. However, they do not take into account any indicators characterizing the impact of the natural resources sector in a resource-based country on ecosystems. It has been revealed that SD indices fail to reflect the specific features of RBEs (such as resource depletion, mining waste accumulation, human impact on the environment, and environmental costs).

None of the indices discussed reflect the complex impact of the mining sector, the amount of mining waste and its impact on the environment, and investments in environmental protection. They only reflect the situation in social, environmental, and economic spheres, which makes it impossible to evaluate the contribution of the mining industry both to pollution and natural resource restoration and means that specific studies are required to analyze these aspects.

The mining sector strongly influences all elements of the natural environment. In terms of impact complexity, this sector ranks first among all other sectors of the economy. The mining sector is a leader in air pollutant emissions and the generation of waste. More than 90% of waste is accounted for by the mining sector, more than half of the waste produced in the mining sector is recycled, and the rest goes to spoil tips or tailings dams, which can act as secondary sources of pollution.

The main environmental expenses were aimed at protecting the atmosphere and water resources. The share of investment in the GDP of the Russian Federation was 0.15–0.16% which is clearly not enough to compared to the level for developed countries, which is1.25–1.5% of GDP. RBEs make the biggest contribution to environmental impact, as a result of which environmental costs are increasing. This has an impact on social well-being.

Author Contributions: Conceptualization, T.P. and M.N.; methodology, T.P. and M.N.; software, M.N.; validation, T.P., M.N., and O.M.; formal analysis, M.N. and O.M.; investigation, T.P. and M.N.; resources, O.M.; writing—original draft preparation, T.P., M.N., and O.M.; writing—review and editing, M.N. and O.M.; visualization, M.N. and O.M.; funding acquisition, T.P., M.N., and O.M. All authors have read and agreed to the published version of the manuscript.

Funding: This research was funded by RFBR and MCESSM, grant number № 19–510–44013\19.

Acknowledgments: Special thanks to Natalia Savelyeva for the administrative and technical support.

Conflicts of Interest: The authors declare no conflict of interest involving the results.

References

1. Costanza, R.; Daly Herman, E. Natural Capital and Sustainable Development. *Conserv. Biol.* **1992**, *6*, 37–46. [CrossRef]
2. Tolba, M.K. Perceptions and attitudes. *Sav. Our Planet.* **1992**, 221–230. [CrossRef]
3. Meadows, D.H.; Randers, J.; Meadows, D.L.; Behrens, W.W. *The Limits to Growth: A Report for the Club of Rome's Project on the Predicament of Mankind*; Universe Books: New York, NY, USA, 1972; p. 211.
4. Biely, K.; Maes, D.; Passel, S.V. The idea of weak sustainability is illegitimate Environment. *Dev. Sustain.* **2018**, *20*, 223–232. [CrossRef]

5. Adrianov, V. Konceptual'nye podhody k razrabotke strategii ustojchivogo razvitiya ekonomiki Rossii do 2030 g. *Obshchestvo Ekon.* **2016**, *7*, 5–36.

6. Sharachchandra, L. Sustainable Development: A Critical Review. *World Dev.* **1991**, *19*, 607–621.

7. Pozdnyakova, T.S. Ocenka posledstvij syr'evoj zavisimosti dlya ekonomiki Rossii. *Akad. Vestn.* **2014**, *2*, 424–432.

8. Majburov, I. Ustojchivoe razvitie kak koevolyucionnyj process. *Obshchestvo Ekon.* **2004**, *4*, 124–143.

9. Saisana, M.; Philippas, D. Sustainable Society Index (SSI): Taking societies' pulse along social, environmental and economic issues. The Joint Research Centre audit on the SSI. *Environ. Impact Assess. Rev.* **2012**, *32*, 94–106. Available online: http://composite-indicators.jrc.ec.europa.eu (accessed on 19 May 2020).

10. Ikonnikova, O.V. Osnovnye podhody k opredeleniyu ponyatiya «Ustojchivoe razvitie sel'skih territorij». *Probl. Sovrem. Ekon.* **2012**, *1*, 349–352.

11. Daly, H.E. *Beyond Growth: The Economics of Sustainable Development*; Beacon Press: Boston, MA, USA, 1996.

12. Davies, G.R. Appraising weak and strong sustainability: Searching for a middle ground. *Cons. J. Sustain. Dev.* **2013**, *10*, 111–124.

13. Ekins, P.; Simon, S.; Deutsch, L.; Folke, C.; De Groot, R. A framework for the practical application of the concepts of critical natural capital and strong sustainability. *Ecol. Econ.* **2003**, *44*, 165–185. [CrossRef]

14. Pearce, D. Green Economics. *Environ. Values* **1992**, *1*, 3–13. [CrossRef]

15. William, C.C.; Nancy, M. Dickson. Sustainability science: The emerging research program. *Proc. Natl. Acad. Sci. USA* **2003**, *100*, 8059–8061. [CrossRef]

16. Starikova, E.A. Sovremennye podhody k traktovke koncepcii ustojchivogo razvitiya. Vestnik RUDN. *Seriya Ekon.* **2017**, *1*, 7–17.

17. Alcott, B. The sufficiency strategy: Would rich-world frugality lower environmental impact? *Ecol. Econ.* **2008**, *64*, 770–786. [CrossRef]

18. Dok. YUNEP. Zelenaya Ekonomika: Sprav. Available online: http://old.ecocongress.info/5_congr/docs/doklad.pdf (accessed on 19 May 2020).

19. Budanov, I.A.; Terent'ev, N.E. Problemy i napravleniya tekhnologicheskoj modernizacii metallurgicheskogo kompleksa Rossii v kontekste «Zelenogo» rosta ekonomiki. Nauchnye trudy: Institut narodnohozyajstvennogo prognozirovaniya. *RAN* **2017**, *15*, 76–91.

20. Batova, N.; Sachek, P.; Tochitskaya, I. On the Path to Green Growth: A Window of Opportunity for a Circular Economy. BEROC Green Economy Policy Paper Series. 2018. Available online: http://www.beroc.by/webroot/delivery/files/GE_1.pdf (accessed on 19 May 2020).

21. Pakhomova, N.V.; Richter, K.K.; Vetrova, M.A. Transition to circular economy and closedloop supply chains as driver of sustainable development. *St Petersburg Univ. J. Econ. Stud.* **2017**, *33*, 244–268. [CrossRef]

22. Di Maio, F.; Rem, P. A Robust Indicator for Promoting Circular Economy through Recycling. *J. Environ. Prot.* **2015**, *6*, 1095–1104. [CrossRef]

23. YAkovleva-CHernysheva, A.Y.U. Teoreticheskie osnovy upravleniya ustojchivym razvitiem kommercheskoj organizacii. *Nauchno-prakticheskij zhurnal Upr. Ekon. XXI Veke* **2015**, *2*, 5–11.

24. Kanaeva, O.A. Social'nye imperativy ustojchivogo razvitiya. *Vestn. SPbGU Ekon.* **2018**, *34*, 26–58. [CrossRef]

25. *Where Is the Wealth of Nations? Measuring Capital for the 21st Century*; The World Bank: Washington, DC, USA, 2006; Available online: http://documents.worldbank.org/curated/en/287171468323724180/pdf/348550REVISED0101Official0use0ONLY1.pdf (accessed on 19 May 2020).

26. Kaznacheev, P. *Prirodnaya Renta i Ekonomicheskij Rost*; RANHiGS: Saint Petersburg, Russia, 2013; p. 102.

27. Abramovitz, M. Simon Kuznets (1901–1985). *J. Econ. Hist.* **1986**, *46*, 241–246. [CrossRef]

28. Kenny, D.C.; Costanza, R.; Dowsley, T.; Jackson, N.; Josol, J.; Kubiszewski, I.; Narulla, H.; Sese, S.; Sutanto, A.; Thompson, J. Australia's Genuine Progress Indicator Revisited (1962–2013). *Ecol. Econ.* **2019**, *158*, 1–10. [CrossRef]

29. Costanza, R.; Hart, M.; Posner, S.; Talberth, J. *Beyond GDP: The Need for New Measures of Progress*; THE PARDEE PAPERS No. 4; Boston University: Boston, MA, USA, 2009; Available online: https://www.oecd.org/site/progresskorea/globalproject/42613423.pdf (accessed on 19 May 2020).

30. Kaivo-oja, J.; Panula-Ontto, J.; Vehmas, J.; Luukkanen, J. Relationships of the dimensions of sustainability as measured by the sustainable society index. *Int. J. Sustain. Dev. World Ecol.* **2013**, *21*, 39–45. [CrossRef]

31. Porter, M.E.; Stern, S.; Green, M. *Social Progress Index 2015*; Social Progress Imperative: Washington, DC, USA, 2015; Available online: https://www.socialprogress.org/assets/downloads/resources/2015/2015-Social-Progress-Index-Exec-Summary.pdf (accessed on 19 May 2020).

32. Lior, N.; Radovanović, M.; Filipović, S. Comparing sustainable development measurement based on different priorities: Sustainable development goals, economics, and human well-being—Southeast Europe case. *Sustain. Sci.* **2018**, *13*, 973–1000. [CrossRef]

33. Blanc, I.; Friot, D.; Margni, M.; Jolliet, O. Towards a new index for environmental sustainability based on a DALY weighting approach. *Sustain. Dev.* **2008**, *16*, 251–260. [CrossRef]

34. Kaly, U.; Briguglio, L.; McLeod, H.; Schmall, S.; Pratt, C.; Pal, R. *Environmental Vulnerability Index (EVI) To Summarise National Environmental Vulnerability Profiles*; SOPAC Technical Report 275; SOPAC: Suva, Fiji, 1999; p. 66.

35. Böhringer, C.; Jochem, P.E.P. Measuring the Immeasurable—A Survey of Sustainability Indices Discussion. *Ecol. Econ.* **2007**, *63*, 1–8. Available online: ftp://ftp.zew.de/pub/zew-docs/dp/dp06073.pdf (accessed on 19 May 2020). [CrossRef]

36. Angelakoglou, K.; Gaidajis, G.A. Conceptual Framework to Evaluate the Environmental Sustainability Performance of Mining Industrial Facilities. *Sustainability* **2020**, *12*, 2135. [CrossRef]

37. Alekseeva, N.N.; Arshinova, M.A.; Bancheva, A.I. Polozhenie Rossii v mezhdunarodnyh ekologicheskih rejtingah. Ekologiya i bezopasnost' zhiznedeyatel'nosti. Vestnik. *RUDN* **2018**, *26*, 134–152. [CrossRef]

38. Hezri, A.; Dovers, S. Sustainability indicators, policy and governance: Issues for ecological economics. *Ecol. Econ.* **2006**, *60*, 86–99. [CrossRef]

39. World Development Indicators. Available online: https://mydata.biz/ru/catalog/databases/wdi (accessed on 19 May 2020).

40. OECD. Key Environmental Indicator. 2008. Available online: https://www.oecd.org/env/indicators-modelling-outlooks/37551205.pdf (accessed on 19 May 2020).

41. World Bank Open Data. Available online: http://docs.cntd.ru/document/gost-r-7-0-5-2008 (accessed on 19 May 2020).

42. Natural Resource Governance Institute (NRGI). *Resource Governance Index*; NRGI: New York, NY, USA; Washington, DC, USA, 2014; Available online: https://resourcegovernance.org (accessed on 19 May 2020).

43. Gylfason, T. Lessons from the Dutch disease: Causes, treatment, and cures. INSTITUTE OF ECONOMIC STUDIESWORKING PAPERS 1987-2001. Formerly Iceland Economic Papers Series. 2001. Available online: https://www.researchgate.net/profile/Thorvaldur_Gylfason (accessed on 19 May 2020).

44. Crivelli, E.; Gupta, S. Resource blessing, revenue curse? Domestic revenue effort in resource-rich countries. *Eur. J. Political Econ.* **2014**, 35. [CrossRef]

45. International Monetary Fund (IMF). *Guide on Resource Revenue Transparency*; IMF: Washington, DC, USA, 2007; Available online: https://www.imf.org/external/np/pp/2007/eng/101907g.pdf (accessed on 19 May 2020).

46. Ahrend, R. Sustaining growth in a resource-based economy: The main issues and the specific case of Russia. *ECE Discuss. Papers Ser. UNECE* **2005**, *3*, 24.

47. Teksoz, U.; Kalcheva, K. *Institutional Differences across Resource-Based Economies*; WIDER Working Paper, No. 2016/63; The United Nations University World Institute for Development Economics Research (UNUWIDER): Helsinki, Finland, 2016; Available online: http://dx.doi.org/10.35188/UNU-WIDER/2016/106-2 (accessed on 19 May 2020).

48. Selishcheva, T.A. Problemy ustojchivogo razvitiya ekonomiki v stranah Evrazijskogo ekonomicheskogo soyuza. *Evrazijsk. Ekon. Perspekt. Probl. Resheniya* **2018**, *2*, 15–21.

49. Sustainable Society Foundation: The Hague, The Netherlands. 2014. Available online: http://www.ssfindex.com/ssi2016/wp-content/uploads/pdf/SSI2014.pdf (accessed on 19 May 2020).

50. Sustaining Human Progress Reducing Vulnerabilities and Building Resilience. *Human Development Report 2014*; Technical Notes; The United Nations Development Programme: New York, NY, USA, 2014; p. 10. Available online: http://hdr.undp.org/sites/default/files/hdr14_technical_notes.pdf (accessed on 19 May 2020).

51. Wendling, Z.A.; Emerson, J.W.; Esty, D.C.; Levy, M.A.; de Sherbinin, A. *The Environmental Performance Index*; Yale Center for Environmental Law & Policy: New Haven, CT, USA, 2018; Available online: https://epi.yale.edu (accessed on 19 May 2020).

52. Index, E.P. *2018 EPI Policymakers' Summary. Global Metrics for the Environment: Ranking Country Performance on High.-Priority Environmental Issues*; Yale Center for Environmental Law & Policy; Yale University Center for International Earth Science Information Network; Columbia University: New Haven, CT, USA, 2018; Available online: https://epi.envirocenter.yale.edu/downloads/epi2018policymakerssummaryv01.pdf (accessed on 19 May 2020).

53. Robinson, J.; Torvik, R.; Verdier, T. Political foundations of the resource curse. *J. Dev. Econ.* **2006**, *79*, 447–468. [CrossRef]

54. World Bank. *The Little Green Data Book 2017.World Development Indicators*; World Bank: Washington, DC, USA, 2017; Available online: https://openknowledge.worldbank.org/handle/10986/2134/focus?filtertype= author&filter_relational_operator=equals&filter=World+Bank (accessed on 19 May 2020).

55. World Bank. *The Little Data Book on Financial Inclusion*; World Bank Group: Washington, DC, USA, 2018; Available online: https://openknowledge.worldbank.org/handle/10986/2134/focus?filtertype=author&filter_ relational_operator=equals&filter=World+Bank (accessed on 19 May 2020).

56. Human Development. United Nations Development Programme. Available online: http://hdr.undp.org/en/ data (accessed on 19 May 2020).

57. Sustainable Society Index—Your Compass to Sustainability. Available online: http://www.ssfindex.com/ data-all-countries/ (accessed on 19 May 2020).

58. Indeks Ekologicheskoj Effektivnosti. Gumanitarnyj Portal. Available online: https://gtmarket.ru/ratings/ environmental-performance-index/info (accessed on 19 May 2020).

59. Rejting Stran Mira Po Urovnyu Ustojchivosti Obshchestva. Gumanitarnyj Portal. Available online: https://gtmarket.ru/ratings/sustainable-society-index/info (accessed on 19 May 2020).

60. Rosstat Data on Environmental Issues. Available online: https://www.gks.ru/folder/11194 (accessed on 19 May 2020).

61. Mingaleva, Z.H.A.; Deputatova, L.N.; Starkov, Y.U.V. Primenenie rejtingovogo metoda ocenki effektivnosti gosudarstvennoj ekologicheskoj politiki: Sravnitel'nyj analiz Rossii i zarubezhnyh stran. *Ars Iskusstv. Upr.* **2018**, *10*, 419–438. [CrossRef]

62. Yurak, V.; Dushin, A.; Mochalova, L. Vs sustainable development: Scenarios for the Future. *J. Min. Inst.* **2020**, *242*, 242–247. [CrossRef]

63. Hatkov, V.; Boyarko, G. Administrative methods of import substitution management scarce types of mineral raw materials. *J. Min. Inst.* **2018**, *234*, 683–692. [CrossRef]

64. Cherepovitsyn, A.; Ilinova, A.; Evseeva, O. Stakeholders management of carbon sequestration project in the state – business – society system. *J. Min. Inst.* **2019**, *240*, 731–742.

65. Ulanov, V.; Ulanova, E. Influence of external factors on national energy security. *J. Min. Inst.* **2019**, *238*, 474–480. [CrossRef]

66. Statistical Review of World Energy. BP. Available online: https://www.bp.com/content/dam/bp/business-sites/ en/global/corporate/pdfs/energy-economics/statistical-review/bp-stats-review-2018-full-report.pdf (accessed on 19 May 2020).

Article

A Combined Analysis of Sociological and Farm Management Factors Affecting Household Livelihood Vulnerability to Climate Change in Rural Burundi

Risper Nyairo *, Takashi Machimura * and Takanori Matsui

Graduate School of Engineering, Osaka University, Suita 565-0871, Japan; matsui@see.eng.osaka-u.ac.jp
* Correspondence: risper@ge.see.eng.osaka-u.ac.jp (R.N.); mach@see.eng.osaka-u.ac.jp (T.M.)

Received: 24 April 2020; Accepted: 21 May 2020; Published: 24 May 2020

Abstract: This paper analyzed the livelihood vulnerability of households in two communes using socio-economic data, where one site is a climate analogue of the other under expected future climate change. The analysis was undertaken in order to understand local variability in the vulnerability of communities and how it can be addressed so as to foster progress towards rural adaptation planning. The study identified sources of household livelihood vulnerability by exploring human and social capitals, thus linking the human subsystem with existing biophysical vulnerability studies. Selected relevant variables were used in Factor Analysis on Mixed Data (FAMD), where the first eight dimensions of FAMD contributed most variability to the data. Clustering was done based on the eight dimensions, yielding five clusters with a mix of households from the two communes. Results showed that Cluster 3 was least vulnerable due to a greater proportion of households having adopted farming practices that enhance food and water availability. Households in the other clusters will need to make appropriate changes to reduce their vulnerability. Findings show that when analyzing rural vulnerability, rather than broadly looking at spatial climatic and farm management differences, social factors should also be investigated, as they can exert significant policy implications.

Keywords: questionnaire survey; climate analogues; Factor Analysis on Mixed Data (FAMD); clustering; education; farm management

---

1. Introduction

The Inter-governmental Panel on Climate Change (IPCC) has described vulnerability to climate change as the degree to which a system is susceptible to, and unable to cope with, adverse impacts of climate change [1]. Vulnerability is a combination of the risks people are exposed to and their social, economic, and cultural abilities to cope with the damages incurred [2,3]; therefore, the potential for adaptation is one criterion that may be used to identify key vulnerability of a system [1]. Yet, studies [4,5] have mainly analyzed the biophysical factors that contribute to vulnerability, without accounting for the role played by socio-economic dynamics. Socio-economic conditions have, for instance, been found to profoundly affect food systems through drivers such as soil fertility, irrigation, fertilizer use, demography, and socio-politics [6–8]. Local governments are responsible for providing infrastructure such as water and energy, and hence can play a proactive role in climate change adaptation. The article by [9] elaborates how the development of irrigation farming made possible by federal and provincial governments was a major contributor to community adaptation to climate variability in rural Saskatchewan. Such factors as wealth [10], community organization [11], and access to technology differentiate vulnerability across societies facing similar exposure to climate change [1,7].

On matters scale, while there may be cross-scale interactions due to the interconnectedness of economic and climate systems, local social, cultural, and geographic features may often differ and differentially affect vulnerability levels [12–14]. In fact, social vulnerability has been shown to be a partial product of both social inequalities and place inequalities [15]. IPCC findings [1] clearly illustrate this point by noting that low-latitude less-developed areas are generally at higher risk of climate change impacts and vulnerability due to higher sensitivity and lower adaptive capacity [1]. As [16,17] also noted, attempts to adapt to climate change impacts differed among communities based on geographical location, community attributes, and industrial sectors. IPCC revealed that the impacts of climate variability and extremes are most acutely experienced at the local level. Local level assessment of vulnerability therefore provides a better understanding of where and when to invest and who should make the investment [13,18–21].

The IPCC report further revealed that the African continent is the most vulnerable continent to climate change, given the expected significant reduction in food security and agricultural productivity [1]. Food supply and water resources are some of the sectors in low-latitude areas that are vulnerable to temperature and precipitation changes which result in droughts, decreases in food productivity (especially cereals), and water shortage. There is a general consensus that the sources of vulnerability in Africa are socio-economic in nature and include demography, governance, conflict, and inadequate resources [22]. However, the persistence of drought, which may lead to land cover change, is a potential key impact of climate change [1]. In many parts of the African continent droughts have been experienced more frequently in the last 30 years, and in eastern and southern Africa there is medium confidence that droughts will intensify in some seasons due to reduced precipitation and high evapotranspiration [22]. East Africa, in particular, has experienced temperature increases since the early 1980s. Precipitation in this region is also highly variable in both spatial and temporal terms, but trends show a decrease between the months of March and June [22].

Similarly, Burundi, a country in East Africa, has often experienced the negative impacts of climate variability, especially droughts and flooding. Multi-year droughts have been registered in the periods of 1999–2005, 2007–2008, 2010–2011, and 2016–2017 [23] with dire consequences. With almost 90% of its labor force engaged in agriculture [24], and the agricultural sector making up to 30% of the country's GDP [25], dependence on rain-fed crop production significantly increases the vulnerability of Burundian communities to the negative impacts caused by vagaries of variable weather and climate. Dependence on rain-fed agriculture is a dominant practice among several other countries [26], thus exposing them to drought vulnerability. Apart from the dependence on rain-fed agriculture, the profile of Burundi in terms of economic capacity (agricultural GDP), human resource, and technology is similar to a number of other countries on the African continent. This was illustrated by [27], who ranked Burundi, Somalia, Mali, Chad, Ethiopia, and Niger as countries with high relative vulnerability to drought, based on these similarities.

This paper builds on earlier work by [28], to assess levels of household livelihood vulnerability generated by social processes interacting across geographic scales in two rural communes of Bubanza (Bubanza Province) and Bugabira (Kirundo Province) in Burundi. The objectives are to identify the sociological drivers of household livelihood vulnerability and determine the vulnerability levels among clusters of households in the study area. The aim is to present a human dimension to the analysis of vulnerability of rural livelihoods, which links with existing biophysical vulnerability studies in the two locations.

2. Study Background

2.1. Climate Analogues Approach

The spatial climate analogues approach is one technique used in aiding climate change adaptation planning by assessing local level social vulnerability of target regions to future climate change. The approach was proposed to overcome the challenges of Global Climate Models, crop models, and farm system models by presenting field-based realities of the anticipated novel climates using the current spatial variability in climates [29]. Climate analogues are regions whose present climate resembles the predicted climate of another region [30]. In spatial analogues, the analogue region is expected to have climatic conditions similar to those projected for the target region, and is also expected to have similar socio-economic and political conditions [31,32]. In essence, the analogue region is expected to have developed systems adaptable to its climate that the target region can learn from. The spatial climate analogues approach thus resembles contextual vulnerability assessment [32], where it is assumed that nearer locations are more similar than distant locations. Analogue methodologies significantly improve the identification of determinants of vulnerability.

2.2. Climate Analogue Analysis in Burundi

In a previous study in Burundi, [28] applied the spatial climate analogue approach in Bubanza Province as the analogue location of Kirundo Province, located approximately 97 km apart. The study showed that farming systems may remain largely similar in the two areas with households keeping similar types of livestock and growing similar kinds of crops despite changed climatic conditions. Slight differences could only be noted in the adoption rates of some improved farming techniques, but these could not be attributed to the difference in climate between the regions. Similar results were reported by [33], who found that factors other than climate were the drivers in farm characteristics. The research by [28] provided useful insights on the expected future of the communes in Burundi but did not account for social variability among households in the two locations, which may serve to exacerbate or ameliorate the predicted negative impacts.

3. Materials and Methods

3.1. The Study Area

Kirundo province is the northern-most province of Burundi, bordering Rwanda. Geographically, it consists of hills and depressions of Bugesera (88% of total territory), Northeast Bweru (7%), and Buyenzi (5%) [34] regions with moderate to strong slopes. Since the year 1999, the annual evolution has showed a shortening of the rainy season, but with punctually violent rains and an extended dry season. The principal threat on the wet ecosystems is related to over-silting in lower valleys following intense erosion on strong slopes. Kirundo has more than five lakes, including Rweru, Rwihinda, Cyohoha, Kanzigiri, and Gacamirindi, but little access to underground water sources. The lakes Rwihinda and Cyohoha Sud are the nearest to Bugabira commune, with Rwihinda being threatened by excessive evapotranspiration. Bugabira commune is located at approximately 2.3° S and 30.0° E, with elevation ranging between 1000 m and 1500 m above sea level. Rainfall is irregular and bimodal, with average annual precipitation between 800 mm and 1200 mm [25]; the irregularity and reduction in precipitation has already caused drying up of the shallow water sources and reduction in the agricultural production. The mean annual temperature is 20.5 °C. The climate is classified as "moderate" tropical savanna; forest cover is sparse and the area lacks permanent rivers. The ferralitic lithosols present in the area indicate partly-weathered acidic soils of generally low fertility. More fertile organosols [28] are found in the lower valleys. The 2008 census approximated the population of Bugabira Commune at 89,259 persons. Bugabira is documented for over-cultivation and deforestation.

Bubanza Commune, the analogue used for this study, is in Bubanza province and is located at approximately 3.1° S latitude and 29.4° E longitude. The commune is in the Imbo floodplain region near Lake Tanganyika, which has an annual mean precipitation below 1200 mm and a mean annual temperature of 24.0 °C. The commune experiences two main seasons; the wet season runs from September to April, while the dry season runs from May to August. The area experiences long periods of dryness alternating with heavy rainfall and flooding. Bubanza province is distributed in several natural regions because of its backing to the Mumirwa mountain range. Imbo is the major part, followed by Mumirwa and Mugamba. Elevation stretches from 770 m in the Imbo to 2600 m in the Mugamba region. Soils are lateritic, indicating that they are highly weathered soils with high iron content and low organic matter concentrations. The most common natural vegetation type is savanna. Bubanza is near Nyungwe forest and national park. In terms of hydrology, the Mpanda and Kajeke Rivers and several smaller rivers such as Kidahwe, Nyaburiga, Kadakamwa, and Nyakabingo pass through the commune. The map below shows the two locations (Figure 1). The 2008 census approximated the population of Bubanza Commune at 83,678 persons. Bubanza is known for its paddy fields in the Mitakataka Village.

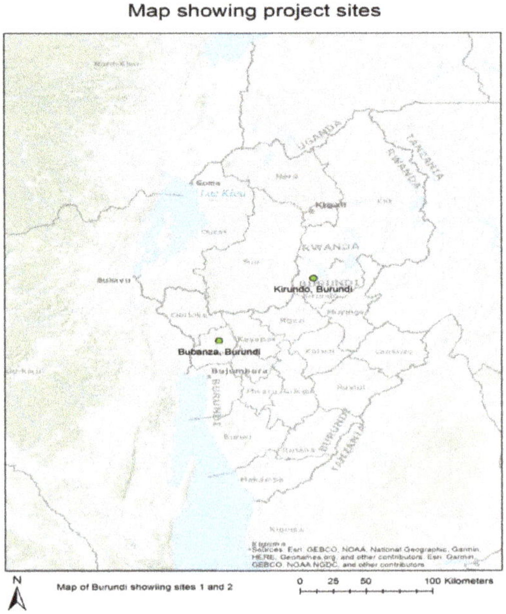

Map showing project sites

Figure 1. Location of the study areas. Kirundo (Target) and Bubanza (Analogue). Source: Environmental Systems Research Institute; National Oceanic and Atmospheric Administration (ESRI and NOAA).

3.2. Data Collection

This study utilized both quantitative and qualitative primary data of 450 households that were collected using a semi-structured questionnaire, as described by [28], for characterizing and clustering the households and also for assessing their vulnerability. A total of 247 households were from Kirundo, the target area, while 203 households were from Bubanza, the analogue. More households were sampled in the target than the analogue location following population estimates of the two locations in order to get representative samples. The questionnaire contained questions on household size, age, education levels, asset ownership, use of farm inputs, food access, land size, whether households identified with community groups and water access and any other changes in resource management and livelihood strategies. This was part of a larger dataset collected for the European Union funded project and implemented by the Consultative Group on International Agricultural Research (CGIAR) program on Climate Change, Agriculture and Food Security (CCAFS). The full list of questions is available as a supplementary table (Appendix A Table A1). The surveys were conducted in April 2012 in the target site and in May 2013 in the analogue location. Households were sampled using the stratified random sampling technique, where they were first sub-divided into two strata according to the administrative unit in which they were located. A number of households were then randomly selected from each unit. Out of the initial more than 1000 questions, including the lower-level dependent ones, this study extracted the 11 most relevant variables, both quantitative and categorical, as described in the data analysis section.

3.3. Data Analysis

The approach for analyzing data was influenced by the objectives and the type of information available. Factor Analysis on Mixed Data (FAMD) [35] and Hierarchical Clustering on Principal Components (HCPC) [36] were combined so as to assess household livelihood vulnerability. FAMD is able to handle both qualitative and quantitative analysis, by applying Principal Component Analysis (PCA) to the quantitative data and Multiple Correspondence Analysis (MCA) to the qualitative data [37]. In this case, MCA was simply a pre-processing step to transform categorical variables into continuous ones. Hierarchical clustering is used in identifying groups of similar observations in a data set, and hence was used to differentiate household profiles.

Out of the total number of original questions obtained by the survey, 107 adaptation-specific variables were first selected (Tables A1 and A2) based on the aspects of demography, infrastructure, household assets, production inputs, and food security [2,38–43]. The variables were then subjected to tests of significance using chi-square [44] and one-way analysis of variance (ANOVA) [45] tests, resulting in only 11 variables (3 quantitative and 8 categorical) for analysis. The 11 selected variables are shown in Table 1. On variable v2 (Lequels), representing rainwater harvesting techniques used, the category "basin" is pools or trenches dig to collect ground water whereas "container" is to store rainwater, and generally "basin" has more capacity. On variable v10 (Monthshung), some farmers suffer food shortage especially in the dry season and the variable represents the severity of food shortage. Variable v11 (Asothe) represents ownership of assets other than crops, animals, radio, cellphone, solar lamp, bicycle, or wheelbarrow.

Table 1. Groups and variables selected from the questionnaire survey of households used in Factor Analysis on Mixed Data (FAMD).

Variable ID	Variable	Description	Type	No. of Levels	Levels (Categorical) Unit (Quantitative)
v1	HHeduc	Highest education attained by household head	Categorical	5	None/Informal/Primary/Secondary/Tertiary
v2	Lequels	Rainwater harvesting technique used	Categorical	4	No/Basin/Container/Others
v3	Ownedland	Amount of land owned	Quantitative	–	Hectares
v4	Ownedfood	Land dedicated to food	Quantitative	–	Hectares
v5	Certseed	Purchase of improved seeds	Categorical	2	Yes/no
v6	Buyfert	Purchase of fertilizers	Categorical	2	Yes/no
v7	Buypest	Purchase of pesticides	Categorical	2	Yes/no
v8	Buyvtmd	Purchase of veterinary medicines	Categorical	2	Yes/no
v9	Crdagact	Getting credit	Categorical	2	Yes/no
v10	Monthshung	No. of months household is hungry	Quantitative	–	Months
v11	Asothe	Ownership of other assets	Categorical	2	Yes/no

To conduct FAMD, data were first imported into R statistics version 3.4.1, and using the FactoMineR package [46], the computation was run with the number of dimensions retained being eight and without any supplementary variables or individuals. The FAMD result was then used for clustering households in order to characterize them, where a hierarchical clustering by Ward's method was employed. The Ward criterion had to be used because it is based on multidimensional variance as well as Principal Component Analysis. The function HCPC, which is also in the FactoMineR package, was used for clustering, where the appropriate number of clusters was decided using a dendrogram. The HCPC function automatically conducted a chi-square test and output p-values of variables that significantly contributed to clustering, with a threshold at 0.05. The ggplot2 [47] and ggpubr packages were used to visualize cluster results.

4. Results

4.1. Factor Analysis on Mixed Data (FAMD) of Households

FAMD results showed that the first eight dimensions had the cumulative contribution to the variability in the data of 67%, as shown in Table 2. Factor loadings of the dimensions to variables are displayed in Figure 2 in pairs of every two dimensions, where axes show partial contribution to total variability of the data.

Table 2. Contributions of the first eight dimensions by FAMD.

Dimension	Eigenvalue	Percentage Contribution	Cumulative Percentage Contribution
1	2.17	13.6	13.6
2	1.79	11.2	24.7
3	1.26	7.9	32.6
4	1.18	7.4	40.0
5	1.12	7.0	47.0
6	1.07	6.7	53.7
7	1.05	6.5	60.2
8	1.02	6.3	66.5

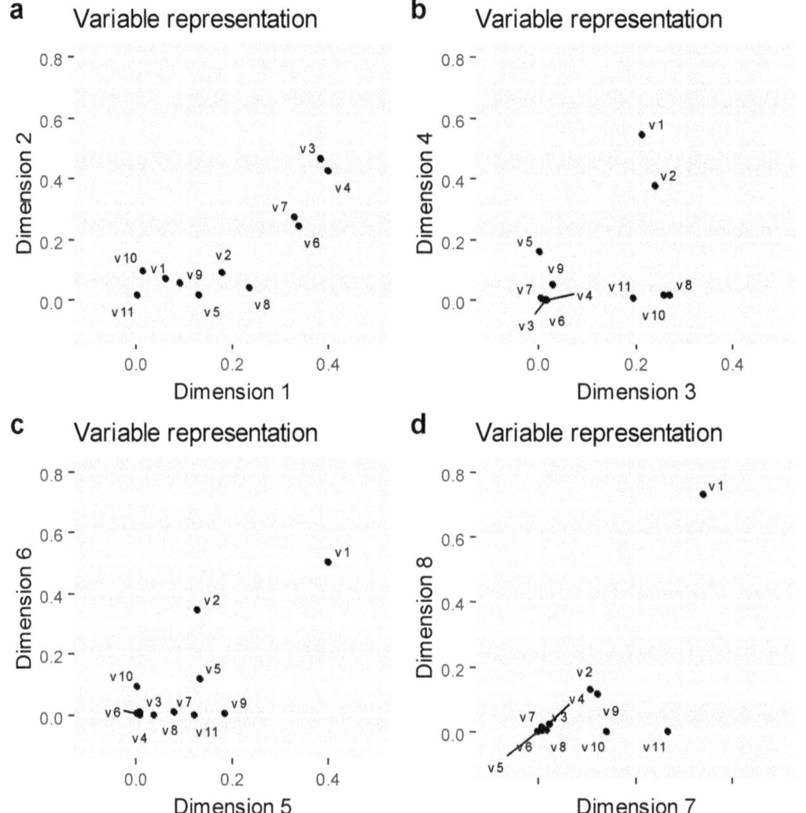

Figure 2. Factor loadings of the top eight dimensions on variable groups by Factor Analysis on Mixed Data (FAMD), showing the variable representations by (**a**) dimensions 1 and 2; (**b**) dimensions 3 and 4 (**c**) dimensions 5 and 6; (**d**) dimensions 7 and 8.

In order to characterize the dimensions of FAMD results as indicators of household livelihood vulnerability, significant variables and categories sensitive to the dimensions were summarized in Table 3. In the table, a "+" means the variables were positively related with the dimension, while a "−" means the variables were negatively related with the dimension. Dimension 1 was highly positively related to land used for growing food (v4), the size of land owned (v3), and the purchase of fertilizers (v6) and pesticides (v7). Dimension 2 was positively related to size of land owned (v3) and land used for growing food (v4), but negatively related to purchase of fertilizers (v6) and pesticides (v7).

Dimension 3 was positively related to hunger period (v10) and rainwater harvesting (v2) using basin, but negatively related to purchase of veterinary medicines (v8) and ownership of extra assets (v11). Dimensions 6 and 8 were both positively related to tertiary education (v1). However, while dimension 6 was negatively related to container water harvesting (v2) and secondary education (v1), dimension 8 was negatively related to no education (v1). Dimension 7 was positively related to asset ownership (v11) and hunger period (v10), which implies that households on positive coordinates of this dimension were hungry for a longer duration. Dimension 4 was positively related to no education and informal education and negatively related to primary school education and other techniques of rainwater harvesting. Dimension 5 was positively related to secondary education and credit access and negatively related to informal education. Dimension 1 represented farmers' action in improving agricultural production, while dimension 2 represented land ownership. Dimensions 3 and 7 represented affluence levels, while dimensions 6 and 8 represented higher education levels and rainwater harvesting.

Table 3. Significant variables and categorical levels significant for the first eight dimensions by Factor Analysis on Mixed Data (FAMD). "+" and "−" denote positive and negative effects on the dimensional scores, respectively.

Dimension	Significant Variables and Levels
1	+: Ownedfood, Ownedland, Buyfert, Buypest −:
2	+: Ownedland, Ownedfood −: Buypest, Buyfert
3	+: Monthshung, Lequels_Basin −: Buyvtmd, Asothe
4	+: HHeduc_Informal, HHeduc_None −: Lequels_Other, HHeduc_Primary
5	+: HHeduc_Secondary, Crdagact −: HHeduc_Informal
6	+: HHeduc_Tertiary −: Lequels_Container, HHeduc_Secondary
7	+: Asothe, Monthshung −:
8	+: HHeduc_Tertiary −: HHeduc_None

4.2. Clustering Households Based on FAMD Dimensions

Clustering based on the first eight dimensions of FAMD yielded five clusters, each with households from both locations, as summarized in Table 4. The number of clusters was decided upon based on a cluster dendrogram. The total number of households in clusters 1 through 5 were 278, 50, 114, 6, and 2, respectively, with clusters 1 and 4 dominated by households from Bugabira (target) and clusters 2 and 3 dominated by households from Bubanza (analogue).

Variables that contributed most to clustering in order of *p*-values were education, purchase of pesticides, purchase of fertilizers, purchase of veterinary medicines, rainwater harvesting, purchase of certified seeds, and access to credit. Among quantitative variables, owned land and land dedicated to food were significant contributors to clustering.

Table 4. Number of households classified to clusters per site.

Cluster	Bugabira (Target)	Bubanza (Analogue)	Total
1	198	80	278
2	12	38	50
3	31	83	114
4	5	1	6
5	1	1	2
Total	247	203	450

Figure 3 shows the distribution of households' score in dimensions with identified clusters, while Table 5 shows the proportions of households per variable within clusters. In order of importance, the significant dimensions for clustering were 1, 2, and 8, followed by 4 and 7. Dimensions 6, 3, and 5 were least important for clustering. Clusters 1 and 3 are almost distinguished in dimensions 1 and 2 (Figure 3a). This result shows that clusters 1 and 3 were defined by dimensions 1 and 2, which represent size of land owned and purchase of fertilizers and pesticides. Extremely high land ownership by only two households that belonged to cluster 5 clearly separated the cluster from other clusters, as shown by panels a and b in Figure 3. Land ownership in cluster 1 was less than the overall average. More than 95% of cluster 1 households did not use pesticides and more than 96% did not use fertilizers, while in cluster 3 more than 84% of households bought pesticides and about 64% used fertilizers (Table 5).

Table 5. Significant proportions of households per variable/category within clusters.

	Variable	Level or Unit	Percent Households or Mean within Clusters				
			1	2	3	4	5
v1	HHeduc	None	0	98	2	0	0
		Informal	17	0	18	0	0
		Primary	66	0	68	0	0
		Secondary	17	2	12	0	100
		Tertiary	0	0	0	100	0
v2	Lequels	No	56	36	18	50	0
		Basin	9	14	25	17	50
		Container	30	48	54	33	50
		Other	5	2	3	0	0
v3	Ownedland	Hectares	1.4	1.0	1.8	4.2	41.0
v4	Ownedfood	Hectares	1.0	0.9	1.3	1.4	23.5
v5	Certseed	Yes	29	56	49	83	50
		No	71	44	51	17	50
v6	Buyfert	Yes	3	22	64	33	50
		No	97	78	36	67	50
v7	Buypest	Yes	4	30	84	17	50
		No	96	70	16	83	50
v8	Buyvtmd	Yes	21	22	68	83	0
		No	79	78	32	17	100
v9	Crdagact	Yes	19	16	39	17	50
		No	81	84	61	83	50
v10	Monthshung	Months	5	6	6	5	6
v11	Asothe	Yes	19	16	16	33	0
		No	81	84	84	67	100

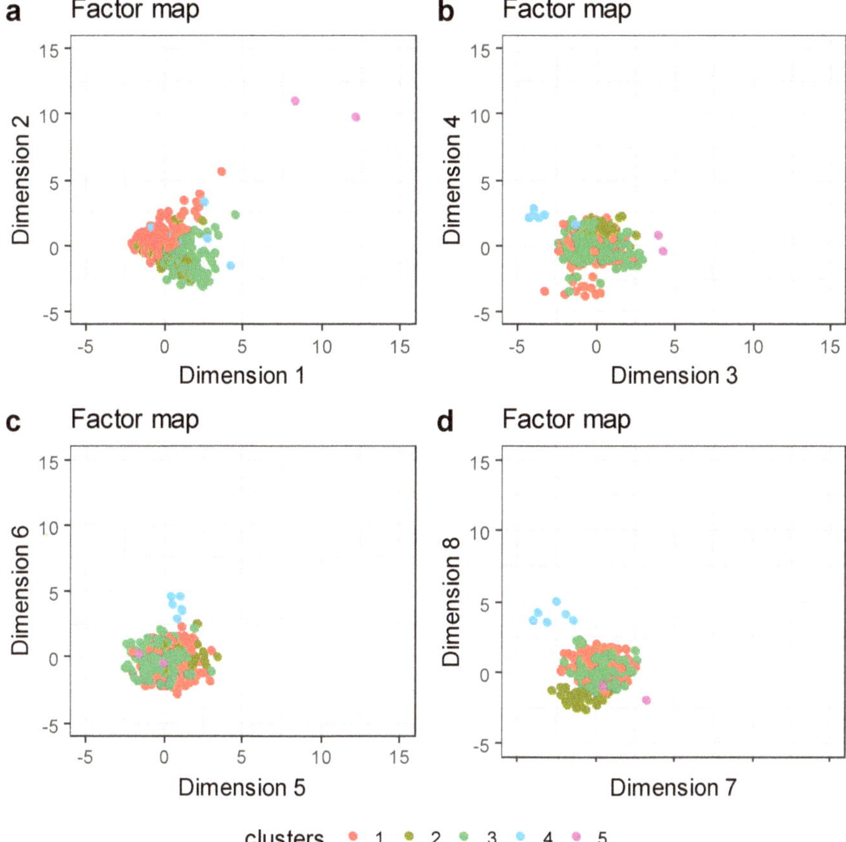

Figure 3. Distribution of households' score in FAMD dimensions with identified clusters, showing cluster scores on (**a**) dimensions 1 and 2; (**b**) dimensions 3 and 4; (**c**) dimensions 5 and 6; (**d**) dimensions 7 and 8.

The table below (Table 6) shows dimensional coordinates of cluster centroids. The centroids of cluster 5 had very high positive values on both dimensions 1 and 2, which explains the positioning of cluster 5 households on the factor map (Figure 3). The centroid with the highest negative value was that of cluster 4 on dimension 3.

Table 6. Dimensional scores of household cluster centroids.

Cluster	Dimension							
	1	**2**	**3**	**4**	**5**	**6**	**7**	**8**
1	−0.68	0.48	0.02	−0.22	−0.01	−0.06	0.10	0.17
2	−0.09	−0.51	0.42	1.30	1.00	0.25	−0.91	−1.61
3	1.42	−1.17	−0.12	−0.17	−0.44	−0.17	0.27	0.10
4	1.71	0.85	−3.5	2.27	0.84	3.93	−2.80	4.07
5	10.2	10.4	4.01	0.25	−0.85	−0.07	1.83	−1.40

Dimension 8 represented households that either had no education (on the negative coordinates) or those that had attained tertiary education (on the positive coordinates). Panels c and d of Figure 3 show cluster 4 households as having the highest positive coordinates on both dimensions 6 and 8.

This is because all households in this cluster had attained tertiary education. Cluster 2 households had negative coordinates on dimension 8 because almost all (98%) had no education. Cluster 2 households also showed the highest score for dimension 5, which represents getting credit. Cluster 4 was also clearly distinguished on the positive coordinates of dimension 6, which represent tertiary education (panel c of Figure 3). Cluster 4 almost entirely lay on the negative coordinates of dimension 3, which represents either purchase of veterinary medicines or asset ownership. The majority (83%) of the households purchased veterinary medicines (Table 5). Cluster 4 was thus defined by dimensions 8, 6, and 3.

Clusters 1 and 3 largely overlapped in dimension 4 because of the almost similar proportions of households that had attained primary school level education. The wide range of coordinates, both positive and negative, suggests large within-cluster variability in education, with some households attaining secondary education, while others having no education or informal education (Table 5).

The overlap of clusters in dimension 7 but with wide variation in coordinates also suggests large within-cluster variation in the duration of hunger period and ownership of assets, with the exception of cluster 4. Cluster 4 almost entirely lay on the negative coordinates of dimension 7 (panel d of Figure 3), suggesting that households in this cluster either had more assets or they suffered shorter hunger periods.

4.3. Important Crop and Livestock Species by Sites and by Target Household Clusters

The questionnaire survey contained two questions on what crop and livestock species were important for households' livelihood. Important crops for livelihood in the target area were wheat, teff, finger millet, peanuts, and sweet potatoes, while in the analogue area the important crops were teff, wheat, banana, barley and sorghum (Figure 4). Important crops in each cluster in the target area are shown in Figure 5, where cluster 5 is excluded because it consisted of only one household. All clusters in the target area except cluster 4 had similar patterns for wheat and teff. Cluster 3 had almost a quarter of the households growing teff and some households growing barley, crops which were more popular in the analogue region.

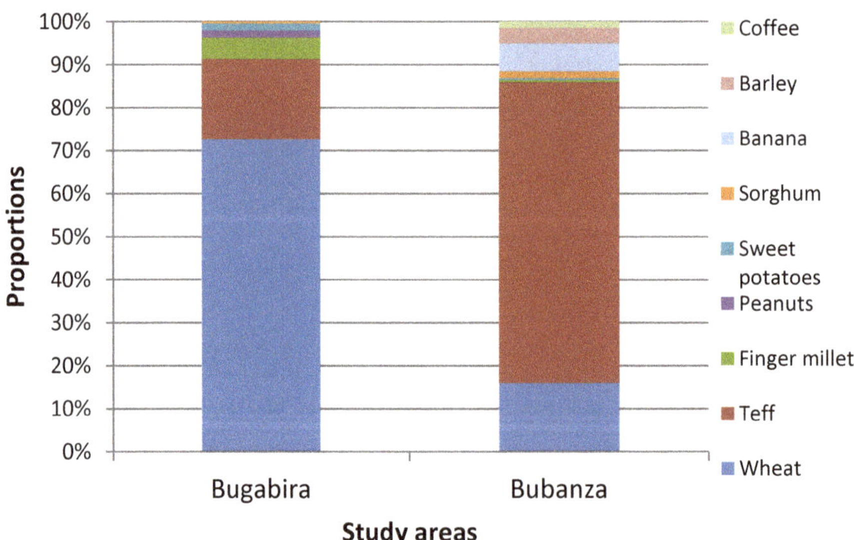

Figure 4. Important crops for livelihood in the target and analogue locations.

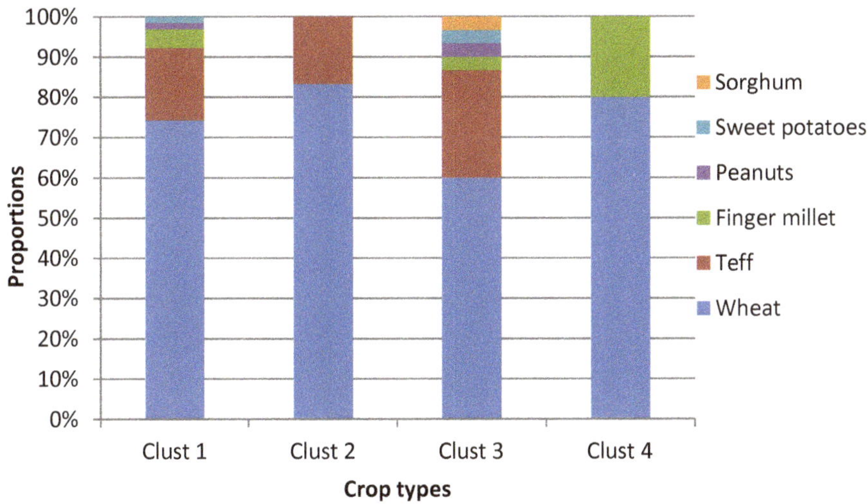

Figure 5. Important crops in Bugabira (target site) per cluster.

As for animals, goats, cattle, hens, and horses were important in both sites (Figure 6). Cattle, hens, and horses were almost of equal proportion in the analogue region. The analogue region thus had more hens and horses compared to the target. Households in the target kept more cattle and goats. The important animals per cluster in the target area are shown in Figure 7. Cluster 2 had only goats. Hens and horses were almost equal in cluster 1, while cattle and goats were almost equal in cluster 3. Cluster 4 was dominated by cattle.

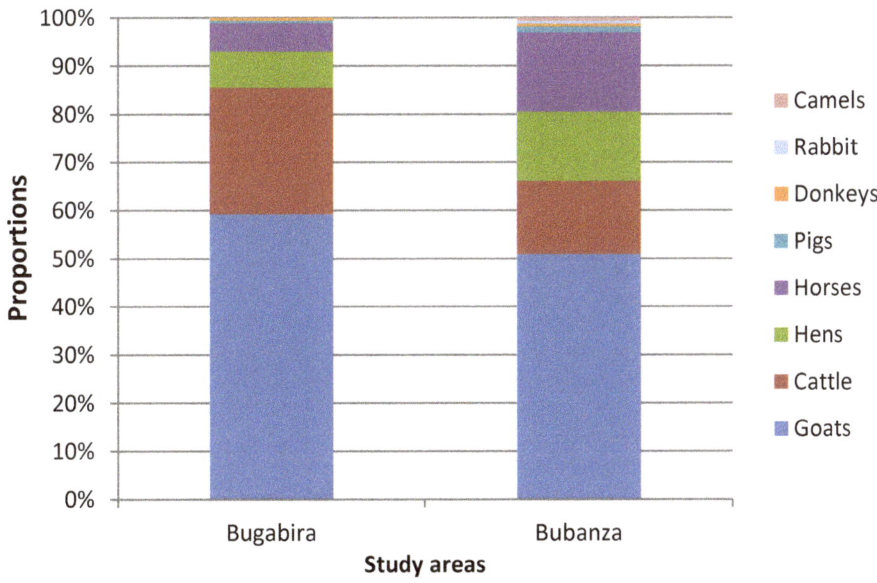

Figure 6. Important animals for livelihood in the target and analogue locations.

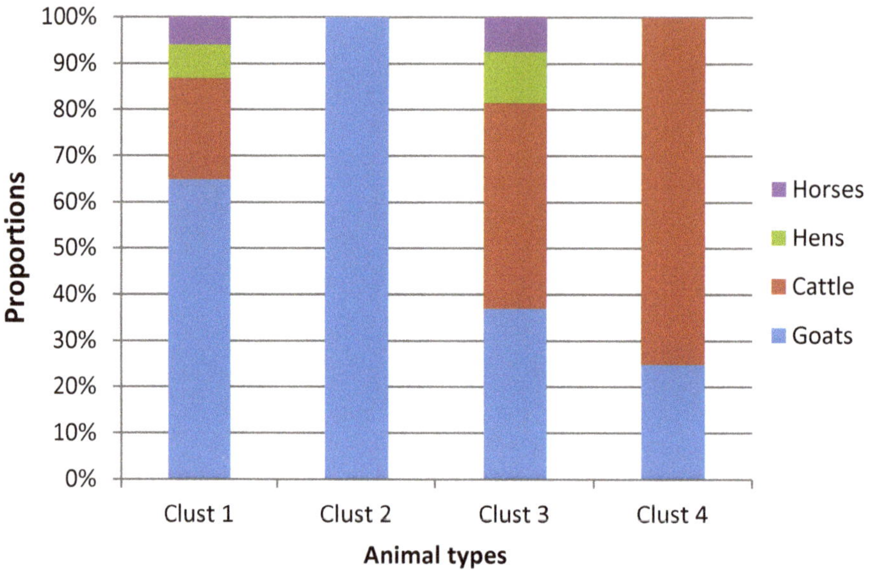

Figure 7. Important animals in Bugabira (the target) per cluster.

5. Discussion

Human and economic capitals are the representative dimensions of rural sustainability [48,49], while the variables represent the factors that can either constrain or enhance the adaptive capacity of rural households to climatic risks; hence, they form the basis for vulnerability assessment. In the FAMD analysis of this study, the first, second, and third dimensions represented affluence and farmers' actions to improve agricultural production (fertilizer, pesticide, veterinary medicines, and rainwater harvesting), while the fourth to sixth and eighth dimensions mainly represented access to education. Dimension 7 represented households without other assets. The household groups' profiles clustered by the FAMD dimensions reflected these two major aspects and summarized as low- and high-input farming groups (clusters 1 and 3, respectively), and least- and most-educated groups (clusters 2 and 4, respectively). Only two households included in cluster 5 were separated from the others by their extremely large land ownership.

The large variation within clusters in dimensions 3, 4, and 7 suggests that education is closely linked with the duration of hunger periods and ownership of assets. This can be deduced from cluster 4, which had the highest education levels and was negatively related to both dimensions 3 and 7. Education was also the main separator between clusters 2 and 3; given that the majority of households in both clusters were drawn from the same (analogue) location and only had minor differences in the use of certified seeds. It can therefore be concluded that cluster 4 is better placed education-wise to learn new technologies [50,51] and possibly to adapt to the predicted future climate. Already, the cluster grows finger millet (Figure 6), a hardy drought-tolerant crop that remains productive even under low-fertility, low-input systems [52,53], and has access to certified seed and veterinary medicine (Table 5). In addition to learning new technologies, educated populations have the capacity to take advantage of various employment opportunities outside of agriculture, thus reducing their vulnerability [54–56]. In their studies, [57,58] found that respondents acknowledged that getting an education and ultimately securing a job was an adaptive measure to the multiple pressures of climatic and socio-economic changes in the island nation of Tuvalu and in Burundi.

Although cluster 2 had only households with no education, 56% could purchase certified seed, as opposed to only 28% in cluster 1 and 49% in cluster 3, both of which had majority households attaining primary-level education. Since the majority of households in clusters 2 and 3 were from the analogue location, it would seem that there are "hidden" factors contributing to the purchase of certified seeds in this area. Possible explanations for this are proximity to urban centers, better access to credit (Table 5), and the types of food crops grown in this area. The results of this study revealed that in cluster 3 the number of households accessing credit was twice that of cluster 1. Across eastern and southern Africa, access to credit enables the adoption of new technologies, including the purchase of drought-tolerant maize [59,60]. Maize had already been established as more commonly grown in the analogue than the target region in the study by [28], and it is common practice for maize farmers to use certified seed whenever accessible. Since cluster 1 was credit-constrained, the availability of rainfall becomes critical, hence increasing its vulnerability.

On another note, despite clusters 1 and 3 having almost similar education proportions, cluster 3 had very high proportions of households harvesting rainwater. The large proportion of cluster 3 households who harvested rainwater used containers, meaning the water was for household use. Lack of access to nearby surface water sources as well as general reduction in rain may be driving this practice. Whereas there are several rivers, the province still suffers inadequate access to water, with less than half of the population having access to safe drinking water. The reasons for this are presently unknown, although some reports have linked this to armed conflicts and irregular time-space distribution of rain [24]. This result shows that issues of water access play a role in household livelihood vulnerability, sometimes to the exclusion of education. It also points to the need to evaluate indigenous knowledge and innovation [61] as a source of information when undertaking vulnerability assessment. Indeed, past research [62] reported wider acceptability of indigenous rainwater harvesting techniques by smallholders as opposed to introduced methods.

Cluster 3 also had a high proportion of households using fertilizers, while the use of fertilizers and pesticides among cluster 1 households was almost nonexistent, which may be attributed to the relatively low average land area (Table 5). Similar results were reported by [63]. Ideally, decreasing farm size should trigger intensification, but it is well-known that socio-economic conditions limit the intensification of smallholder farms in Africa. Households with low incomes and limited land are highly risk-averse [64]. Small land size also often means that fallowing cannot be practiced, even when the land is exhausted. Policies that encourage sedentary lifestyles have led to increased vulnerability of communities [65], since now community members are unable to access communal land or migrate in search of fertile land. Migration helps in decreasing the vulnerability of people to climate change [66,67].

The above findings point to the limitation of only using educational level as a determinant of farm management skill. This is in agreement with [49], who cautioned against using formal education to measure the level of skill in farm management, noting that it should be considered together with other constraints in the natural, physical, financial, and social capitals. The study by [65] found that it was a combination of free primary education, provision of roads, and health centers in Botswana that led to a decrease in vulnerability. Cluster 2, for example, had the second highest proportion of households using certified seed second only to cluster 4, despite not having any education. The cluster also had comparatively reasonable access to credit (Table 5). Past studies show that the role of education in aiding credit access is two-pronged; some findings show that higher education increases the probability for credit access, while others show that education decreases demand for credit [68]. In actual sense, the importance of education in accessing credit varies depending on the source of credit. While education may be important for accessing credit from banks and cooperatives, it is not significant when getting money from local community groups, traders, friends, and relatives [69,70].

From the aspect of farming practice, crop and livestock species choice is one of the measures of climate change adaptability of households in the target site. Teff and wheat were found to be the major crops in the two regions, and this pattern was seen in the clusters as well. Wheat generally does very well at 1200 m above sea level, while teff does well across a range of elevations and is both drought and flood-resistant [71]. Sorghum, which was mainly in the analogue region, is an African grass and its C4 photosynthesis pathway allows it to increase net carbon assimilation at high temperatures [72]. It is this drought tolerance that makes sorghum and finger millet important crops in dry regions. Clusters 1 and 3 had diversified in terms of crop types, with cluster 3 already adopting crops grown in the analogue, such as sorghum, suitable in higher temperatures. Crop diversity at the farm level can be achieved either temporally or spatially and helps to reduce risks of fluctuations in climate [67,73]. Cluster 3 in the target site having a similar household profile to majority in the analogue site showed the highest fraction of drought tolerant species selection (Figure 5).

Animals were important for livelihoods in both regions, which is true of African rural households, where livestock is an important capital serving as both a source of food and income [65,74,75]. Livestock is an especially important source of food during dry seasons, because rain-fed crop cultivation is more sensitive to climatic shocks than livestock production [54]. Goats were found to be the majority, and this could be because goats are browsers rather than grazers, and hence require less pastureland, and being heat-tolerant, can survive in a range of environments. In fact, cluster 2 households in the target only had goats, while goats and cattle were of almost equal proportions in cluster 3. It is likely that in future, the proportion of goats will continue to increase across clusters. Clusters 1 and 3 in the target area had diversified animals compared to clusters 2 and 4. The highest dependency by cluster 4 in the target site on cattle, the most beneficial under present climate (Figure 7), suggests the overfitting to present condition, which may weaken the capacity of adaptation to climate change.

6. Conclusions

The evaluation of household livelihood vulnerability to climate risks such as drought is important for targeted interventions that improve rural livelihoods and build adaptive capacity. This paper set out to analyze the sources of household vulnerability among smallholder farmers in rural Burundi. The study has shown that members of rural communities are not always vulnerable as a whole based on their location, as it is often assumed, but rather that certain social factors differentiate among levels of vulnerability even within the same location. This was clearly illustrated by cluster 3 households in the target area, who have taken steps to improve their agricultural production and reduce livelihood vulnerability through use of inputs, access to credit, rainwater harvesting, and formal education. Given the findings, cluster 3 households can be said to be the least vulnerable to prevailing and predicted climate conditions, even though more still needs to be done to cover the gap in the use of certified seed and to encourage members to go beyond primary school level.

The findings from this study exert a number of policy implications for reducing livelihood vulnerabilities among smallholders in the study area. Firstly, cluster 1 households will need to adopt practices such as rainwater harvesting and use of improved seed to counter the negative effects of predicted longer and hotter dry seasons. This could be done through designating watering sources and directing run-off to these locations. Water is an important driver of rural economy, as it is used for farming and watering animals. Availability of watering resources reduces distances traveled to access it, thus allowing time for more productive activities. The projected increased rainfall in the study areas can be tapped and used for irrigation to fill the water availability gap. Improving the use of farm inputs such as irrigation and application of integrated fertilizer management can address threats to natural resource-based livelihoods and improve ecosystem service provision. Irrigation, for example, supports crop diversification, since farmers do not have to grow only the crops favored by soil moisture content. Such sustainable intensification should be encouraged through sensitization and awareness creation. Improved seeds that are bred to be drought-resistant and to withstand pest attack should be subsidized and access improved through support to agro-dealers and extension agents. In addition, strengthening the entitlements in terms of land area ownership can promote the use of inputs for improving agricultural production. Smallholders often hold back from investing in land because they feel insecure, especially in many African countries where resource rights are poorly defined and almost never enforced. Interventions such as land consolidation seem to be a solution to the farm size challenge, although these may be difficult to apply in areas of rapid population growth and lack of alternative off-farm income-generating activities.

Secondly, the most immediate response for cluster 2 should be to offer a mix of education and extension services to the households, as well as cover the gap in the use of improved seed through subsidies. The concept of model farms to demonstrate farming techniques to local farmers, which has been done in countries like Canada and Kenya, can greatly enhance extension service provision and overcome the difficulty of farmers to transition to sustainable practices. In addition, vocational training may be implemented based on locally available resources in conjunction with charity organizations to equip the household members with practical skills that do not require high levels of education. Training can be integrated with the existing credit systems so as to enable sustainable outcomes for the households that access them. Such training should build on existing indigenous knowledge so as to enhance participation and strengthen local capacity.

Finally, reducing household livelihood vulnerability requires multi-faceted, integrated approaches—combining infrastructure, education, and improved land policies to enable adaptive capacity in rural agrarian-based communities. Household characteristics such as affluence and educational status can be seen as buffers that reduce household livelihood vulnerability, but in addition, governments have a part to play. Unequal development may result in certain people benefiting more than others. Lessons can be drawn from the success of Botswana (see [65]) and other African countries. It is the responsibility of the government through the disaster preparedness and management department to put in place contingency plans for eventualities such as drought. Government investment in smallholder crop insurance schemes, for example, has proved important in reducing vulnerability, especially where farms are managed corporately. Doing this will help to achieve one of the targets of one of the United Nations' (UN) Sustainable Development Goals (SDG 2)—end hunger, achieve food security and improved nutrition, and promote sustainable agriculture—in the SDG global indicator framework, which is to double the agricultural productivity and incomes of small-scale farmers by 2030 through secure and equal access to land, inputs, knowledge, financial services, and markets.

Author Contributions: Conceptualization, R.N.; T.M. (Takashi Machimura) and T.M. (Takanori Matsui); methodology, R.N. and T.M. (Takashi Machimura); formal analysis, R.N. and T.M. (Takashi Machimura); data curation, R.N.; writing—original draft preparation, R.N.; writing—review and editing, T.M. (Takashi Machimura) and T.M. (Takanori Matsui); supervision, T.M. (Takashi Machimura). All authors have read and agreed to the published version of the manuscript.

Funding: This research received no external funding.

Acknowledgments: The authors wish to acknowledge Eike Luedeling and the CGIAR Research Program on Climate Change, Agriculture and Food Security (CCAFS) for support in data acquisition and Institut des Sciences Agronomiques du Burundi (ISABU) for logistic support. Special thanks go to Concern Worldwide, Burundi and all households who participated in the survey. The authors also thank all 3 anonymous reviewers whose comments led to the improvement of the manuscript.

Conflicts of Interest: The authors declare no conflict of interest.

Appendix A

Table A1. Full questionnaire design.

Questionnaire Section	Description	Number of Questions	Number of Variables
1	Household respondent and type	6	6
2	Demography	4	4
3	Sources of livelihood security	3	128
4	Crop, farm animals, and tree management changes	7	607
5	Food security	2	24
6	Land and water	11	42
7	Input and credit	11	32
8	Climate and weather information	1	50
9	Community groups	5	44
10	Assets	1	56
11	Constraints to production	2	18

Table A2. Selected questions for analysis.

Section	Variable Group	Variable	Levels or Unit
1	Age	Age of household head; Age of respondent	Age in years
1	Sex/Education	Sex of the household head;	Male/Female
		Highest education attained by household head	No education/Informal/ Primary/Secondary/Tertiary
2	Household size	Total household size Total household males Total household females Males below 5 years Females below 5 years Males between 5 and 15 years Females between 5 and 15 years	Number
6	Rainwater harvesting	Practice rainwater harvesting	Yes/No
		Use of harvested rainwater	Irrigation/Household/ Livestock
		Practice irrigation	Yes/No
		Type of irrigation practiced	Basin/Container/Dam
10	Assets	Do you own a radio? Cellphone? Bicycle? Motorcycle? Solar panel? Machete? Wheelbarrow? Spade? Watering cans? Oil lamp? Large battery? Fishing net? Bank account? Cattle? Goats? Sheep? Poultry? Other assets? Water storage tank? Improved house, e.g., brick/concrete Improved roofing Separate animal structure	Yes/No
6	Land access	Size of land owned Land rented Communal land Total land access Land owned dedicated to food Land rented dedicated to food Communal land dedicated to food Land owned for industrial purpose Land owned dedicated to grazing Land owned dedicated to trees Land rented dedicated to trees Land owned for aquaculture Degraded land	Hectares
6	Water sources	Tanks for water harvesting Dams or water ponds Water inlet	Yes/No
7	Inputs	Purchase of certified seed Purchase of fertilizers Purchase of pesticides Purchase of veterinary medicine Access to agricultural credit	Yes/No
9	Group membership	Tree planting, Fish pond, Fishing, Forest product collection, Water catchment management, Soil improvement activities, Crop substitution, Irrigation, Savings and credit, Marketing products, Productivity enhancement, Seed production, Vegetable production Any other group	Yes/No
9	Support groups	Help from friends From government From politicians, MPs From NGOs From religious organizations/church From local community group	Yes/No
9	Association membership	Purpose of association you are a member of Association benefits	Agriculture/Marketing/Savings and credit/ Other support Access to credit/ Shared manpower/ Common purchase of inputs/ Easier access to information/ Coordinated agricultural sales
9	Association size	Total association membership Number of men Number of women	Number
5	Food source/scarcity	Source of food in January, February, March, April, May, June, July, August, September, October, November, December	Mainly from own farm/ Mainly from off-farm (purchase, aid)
		Shortage of food in January, February, March, April, May, June, July, August, September, October, November, December	Yes/No

References

1. Schneider, S.H.; Semenov, S.; Patwardhan, A.; Burton, I.; Magadza, C.H.D.; Oppenheimer, M.; Yamin, F. Assessing key vulnerabilities and the risk from climate change. In *Climate Change 2007. Impacts, Adaptation and Vulnerability*; Contribution of Working Group II to the Fourth Assessment Report of the Intergovernmental Panel on Climate, Change; Parry, M.L., Canziani, O.F., Palutikof, J.P., van der Linden, P.J., Hanson, C.E., Eds.; Cambridge University Press: Cambridge, UK, 2007; pp. 779–810.

2. Coulibaly, J.Y.; Mbow, C.; Sileshi, G.W.; Beedy, T.; Kundhlande, G.; Musau, J. Mapping Vulnerability to Climate Change in Malawi: Spatial and Social Differentiation in the Shire River Basin. *Am. J. Clim. Chang.* **2015**, *4*, 282–294. [CrossRef]

3. Yusuf, A.A.; Francisco, H. *Climate Change Vulnerability Mapping for Southeast Asia*; Economy and Environment Program for Southeast Asia, South Bridge Court: Singapore, 2009.

4. Schroth, G.; Läderach, P.; Martinez-Valle, A.I.; Bunn, C.; Jassogne, L. Vulnerability to climate change of cocoa in West Africa: Patterns, opportunities and limits to adaptation. *Sci. Total. Environ.* **2016**, *556*, 231–241. [CrossRef] [PubMed]

5. Gan, T.Y.; Ito, M.; Hülsmann, S.; Qin, X.; Lu, X.X.; Liong, S.Y.; Rutschman, P.; Disse, M.; Koivusalo, H.; Huelsmann, S.; et al. Possible climate change/variability and human impacts, vulnerability of drought-prone regions, water resources and capacity building for Africa. *Hydrol. Sci. J.* **2016**, *61*, 1209–1226. [CrossRef]

6. Huq, N.; Hugé, J.; Boon, E.; Gain, A.K. Climate Change Impacts in Agricultural Communities in Rural Areas of Coastal Bangladesh: A Tale of Many Stories. *Sustainability* **2015**, *7*, 8437–8460. [CrossRef]

7. Thornton, P.K.; Ericksen, P.J.; Herrero, M.; Challinor, A.J. Climate variability and vulnerability to climate change: A review. *Glob. Chang. Biol.* **2014**, *20*, 3313–3328. [CrossRef] [PubMed]

8. Barbier, B.; Yacouba, H.; Karambiri, H.; Zoromé, M.; Somé, B. Human vulnerability to climate variability in the Sahel: Farmers' adaptation strategies in northern Burkina Faso. *Environ. Manag.* **2009**, *43*, 790–803. [CrossRef] [PubMed]

9. Pittman, J.; Wittrock, V.; Kulshreshtha, S.; Wheaton, E. Vulnerability to climate change in rural Saskatchewan: Case study of the Rural Municipality of Rudy No. 284. *J. Rural. Stud.* **2011**, *27*, 83–94. [CrossRef]

10. Wu, C.-C.; Jhan, H.-T.; Ting, K.-H.; Tsai, H.-C.; Lee, M.-T.; Hsu, T.-W.; Liu, W.-H. Application of Social Vulnerability Indicators to Climate Change for the Southwest Coastal Areas of Taiwan. *Sustainability* **2016**, *8*, 1270. [CrossRef]

11. Williams, P.A.; Crespo, O.; Abu, M.; Simpson, N.P. A systematic review of how vulnerability of smallholder agricultural systems to changing climate is assessed in Africa. *Environ. Res. Lett.* **2018**, *13*, 103004. [CrossRef]

12. Grothmann, T.; Petzold, M.; Ndaki, P.; Kakembo, V.; Siebenhüner, B.; Kleyer, M.; Yanda, P.; Ndou, N. Vulnerability Assessment in African Villages under Conditions of Land Use and Climate Change: Case Studies from Mkomazi and Keiskamma. *Sustainability* **2017**, *9*, 976. [CrossRef]

13. Malone, E. *Vulnerability and Resilience in the Face of Climate Change: Current Research and Needs for Population Information (No. PNWD-4087)*; Battelle Memorial Institute: Washington, USA, 2009.

14. Turner, B.L.; Kasperson, R.E.; Matson, P.A.; McCarthy, J.J.; Corell, R.W.; Christensen, L.; Eckley, N.; Kasperson, J.X.; Luers, A.; Martello, M.L.; et al. A framework for vulnerability analysis in sustainability science. *Proc. Natl. Acad. Sci. USA* **2003**, *100*, 8074–8079. [CrossRef] [PubMed]

15. Boko, M.; Niang, I.; Nyong, A.; Vogel, C.; Githeko, A.; Medany, M.; Osman-Elasha, B.; Tabo, R.; Yanda, P.Z. Africa. In *Climate Change 2007: Impacts, Adaptation and Vulnerability*; Contribution of Working Group II to the Fourth Assessment Report of the Intergovernmental Panel on Climate Change; Cambridge University Press: Cambridge, UK, 2007; pp. 433–467.

16. Ludeña, C.E.; Yoon, S.W. *Local Vulnerability Indicators and Adaptation to Climate Change: A Survey*; Technical Note No. 857 (IDB-TN- 857); Inter-American Development Bank: Washington, DC, USA, 2015.

17. Westerhoff, L.; Smit, B. The rains are disappointing us: Dynamic vulnerability and adaptation to multiple stressors in the Afram Plains, Ghana. *Mitig. Adapt. Strateg. Glob. Chang.* **2009**, *14*, 317. [CrossRef]

18. IPCC. *Managing the Risks of Extreme Events and Disasters to Advance Climate Change Adaptation*; A Special Report of Working Groups I and II of the Intergovernmental Panel on Climate, Change; Field, C.B., Barros, V., Stocker, T.F., Qin, D., Dokken, D.J., Ebi, K.L., Mastrandrea, M.D., Mach, K.J., Plattner, G.K., et al., Eds.; Cambridge University Press: Cambridge, UK; New York, NY, USA, 2012; p. 582.

19. Shameem, M.I.M.; Momtaz, S.; Rauscher, R. Vulnerability of rural livelihoods to multiple stressors: A case study from the southwest coastal region of Bangladesh. *Ocean. Coast. Manag.* **2014**, *102*, 79–87. [CrossRef]

20. Ribot, J. Cause and response: Vulnerability and climate in the Anthropocene. *J. Peasant Stud.* **2014**, *41*, 667–705. [CrossRef]

21. Downing, T.E.; Patwardhan, A.; Klein, R.J.; Mukhala, E.; Stephen, L.; Winograd, M.; Ziervogel, G. Assessing vulnerability for climate adaptation. In *Adaptation Policy Frameworks for Climate Change: Developing Strategies, Policies, and Measures*; Lim, B., Spanger-Siegfried, E., Eds.; Cambridge University Press: Cambridge, UK, 2004; Chapter 3.

22. IPCC. *Climate Change 2014: Impacts, Adaptation, and Vulnerability: Part B: Regional Aspects; Contribution of Working Group II to the Fifth Assessment Report of the Intergovernmental Panel on Climate Change*; Barros, V.R., Field, C.B., Dokken, D.J., Mastrandrea, M.D., Mach, K.J., Bilir, T.E., Chatterjee, M., Ebi, K.L., Estrada, Y.O., et al., Eds.; IPCC: Geneva, Switzerland, 2014; p. 151.

23. Vinck, P.; Bizuneh, M.; Rubavu, M.; Tahirou, L. *VAM WFP Food Security Analysis*; Burundi. United Nations World Food Programme: Rome, Italy, 2008.

24. ETDELE DU TOURISME. National Adaptation Plan of Action to Climate Change. 2007. Available online: http://unfccc.int/resource/docs/napa/bdi01e.pdf (accessed on 3 April 2018).

25. Minani, B.; Rurema, D.-G.; Lebailly, P. Rural resilience and the role of social capital among farmers in Kirundo province, Northern Burundi. *Appl. Stud. Agribus. Commer.* **2013**, *7*, 121–125. [CrossRef]

26. Rockström, J.; Karlberg, L.; Wani, S.P.; Barron, J.; Hatibu, N.; Oweis, T.; Bruggeman, A.; Farahani, J.; Qiang, Z. Managing water in rainfed agriculture–the need for a paradigm shift. *Agric. Water Manag.* **2010**, *97*, 543–550. [CrossRef]

27. Naumann, G.; Barbosa, B.; Garrote, L.; Iglesias, A.; Vogt, J. Exploring drought vulnerability in Africa: An indicator based analysis to be used in early warning systems. *Hydrol. Earth Syst. Sci.* **2014**, *18*, 1591–1604. [CrossRef]

28. Nyairo, R. Applicability of Climate Analogues for Climate Change Adaptation Planning in Bugabira Commune, Burundi. Master's Thesis of Science in Agroforestry, University Of Nairobi, Nairobi, Kenya, 2014.

29. Ramírez-Villegas, J.; Lau, C.; Köhler, A.-K.; Signer, J.; Jarvis, A.; Arnell, N.; Osborne, T.; Hooker, J. *Climate Analogues: Finding Tomorrow's Agriculture Today*; Working Paper No. 12; CGIAR Research Program on Climate Change, Agriculture and Food Security: Cali, Colombia, 2011.

30. Hallegatte, S.; Hourcade, J.-C.; Ambrosi, P. Using climate analogues for assessing climate change economic impacts in urban areas. *Clim. Chang.* **2007**, *82*, 47–60. [CrossRef]

31. Arnbjerg-Nielsen, K.; Funder, S.G.; Madsen, H. Identifying climate analogues for precipitation extremes for Denmark based on RCM simulations from the ENSEMBLES database. *Water Sci. Technol.* **2015**, *71*, 418–425. [CrossRef]

32. Ford, J.D.; Keskitalo, E.C.H.; Smith, T.; Pearce, T.; Berrang-Ford, L.; Duerden, F.; Smit, B. Case study and analogue methodologies in climate change vulnerability research. *Wiley Interdiscip. Rev. Clim. Chang.* **2010**, *1*, 374–392. [CrossRef]

33. Bos, S.P.M.; Pagella, T.; Kindt, R.; Russell, A.J.M.; Luedeling, E. Climate analogs for agricultural impact projection and adaptation—A reliability test. *Front. Environ. Sci.* **2015**, *3*, 65. [CrossRef]

34. Republic of Burundi. *Ministère de l'Eau, de l'Environnement, de l'Aménagement du Territoire et de l'Urbanisme–Plan régional de mise en oeuvre de la Stratégie Nationale et Plan d'Action sur la Biodiversité dans la dépression de Bugesera*; National Institute for Nature Conservation and the Environment (INECN): Bujumbura, Burundi, 2013; p. 46.

35. Thurstone, L.L. *Multiple-Factor Analysis: A Development and Expansion of The Vectors of Mind*; University of Chicago Press: Chicago, IL, USA, 1947.

36. Argüelles, M.; Benavides, C.; Fernández, I. A new approach to the identification of regional clusters: Hierarchical clustering on principal components. *Appl. Econ.* **2014**, *46*, 2511–2519. [CrossRef]

37. Abdi, H.; Williams, L.J.; Valentin, D. Multiple factor analysis: Principal component analysis for multitable and multiblock data sets. *Wiley Interdiscip. Rev. Comput. Stat.* **2013**, *5*, 149–179. [CrossRef]

38. Tiepolo, M.; Bacci, M. Tracking Climate Change Vulnerability at Municipal Level in Rural Haiti Using Open Data. In *Renewing Local Planning to Face Climate Change in the Tropics*; In Green Energy and Technology; Springer: Berlin/Heidelberg, Germany, 2017; pp. 103–131. [CrossRef]

39. Žurovec, O.; Čadro, S.; Sitaula, B.K. Quantitative Assessment of Vulnerability to Climate Change in Rural Municipalities of Bosnia and Herzegovina. *Sustainability* **2017**, *9*, 1208. [CrossRef]

40. Simane, B.; Zaitchik, B.F.; Foltz, J.D. Agroecosystem specific climate vulnerability analysis: Application of the livelihood vulnerability index to a tropical highland region. *Mitig. Adapt. Strat. Glob. Chang.* **2016**, *21*, 39–65. [CrossRef]

41. Tibesigwa, B.; Visser, M.; Collinson, M.; Twine, W. Investigating the sensitivity of household food security to agriculture-related shocks and the implication of social and natural capital. *Sustain. Sci.* **2016**, *11*, 193–214. [CrossRef]

42. Lee, Y.-J.; Lin, S.-C.; Chen, C.-C. Mapping Cross-Boundary Climate Change Vulnerability—Case Study of the Hualien and Taitung Area, Taiwan. *Sustainability* **2016**, *8*, 64. [CrossRef]

43. Christiaensen, L.J.; Subbarao, K. Towards an Understanding of Household Vulnerability in Rural Kenya. *J. Afr. Econ.* **2005**, *14*, 520–558. [CrossRef]
44. McHugh, M.L. The Chi-square test of independence. *Biochem. Medica.* **2013**, *23*, 143–149. [CrossRef]
45. Kim, H.-Y. Analysis of variance (ANOVA) comparing means of more than two groups. *Restor. Dent. Endod.* **2014**, *39*, 74–77. [CrossRef]
46. Lê, S.; Josse, J.; Husson, F. FactoMineR: An R package for multivariate analysis. *J. Stat.* **2008**, *25*, 1–18. [CrossRef]
47. Hadley, W. ggplot2: Elegant Graphics for Data Analysis. 2016. Available online: https://cran.r-project.org/web/packages/ggplot2/ggplot2.pdf (accessed on 18 October 2019).
48. Serrat, O. The Sustainable Livelihoods Approach. In *Knowledge Solutions*; Springer: Singapore, 2017; pp. 21–26. [CrossRef]
49. Nelson, R.; Kokic, P.; Crimp, S.; Martin, P.; Meinke, H.; Howden, S.M.; de Voil, P.; Nidumolu, U. The vulnerability of Australian rural communities to climate variability and change: Part II—Integrating impacts with adaptive capacity. *Environ. Sci. Policy.* **2010**, *13*, 18–27. [CrossRef]
50. Blankespoor, B.; Dasgupta, S.; Laplante, B.; Wheeler, D. *The Economics of Adaptation to Extreme Weather Events in Developing Countries*; CGD Working Paper 198; Center for Global Development: Washington, DC, USA, 2010; Available online: http://www.cgdev.org/content/publications/detail/1423545 (accessed on 6 February 2019).
51. Ulrich, A.; Speranza, I.C.; Roden, P.; Kiteme, B.; Wiesmann, U.; Nüsser, M. Small-scale farming in semi-arid areas: Livelihood dynamics between 1997 and 2010 in Laikipia, Kenya. *J. Rural. Stud.* **2012**, *28*, 241–251. [CrossRef]
52. Hittalmani, S.; Mahesh, H.B.; Shirke, M.D.; Biradar, H.; Uday, G.; Aruna, Y.R.; Lohithaswa, H.C.; Mohanrao, A. Genome and Transcriptome sequence of Finger millet (Eleusine coracana (L.) Gaertn.) provides insights into drought tolerance and nutraceutical properties. *BMC Genom.* **2017**, *18*, 465. [CrossRef]
53. Parvathi, M.S.; Nataraja, K.N.; Yashoda, B.K.; Ramegowda, H.V.; Mamrutha, H.M.; Rama, N. Expression analysis of stress responsive pathway genes linked to drought hardiness in an adapted crop, finger millet (Eleusine coracana). *J. Plant. Biochem. Biot.* **2013**, *22*, 193–201. [CrossRef]
54. Mertz, O.; Mbow, C.; Reenberg, A.; Genesio, L.; Lambin, E.F.; D'haen, S.; Zorom, M.; Rasmussen, K.; Diallo, D.; Barbier, B.; et al. Adaptation strategies and climate vulnerability in the Sudano-Sahelian region of West Africa. *Atmos. Sci. Lett.* **2011**, *12*, 104–108. [CrossRef]
55. Younus, M.A.F.; Kabir, M.A. Climate Change Vulnerability Assessment and Adaptation of Bangladesh: Mechanisms, Notions and Solutions. *Sustainability* **2018**, *10*, 4286. [CrossRef]
56. Sietz, D.; Choque, S.E.M.; Lüdeke, M.K.B. Typical patterns of smallholder vulnerability to weather extremes with regard to food security in the Peruvian Altiplano. *Reg. Environ. Chang.* **2012**, *12*, 489–505. [CrossRef]
57. McCubbin, S.; Smit, B.; Pearce, T. Where does climate fit? *Vulnerability to climate change in the context of multiple stressors in Funafuti, Tuvalu. Glob. Environ. Chang.* **2015**, *30*, 43–55. [CrossRef]
58. Sommers, M.; Uvin, P. *Youth in Rwanda and Burundi: Contrasting Visions*; Special Report #293; United States Institute of Peace: Washington, DC, USA, 2011.
59. Shiferaw, B.; Kebede, T.; Kassie, M.; Fisher, M. Market imperfections, access to information and technology adoption in Uganda: Challenges of overcoming multiple constraints. *Agric. Econ.* **2015**, *46*, 475–488. [CrossRef]
60. Fisher, M.; Abate, T.; Lunduka, R.W.; Asnake, W.; Alemayehu, Y.; Madulu, R.B. Drought tolerant maize for farmer adaptation to drought in sub-Saharan Africa: Determinants of adoption in eastern and southern Africa. *Clim. Chang.* **2015**, *133*, 283–299. [CrossRef]
61. Altieri, M.A.; Koohafkan, P. *Enduring Farms: Climate Change, Smallholders And Traditional Farming Communities*; Third World Network (TWN): Penang, Malaysia, 2008; Volume 6.
62. Biazin, B.; Sterk, G.; Temesgen, M.; Abdulkedir, A.; Stroosnijder, L. Rainwater harvesting and management in rainfed agricultural systems in sub-Saharan Africa—A review. *Phys. Chem. Earth* **2012**, *47*, 139–151. [CrossRef]
63. Waithaka, M.M.; Thornton, P.K.; Shepherd, K.D.; Ndiwa, N.N. Factors affecting the use of fertilizers and manure by smallholders: The case of Vihiga, western Kenya. *Nutr. Cycl. Agroecosystems* **2007**, *78*, 211–224. [CrossRef]

64. Molua, E.L. Farm income, gender differentials and climate risk in Cameroon: Typology of male and female adaptation options across agroecologies. *Sustain. Sci.* **2011**, *6*, 21–35. [CrossRef]
65. Dougill, A.J.; Fraser, E.D.G.; Reed, M.S. Anticipating Vulnerability to Climate Change in Dryland Pastoral Systems: Using Dynamic Systems Models for the Kalahari. *Ecol. Soc.* **2010**, *15*, 17. [CrossRef]
66. Grecequet, M.; DeWaard, J.; Hellmann, J.J.; Abel, G.J. Climate Vulnerability and Human Migration in Global Perspective. *Sustainability* **2017**, *9*, 720. [CrossRef]
67. Ayeb-Karlsson, S.; van der Geest, K.; Ahmed, I.; Huq, S.; Warner, K. A people-centred perspective on climate change, environmental stress, and livelihood resilience in Bangladesh. *Sustain. Sci.* **2016**, *11*, 679–694. [CrossRef]
68. Wamsler, C.; Brink, E.; Rentala, O. Climate change, adaptation, and formal education: The role of schooling for increasing societies' adaptive capacities in El Salvador and Brazil. *Ecol. Soc.* **2012**, *17*, 2. [CrossRef]
69. Mpuga, P. Constraints in Access to and Demand for Rural Credit: Evidence from Uganda. *Afr. Dev. Rev.* **2010**, *22*, 115–148. [CrossRef]
70. Njeru, T.N.; Mano, Y.; Otsuka, K. Role of Access to Credit in Rice Production in Sub-Saharan Africa: The Case of Mwea Irrigation Scheme in Kenya. *J. Afr. Econ.* **2016**, *25*, 300–321. [CrossRef]
71. Roseberg, R.J.; Norberg, S.; Smith, J.; Charlton, B.; Rykbost, K.; Shock, C. Yield and quality of teff forage as a function of varying rates of applied irrigation and nitrogen. *Klamath Exp. Station* **2005**, *1069*, 119–136.
72. Paterson, A.H.; Bowers, J.E.; Bruggmann, R.; Dubchak, I.; Grimwood, J.; Gundlach, H.; Haberer, G.; Hellsten, U.; Mitros, T.; Poliakov, A.; et al. The Sorghum bicolor genome and the diversification of grasses. *Nature* **2009**, *457*, 551–556. [CrossRef] [PubMed]
73. Lin, B.B. Resilience in Agriculture through Crop Diversification: Adaptive Management for Environmental Change. *J. Biosci.* **2011**, *61*, 183–193. [CrossRef]
74. Thiede, B.C. Rainfall Shocks and Within-Community Wealth Inequality: Evidence from Rural Ethiopia. *World Dev.* **2014**, *64*, 181–193. [CrossRef]
75. Vandamme, E.; D'Haese, M.; Speelman, S.; D'Haese, L. Livestock against risk and vulnerability: Multi-functionality of livestock keeping in Burundi. In *The Role of Livestock in Developing Communities: Enhancing Multifunctionality*; Swanepoel, F.J.C., Stroebel, A., Moyo, S., Eds.; The Technical Centre for Agricultural and Rural Cooperation (CTA): Cape Town, South Africa, 2010; pp. 107–121.

Article

How do Clusters Foster Sustainable Development? An Analysis of EU Policies

Niki Derlukiewicz [1,*], Anna Mempel-Śnieżyk [2], Dominika Mankowska [3], Arkadiusz Dyjakon [4], Stanisław Minta [3] and Tomasz Pilawka [3]

[1] Department of Macroeconomics, Wroclaw University of Economics and Business, 53-345 Wroclaw, Poland
[2] Department of Spatial Economy and Local Government Administration, Wroclaw University of Economics and Business, 53-345 Wroclaw, Poland; anna.sniezyk@ue.wroc.pl
[3] Institute of Economics Sciences, Wroclaw University of Environmental and Life Science, 50-375 Wroclaw, Poland; dominika.mankowska@upwr.edu.pl (D.M.); stanislaw.minta@upwr.edu.pl (S.M.); tomasz.pilawka@upwr.edu.pl (T.P.)
[4] Institute of Agricultural Engineering, Wroclaw University of Environmental and Life Sciences, 50-375 Wroclaw, Poland; arkadiusz.dyjakon@upwr.edu.pl
*   Correspondence: niki.derlukiewicz@ue.wroc.pl; Tel.: +48-601-541-144

Received: 17 December 2019; Accepted: 6 February 2020; Published: 11 February 2020

Abstract: Sustainable development is one of the fundamental and most important objectives of the worldwide policy. The conducted research shows that sustainable development (SD) is increasingly important in the consciousness of the EU countries, which can be viewed through a prism of the undertaken projects. This paper raises the issue of clusters and their significance in the development of a sustainable economy. The article explores trends in the European Union policy related to sustainable development and clusters. The purpose of this study is to find an answer to the following questions: How can clusters contribute to sustainable development and what are the key factors that ensure this process? To achieve the goal of the article a systematic study of the literature and reports was carried out. Moreover, the analysis of the activity of European clusters in the context of sustainable development was performed. Next, the examples of cluster projects focused on sustainable development were presented. It was shown that the clusters contribute a smarter and sustainable development by succeeding in technological and scientific results, developing new technologies for emerging industries, creating new business activities, enticing major technology companies, and connecting local firms into world-class value systems. Furthermore, the clusters participate actively in sustainable development as they promote knowledge creation, joint learning, technology transfer, as well as collaboration, and sustainable innovations. Finally, clusters facilitate the sustainable upgrading of small and medium enterprises and encourage the participation of stakeholders in the process of sustainable development.

Keywords: cluster; sustainable development; sustainable economy; SDG implementation

---

1. Introduction

Recently, the possibility of spreading the idea of sustainable development (SD) through the development of cluster initiatives was recognized [1]. Connections between clusters and sustainable development are not so precise and the increased interest in SD vs. clusters by academics has been observed in the last decade [2–5].

The basic factors that determine the potential of clusters to implement SD include [6,7]: the ability to influence the strategic goals of enterprises collected in clusters by local governments, building social trust, creating favorable conditions of social acceptance for external effects of enterprise development, increasing interest in innovative activities, including eco-innovation, knowledge transfer between cluster members and business environment institutions.

One of the mentioned priorities is a global partnership, which should result in cooperative and strong partnerships at all levels and between different governments, the private sector, civil society and other parties [8]. To achieve cooperation at the international level and to increase innovation, resulting in new products and technologies aiming at achieving sustainable development groups (SDGs) (e.g., reducing water waste, decrease of environment pollution, poverty reduction, new jobs creation, activation of local community, improving economic indicators) the paper refers to clusters and their relation with sustainable development.

As the clusters are the most common places where one can find education, research, innovation, environmental solutions, and sustainable technologies, the paper raises the issue of clusters and their significance in the development of a sustainable economy as well as explores trends in the European Union (EU) policy related to SD and clusters. As a result, this paper aims to determine the potential of clusters in promoting sustainable development. To achieve the main goal of the paper the following specific objectives will be realized: (i) analysis of the effects of clusters on sustainable development according to the state of the art, (ii) analysis of the directions of EU cluster policy undertaken in last two decades relating to sustainability—EU projects directly or indirectly related to sustainable development, and (iii) an indication of examples of the activities performed by clusters to support sustainable development.

2. Theoretical Framework

2.1. Cluster Concept

Since the 1990s, researchers have emphasized the importance of clusters in economic development in different aspects [9–16]. Such structures became an important part of almost every national and regional economy in countries all over the world [17,18]. The concept of clusters is built on traditional theories of localization and integrates other concepts such as industrial districts, growth poles, systems of production, regional innovation systems, or learning and creative regions. The original concept of the territorially concentrated enterprises was established by Marshall [19] (industrial district) [20]. The concept was developed by researchers who emphasized the results of the interaction between participants who enhance innovation capacity, increase the level of competitiveness and help to achieve a beneficial coefficient of socio-economic development [21–25]. Historically, clusters have been found in traditional industries such as textiles in northern Italy, steel in Pittsburgh or car manufacture in Detroit [26]. Porter [27] merely popularized the theoretical achievements of Marshall; nevertheless, Porter is seen as the precursor of the economic aspect of clusters. According to his definition a cluster is a geographically concentrated, competitive, cooperative group of interrelated companies, specialized suppliers, service providers, and finally, companies operating in related sectors and associated institutions [27,28]. The popularity of cluster structures was also increased by organizations such as UNIDO and OECD after their attempts to use the concept of clusters as a tool for development [29,30] Creating networks and enterprise cooperation contributes to positive effects for entities belonging to this network [31]. The academics highlight that clusters as networks generate benefits for enterprises located in this structure, e.g., easier and affordable access to means of production, distribution channels, human resources, or knowledge and innovation [32]. Benefits for companies and institutions within a cluster comprise profit increase due to lower costs incurred by companies operating within the network, export increase, higher innovativeness, better expansion of knowledge and technological progress, enhanced competitive advantage, faster productivity growth related to the concentration of the resources of innovation absorption capacity [33]. The existence of networks also ensures risk sharing, joint analysis of ideas and initiatives, sharing costs of introducing innovations, availability and possibility to exchange experienced and specialized employees. This confidence of partners contributes to frequent formal and informal contacts between them and the exchange of experience.

The role of the government in creating a cluster is very important. It is worth noting that clusters are more willingly created when there is financial and institutional support from the government. The most popular kinds of support include such incentives as:

(1)   Creation of the offices or the employment of the specialists–advisers financed by the authorities in establishing and running clusters,

(2)   Subsidies for establishing the cluster and the costs of its management (especially in the initial period of operation),

(3)   Facilitating the application for public funds for research and development (in the form of additional points for the mere fact of being a cluster), which gives an advantage compared to non-cluster enterprises,

(4)   Promotion and information activities financed by public authorities to increase the intensity of cluster creation.

2.2. Sustainable Development Concept

The concept of sustainable development (SD) is a concept of development directed towards achieving a balance between social aspects, economic activities, and the environment. The term SD is widely discussed in present political and environmental discourses [34]. Currently, most countries in the world face many sustainability challenges from youth unemployment to the aging population, climate change, pollution, sustainable energy, international migration, and rural area depopulation. The concept of sustainability [35] is not new and has both enthusiasts and opponents, and it is deliberated in various aspects [36,37]. SD is one of the most critical challenges and priorities of the modern world. It is also observed as a strategic trend in global environment protection policy and socio-economic development [38]. Challenges of the present day and willingness to strive for sustainable development result in the creation of programs and projects supporting research and implementation of solutions that take into account economic, environmental and social aspects, and which may be motivators for other market participants (the examples are RUBIZMO [39] and EuroSea [40] projects implemented under the Horizon 2020 program).

SD is more and more often identified in the pro-ecological context [41]. However, it should be remembered that the area of the environment is not the only pillar of this concept. The other two pillars—society and economy—are equally important. The three pillars of SD are fundamental and occur across all sectors of the economy [42]. It should be highlighted that SD recommends that the needs of the future can be met depending on how well social (equity, participation, empowerment, social mobility, and cultural preservation), economic (services, household needs, industrial growth, agriculture growth, and efficient use of labor), and environmental (biodiversity, natural resources, carrying capacity, ecosystem integrity, and clean air and water) objectives or needs are balanced. Very often needs are incompatible; for instance, industrial growth may conflict with the protection of natural resources [43]. Besides, sustainable development can be seen as equalizing opportunities between regions with high development potential (usually large urban agglomerations) and areas with lower development potential (e.g., rural areas, less industrialized regions, etc.) [44]. Moreover, SD is understood as growth based on education, research, and innovation, digital society as well as growth based on a more competitive low-carbon economy. SD makes/involves efficient and sustainable use of resources, protection of the environment (by reducing emissions and preventing biodiversity loss), capitalization on Europe's leadership in developing new green technologies and production methods, as well as the introduction of efficient smart electricity grids [45,46]. From the EU point of view, sustainability should be considered by companies as attractive business opportunities which result in profits [47]. Increased resource efficiency, circular economy solutions, and participation in green markets represent essential opportunities for European small and medium enterprises (SMEs) and their improvement in productivity and boosting their competitiveness.

SD is characterized by various dimensions and results in complex interactions. To follow the changes referring to sustainability in particular areas SD indicators were developed and applied.

They refer to the main areas of SD—social, economic, and environmental [48]—and are based on selected indicators from the global indicator framework and reflect the SD goals (SDGs) [49]. SDGs are presented in the United Nations *2030 Agenda for Sustainable Development*: no poverty; zero hunger; good health and well-being; quality education; gender equality; clean water and sanitation; affordable and clean energy; decent work and economic growth; industry, innovation, and infrastructure; reduced inequalities; sustainable cities and communities; responsible consumption and production; climate action; life underwater, life on land; peace, justice, and strong institutions; and finally, partnership for the goals [8].

3. Material and Methods

To elaborate on the necessary information about clusters in the context of SD, the authors conducted desk research. The identification of the papers concerning SD and the cluster concept was performed. This part of desk research was devoted to establishing the theoretical framework of the EU cluster policy directions and to broadening the knowledge related to the dependence between clusters and SD. The desk research was supported by Google Scholar, Emerald, EBSCO databases and websites dedicated to clusters and SD concepts. The analysis allowed the authors to systematize the knowledge and definitions related to the examined phenomenon. To research SD in the cluster's context it was decided to perform Boolean keyword and subject term searches in EBSCO, Emerald, and Google Scholar databases using search phrases reflecting the phenomena and Boolean search operators such as AND and NEAR, with the phrases: sustainable development, environmental development. The searching was realized on 10–12 July 2019.

Moreover, the research was also conducted based on the authors' expert knowledge resulting from research and analyses of reports, case studies, and the following interactive maps of the EU e-platforms dedicated to clusters: European Cluster Collaboration Platform, Excellence Cluster for Regional Improvement Website, Green and Cluster Excellence Programme Website. The platforms were analyzed to check the directions of the EU's activities in the context of SD, to provide knowledge on actions and types of activities resulting in the improvement of sustainable development and to enable the assessment of activities carried out in the context of clusters vs. SD.

To complete the research and to achieve the goal of the paper, examples of the activities performed by clusters to support sustainable development were selected. Moreover, all cluster partnerships were analyzed that were realized under the European Strategic Cluster Partnerships for Smart Specialization Investments (ESCP-S3) and European Strategic Cluster Partnerships for Going International (ESCP-4i). Then, using the method of deduction and inference, a list of cluster activities that contribute to sustainable development was indicated. Finally, the analysis of the Cluster Partnership recently created in the EU and an assessment of the activity of chosen clusters and its contribution to sustainable development was performed.

4. Results

The desk research proved that clusters are an essential part of the contemporary economic development of most countries and regions. Clusters are considered as engines of growth and have become a popular policy tool for boosting economic growth and innovation.

The coexistence of clusters with sustainable development/sustainable economy and sustainability in scholarly publications in the years 1990–2019 was checked (Table 1).

As seen from conducted research, in the last two decades a growing number of the papers referring to the issue of clusters concerning SD was observed. Therefore, a more detailed analysis was conducted for the years 2000–2019, as confirmed in the research included in the EBSCO, EMERALD, and Google Scholar databases. The used operator AND allowed us to distinguish the papers related to clusters and SD in the content. In the EBSCO database the number of results of "cluster AND sustainable development" was significantly lower (one in 2002, 47 in 2017, and 13 in 2018–2019) than in EMERALD and Google Scholar. The results of the search in EMERALD and Google Scholar databases are presented in Figures 1 and 2, respectively.

Table 1. Cluster and sustainable development—literature analysis.

Keyword Boolean Operators	EBSCO	Emerald	Google Scholar
Cluster AND sustainable development	159	9511	847,000
Cluster NEAR sustainable development	2695	309	19,700
Cluster AND sustainable economy	9	6852	334,000
Cluster NEAR sustainable economy	1031	254	239,000
Cluster AND sustainability	182	6519	739,000
Cluster NEAR sustainability	195	241	334,000

Source: own research.

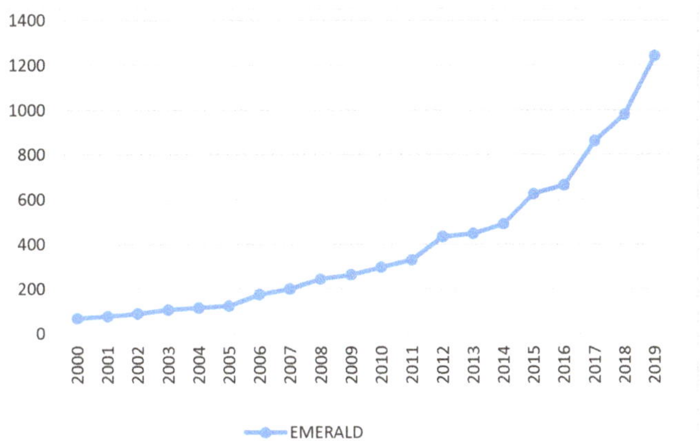

Figure 1. Number of articles containing the search "cluster" and "SD" (2000–2019). Source: Own elaboration based on EMERALD (between 1 January 2000 and 31 December 2019).

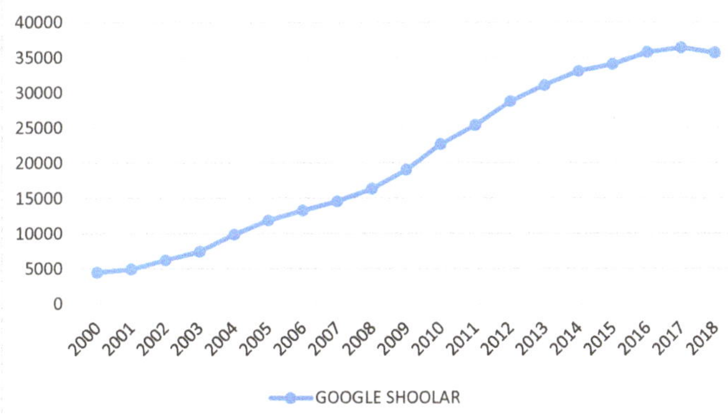

Figure 2. Number of articles containing the search "cluster" and "SD" (2000–2019). Source: Own elaboration based on Google Scholar database (between 1 January 2000 and 1 December 2019).

In the Emerald database, a constant increase in the number of analyzed papers was observed: from 69 in 2000 to 1243 in 2019, while in the Google Scholar database, the number of papers increased from 4460 in 2000 to 36,400 in 2017. In 2018 the upward trend stopped, and in 2019 there were 35,700 papers.

The issue of clusters and their significance in SD is the subject of theoretical and empirical studies [50–52]. The research also concerns the role of clusters in regional and economic development [53, 54]. It is also recognized as an extensive solution to the existing urban planning issues, the aim of which is providing a better quality of life. The cluster approach is also used in connection with the concept of smart specialization to increase the efficiency and effectiveness of economic systems contributing to sustainable development simultaneously [55]. In smart specialization, strategies of countries and regions clusters are a fundamental component and they bring together universities, local authorities, and businesses. For business, clusters are a platform through which they can have many benefits, i.e., easier access to the appropriate information, participation in research projects, better access to financing sources, knowledge exchange, cooperation with R&D and public sector institutions [56]. Universities and research centers take advantage of being a part of clusters as they have great opportunities to realize projects in cooperation with the business sector. Finally, the public sector uses clusters to communicate more efficiently with private actors/players and design cluster policies more competently [57]. Thus, there are many benefits of being in a cluster for all entities in general.

However, also some disadvantages of clusters are indicated. For example, in a situation of economic downturn in a given industry, with excessive concentration and scale of cluster activity, there is a risk that a given region will be overly dependent on one industry/sector, i.e., there will be an imbalance in the structure of the region's economy (too high a share of one industry and too small a share of other industries). Usually, the cluster helps in the development of the region, but the collapse in the market, where it operates, may lead to a regional crisis. A good example is the city of Detroit in the U.S.A. When the car industry in the U.S.A. collapsed, the city was depopulated and the industries dependent on the automotive industry collapsed. As a result, the population's wealth declined and the unemployment index increased significantly, and a problem with the entire city's budget for its development appeared.

Nonetheless, the awareness of the European Commission of the importance of the cluster in sustainable economic development forced them to undertake different projects indirectly related to the improvement of the environment. As the effect of the EU activities, the European Cluster Alliance was established, constituting a network of institutions representing central and regional authorities and development agencies that support transnational cooperation between regions in the field of cluster policy development. The main aim of EU actions is to intensify cluster and business network collaboration across borders and sectoral boundaries [58]. The EU actions enable European cluster cooperation, exchange of knowledge and experience by participation in international projects. As the result of undertaking activities in 2014, the European Strategic Cluster Partnerships (ESCPs) under the COSME program was established and two projects were undertaken to boost industrial competitiveness and investment within the EU: the European Strategic Cluster Partnerships for Smart Specialization Investments (ESCP-S3) and European Strategic Cluster Partnerships for Going International (ESCP-4i). Moreover, the EU promoted projects for cluster excellence and new industrial value chains (13 INNOSUP-1 projects) [59].

In 2018 the nine European Strategic Cluster Partnerships for Smart Specialization Investments was launched and they enabled cluster cooperation in thematic areas related to regional smart specialization strategies and increased the participation of the industry. All cluster partnerships are presented in Table 2.

Most cluster partnerships activities (Table 2) stress the importance of the environment and concentrate on increasing knowledge and cooperation, technology transfer, new business collaborations, and what is the most important industrial modernization. It is expected that all planned activities will contribute not only to increased international cooperation between companies but also to more sustainable development in the EU.

Table 2. European Strategic Cluster Partnerships for Smart Specialization Investments (ESCP-S3) and its contribution to sustainable development.

Acronym	Sector	S3 EU Priority Areas	Expected Results on Sustainable Development
TRACK	Agricultural Inputs and Services	Sustainable innovation, Sustainable agriculture	Adoption of new technologies for improving efficiency and traceability in various vegetal-based agri-food chains, modernization of the vegetal-based agri-food chain.
TEX4IM	Textile Manufacturing	Manufacturing and Industry, Textiles, wearing apparel and leather and related products	Generation of a long-term partnership for boosting industrial competitiveness and investment in European textile and clothing sector, set up a sustainable incubator and accelerator system for the generation of joint investment projects in the textile and clothing sector. Joint strategy for industrial modernization of the telecommunication (TC) sector of Europe.
S3martMed	Biopharmaceuticals	Human health and social work activities, Human health activities (medical services)	Foster cooperation between European clusters and their small and medium enterprise (SME) members in the field of medical technologies.
EACP-EUROSME	Aerospace Vehicles and Defence	Aeronautics and Space, Aeronautics	Reconsideration of the present sector conditions, stressing the importance of the environment with new cluster strategies that support the circular economy across the entire aerospace value chain and life cycle.
EACN	Automotive	Advanced manufacturing systems	Industrial modernization in the automotive industry, investing in a smart, innovative and sustainable industry.
DIGICLUSTERS	Food Processing and Manufacturing	Information and Communication Technologies (ICT), Computer programming, consultancy and related activities	Prototyping new value chains and emerging industries based on combined competencies of consortium partners, focusing on developing competitive next generation and added-value products and services through the innovative approach of intra-regional and interregional hackathons.
CYBER SECURE LIGHT	Lighting and Electrical Equipment	Digital Agenda, ICT trust, cybersecurity and network security	Knowledge and cooperation assets concerning the cybersecurity implementation, strengthen with insights on cybersecurity in smart building and related innovation investments, investment for pilot business initiatives/technology transfer projects.
CONNSENSYS	Food Processing and Manufacturing	Manufacturing and Industry, Food, beverage and tobacco products	More interconnected, resilient and smart agri-food system in Europe, validation and implementation of IoT solutions based on smart electronic systems and embedded technology in the food industry on a wide scale (from multinationals to locally operating SMEs).
AI4Diag	Biopharmaceuticals	Manufacturing and Industry, Biotechnology	Industrial modernization of diagnostics companies to reinforce their international leadership.

Source: own elaboration based on [59].

As the first example, the European Automotive Cluster Network for Joint Industrial Modernisation (EACN) is presented [60]. Six partnerships clusters are represented: Poland, Spain, Bulgaria, France, and Serbia. The partnerships resulted in the engagement of 560 enterprises and were aimed at initiating joint R&D projects concerning the virtualization of processes, robotics and artificial intelligence, the elasticity of production, and skills and competencies. EACN's contributions are directed towards areas such as efficient and sustainable manufacturing or entrepreneurs' integration in Industry 4.0.

The next partnership [61] refers to Cyber Secure IoT Lighting and Home Automation systems for Smart Building (CYBER SECURE LIGHT) and refers to ICT/Digitalization—Lighting and Electrical Equipment sector. The partnership enables seven clusters from Italy, Poland, France, Spain, Hungary, and Slovenia to engage their 1115 representatives. The most important CYBER SECURE LIGHT partnership's aims, from the SD point of view, are: fostering interregional business-to-business collaboration deals for innovation and smart investments, cooperation ensuring company growth in the smart building/IT/cyber, technology transfer projects and cooperative agreements, co-learning and knowledge sharing, supporting the development of cluster bridges with other complementary ecosystem actors to enlarge the CYBER SECURE LIGHT consortium and investigate new development and cooperation opportunities [61].

To strengthen clusters, to internationalize them, to enable their transfer of knowledge, and to make SDGs achievable for them the Cluster Excellence Programme [62] was introduced. The project aimed also at strengthening cluster management excellence in the EU. The program was organized as two editions and focused among other things on providing top quality services to cluster enterprises, facilitating internationalization, developing strategic plans and action plans for the sustainable development of clusters. As the result of the Cluster Excellence Programme 11 projects were composed by partnerships of clusters representing various countries (the proposal has been continued). Thanks to this project, companies gain an opportunity for knowledge and experience exchange, spreading innovation, participation in workshops, study visits and training, etc. Moreover, partners improve and test new management skills, and SMEs strengthen their competitiveness [62].

The next initiative of the EU contributing to knowledge transfer and skills improvement is the Smarter Cluster Policies for southeast Europe (ClusterPoliSEE) [63]. The project aims at developing and implementing effective smart specialization strategies for cluster development and concentrates on eco-innovation. This project gives opportunities for sustainable business through collaborative R&D projects leading to greater sustainability and economic diversification [64].

Moreover, to present the activity of European clusters in the context of SD, an analysis of European clusters referring to their activities in sustainable development was conducted. As results from the conducted research show, clusters are conscious of SD importance and pay attention to the development in accordance with sustainability. To specify the cluster activities resulting in SD, it is worth presenting particular examples of clusters and their undertaken initiatives to strengthen, promote and ensure sustainable socio-economic development. Clusters are networking engines which facilitate the interactions among their members at regional, national, trans-national, international or sectoral level. Participation in clusters facilitates the exchange of new knowledge, liaising and offers various opportunities for the members. But what is most important is that more and more often clusters contribute to the development of a sustainable economy. In the EU there are clusters that directly and indirectly foster SD. Clusters that directly influence SD operate in sectoral industries such as sustainable energy, automotive and environmental services. Many clusters operate in the EU but only some of them are presented in this paper. For example, TWEED cluster [65] is a sustainable energy cluster located in Belgium. The cluster aims to set up high-quality projects in the fields of production and exploitation of sustainable energy: renewable energy sources and the implementation of new processes to achieve energy savings, energy efficiency or the reduction of greenhouse gas emissions, including CO_2. This cluster consists of companies and R&D institutions cooperating in the following sectors: solar energy, wind energy, biomass, energy efficiency, smart grid, and green products and services. The activities of TWEED include [65]:

- Networking between industrial or commercial companies and other actors of sustainable energy sectors,

- Developing synergies with other actors of sustainable energy sectors (including other clusters),
- Promoting cluster members locally and internationally,
- Carrying out industrial, technical, market and economic studies on the sustainable energy sector.

The Electric Vehicles Industrial Cluster [66] of Bulgaria is the next example. The cluster constitutes a good practice example of a cluster initiative directly promoting sustainability aimed at achieving SD, improved competitiveness of the cluster members and a sustainable and clean environment. The cluster aims among other things at industrial investment projects for technological renewal and introduction of innovations to reduce the energy intensity of transport schemes, implementation of models of mass urban transport by electric buses, building sustainable educational models, etc. [66].

There are also clusters in which activity indirectly contributes to SD. For example, the Upper Rhine Cluster for Sustainability Research (URCforSR) [67] is a cluster located in Germany that engages various universities of applied sciences and research institutes. It aims to establish a research association with European significance in such fields of study as [67] governance, energy, infrastructure, societal change, transformation processes and technology, resource management, and multiculturalism and multilingualism. Cooperation in the clusters allows the exchange and transfer of knowledge as well as leads to interdisciplinary research on the governance of sustainable growth. Moreover, the cluster conducts different projects for protection against pollution or to increase renewable energies [67]:

- NAVEBGO—Sustainable reduction of biocide input into the groundwater of the Upper Rhine,
- SMI: Inclusive Smart Meter—artificial intelligence to support proactive control of energy consumption by end-users,
- SuMo-Rhine—promotion of Sustainable Mobility in the Upper Rhine Region,
- The cluster also promotes SD in its book *Sustainability Governance and Hierarchy* [68]. The book analyzes the concept of sustainability governance through sustainability and environmental studies.

Another example of a cluster that can contribute indirectly to SD is the Green and Sustainable Finance Cluster in Germany (GSFCG) [69]. It aims at developing user-oriented concepts for the implementation of sustainable criteria in the business models of the various stakeholders operating in the financial center. This cluster also promotes investing in private capital in accordance with sustainability [69].

Studies have shown the increasing trend in the EU of activities referring to the importance of clusters in the context of sustainability. The directions undertaken by the EU have a positive effect on the awareness of companies operating in clusters by broadening their knowledge concerning sustainable development, as well as encouraging them to sustainable activities. However, the main idea is wider and, next to the international projects, worldwide actions are promoted concerning third world countries.

5. Discussion

Among numerous benefits that clusters bring [70] this study focused on their influence on the development of a sustainable economy [51]. Clusters can lead to smarter and sustainable development by succeeding in technological and scientific results, developing new technologies for emerging industries, enticing major technology companies, and connecting local firms with world-class value systems [46]. The literature also provides cluster activity in the field of environment protection [71]: technological activities (e.g., energy, biomass, water treatment, innovative solutions in reducing storage, recovery, and recycling of rain), educational activities (e.g., training in the use of renewable energy issues, energy efficiency, biomass, photovoltaic, wind and water turbines, geothermal energy and biogas plants) and research activities (e.g., environmental studies). Dyrda-Muskus [71] underlines the environmental advantages of clusters activities. However, there are more factors by which clusters can support achievements in the area of sustainable development. Interesting research is presented by

Borkowska-Niszczota [72], i.e., classification and analysis of factors stimulating the impact of a tourist cluster on SD. Some research referring to SD concerns SMEs belonging to clusters and their further development, as well as their corporate social responsibility (CSR) [31,52].

Moreover, the research related to the investigated issue is presented by Bankova and Slavova-Georgieva [73] on a Bulgarian example. The research was based on the indicators (ecological sustainability index, social progress index, and state of cluster development). The possibilities of clusters in contributing to sustainable development were checked in the context of cluster development and CSR of enterprises belonging to clusters [73].

The EU policymakers, aware of the role of clusters in sustainable development, decided to make an effort to assist the growth of clusters. Cluster policies are widely encouraged by international authorities, such as the European Union and OECD, and have grown worldwide both at regional and national levels [74,75]. The policy actions used by policymakers comprise direct and indirect financial support, start-up support, aid for administration, networks, and cooperation, and general assistance for cluster activities [74]. The models of cluster policy in particular countries vary significantly. Generally, this policy is based on activities focused on the creation of an environment favorable for cluster development, such as ensuring appropriate financial instruments and improvement of coordination channels, as well as supporting science-business cooperation [76].

However, these activity costs are financed from public funds, i.e., taxes from the whole society and other state or regional revenues, which means that there may be a shortage of funds or the reduction of other social or development-oriented activities. Moreover, supporting clusters from public funds gives them an advantage over other enterprises also wanting/expecting to develop, but not being a part of cluster networks. So, there is a doubt about the equal treatment of all entities operating in a given sector/region. Another question is connected with the cost-benefit analysis, i.e., whether society has measurably more benefits from creating clusters than publicly funded expenditures.

Nevertheless, the EU underlines the trend of internationalization of clusters in accordance with Agenda 2030 development goals and global partnerships. As a result of changes taking place in the modern world and economy, a new stage in the development of clusters is promoted—the stage of foreign expansion and international cooperation. Several activities aimed at accelerating the process of increasing innovation and competitiveness of the European economy are realized in particular countries and regions in the EU.

Connections between clusters and sustainable development are not so precise and the increased interest in SD vs. clusters by academics is observed in the last decade. The links between business and SD refer to different types of innovations (sustainable innovation or eco-innovation) that have a reduced negative impact on the environment [77].

6. Conclusions

As a result of the undertaken actions in 2016–2017, about 150 cluster organizations across 23 European countries were involved. They developed and implemented joint strategies to support the internationalization of enterprises. Moreover, collaboration projects were implemented among EU clusters, as well as clusters and international organizations from outside the EU. Thanks to international cooperation and exchange of knowledge and experience, more and more clusters are aware of how they can influence the SD.

The selected examples of the clusters' contributions to sustainable development were presented. It is worth noting that clusters are the places where sustainability can be achieved. As evidenced from the research, clusters create synergies between businesses and universities, and research and reinforce links between them using existing strengths and opportunities. Clusters ensure the continuous development of new technologies resulting, first of all, from the possibility of cooperation among different entities and involvement in research of scientific units [78]. Clusters constitute an integral part of the innovative environment and play a significant function by providing a fast flow of information and effective acquisition of knowledge in the process of absorption of innovation. Functioning in a network allows the use of highly specialized knowledge that is not available directly on the market. The development of high-tech sectors is one of the possible methods of strengthening the competitive position. Clusters are characterized by highly innovative entities and institutions, mainly due to the specific conditions of the high technology sector and the created environment. By participating in EU projects clusters meet SDGs [79,80], increasing their influence on the internal and external market.

Thus, the clusters can foster SD in various direct or indirect manners (promoting activities and education). The analysis carried out allowed the authors to formulate the following conclusions and recommendations:

- Clusters can play an important role in supporting the SD process,
- The last decade provided a lot of research on clusters and some of it refers to SD,
- The cluster policy in the EU contributes to strengthening the position of clusters in achieving SD,
- There are many initiatives in the EU aimed at cluster activity enhancement, and consequently at SD.

However, there are some challenges associated with clusters. For example, cluster companies gain a competitive advantage over other companies outside of clusters, which weakens the other companies and may lead to their collapse or industry change. This may contribute to an increase in the concentration of production, income, and resources in a particular place (cluster). Excessive concentration of capital and resources in an ever-smaller group of people and entities may lead to negative stratification of income and social inequalities due to the wealth, and this is contrary to the assumptions of sustainable development. There is also a risk that some clusters are created only to collect grants, and then their activity ends (the question that arises here is how many clusters are formed due to the desire to network activities, and how many to collect grants). Besides, not all clusters contribute to sustainable development (i.e., heavy industrial clusters) and it is a challenge for governments and management of clusters to encourage them to undertake sustainable activities.

It seems that future cluster policy should support particularly the development of those clusters that contribute to SD and energy efficiency through pro-ecological solutions, ensuring the effects on SD. It is important to elaborate on a cluster development strategy focused on SD by including a document/agreement for the entire structure, involving all members and referring to the principles of SD. Only in this way will SD become an objective of every particular company or organization in the cluster. The performed analysis indicated that there is a research gap in the field of the constant measurement of clusters in the context of SD.

Author Contributions: Conceptualization, N.D., A.M.-Ś., D.M.; methodology, N.D. and A.M.-Ś.; validation, N.D., A.M.-Ś., A.D., S.M., and T.P.; formal analysis, A.D., S.M., D.M.; investigation, N.D. and A.M.-Ś.; resources, A.D., S.M., D.M.; data curation T.P.; writing—original draft preparation, N.D. and A.M.-Ś. All authors have read and agreed to the published version of the manuscript.

Funding: The project is co-financed by the Ministry of Science and Higher Education in Poland under the program Regional Initiative of Excellence 2019–2022, project number 015/RID/2018/19, total funding amount 10 721 040,00 PLN.

Conflicts of Interest: The authors declare no conflict of interest.

Abbreviations

SD	sustainable development
SDG	sustainable development group
CSR	corporate social responsibility
SMEs	small and medium enterprises
EU	European Union
TC	Telecommunication

References

1. Yang, J.; Černevičiūtė, J. Cultural and Creative Industries (CCI) and sustainable development: China's cultural industries clusters. *Entrep. Sustain. Issues Entrep. Sustain. Cent.* **2017**, *5*, 231–242. [CrossRef]
2. Glinskiy, V.; Serga, L.; Chemezova, E.; Zaykov, K. Clusterization economy as a way to build sustainable development of the region. *Procedia CIRP* **2016**, *40*, 324–328. [CrossRef]
3. Kuraś, P. Cluster management standards in Poland in the context of sustainable development. *Glob. J. Environ. Sci. Manag. (GJESM)* **2019**, *5*, 41–50.
4. Zaburanna, L.; Yarmolenko, Y.; Kozak, M.; Artyukh, T. Modelling of regional clusters considering sustainable development. *Adv. Econ. Bus. Manag. Res.* **2019**, *99*, 222–226.
5. Sonetti, G.; Lombardi, P.; Chelleri, L. True green and sustainable university campuses? Toward a clusters approach. *Sustainability* **2016**, *8*, 83. [CrossRef]
6. Scheller, D.; Thörn, H. Governing 'Sustainable Urban Development' through self-build groups and co-housing: The cases of Hamburg and Gothenburg. *Int. J. Urban Reg. Res.* **2018**, *42*, 914–933. [CrossRef]
7. Raszkowski, A.; Bartniczak, B. Towards sustainable regional development: Economy, society, environment, good governance based on the example of Polish regions. *Transform. Bus. Econ.* **2018**, *17*, 225–245.
8. European Commission (EC). *Sustainable Development in the European Union. Monitoring Report on Progress Towards the SDGs in an EU Context*; Eurostat: Brussels, Belgium, 2019. Available online: https://ec.europa.eu/eurostat/documents/3217494/9940483/KS-02-19-165-EN-N.pdf/1965d8f5-4532-49f9-98ca-5334b0652820 (accessed on 2 December 2019).
9. Swann, G.M.P.; Prevezer, M.; Stout, D. (Eds.) *The Dynamics of Industrial Clustering: International Comparisons in Computing and Biotechnology*; Oxford University Press: Oxford, UK, 1998.
10. Meyer-Stamer, J. Strategien lokaler/regionaler Entwicklung: Cluster, Standortpolitik und systemische Wettbewerbsfähigkeit. *Nord-Süd Aktuell* **1999**, *13*, 447–460. Available online: http://www.meyer-stamer.de/1999/nsa.pdf (accessed on 8 December 2019).
11. Voyer, R. Knowledge-based industrial clustering: International comparisons. In *Local and Regional Systems of Innovation*; De la Mothe, J., Paquet, G., Eds.; Economics of Science, Technology and Innovation; Springer: Boston, MA, USA, 1998. [CrossRef]
12. Simmie, J. Innovation and clustering in the globalised international economy. *Urban Stud.* **2004**, *41*, 1095–1112. [CrossRef]
13. Martin, R.; Sunley, P. Deconstructing clusters: Chaotic concept or policy panacea? *J. Econ. Geogr.* **2003**, *3*, 5–35. [CrossRef]
14. Njos, R.; Jakobsen, S.E. Cluster policy and regional development: Scale, scope and renewal. *Reg. Stud. Reg. Sci.* **2016**, *3*, 146–169. [CrossRef]
15. Vlasceanu, C. Impact of clusters on innovation, knowledge and competitiveness in the Romanian economy. *Econ. Ser. Manag.* **2014**, *17*, 50–60.
16. Breschi, S.; Malerba, F. The geography of innovation and economic clustering: Some introductory notes. *Ind. Corp. Chang.* **2001**, *10*, 817–833. [CrossRef]
17. Ahuja, G. Collaboration Networks, Structural Holes, and Innovation: A Longitudinal Study. *Adm. Sci. Q.* **2000**, *45*, 425–455. [CrossRef]
18. OECD. Innovative Clusters. Drivers of National Innovation System. 2000. Available online: www.oecd.org (accessed on 19 October 2019).
19. Marshall, A. *Principles of Economics*, 8th ed.; Macmillan: London, UK, 1920.
20. Li, J.; Webster, D.; Cai, J.; Muller, L. Innovation clusters revisited: On dimensions of agglomeration, institution, and built-environment. *Sustainability* **2019**, *11*, 3338. [CrossRef]
21. Bagnasco, A. *Tre Italie. La Problematica Territoriale Dello Sviluppo Italiano*; Il Mulino: Bologna, Italy, 1977.
22. Becattini, G. Dal settore industriale al distretto industriale: Alcune considerazioni sullunità di indagine delleconomia industrial. *Riv. Di Econ. E Politica Ind.* **1979**, *5*, 7–21.
23. Dahmén, E. *Svensk Industriell Företagsverksamhet, Kausalanalys av den Industriella Utvecklingen 1919–1939*; Industrins Utredningsinstitut, IUI: Stockholm, Sweden, 1950.
24. Perroux, F. Economic Spaces: Theory and application. *Q. J. Econ.* **1950**, *64*, 90–97. [CrossRef]

25. Pyke, F.; Sengenberger, W. Industrial districts and local economic regeneration: Research and policy issue. In *Industrial Districts and Local Economic Regeneration*; Pyke, F., Sengenberger, W., Eds.; International Institute for Labor Studies: Geneva, Switzerland, 1992.

26. Kuah, A.T.H. Cluster theory and practice: Advantages for the small business locating in a vibrant cluster. *J. Res. Mark. Entrep. Bus.* **2002**, *4*, 206–228. [CrossRef]

27. Porter, M.E. *The Competitive Advantage: Creating and Sustaining Superior Performance*; Free Press: New York, NY, USA, 1985.

28. Porter, M.E. Location, Competition, and Economic Development: Local Clusters in a Global Economy. *Econ. Dev. Q.* **2000**, *14*, 15–34. [CrossRef]

29. Simmie, J.; Sennett, J. Innovative Clusters: Global or Local Linkages? *Natl. Inst. Econ. Rev.* **1999**, *170*, 87–98. [CrossRef]

30. UNIDO. Industrial Development Report 2009. *Development* **2009**, *12*, 99. [CrossRef]

31. Ginevicius, R. The effectiveness of cooperation of industrial enterprises. *J. Bus. Econ. Manag.* **2010**, *11*, 283–296. [CrossRef]

32. Lin, H.P.; Hu, T.S. Knowledge interaction and spatial dynamics in industrial districts. *Sustainability* **2017**, *9*, 1421. [CrossRef]

33. Li, S.; Han, S.; Shen, T. How can a firm innovate when embedded in a cluster? Evidence from the automobile industrial cluster in China. *Sustainability* **2019**, *11*, 1837. [CrossRef]

34. Emma Pravitasari, A.; Rustiadi, E.; Pratika Mulya, S.; Nursetya Fuadina, L. Developing regional sustainability index as a new approach for evaluating sustainability performance in Indonesia. *Environ. Ecol. Res.* **2018**, *6*, 157–168. [CrossRef]

35. Zulfiquar, A.N.; Kant, R. A state-of-art literature review reflecting 15 years of focus on sustainable supply chain management. *J. Clean. Prod.* **2017**, *142*, 2524–2543. [CrossRef]

36. Herrmann, M. The challenge of sustainable development and the imperative of green and inclusive economic growth. *Mod. Econ.* **2014**, *5*, 42983. [CrossRef]

37. Robert, K.W.; Parris, T.M.; Leiserowitz, A.A. What is sustainable development? Goals, indicators, values, and practice. *Environ. Sci. Policy Sustain. Dev.* **2005**, *47*, 8–21. [CrossRef]

38. Megyesiova, S.; Lieskovska, V. Analysis of the sustainable development indicators in the OECD countries. *Sustainability* **2018**, *10*, 4554. [CrossRef]

39. RUBIZMO (Replicable Business Models for Modern Rural Economies—Grant Agreement ID: 773621). Available online: www.rubizmo.eu (accessed on 25 January 2020).

40. EUROSEA (Improving and Integrating European Ocean Observing and Forecasting Systems for Sustainable Use of the Oceans—Grant Agreement ID: 862626). Available online: https://www.eurosea.eu/ (accessed on 25 January 2020).

41. Socińska, J. Klasty jako czynnik zrównoważonego rozwoju (Clusters as a factor of sustainable development). *J. Agribus. Rural Dev.* **2012**, *3*, 251–259. (In Polish)

42. Fischer, J.; Helman, K.; Zeman, J. Sustainable development indicators at the regional level in the Czech Republic. *Statistika* **2013**, *93*, 5–19.

43. Hopwood, B.; Mellor, M.; O'Brien, G. Sustainable development: Mapping different approaches. *Sustain. Dev.* **2005**, *13*, 38–52. [CrossRef]

44. Minta, S.; Tańska-Hus, B.; Nowak, M. Koncepcja wdrożenia produktu regionalnego "Wołowina Sudecka" w kontekście ochrony środowiska (The implementation of the concept of a regional product "Sudeten beef" in the context of environment protection). *Annu. Set Environ. Prot. (Rocz. Ochr. Środowiska)* **2013**, *15*, 2887–2898.

45. Derlukiewicz, N.; Mempel-Śnieżyk, A. European cities in the face of sustainable development. *Ekon. I Prawo* **2018**, *17*, 125–135. [CrossRef]

46. Scheel, C. Knowledge clusters of technological innovation systems. *J. Knowl. Manag.* **2002**, *6*, 356–367. [CrossRef]

47. Bumberova, V.; Milichovsky, F. Sustainability development of KIBS: Strategic actions and business performance. *Sustainability* **2019**, *11*, 5136. [CrossRef]

48. United Nations. *Indicators of Sustainable Development: Guidelines and Methodologies*; United Nations Publications: New York, NY, USA, 2007. [CrossRef]

49. United Nations. *Transforming Our World: The 2030 Agenda for Sustainable Development*; 2015; Available online: https://sustainabledevelopment.un.org/content/documents/21252030%20Agenda%20for%20Sustainable%20Development%20web.pdf (accessed on 9 September 2019).

50. Abrahams, G. Constructing definitions of sustainable development. *Smart Sustain. Built Environ.* **2017**, *6*, 34–47. [CrossRef]

51. Sen, S.K.; Ongsakul, V. Clusters of sustainable development goals: A metric for grassroots implementation. *Sustainability* **2018**, *11*, 118–126. [CrossRef]

52. Srovnalikova, P.; Haviernikova, K.; Guscinskiene, J. Assessment of reasons for being engaged in clusters in terms of sustainable development. *J. Secur. Sustain. Issues* **2018**, *8*, 103–112. [CrossRef]

53. Ketels, C.; Lindqvist, G.; Solvell, O. Strengthening clusters and competitiveness in Europe. In *The Role of Cluster Organisations*; The Cluster Observatory, Stockholm School of Economics: Stockholm, Sweden, 2012.

54. Knauseder, J. Business Clusters as Drivers of Sustainable Regional Development? An Analysis of Cluster Potentials for Delivering Sustainable Development in Regions-With a Case Study of the Mexican Automotive Cluster Saltillo-Ramos Arizpe. 2009. Available online: https://pdfs.semanticscholar.org/b042/709331ed9341be31d9181b012226c4e0a5d0.pdf?_ga=2.26568765.1242069040.1580225972-1718181064.1580225972 (accessed on 15 October 2019).

55. Del Castillo, J.; Paton, J.; Saez, A. Smart Specialisation and Clusters: The Basque Country Case. 2013. Available online: http://www.reunionesdeestudiosregionales.org/Oviedo2013/htdocs/pdf/p880.pdf (accessed on 25 October 2019).

56. Pasha, A. Role of entrepreneurial universities, research centers and economic zones in driving entrepreneurship and innovation in cluster ecosystems. *SSRN Electron. J.* **2019**. [CrossRef]

57. Anic, I.D.; Corrocher, N.; Morrison, A.; Aralica, Z. The development of competitiveness clusters in Croatia: A survey-based analysis. *Eur. Plan. Stud.* **2019**, *27*, 2227–2247. [CrossRef]

58. Slaper, T.F.; Harmon, K.M.; Rubin, B.M. Industry clusters and regional economic performance: A study across U.S. Metropolitan Statistical Areas. *Econ. Dev. Q.* **2018**, *32*, 44–59. [CrossRef]

59. EU Cluster Partnerships. 2019. Available online: https://www.clustercollaboration.eu/eu-cluster-partnerships (accessed on 9 October 2019).

60. European Automotive Cluster Network for Joint Industrial Modernisation (EACN). Available online: http://silesia-automotive.pl/index/?id=44ac09ac6a149136a4102ee4b4103ae6 (accessed on 9 June 2019).

61. Cyber Secure IoT Lighting and Home Automation Systems for Smart Building (CYBER SECURE LIGHT). Available online: https://www.clustercollaboration.eu/partner-search/s3-partnership-cyber-secure-iot-lighting-and-home-automation (accessed on 9 October 2019).

62. Cluster Excellence Programme. Available online: https://www.clustercollaboration.eu/eu-initiative/cluster-excellence-calls (accessed on 9 June 2019).

63. Smarter Cluster Policies for South East Europe (ClusterPoliSEE). Available online: http://www.southeast-europe.net/en/projects/approved_projects/?id=168 (accessed on 9 June 2019).

64. Excellence Cluster for Regional Improvement Website. Available online: https://www.clustercollaboration.eu/eu-project-profile/ecri-excellence-cluster-regional-improvement (accessed on 9 November 2019).

65. Technology of Wallonia Energy, Environment and Sustainable Development Website. Available online: https://clusters.wallonie.be/tweed-en/ (accessed on 11 November 2019).

66. Electric Vehicles Industrial Cluster. Available online: http://www.emic-bg.org/?lang_id=2/ (accessed on 11 November 2019).

67. The Upper Rhine Cluster for Sustainability Research. Available online: https://www.sustainability-upperrhine.info/de/home/ (accessed on 20 September 2019).

68. Hamman, P. *Sustainability Governance and Hierarchy*; Routledge: London, UK, 2019. [CrossRef]

69. Green and Sustainable Finance Cluster in Germany (GSFCG) Website. Available online: https://gsfc-germany.com/sustainable-finance-status-quo-und-innovation/ (accessed on 8 November 2019).

70. Nallari, R.; Griffith, B. *Clusters of Competitiveness*; World Bank: Washington, DC, USA, 2013. [CrossRef]

71. Dyrda-Muskus, J. Polish economic clusters and their efforts to protect the environment—Selected examples. *J. Perspect. Econ. Political Soc. Integr.* **2012**, *20*, 25–36. [CrossRef]

72. Borkowska-Niszczota, M. Produkt turystyczny. Innowacje-marketing-zarządzanie. In *Przedsiębiorczość I Zarządzanie*; Dębski, M., Żuławska, U., Eds.; Wydawnictwo SAN: Łódź-Warszawa, Poland, 2019; Volume 20, pp. 155–170. (In Polish)

73. Slavova-Georgieva, I.; Bankova, Y. The role of clusters for sustainable development: Socially responsible practices, limitations and challenges. Case Study of a Bulgarian Industrial Cluster. In Proceedings of the 34th International Academic Conference, Florence, Italy, 13–16 September 2017; 2017. [CrossRef]

74. Vernay, A.L.; D'Ippolito, B.; Pinkse, J. Can the government create a vibrant cluster? Understanding the impact of cluster policy on the development of a cluster. *Entrep. Reg. Dev.* **2018**, *30*, 901–919. [CrossRef]

75. Fang, L. The dual effects of information technology clusters: Learning and selection. *Econ. Dev. Q.* **2018**, *32*, 195–209. [CrossRef]

76. Haviernikova, K.; Kordas, M.; Vojtovic, S. Cluster Policy Report. 2016. Available online: Cluster-portal.cz/Resources/Upload/Home/ke-stazeni/v4clusterpol/v4clusterpol-reports//v4cp_cluster-policy_slovakia.pdf (accessed on 13 November 2019).

77. Klewitz, J.; Zeyen, A.; Hansen, E.G. Intermediaries driving eco-innovation in SMEs: A qualitative investigation. *Eur. J. Innov. Manag.* **2012**, *15*, 442–467. [CrossRef]

78. Derlukiewicz, D. Application of a design and construction method based on a study of user needs in the prevention of accidents involving operators of demolition robots. *Appl. Sci.* **2019**, *9*, 1500. [CrossRef]

79. Szewrański, S.; Bochenkiewicz, M.; Kachniarz, M.; Kazak, J.K.; Sylla, M.; Świąder, M.; Tokarczyk-Dorociak, K. Location support system for energy clusters management at regional level. *IOP Conf. Ser. Earth Environ. Sci.* **2019**, *354*, 012021. [CrossRef]

80. Kazak, J.K.; van Hoof, J. Decision support systems for a sustainable management of the indoor and built environment. *Indoor Built Environ.* **2018**, *27*, 1303–1306. [CrossRef]

Article

Eco-Environmental Risk Evaluation for Land Use Planning in Areas of Potential Farmland Abandonment in the High Mountains of Nepal Himalayas

Suresh Chaudhary [1,2,3,4,*], **Yukuan Wang** [1,3,*], **Amod Mani Dixit** [4], **Narendra Raj Khanal** [5], **Pei Xu** [1,3], **Kun Yan** [1,3], **Qin Liu** [1,3], **Yafeng Lu** [1,3] and **Ming Li** [1,3]

[1] Institute of Mountain Hazards and Environment, Chinese Academy of Sciences, Chengdu 610041, China; xupei@imde.ac.cn (P.X.); yankun@imde.ac.cn (K.Y.); liuqin@imde.ac.cn (Q.L.); luyafeng@imde.ac.cn (Y.L.); liming@imde.ac.cn (M.L.)

[2] University of Chinese Academy of Sciences (UCAS), Beijing 100049, China

[3] Wanzhou Key Regional Ecology and Environment Monitoring Station of Three Gorges Project Ecological Environmental Monitoring System, Wanzhou 404020, China

[4] National Society for Earthquake Technology-Nepal, Kathmandu Lalitpur, P.O. Box 13667, Nepal; adixit@nset.org.np

[5] Central Department of Geography, Tribhuvan University, University Campus, Kirtipur, P.O. Box 44613 Nepal; nrkhanal.geog@gmail.com

[*] Correspondence: suresh.nset@gmail.com (S.C.); wangyukuan@imde.ac.cn (Y.W.); Tel.: +86-028-8523-0627 (Y.W.)

Received: 2 November 2019; Accepted: 3 December 2019; Published: 5 December 2019

Abstract: Land use change, especially that due to farmland abandonment in the mountains of Nepal, is being seen as a major factor contributing to increasing eco-environmental risk, undesirable changes in the socio-cultural landscape, biodiversity loss, and reduced capacity of the ecosystem to provide key services. This study aims to: i) evaluate eco-environmental risk for one of the high mountain river basins, the Dordi river basin in Nepal, that has a growing potential of farmland abandonment; and ii) develop a risk-based land use planning framework for mitigating the impact of risk and for enhancing sustainable management practices in mountain regions. We employed a multi-criteria analytic hierarchy process (AHP) to assign risk weightage to geophysical and socio-demographic factors, and performed spatial superposition analysis in the model builder of a geographic information system (GIS) to produce an eco-environmental risk map, which was subjected to a reliability check against existing eco-environmental conditions by ground truthing and using statistical models. The result shows that 22.36% of the basin area has a high level of risk. The very high, extreme high, moderate, and low zones accounted 17.38%, 7.93%, 28.49%, and 23.81%, respectively. A high level of eco-environmental risk occurs mostly in the north and northwest, but appears in patches in the south as well, whereas the level of moderate risk is concentrated in the southern parts of the river basin. All the land use types, notably, forest, grassland, shrub land, and cultivated farmland, are currently under stress, which generally increases with elevation towards the north but is also concentrated along the road network and river buffer zones where human interference with nature is the maximum. The risk map and the framework are expected to provide information and a scientific evidence-base for formulating and reasonable development strategies and guidelines for consensus-based utilization and protection of eco-environmental resources in the river basin. As an awareness raising tool, it also can activate social processes enabling communities to design for and mitigate the consequences of hazardous events. Moreover, this risk assessment allows an important link in understanding regional eco-environmental risk situation, land use, natural resources, and environmental management.

Keywords: eco-environmental risk assessment; land use planning; analytical hierarchy process; GIS; mountain region; Nepal

---

1. Introduction

The Nepal mountains are characterized by a richly diverse ecological and socio-cultural landscape, and have been shaped by traditional land use activities [1–3]. The land management systems in practice have developed over the years in accordance with the prevailing environmental factors, including topography, climate, relief, soil, and proximity to roads and settlements [4]. In particular, cultivated farmland is located in the most fertile gentle slopes at a lower elevation, whereas forest and grassland are scattered, mostly occupying the steep slopes that are relatively farther from the villages. Thus, a balance between people, settlement, farmland, forest, and grassland was achieved in the traditional land use practice. This practice contributed to the ecological and environmental diversity in the mountains and to the socio-cultural values of the landscape. However, recent farmland abandonment has triggered the process of land degradation as well as deterioration of mountain ecosystem services [5–8]. The resulting patterns have also led to the problems of soil erosion and siltation of rivers and reservoirs, which in turn have increased the frequency and severity of landslides, debris flow, rock falls, and sinkhole formation [9,10]. Therefore, in order to manage the ever-increasing risk and to promote sustainable mountain development and eco-environmental protection, proper eco-environmental risk-based land use planning is essential [11].

Eco-environmental risk is the likelihood of harmful consequences occurring as a result of exposure to biological, physical, chemical, and social stressors [12]. The risk can exert harmful or fatal effects on humans and the natural environment [13]. Thus, eco-environmental risk assessment (ERA) is performed to identify, analyze, and evaluate risk to the environment, ecosystem services, living organisms, and human beings [14]. It has been applied in many fields including environmental sciences and health sciences. In particular, in 1987, the eco-environmental risk assessment (ERA) emerged as a way to assess the likelihood of adverse effects to ecosystems rather than only to human health [15]. Since then, ERA has been applied to evaluate eco-environmental risks to ecosystems and thus identify opportunities for regulating sustainable land use [16]. Surrogate information contributing to the distribution of risks across the mountain ecosystem such as topographic condition (elevation, slope, and aspect), types of land use, rainfall, geology, soil, and socio-economic condition, road distance, and settlement location are considered to develop risk assessment processes [17]. Mainly the risk analysis consists of characterizing both the exposure and effects to the environment [18]. Such understanding and mapping are important especially for eco-environmentally fragile mountainous regions, where the relationship between the natural and human built-up environment is crucial for sustainable land use [19,20].

A variety of approaches and methods are available and used to evaluate and identify eco-environmental risks [21,22]. These were designed from diverse contexts and performed at different spatial and temporal scales [23,24]. For example, Liu et al. conducted and viewed eco-environmental risk from the perspective of watershed management [25] and agricultural activities [26]. These were carried out using a geographic information system (GIS), remote sensing or a combination of both, accompanied with field evaluation [27], and sometimes coupled with other environmental models or methods [28]. These techniques are cost-effective, easily-compared, time-efficient, and ideal for mapping and monitoring the trends of eco-environmental degradation and spatial changes [29,30]. Among the methods, the analytical hierarchy process (AHP) remains the most common and widely used multi-criteria decision aid (MCDA) method to date and has appeared in many research reports [31–33]. It can be used on risk assessment accounting or for overlaying for a multitude of criteria such as soil, climate, vegetation, and their sub-criteria [34]. Subjective opinions from decision makers (DMs), stakeholders, and experts in a given field also can be used in this method [35]. Moreover, the AHP method allows the integration of qualitative and quantitative factors and can solve complex risk assessment problems where data and information are lacking or ambiguous [36]. We used this AHP method. With this background, the present study aims to: i) evaluate eco-environmental risk for one of the high mountain river basins in Nepal that has a growing potential for farmland abandonment; and ii) develop a risk-based land use planning framework for mitigating the impact of risk and for enhancing sustainable management practices in mountain regions. The study at first identifies and analyzes the role of different natural and social factors together with other interdependent factors that influence the eco-environmental risk processes to evaluate the level of risks. Based upon the risk assessed, we propose a draft framework of ERA-based land use planning considering the principles of best practices in sustainable land management. The risk-based land use planning thus prepared is expected to serve as a tool that can help local governance to improve economic opportunities and develop land use options that reconcile conservation and development objectives. The use of this framework thereby can activate social processes of decision making and consensus building among stakeholders on matters concerning the utilization and protection of local resources. The risk map also enables communities to design for and mitigate the consequences of hazardous events. Moreover, this risk assessment allows an important link in understanding regional eco-environmental risk situation, making land use, natural resources, and environmental management.

2. Materials and Methods

2.1. Study Area

The assessment was carried out in the Dordi river basin that lies in the high mountainous region of western Nepal. It is located within the coordinates of 28°8′ N–28°27′ N latitude, and 84°24′ E–84°42′ E longitude, along the southern faces of the Himalayan range, with elevations ranging from 546 m above mean sea level (ma.s.l.) to more than 7746 m a.s.l (see Figure 1). The basin covers a land area of 496 km^2 approximately.

The region has a pleasant mountainous subtropical climate with concentrated monsoon rainfall during the period between June and September with an average rainfall of 2600 mm per year. Chir pine (Pinusrox burghii) and broad-leaf trees, such as *Chilaune (Schimawal lichii), oak (Quercussemi carpifolia), sal (Shorearobusta), Schima Castanopsis,* and *Engelhardtia* are the dominant vegetation types [37]. The basin has a long history of human habitation which has evolved the present day landscapes and land use patterns [37]. High altitude, steep slopes, undulating terrain as well as several major rivers and streams running north to south define the Dordi river basin which is sensitive to land use change and the ongoing risk processes. The rivers/streams are ephemeral in character with high flows when it rains, causing frequent flash floods. The basin was severely affected by the 2015 earthquake. Landslides, debris flow, and rock falls frequently occur over the basin. Many cultivated farmlands and settlements, especially those adjacent to the rivers and secondary streams face landslide hazards. Many landslides, debris flow, and ground settlements are directly associated with the abandonment of farmland terraces located on the hill slopes [6]. Additionally, gully formation, soil erosion, and siltation of soil on water bodies (river water, drinking water supply, and irrigation canal) are common. Moreover, river/stream bank cutting also poses a threat to houses and other assets along the bank of the Dordi River.

The main coping strategy of the communities is to rebuild, clean, and maintain the infrastructure after landslides, debris flow, rock falls or soil erosion events. Communities erected at places gabion walls for landslide or debris flow protection, but these are not entirely effective: most of the household respondents claimed that they have been affected by the geologic and hydrologic hazards several times every year. Due to this, some residents permanently migrated, leaving their farmlands uncultivated. The remaining residents are very worried about these eco-environmental hazards and concerned with the high probability of repeated landslides, debris flow, and soil erosion. While the people's priority is to improve education, employment, and road development, their day-to-day concerns include mitigation of landslides, floods, and soil erosion. This kind of deterioration in the Dordi river basin arrested our attention and decision to conduct a systematic eco-environmental risk sensitivity mapping and assessment of associated risks, and to draft a risk-based land use planning framework that could assist in the peoples' efforts to mitigate and control the problems caused by geologic, hydrological, and natural hazards, or those induced due to the abandonment of farmlands.

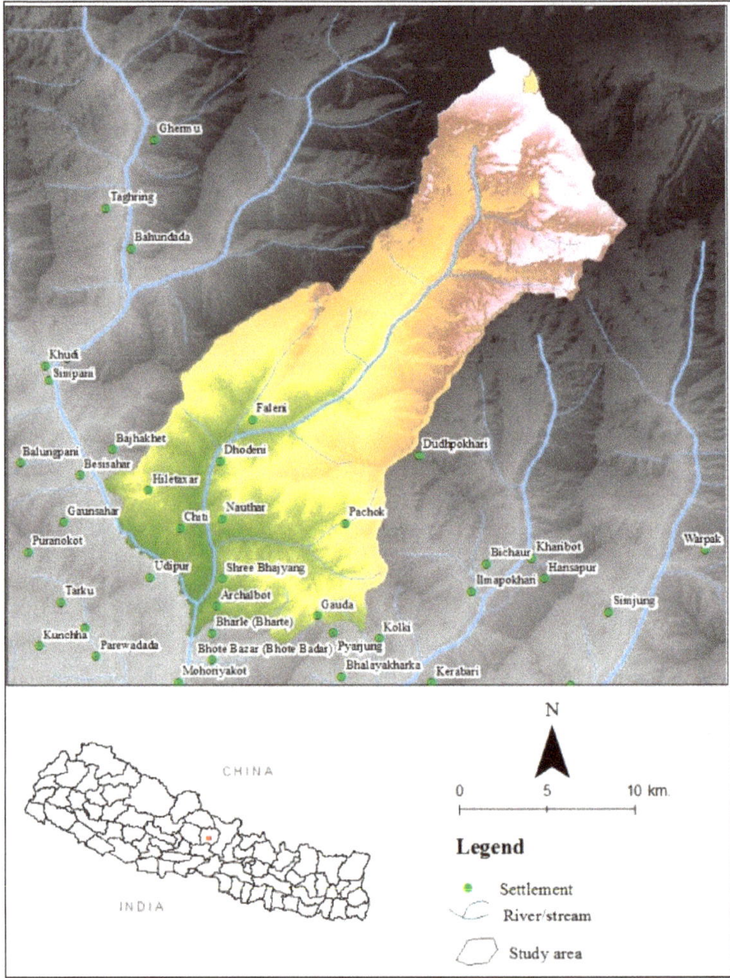

Figure 1. Location map of the study area.

2.2. Data Collection and Processing

The dataset used for this study includes a series of geographic information prepared from satellite imagery, secondary data, and field observation. The dataset used includes elevation, slope, aspect, soil, geology, rainfall, existing land use, normalized difference vegetation index (NDVI), and distances from settlements to the nearest rivers/streams, road network, and settlement location. A digital elevation model (DEM) at 30 m resolution, prepared using data obtained from the geospatial data cloud (http://www.glovis.com) was used for elevation, slope, and aspect mapping. Soil type (scale = 1:1,000,000) and geology (scale = 1:50,000) maps were derived from the soil and terrain database for Nepal [38] and Department of Mines and Geology, Government of Nepal, respectively. Rainfall data were collected from the Hydrology and Meteorology Department of Nepal. They were interpolated by using the inverse distance weighting (IDW) method in ArcGIS 10.5 software. The positions of rivers/streams, road networks, and settlement locations were taken from the Department of Survey, Government of Nepal.

Land use/land cover map and NDVI were prepared from the open access Landsat 7 Enhanced Thematic Mapper Plus (30 m resolution, LANDSAT 7 ETM SLC-off, acquisition date: 22-08-2019) images. Two scenes of Landsat 7 ETM data (path 141, 142 and row 41) for August 2019 were downloaded from Geospatial Data Cloud (http://www.gscloud.cn; National Basic Science Data Sharing Service Platform, Chinese Academy of Sciences, Beijing, China). A false color image map produced by the computer software ArcGIS Desktop 10.5, version 10.5.0.6491 (Esri ®ArcMap^{TM} 380 New York Street Redlands, California 923738100 USA) was then analyzed for different land units. These thematic factors were converted into raster data with 30 × 30 m resolution (WGS_1984_Universal Transverse Mercator System, false easting = 500,000, central meridian = 84, sale factor = 0.9996 and Datum – D_WGS_1984) in an ArcGIS 10.5 environment.

2.3. Selection of Criteria and Construction of Assessment Indicator System

We adopted eleven criteria including topographic characteristics (elevation, slope, and aspect), soil, geology, rainfall, rivers/streams, existing land use/land cover, normalized difference vegetation index, and socio-economic conditions such as distance to roads and settlements, which revealed the major natural and built-up environments of the river basin. The description of all the criteria and their sub-criteria are described in the following paragraphs.

2.3.1. Elevation

A topographic characteristic, elevation has potent influence on land use, surface runoff, as well as development and distribution of accidental events such as landslides and debris flow [39,40]. The elevation directly reflects the terrain ruggedness. In this study, the elevation variation was generated from the digital elevation model and classified into classes as: (i) <1000, (ii) 1000–1500, (iii) 1500–2000, (iv) 2000–2500, and (v) >2500.

2.3.2. Slope

Slope gradient mainly controls the infiltration of groundwater into subsurface and surface flow. In general, steeper slopes result in a faster speed of flow and are not suitable for the maintenance of eco-environmental conditions [41]. In this study, the slope angles were generated from the digital elevation model, and classified into five classes: flat to gentle slope (<15°), moderate slope (15°–25°), fairly moderate slope (25°–35°), steep slope (35°–45°), and very steep slope (>45°).

2.3.3. Aspect

Topographic aspect determines the maximum slope of the terrain surface as well as the relative amounts of sunshine and atmospheric moisture it receives [42]. In this study, the topographic aspects were generated from the digital elevation model, and classified into classes: (i) north (N), (ii) northeast (NE), (iii) east, (iv) southeast (SE), (v) south (S), (vi) southwest (SW), (vii) west (W), (viii) northwest (NW), and (ix) flat.

2.3.4. Geology

Geological formation plays an important role in the production of loose materials on the slope by weathering and slope movements such as landslide, debris flow, etc. [43]. The river basin geology is mainly composed of easily weathered weak rocks such as shale and phyllite of Ranimatta formation that produce loose clayey soils, moderately hard slates of Ghanapokhara formation, and granite and augen gneiss of Ulleri formation, that give rise to gravelly soil; whereas the hard quartzite of the Naudanda formation produces sandy or boundary slopes not very suitable for vegetative growth [37,44].

2.3.5. Soil

Soil structures play an important role in the definition of land use and land cover, as well as in the production on the slope of loose materials by weathering and slope movements [38,43]. The soil types in the basin include red soil, sandy, cobbly, loamy, loamy-boulder, and loamy-skeletal, that have been classified into: (i) chromic cambisols, (ii) eutric cambisols, (iii) eutric regosols, (iv) gelic laptosols, (v) gleyic cambisols, and (vi) humic Cambisols [37].

2.3.6. Rainfall

Rainfall is a critical factor for constituting the natural environment as well as the energy bases of ecosystem services (e.g., agricultural land, forests, rivers, etc.) [45]. When the energy transmission and transformation of rainfalls are not in accordance with other environmental factors in time and space, it will lead to ecological degradation [46]. On one hand, lack of precipitation disturbs the entire ecosystem with strong erosion along the numerous tributaries on the other; strong rainfall induces heavy soil erosion, giving rise to geological hazards such as landslips and mudslides. Therefore, eco-environmental risk bears strong and positive correlations between rainfall concentrations. Here, we used the inverse distance weighting (IDW) interpolation method to generate a rainfall map that was classified into five classes: (i) <2613 mm/year, (ii) 2613–2614 mm/year, (iii) 2614–2615 mm/year, (iv) 2615–2616 mm/year, and (v) >2616 mm/year.

2.3.7. Land Use/Land Cover (LULC)

Land use/land cover is one of the most important forms of eco-environmental landscape that serves as also the source of hazardous events in mountain regions [47]. In general, the forest area provides stability to the slope and it is widely accepted that vegetation cover has a positive influence on slope stability [48]. Geological hazards get triggered frequently in the barren mountains and abandoned farmlands [7]. In this study, the land use/cover was mapped into eight classes: (i) agricultural land, (ii) forest, (iii) grass land, (iv) bare land, (v) shrub land, (vi) water body, and (vii) snow/glacier. Agriculture land referred to land used for growing crops, even if it was cultivated every year. Forest is referred to as any vegetated land that is dominated by trees or shrubs. Grassland is defined as any land used to grow grasses for the purpose of grazing. Shrub land is categorized when forest does not entirely apply.

2.3.8. Normalized Difference Vegetation Index (NDVI)

NDVI is an important in that it measures the ground vegetation condition. The vegetation cover condition directly affects and even determines the eco-environmental conditions such as the amount of primary biological production, ecological carrying capacity, and soil erosion [49]. The higher the NDVI value, the healthier the vegetation; while a lower NDVI value represents less or no vegetation. Theoretically, NDVI values range from −1 to 1. In practice, extreme negative values represent water, values around zero represent bare soil, and values over 0.6 represent dense green vegetation. Here, NDVI values have been divided into five classes: (−0.17)–(0.02), (−0.02)–0.10, 0.10–0.20, 0.20–0.29, and 0.29–0.50.

2.3.9. Distance to Rivers/Streams

The hydrological system with rivers, streams, ponds, and natural springs as its components has importance not only because it assists in draining rain water but also because it can induce landslides, debris flow, and flooding [50]. Most of the time, the occurrence of flooding is related to the distribution of the drainage system. Concave stream banks usually host soil fall or landslides. In this study, a map for distance to river and streams was prepared using the buffer distance analysis (BDA) method, and classified into five classes: (i) close (<50 m), (ii) nearby (50–100 m), (iii) distant (100–200 m), (iv) little distance (200–500 m), and (v) far (>500 m).

2.3.10. Distance to Road Networks

Road networks and mountain slopes are important factors that influences the geomorphic processes affecting the surrounding eco-environment [51]. Road construction often leads to high runoff coefficients and soil losses from mountainous slope surfaces. In this study, motorable earthen roads were considered for network, and a distance map was prepared using the buffer distance analysis (BDA) method. These were classified into five classes: (i) close (<50 m), (ii) nearby (50–100 m), (iii) nearby distant (100–200 m), (iv) far (200–500), and (v) more far (>500 m).

2.3.11. Distance to Settlements

Population density, settlement distribution, and socio-economic activities acting separately or interacting together influence the regional eco-environmental quality [52]. In particular, human activities such as grazing, mining, and soil digging are associated with the location of settlements, that can accelerate surface geomorphic processes indicating geomorphic hazards [53–56]. This study considered a settlement as one that includes at least five buildings located in a cluster. Digital thematic distances from each settlement were prepared in the form of a map using the buffer distance analysis (BDA) interpolation method. These were classified into three classes: (i) close (<1000 m), (ii) nearby (1000–2000 m), (iii) nearby distant (2000–3000 m), and far (>3000).

2.4. Determining Relative Importance of the Different Criteria

Each of the criteria was weighted according to its relative contribution to eco-environmental risk and quality. A pairwise comparison matrix based on the AHP method was used to determine the weight of each criterion (see Figure 2). A numerical value from 1 to 9 was assigned, indicating how many times a criterion was more important or how dominant one criterion was over another criterion. This was determined based upon a consensus of five experts (1 geologist, 2 geographers, 1 soil scientist, and 1 environmental scientist) experienced and well-versed in the relative influence of each criterion on eco-environmental risk, land use, ecosystem health, and ecological quality [57]. Specifically, higher weights were given to higher elevations while lower elevations were assigned lower weights for the elevation factor: steeper hill slopes received higher weightage values. Similarly, higher weights were given to north, northwest, and northeast aspects because northern aspects are mostly devoid of vegetation or have scant vegetation in the basin. For land use types, bare land was given a high weight because of its high influence on geomorphic processes; grassland and scrub were given medium weights as grasses become prone to landslides and other events, thus becoming medium susceptible to environmental hazards (e.g., landslides, debris flow). The forest classes were given less weight because forests hinder environmental hazards. The major rock type of the area was rated according to their relative importance for landslides and debris flow. Accordingly, proximity to rivers/streams, road networks, and settlements was assigned higher scores. It was assumed that the possibility of a landslide and other geomorphological processes was more frequently near streams because of undercutting.

After assigning values to all thematic layers, the validity of comparisons was evaluated through the consistency ratio (CR).

$$\text{Consistency ratio (CR)} = \frac{Consistency\ index\ (CI)}{Random\ index\ (RI)} \tag{1}$$

where CI is the consistency index, expressed as,

$$\text{Consistency index (CI)} \ = \ \frac{\lambda\text{max} - \text{n}}{n - 1} \tag{2}$$

where λmax is the largest eigenvalue and n is the number of the criteria. RI is the random consistency index value. Table 1 shows the values for the RI of the matrix dimensions of 1–15. In general, the acceptable value of CR depends on the size of the matrix (0.1 for matrices n ≥5). If the CR value is equal to or less than the specified value, this indicates that the evaluation within the matrix is acceptable and close to ideal values. However, if the CR is higher than the acceptable value, an evaluation process is needed to be improved [60]. Here, the CR of the matrix that derived from criteria weights is 0.1, which is equal to <0.1, indicating the judgment is consistent, and the calculated weights can be used (Table S1).

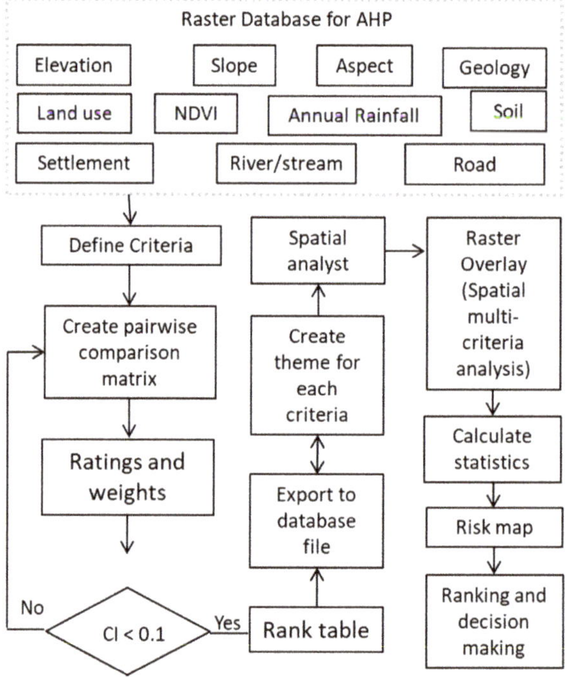

Figure 2. Determining criteria weight in the analytic hierarchy process (AHP) model. NDVI: normalized difference vegetation index; CI: consistency index.

Table 1. Random consistency index [58,59].

n	1	2	3	4	5	6	7	8	9	10	11	12	13	14	15
RI	0	0	0.58	0.90	1.12	1.24	1.32	1.41	1.45	1.49	1.51	1.53	1.56	1.57	1.59

2.5. Risk Calculation and Classification of Results

The eco-environmental risk was calculated by overlaying weightage value of each criterion for all unit areas. The defined criteria of elevation, slope, aspect, geology, soil, land use/land cover, NDVI, annual rainfall, distance from rivers/streams, roads, and settlements, were imported into different GIS layers at first, then arranged as the input of a model builder module of ArcGIS 10.5. This model enables running workflows containing a sequence of geo-processing tools to obtain assessment results efficiently. Secondly, all these criteria were incorporated using the raster calculator function in ArcGIS employing the following linear additive model:

$$\text{Eco} - \text{environmental risk index (ERI)} \ = \ \sum_{j=1}^{n} Wjwij \qquad (3)$$

where *Wj* is the weight value of causative criteria's *j*, *wij* is the weight value of class *i* of each causative criterion's *j*, and *n* is the number of the causative factor.

The resulting ERI map has categories according to expert opinions as there are no general rules to categorize such continuous data automatically [61]. In this study, the resulting ERI map is classified into five levels, namely, low, moderate, high, very high, and extreme high risk zones according to the methods of natural breaks classification. The definition of risk level is shown in Table 2.

Table 2. Eco-environmental risk classification in the Dordi river basin.

Risk classification	Eco-environmental risk index(ERI)	Character description
Low risk	>0.15	The eco-environmental system is stable including strong risk resistance, fertile soil, relatively low altitude, and great vegetation coverage.
Moderate risk	0.15–0.20	The eco-environmental system is relatively stable including risk resistance, fertile soil, relatively low altitude, and better vegetation coverage.
High risk	0.20–0.25	The eco-environmental system is relatively unstable including relatively poor risk resistance, relatively barren soil, and relatively complicated vegetation types.
Very high risk	0.25–0.30	The eco-environmental system is unstable including poor risk resistance, barren soil, and few vegetation types.
Extreme high risk	>30	The eco-environmental system is extremely unstable including poor risk resistance, relatively high altitude, barren soil, and sparse vegetation that are mainly hardy plants.

2.6. Development of aFramework for Land Use Planning

We developed a framework for land use planning (LUP) that provides an important scientific basis for supporting the strategies of ecological construction, regional development, and planning of management options. LUP integrates the conditions in the five categories of risks (extreme high, very high, high, moderate, and low) with four categories of land use types: forest, shrub land, grassland, agricultural farmland (see Figure 3). In this way, 20 land use types, for example forest with high risk (FH) and agricultural land with low risk (AL), were categorized. Secondly, all these categories were assigned into four different types of scenarios, namely, (i) restricted areas, (ii) priority control areas, (iii) control areas, and (iv) monitored areas. For the different level of scenarios, we provide important information to support the strategy of ecological construction and risk mitigation measures, including land conservation, land replacement, and low impact development (LID) practices. Land conservation measures serve to prevent human disturbances (protecting land of agricultural significance, protecting natural capital from encroachment), and land replacement (e.g., rehabilitation and/or avoidance of sites) serves to eliminate risk sources in eco-environmentally sensitive areas or transfer the sources to other sites. Low impact practices include protection of the quality and quantity of natural resources and minimization of land degradation establishing appropriate buffers between socio-economic development activities such as road network facilities, permeable pavements, and storage tanks, which can control risks related to surface runoff at their source through natural processes of infiltration, detention, storage, and purification. As the last step, land-use measures were validated, or adjusted as required, to ensure that environmental concerns were effectively addressed.

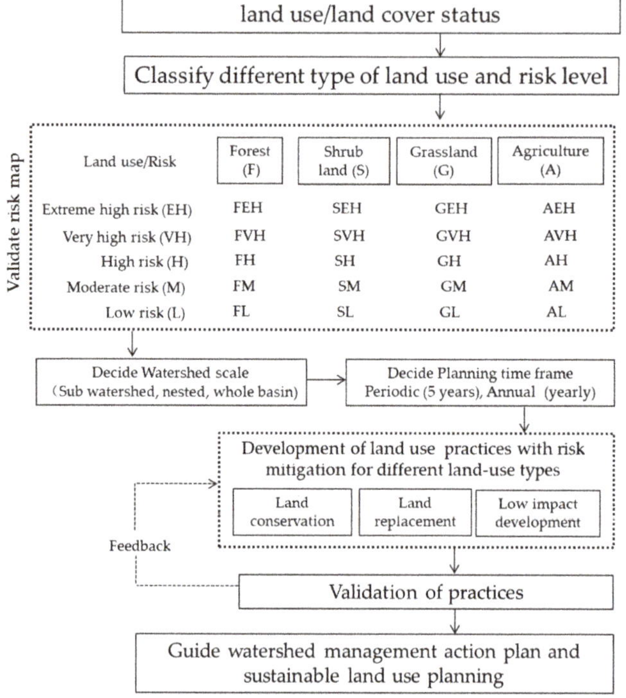

Figure 3. A framework for guiding land use practices.

3. Results and Discussions

3.1. Spatial Distribution of Eco-Environmental Risk

The spatial distribution of eco-environmental risk levels for the Dordi river basin were calculated, using the level of calculated risk at each grid number multiplied by the grid area. An area of 111.10 km², accounting for 22.36% of the total Dordi river basin, belongs to the high level of risk. Approximately 86.35 km² (17.38%) and 39.43 km² (7.93%) belong to very high level of risk and extreme high level of risk, respectively. This means that nearly half (>47.67%) of the total basin area is highly risky. The low and moderate zones accounted for 23.81% (118.29km²) and 28.49% (141.51km²), respectively (Table 3). A high level of eco-environmental risk occurs mostly in the north and northwest, but appears in patches in the south as well, whereas the level of moderate risk is concentrated in the southern parts of the river basin. Areas with a lower level of risk are scattered throughout the basin. In general, the risk is relatively light in center parts and heavy in northern parts (see Figure 4).

The eco-environmental risk level also closely correlates with altitude: high and very high levels of risk are mostly distributed in areas with an elevation above 2500 m. The extreme level of risk is distributed entirely above the elevation of 2500 m, whereas the moderate and light levels are mostly distributed in the lower elevation range of 1000 and 2500 m (see Figures 5 and 6). It is interesting to note that all land use types contain all categories of risk, albeit in different proportions. Of the total 207.2 km² of forest, 44.17% represent low risk, 34.50% moderate risk, and 21.31% of the high risk, very high, and extreme high risk. The high risk characteristics of these areas can be attributed both to their importance in providing ecosystem services and to their locations at higher elevations and steep slopes especially for the forest reserves. Shrub lands occupy12.88 km² area, of which 77.12% is at high risk. The grass land use type is widely distributed in the study area, but 80.42% of its area is in high risk zones, mostly around the forest as well as in the buffer zones of the river. Additionally, one-fourth of the cultivated farmlands are at high risk mainly along the road network and river buffer zones. Table 4 provides the details.

Table 3. Area and proportion of eco-environmental risk in the Dordi river basin.

Risk level	ERI	Number of Grid	Area (km^2)	Percentage
Low risk	<0.15	112,762	118.29	23.81
Moderate risk	0.15–0.20	134,889	141.51	28.49
High risk	0.20–0.25	105,900	111.10	22.36
Very high risk	0.25–0.30	82,311	86.35	17.38
Extreme high risk	>0.30	37,590	39.43	7.93

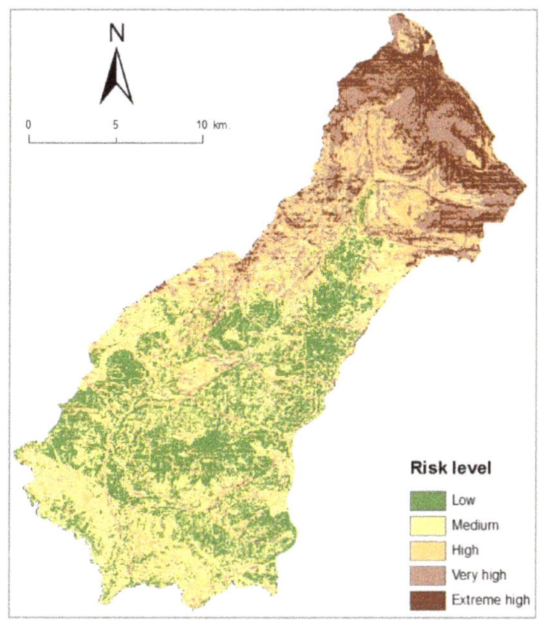

Figure 4. Eco-environmental risk map of Dordi river basin.

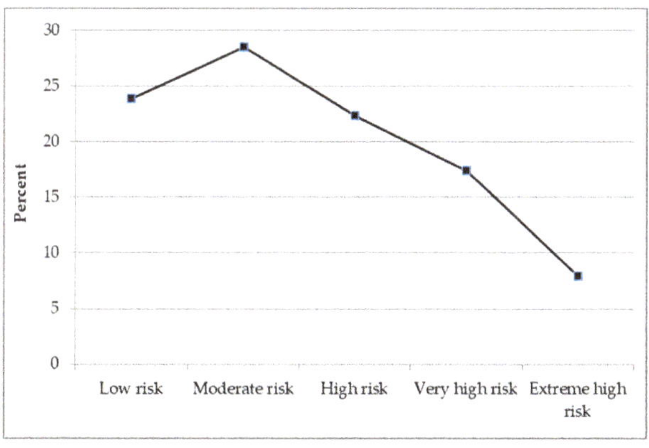

Figure 5. Distribution of the eco-environmental risk.

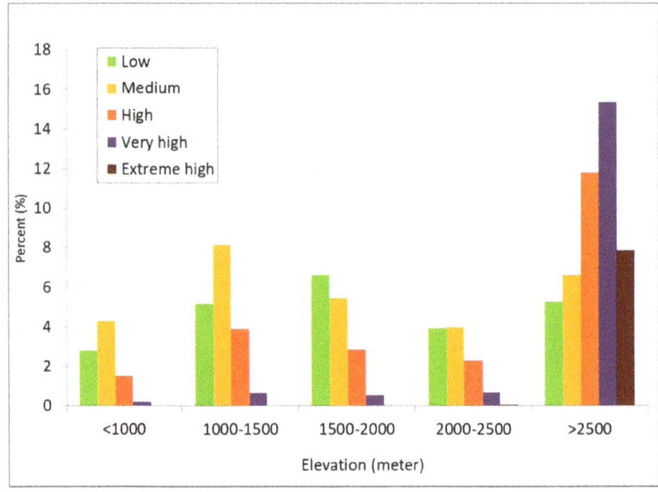

Figure 6. Distribution of eco-environmental risk in relation to altitude.

Table 4. Extent of risk on different land use type.

Land use	Forest	Shrub land	Grassland	Agriculture
Area (km^2)	207.2 (46.20%)	12.88(2.88%)	61.16 (13.64%)	71.25 (15.89%)
Low (km^2)	91.53 (44.17%)	0.41(3.25%)	0.18 (0.30%)	14.53(20.39%)
Medium(km^2)	71.49 (34.50%)	2.52(19.62%)	11.78(19.27%)	38.33(53.80%)
High(km^2)	38.29 (18.47%)	5.96(46.30%)	24.63(40.28%)	15.48(21.73%)
Very high(km^2)	5.81(2.80%)	3.72(28.91%)	20.19(33.02%)	2.87(4.03%)
Extreme high (km^2)	0.067 (0.03%)	0.24(1.90%)	4.35(7.11%)	0.02(0.03%)

Eco-environmental risk map especially for the high, very high, and extreme high areas requires verification [62]. The best way to physically validate the research findings is by conducting field observations, which is rather a difficult job logistics-wise [63]. However, this can be done on the basis of field information and past geo-meteorological events (e.g., landslides, rock falls, debris flows, etc.) [64]. In this study, we observed that the large number of eco-environmental risks such as landslides, debris flow, rock falls, and gully formation processes are clearly marked in the areas of abandoned farmlands and/or in the areas of high, very high, and extreme high areas of eco-environmental risk zones.

A large number of landslides, debris flows, and rock falls were observed along the bank of the river stream, and in areas influenced by development infrastructure constructions such as road networks, irrigation canals, water supply and hydropower project, and on the abandoned farmlands [6]. Sometimes landslides were observed spreading out in adjacent agricultural fields. Slope land plants in several places were severely affected by landslides, debris flows, and rock falls; sometimes these were destroyed heavily, exacting an adverse effect on the ecological and environmental conditions of the region. In all such locations, soil is exposed with an increasing extent of risk of soil erosion. These hazards also affect the fertility of the soil and may cause habitat destruction for wild animals to a large extent. Landslides were found in conditions of poor soil structure and poor vegetation resulting in high risk to nearby villages. A very large newly triggered landslide was observed in the slope uphill of Basnetgaun and Hiletaksar road. This slide was due to severe undercutting of the slope by the road construction in hill slopes covered by thin grasses and some trees. Another large debris flow with a wide debris fan was situated at the uphill slope of the settlement called Hile. Construction, poor vegetation, and the presence of soft weathered mudstone and sandstone were the main causes of this slide. An old rotational slide was also observed along the bank of the river and stream (see Figure 7). Although the entire slope is covered by moderately dense forest, the landslide scars are still visible. This slide occurred due to river incision. We checked the risk map prepared against all these conditions and found that the map reflected the realities on the ground. We conclude that our eco-environmental risk map represents the ground conditions and, therefore can be used as a scientific basis to guide land use planning in the study area.

(a)   (b)
(c)   (d)

Figure 7. Occurrence of landslides with or without abandoned farmland in Dordi river basin; (**a**) A large landslide near the settlement of Chiti; (**b**) debris flow near Basnetgaun; (**c**) landslide occurrence along the road of Hile-Nauthar; (**d**) large landslides at Basnetgaun-Hile road. Source: field survey, 2018.

The eco-environmental risk map was also validated using mathematical and statistical tools such as computation of geological hazards (e.g., landslides), density, and success rate curve [65]. In particular, the geological hazard density of each risk level, a ratio of observed landslide occurrences in a respective risk, gives the overall quality of the eco-environmental risk map [66]. Our results of such treatments are given in Table 5; landslide density for the extreme high risk zone is 0.0796, which is distinctly larger than for other risk zones. The landslide densities for very high, high, moderate, and low level of risk zones are respectively 0.0299, 0.0207, 0.0090, and 0.0033. These results show that there is gradual decrease in landslide density from the extreme high to low risk zones and there is also considerable separation in landslide density values between the different risk zones. A similar kind of success rate (85.3%) was found for one of the landslide susceptibility mappings conducted along the Dordi river basin [67]. This finding indicates that the eco-environmental risk increases with increasing elevation, which could reflect the harsh eco-environmental conditions at higher elevations which is in accordance with similar conclusions arrived at by other researchers in other countries such as China [17], Ethiopia [68], and Slovenia [69]. The study clearly indicates that landslides are frequently occurring natural hazards especially in high and very high risk areas that lead to massive destruction of life and property and sometimes lead to large-scale landscape transformations. Therefore, such concurrence of results from both methodologies allows concluding that the calculated eco-environmental risk map and classified risk zones are found to be in good agreement with occurrences of hazardous processes.

Table 5. Observed geological hazard density in the different risk zones of the eco-environmental risk map.

Eco-environmental risk zones	Area		Hazard area		Landslide density
	(km^2)	(%)	(km^2)	(%)	
Low risk	118.29	23.81	0.40	4.11	0.0033
Moderate risk	141.51	28.49	1.28	13.17	0.0090
High risk	111.10	22.36	2.31	23.77	0.0207
Very high risk	86.35	17.38	2.59	26.64	0.0299
Extreme high risk	39.43	7.93	3.14	32.30	0.0796

3.2. Land Use Planning Framework

We implemented a land use planning framework combining the risk zones with the existing four major land use types. This created 20 conditions that we assigned alphabetic nomenclature that appears as a matrix in Figure 8. A logical reasoning then followed: classification with nomenclatures such as "EFH, SEH, GEH, AEH, FVH, SVH, GVH, AVH" were considered as "restricted areas". These zones are extremely eco-sensitive and vulnerable to natural and human disturbances. The eco-environmental conditions are severely polluted, and the ecosystems are crippled. In particular, risks in these areas are considered as unacceptable and thus require corrective actions.

Risk	Forest (F)	Shrub land (S)	Grassland(G)	Agriculture (A)
Extreme high risk(EH)	FEH	SEH	GEH	AEH
Very high risk (VH)	FVH	SVH	GVH	AVH
High risk (H)	FH	SH	GH	AH
Moderate risk (M)	FM	SM	GM	AM
Low risk (L)	FL	SL	GL	AL

Scenario I – Restricted areas "FEH, SEH, GEH, AEH, FVH,SVH, GVH, AVH" risks are unacceptable and must be mitigated through land conservation and replacement

Scenario II - Priority control areas "FH, SH, GH, AH ", risks are unacceptable, urgent actions including land replacement or LID practices are required

Scenario III - Control areas "FM, SM, GM, AM", risks are undesirable, actions including LID practices can be applied

Scenario IV - Monitoring areas "FL, SL, GL, AL", risks are acceptable, actions are optional

Figure 8. Land use planning framework combining land uses and risk level.

The high risk zones with nomenclatures of "FH, SH, GH, AH" are assigned as "priority control areas". These zones are highly eco-sensitive and vulnerable to natural and human disturbances. Therefore, development activities in these zones should be strictly restricted by the principal functional orientation. By taking ecological and resourceful advantage, the land use in such "priority control areas" should focus on the (i) development of ecological forestry, (ii) promotion of biodiversity, natural beauty, and physical endowments, (iii) promotion of suitable eco-tourism development, and (iv) preservation of cultural and natural heritages, while constructing an ecological tourism demonstration that could serve as world class tourist destinations with green tourism products and services.

The moderate risk land areas "FM, SM, GM, AM" areas are classified as "control areas" which are envisioned as optimized development zones in ecological planning. Development orientation of these areas should be adjusted to alleviate the impacts of production and construction on the ecological environment and act as a buffer for human activities. Specially in high altitudes, management should be transformed into ecological construction as integrative water protection and the establishment of a shelter belt for preventing land desertification and soil erosion and developing ecological agriculture. Suggested actions in these areas include low impact development (LID) practices and conservation of forest and farmland. LID-based land use practices such as strict control of industrial environmental pollution, and the presence or increase of vegetation coverage, restoration of lakes from farmland, and control of water loss and soil erosion are the chief tasks in ecological construction.

In the zones of low risk scenarios "FL, SL, GL, AL" are assigned as "monitoring" areas. Risks are acceptable and thus socio-economic development actions are optional. The ecosystem has a lower sensitivity degree to outside interferences, and the land resources and environment can support the demands of exploitation and construction. Thus, all areas of human settlement development that can be devoted to major infrastructure and utility systems can be developed. However, the important restrictive factor, that is shortage of water resources in the river basin, should be taken into full consideration in economic and social activities insistently, and the water-saving construction system is in urgent need of construction. In the southern and riverbed plain area, there is a favorable natural condition with a pleasant subtropical humid climate, fertile soil condition, and plentiful wetland resource. The modern agricultural production system should be developed considering timely ecological restoration to prevent destruction of landscapes and large areas of water loss and soil erosion.

Further, the areas suggested for monitoring can be divided into two classes namely, farmlands and built-up areas. The settlements of Tillar, Ramchokbesi, Basnetgaun, Hile, Majhgaun, Chisapani, Tiwari danda, Karki Danda, Nauthar, and Sera are potentially the most important bases for the socio-economic development of the whole Dordi river basin. These areas are suitable primarily for the residential portion of the built-up environment. Thus the emphasis should be made to ensure the following achievements: (a) integration of activities within and among settlements and efficient production and movement of people and commodities, and (b) access of the population to housing, education, health care, recreation, transportation and communication, sanitation, and basic utilities such as water, power, waste disposal, and other services. However, farmlands of these prefectures are highly threatened by abandonment. Thus, in areas where water is available for irrigation, trees, shrubs, and grass belts should be planted as barrier fences or small plots in order to reduce the rate of desertification and reclamation of lost lands.

In the mountainous regions of Nepal, there are mainly four types of land use, forest, shrub land, grassland, and farmland, that contain many valuable areas of conservation, despite significant transformations to the landscape [70]. Primarily, the vegetation barrier could break the wind velocity preventing sands from blowing laterally up the slope land. These infrastructures form a network of natural and semi-natural areas, such as existing forests and plantations; they enhance ecosystem health and resilience, minimizing natural disaster risks, including lowering surface water runoff which reduces the risk of flooding, connecting habitats, and mitigating mountain disaster effects; they also contribute to biodiversity conservation in an integrated manner, improving flora and fauna, and human well-being. They can also protect farmlands, conserve water and soils, and provide wood for fuel. In addition, green infrastructure offers a promising way to integrate biodiversity and ecosystem services in the mountain landscape. However, maintaining and enhancing these areas requires a policy objective from Nepalese governments at all levels of the governing system [71]. Therefore, this study incorporated forests, shrubs, grasslands, and farmlands at the land use planning stage.

The current conservation areas in the greater mountain areas are in place for a range of historical reasons, including their scenic and recreational values, their non-suitability for various land use, the influence of lobby groups, and the tenure of the land. Since the Third Five-Year Plan (1965–1970), the Nepalese governments have established a holistic approach to managing watershed resources that integrate forestry, agriculture, pasture, and water management, with an objective of sustainable management of natural resources. This approach seeks to promote interactions among multiple stakeholders within and between the upstream and downstream locations of a watershed. However, these experiences suggest that these ideals of watershed management do not appear to be strongly linked with the eco-environmental risk assessment and/or developed solely for conservation values [72]. Regarding the eco-environmental risk level, we categorized these four land use types as restricted areas, priority control areas, control areas, and monitoring areas. This can be used to assess the efficacy of current conservation areas and to identify land that is more likely to protect areas. Comparison of different risk scenarios and a mechanism for a constant dialogue between policymakers, practitioners, and communities at the landscape level would help in linking the upstream and downstream ecology to improve the livelihoods of the local people and sustainable watershed resource management.

4. Conclusions and Recommendations

This study evaluates eco-environmental risk conditions for the Dordi river basin located in the high mountains of the Nepalese Himalayas, produces a risk map, and develops a land planning framework. The proposed land use planning framework can help to achieve better outcomes for preventing further eco-environmental changes and loss of biodiversity. This can be used as a scientific basis to carry forward plans and enforce them, and/or for mobilizing public participation in land and to create a political leverage to support the planning processes. Likewise, the maps produced by this research may also help in facilitating improved collaboration between agencies responsible for land use planning, environmental management, and the private sector and non-governmental organization (NGO/INGOs) who work on various aspects of community development, serving as a common understanding of the situation. In addition, the land use planning tools demonstrated in this study can serve land managers and conservationists as a sophisticated analytical tool beyond simple rules of thumb. It can be used for designing and implementing the much-needed innovative financial incentives, and for discouraging risk-prone development. Moreover, this risk assessment allows an important link in understanding regional eco-environmental risk situation, making land use, natural resources, and environmental management. However, the analysis presented here does have limitations: reasonable data for the specific task of interest are not available in Nepal. For this study, the risk distribution maps were derived from secondary data on elevation, slope, aspect, soil, geology, land use, NDVI, rainfall, distance to river, settlements, and roads, and using expert opinions and field visits. More detailed field-surveyed data could yield better granularity and accuracy. Nonetheless, the results convincingly allow us to make the following suggestions to the national and local governmental authorities to take steps for the effective and practical application to the recovery of the local ecosystem.

- Region of strict protection: the region where eco-environmental risk is high, very high, and extreme is identified as the region of strict protection. This area constitutes nearly half (48%) of the basin. Considering the status of the area, all the development activities must be effectively monitored by the local government authority, and a proper reclamation plan for ecological recovery should be immediately put in place. Comprehensive strategy for combating hazard risks should be implemented. Also, human activities should be reduced as much as possible and eco-restoration activities should be initiated immediately.

\- Region of priority control: area under moderate and low risk constitutes more than half (52%) of the total area. It is suggested for focal protection. In this region the improved implementation of conservation measures is needed. This can be achieved by providing alternative sources of income to local people. Active participation of the local people in eco-restoration is recommended. Awareness of these trade-offs can underpin effective land use allocation that promotes sustainable land management and multifunctional land system through the efficient supply of multiple ecosystem services.

Supplementary Materials: The following is available online at http://www.mdpi.com/2071-1050/11/24/6931/s1, Table S1: AHP method, pairwise comparison matrix and normalized principal eigenvector for eco-environmental risk assessment criteria's, and for the classes within each criterion.

Author Contributions: Conceptualization, S.C.; field survey conduction, S.C.; data curation, S.C.; writing—original draft preparation, S.C.; writing—review and editing, S.C., YW, A.M.D., and N.R.K.; supervision, Y.W., A.M.D., and N.R.K.; design guidance, A.M.D., N.R.K., P.X., K.Y., Q.L., Y.L., and M.L.; discussion, A.M.D., N.R.K., P.X., K.Y., Q.L., Y.L., and M.L.

Funding: This research was funded by the 135 Strategic Program of the Institute of Mountain Hazards and Environment, Chinese Academy of Science, grant number SDS–135-1703; International talent program of the Chinese Academy of the Sciences, grant number 2019 VCA 0026; and the Chinese Academy of Sciences-the World Academy of Sciences (CAS-TWAS) Presidents Fellowship program for international PhD study.

Acknowledgments: We are extremely grateful to the Central Department of Geography (CDG) and National Society for Earthquake Technology (NSET), for sharing their valuable dataset and review of the manuscript. The authors are grateful for the constructive comments from the editor and anonymous reviewers.

Conflicts of Interest: The authors declare no conflicts of interest.

References

1. Latocha, A. Land-use changes and longer-term human–environment interactions in a mountain region (Sudetes Mountains, Poland). *Geomorphology* **2009**, *108*, 48–57. [CrossRef]
2. Huber, R.; Rigling, A.; Bebi, P.; Brand, F.S.; Briner, S.; Buttler, A.; Elkin, C.; Gillet, F.; Grêt-Regamey, A.; Hirschi, C.; et al. Sustainable land use in mountain regions under global change: Synthesis across scales and disciplines. *Ecol. Soc.* **2013**, *18*, 36. [CrossRef]
3. Awasthi, K.; Sitaula, B.K.; Singh, B.R.; Bajacharaya, R.M. Land-use change in two Nepalese watersheds: GIS and geomorphometric analysis. *Land Degrad. Dev.* **2002**, *13*, 495–513. [CrossRef]
4. Paudel, G.S.; Thapa, G.B. Changing farmers' land management practices in the hills of Nepal. *Environ. Manag.* **2001**, *28*, 789–803. [CrossRef]
5. Paudel, B.; Gao, J.; Zhang, Y.; Wu, X.; Li, S.; Yan, J. Changes in cropland status and their driving factors in the Koshi River basin of the Central Himalayas, Nepal. *Sustainability* **2016**, *8*, 933. [CrossRef]
6. Chaudhary, S.; Wang, Y.; Khanal, N.; Xu, P.; Fu, B.; Dixit, A.; Yan, K.; Liu, Q.; Lu, Y. Social Impact of Farmland Abandonment and Its Eco-Environmental Vulnerability in the High Mountain Region of Nepal: A Case Study of Dordi River Basin. *Sustainability* **2018**, *10*, 2331. [CrossRef]
7. Khanal, N.; Watanabe, T. Abandonment of Agricultural Land and Its Consequences: A Case Study in the Sikles Area, Gandaki Basin, Nepal Himalaya. *Mt. Res. Dev.* **2006**, *26*, 32–40. [CrossRef]
8. Jamshidi, R.; Dragovich, D.; Webb, A.A. Distributed empirical algorithms to estimate catchment scale sediment connectivity and yield in a subtropical region. *Hydrol. Process.* **2014**, *28*, 2671–2684. [CrossRef]
9. Anache, J.A.; Flanagan, D.C.; Srivastava, A.; Wendland, E.C. Land use and climate change impacts on runoff and soil erosion at the hillslope scale in the Brazilian Cerrado. *Sci. Total Environ.* **2018**, *622*, 140–151. [CrossRef]
10. García-Ruiz, J.M.; Lana-Renault, N. Hydrological and erosive consequences of farmland abandonment in Europe, with special reference to the Mediterranean region–A review. *Agric. Ecosyst. Environ.* **2011**, *140*, 317–338. [CrossRef]
11. Lin, W.-T.; Lin, C.; Tsai, J.; Huang, P. Eco-environmental changes assessment at the Chiufenershan landslide area caused by catastrophic earthquake in Central Taiwan. *Ecol. Eng.* **2008**, *33*, 220–232. [CrossRef]
12. Walker, R.; Landis, W.; Brown, P. Developing a regional ecological risk assessment: A case study of a Tasmanian agricultural catchment. *Hum. Ecol. Risk Assess.* **2001**, *7*, 417–439. [CrossRef]

13. Enete, I.; Alabi, M.; Adoh, E. Evaluation of eco-environmental vulnerability in Efon Alaye using remote sensing and Geographic Information System (GIS) techniques. *J. Sustain. Dev. Afr.* **2010**, *12*, 199–212.

14. Jin, X.; Jin, Y.; Mao, X. Ecological risk assessment of cities on the Tibetan Plateau based on land use/land cover changes—Case study of Delingha City. *Ecol. Indic.* **2019**, *101*, 185–191. [CrossRef]

15. Suter, G.W. Endpoints for regional ecological risk assessments. *Environ. Manag.* **1990**, *14*, 9–23. [CrossRef]

16. Nandy, S.; Singh, C.; Das, K.K.; Kingma, N.C.; Kushwaha, S.P.S. Environmental vulnerability assessment of eco-development zone of Great Himalayan National Park, Himachal Pradesh, India. *Ecol. Indic.* **2015**, *57*, 182–195. [CrossRef]

17. Dai, X.; Li, Z.; Lin, S.; Xu, W. Assessment and zoning of eco-environmental sensitivity for a typical developing province in China. *Stoch. Environ. Res. Risk Assess.* **2012**, *26*, 1095–1107. [CrossRef]

18. Jones, R.N. An environmental risk assessment/management framework for climate change impact assessments. *Nat. Hazards* **2001**, *23*, 197–230. [CrossRef]

19. Chambers, R. *Sustainable Livelihoods: An Opportunity for the World Commission on Environment and Development*; Institute of Development Studies, University of Sussex: Brighton, UK, 2011.

20. Giri, S.; Qiu, Z. Understanding the relationship of land uses and water quality in Twenty First Century: A review. *J. Environ. Manag.* **2016**, *173*, 41–48. [CrossRef]

21. Wang, X.; Zhong, X.H.; Liu, S.Z.; Liu, J.G.; Wang, Z.Y.; Li, M.H. Regional assessment of environmental vulnerability in the Tibetan Plateau: Development and application of a new method. *J. Arid Environ.* **2008**, *72*, 1929–1939. [CrossRef]

22. Shao, H.; Sun, X.; Wang, H.; Zhang, X.; Xiang, Z.; Tan, R.; Chen, X.; Xian, W.; Qi, J. A method to the impact assessment of the returning grazing land to grassland project on regional eco-environmental vulnerability. *Environ. Impact Assess. Rev.* **2016**, *56*, 155–167. [CrossRef]

23. Tian, P.; Li, J.; Gong, H.; Pu, R.; Cao, L.; Shao, S.; Shi, Z.; Feng, X.; Wang, L.; Liu, R.; et al. Research on Land Use Changes and Ecological Risk Assessment in Yongjiang River Basin in Zhejiang Province, China. *Sustainability* **2019**, *11*, 2817. [CrossRef]

24. Tran, L.T.; O'Neill, R.V.; Smith, E.R. Spatial pattern of environmental vulnerability in the Mid-Atlantic region, USA. *Appl. Geogr.* **2010**, *30*, 191–202. [CrossRef]

25. Liu, J.; Li, Y.; Zhang, B.; Cao, J.; Cao, Z.; Domagalski, J. Ecological risk of heavy metals in sediments of the Luan River source water. *Ecotoxicology* **2009**, *18*, 748–758. [CrossRef] [PubMed]

26. Galic, N.; Schmolke, A.; Forbes, V.; Baveco, H.; van den Brink, P.J. The role of ecological models in linking ecological risk assessment to ecosystem services in agroecosystems. *Sci. Total Environ.* **2012**, *415*, 93–100. [CrossRef]

27. Pradhan, B.; Lee, S. Landslide risk analysis using artificial neural network model focussing on different training sites. *Int. J. Phys. Sci.* **2009**, *4*, 1–15.

28. Wang, B.; Yu, G.; Huang, J.; Wang, T.; Hu, H. Probabilistic ecological risk assessment of OCPs, PCBs, and DLCs in the Haihe River, China. *Sci. World J.* **2010**, *10*, 1307–1317. [CrossRef]

29. Anjaneyulu, Y.; Manickam, V. *Environmental Impact Assessment Methodologies*; BS Publications: Hyderabad, India, 2011.

30. Skidmore, A. *Environmental Modelling with GIS and Remote Sensing*; CRC Press: Boca Raton, FL, USA, 2003.

31. Tie, Y.; Tang, C. Application of AHP in single debris flow risk assessment. *Chin. J. Geol. Hazard Control* **2006**, *4*, 79–84.

32. Wong, J.K.; Li, H. Application of the analytic hierarchy process (AHP) in multi-criteria analysis of the selection of intelligent building systems. *Build. Environ.* **2008**, *43*, 108–125. [CrossRef]

33. Ying, X.; Zeng, G.; Chen, G.; Tang, L.; Wang, K.; Huang, D. Combining AHP with GIS in synthetic evaluation of eco-environment quality—A case study of Hunan Province, China. *Ecol. Model.* **2007**, *209*, 97–109. [CrossRef]

34. Ishizaka, A.; Labib, A. Analytic hierarchy process and expert choice: Benefits and limitations. *Or Insight* **2009**, *22*, 201–220. [CrossRef]

35. Aryafar, A.; Yousefi, S.; Ardejani, F.D. The weight of interaction of mining activities: Groundwater in environmental impact assessment using fuzzy analytical hierarchy process (FAHP). *Environ. Earth Sci.* **2013**, *68*, 2313–2324. [CrossRef]

36. Malczewski, J.; Rinner, C. *Multicriteria Decision Analysis in Geographic Information Science*; Springer: New York, NY, USA, 2015.

37. LRMP. *Land Capability Map*; Land Resource Mapping Project: Kathmandu, Nepal, 1986.
38. Dijkshoorn, J.; Huting, J. *Soil and Terrain Databse for Nepal (1.1 million)*; ISRIC—World Soil Information: Wageningen, The Netherlands, 2009.
39. Needelman, B.A.; Gburek, W.J.; Petersen, G.W.; Sharpley, A.N.; Kleinman, P.J.A. Surface runoff along two agricultural hillslopes with contrasting soils. *Soil Sci. Soc. Am. J.* **2004**, *68*, 914–923. [CrossRef]
40. Gritzner, M.L.; Marcus, W.A.; Aspinall, R.; Custer, S.G. Assessing landslide potential using GIS, soil wetness modeling and topographic attributes, Payette River, Idaho. *Geomorphology* **2001**, *37*, 149–165. [CrossRef]
41. Deoja, B.; Dhital, M.R.; Thapa, B.; Wagner, A. *Mountain Risk Engineering Handbook: Vol I*; International Centre for Integrated Mountain Development (ICIMOD): Kathmandu, Nepal, 1991.
42. Li, Z.-W.; Zeng, G.; Zhang, H.; Yang, B.; Jiao, S. The integrated eco-environment assessment of the red soil hilly region based on GIS—A case study in Changsha City, China. *Ecol. Model.* **2007**, *202*, 540–546. [CrossRef]
43. El-Ramly, H.; Morgenstern, N.; Cruden, D. Probabilistic slope stability analysis for practice. *Can. Geotech. J.* **2002**, *39*, 665–683. [CrossRef]
44. Dhital, M.R. *Geology of the Nepal Himalaya: Regional Perspective of the Classic Collided Orogen*; Springer: New York, NY, USA, 2015.
45. Bangash, R.F.; Passuello, A.; Sanchez-Canales, M.; Terrado, M.; López, A.; Elorza, F.J.; Ziv, G.; Acuña, V.; Schuhmacher, M. Ecosystem services in Mediterranean river basin: Climate change impact on water provisioning and erosion control. *Sci. Total Environ.* **2013**, *458*, 246–255. [CrossRef]
46. Oluwasemire, K.; Alabi, S. Ecological impact of changing rainfall pattern, soil processes and environmental pollution in the Nigerian Sudan and northern Guinea savanna agro-ecological zones. *Niger. J. Soil Environ. Res.* **2004**, *5*, 23–31. [CrossRef]
47. Promper, C.; Puissant, A.; Malet, J.; Glade, T. Analysis of land cover changes in the past and the future as contribution to landslide risk scenarios. *Appl. Geogr.* **2014**, *53*, 11–19. [CrossRef]
48. Nisbet, T. The role of forest management in controlling diffuse pollution in UK forestry. *For. Ecol. Manag.* **2001**, *143*, 215–226. [CrossRef]
49. Pettorelli, N.; Ryan, S.; Mueller, T.; Bunnefeld, N.; Jędrzejewska, B.; Lima, M.; Kausrud, K. The Normalized Difference Vegetation Index (NDVI): Unforeseen successes in animal ecology. *Clim. Res.* **2011**, *46*, 15–27. [CrossRef]
50. Collins, A.L.; Walling, D.E. Documenting catchment suspended sediment sources: Problems, approaches and prospects. *Prog. Phys. Geogr.* **2004**, *28*, 159–196. [CrossRef]
51. Nuissl, H.; Haase, D.; Lanzendorf, M.; Wittmer, H. Environmental impact assessment of urban land use transitions—A context-sensitive approach. *Land Use Policy* **2009**, *26*, 414–424. [CrossRef]
52. Cui, E.; Ren, L.; Sun, H. Evaluation of variations and affecting factors of eco-environmental quality during urbanization. *Environ. Sci. Pollut. Res.* **2015**, *22*, 3958–3968. [CrossRef]
53. Marston, R.A.; Miller, M.M.; Devkota, L.P. Geoecology and mass movement in the Manaslu-Ganesh and Langtang-Jugal himals, Nepal. *Geomorphology* **1998**, *26*, 139–150. [CrossRef]
54. Barnard, P.L.; Owen, L.A.; Sharma, M.C.; Finkel, R.C. Natural and human-induced landsliding in the Garhwal Himalaya of northern India. *Geomorphology* **2001**, *40*, 21–35. [CrossRef]
55. Devkota, K.C.; Regmi, A.D.; Pourghasemi, H.R.; Yoshida, K.; Pradhan, B.; Ryu, I.C.; Dhital, M.R.; Althuwaynee, O.F. Landslide susceptibility mapping using certainty factor, index of entropy and logistic regression models in GIS and their comparison at Mugling–Narayanghat road section in Nepal Himalaya. *Nat. Hazards* **2013**, *65*, 135–165. [CrossRef]
56. Cao, Y.; Wu, Y.; Zhang, Y.; Tian, J. Landscape pattern and sustainability of a 1300-year-old agricultural landscape in subtropical mountain areas, Southwestern China. *Int. J. Sustain. Dev. World Ecol.* **2013**, *20*, 349–357. [CrossRef]
57. Landis, W.G.; Wiegers, J.K. Ten years of the relative risk model and regional scale ecological risk assessment. *Hum. Ecol. Risk Assess.* **2007**, *13*, 25–38. [CrossRef]
58. Saaty, T.L. How to make a decision: The analytic hierarchy process. *Eur. J.Oper. Res.* **1990**, *48*, 9–26. [CrossRef]
59. Saaty, T.L. A scaling method for priorities in hierarchical structures. *J.Math. Psychol.* **1977**, *15*, 234–281. [CrossRef]
60. Saaty, T.L. Decision making with the analytic hierarchy process. *Int. J.Serv. Sci.* **2008**, *1*, 83–98. [CrossRef]

61. Ayalew, L.; Yamagishi, H.; Ugawa, N. Landslide susceptibility mapping using GIS-based weighted linear combination, the case in Tsugawa area of Agano River, Niigata Prefecture, Japan. *Landslides* **2004**, *1*, 73–81. [CrossRef]

62. Kwak, B.K.; Kim, J.H.; Park, H.; Kim, N.G.; Choi, K.; Yi, J. A GIS-based national emission inventory of major VOCs and risk assessment modeling: Part II—quantitative verification and risk assessment using an air dispersion model. *Korean J.Chem. Eng.* **2010**, *27*, 121–128. [CrossRef]

63. Roslee, R.; Mickey, A.C.; Simon, N.; Norhisham, M.N. Landslide susceptibility analysis (LSA) using weighted overlay method (WOM) along the Genting Sempah to Bentong Highway, Pahang. *Malays. J. Geosci.* **2017**, *1*, 13–19. [CrossRef]

64. Tehrany, M.S.; Pradhan, B.; Jebur, M.N. Flood susceptibility analysis and its verification using a novel ensemble support vector machine and frequency ratio method. *Stoch. Environ. Res. Risk Assess.* **2015**, *29*, 1149–1165. [CrossRef]

65. Nandi, A.; Shakoor, A. A GIS-based landslide susceptibility evaluation using bivariate and multivariate statistical analyses. *Eng. Geol.* **2010**, *110*, 11–20. [CrossRef]

66. Zou, Q.; Cui, P.; Zhang, J.; Xiang, L. Quantitative evaluation for susceptibility of debris flow in upper Yangtze River basin. *Environ. Sci. Technol.* **2012**, *35*, 159–163.

67. Pokhrel, P.; Pathak, D. Landslide susceptibility mapping of southern part of Marsyangdi River basin, West Nepal using logistic regression method.international journal of geomatics and geosciences. *Int. J. Geomat. Geosci.* **2010**, *1*, 24–32.

68. Seid, A.; Gadisa, E.; Tsegaw, T.; Abera, A.; Teshome, A.; Mulugeta, A.; Herrero, M.; Argaw, D.; Jorge, A.; Kebede, A.; et al. Risk map for cutaneous leishmaniasis in Ethiopia based on environmental factors as revealed by geographical information systems and statistics. *Geospat. Health* **2014**, *8*, 377–387. [CrossRef]

69. Komac, M. A landslide susceptibility model using the analytical hierarchy process method and multivariate statistics in perialpine Slovenia. *Geomorphology* **2006**, *74*, 17–28. [CrossRef]

70. Cowell, R.; Lennon, M. The utilisation of environmental knowledge in land-use planning: Drawing lessons for an ecosystem services approach. *Environ. Plan. C Gov. Policy* **2014**, *32*, 263–282. [CrossRef]

71. Nagendra, H.; Karmacharya, M.; Karna, B. Evaluating forest management in Nepal: Views across space and time. *Ecol. Soc.* **2005**, *10*, 24. [CrossRef]

72. Pandit, B.H.; Wagley, M.P.; Neupane, R.P.; Adhikary, B.R. Watershed management and livelihoods: Lessons from Nepal. *J. For. Livelihood* **2007**, *6*, 67–75.

Article

The Comparative Analysis of the Adaptability Level of Municipalities in the Nysa Kłodzka Sub-Basin to Flood Hazard

Grzegorz Dumieński [1,*], Agnieszka Mruklik [2], Andrzej Tiukało [1] and Marta Bedryj [1]

[1] Institute of Meteorology, Hydrology and Water Management—National Research Institute, 01673 Warsaw, Poland; andrzej.tiukalo@imgw.pl (A.T.); marta.bedryj@imgw.pl (M.B.)

[2] Department of Mathematics, Faculty of Environmental Engineering and Geodesy, Wroclaw University of Environmental and Life Sciences, 50375 Wroclaw, Poland; agnieszka.mruklik@upwr.edu.pl

* Correspondence: grzegorz.dumienski@imgw.pl; Tel.: +48-7132-001-75

Received: 27 February 2020; Accepted: 30 March 2020; Published: 9 April 2020

Abstract: A municipality is a basic local government unit (LGU) in Poland. It is responsible for the safety of its citizens, especially in circumstances of flood hazard. A municipality is a unique social-ecological system (SES), distinguished by its ability to adapt to flood hazard. It is impossible to specify the conditions a municipality has to reach to achieve the highest adaptability to flood hazard, however, it is possible to assign a level of adaptability to a municipality, one that corresponds to the position of a given municipality among the population of assessed municipalities threatened by floods. Therefore, a tool was developed to rank municipalities by their adaptability on the assumption that the assessment of municipal adaptability was influenced by 15 selected features. The research was carried out using data from the period of 2010–2016 for 18 municipalities-SESs located in the Nysa Kłodzka sub-basin. It was indicated that municipalities located in the higher course of the river possess higher levels of adaptability. At the same time, the size of a flood stands for each municipality position with regards to the synthetic adaptability index (SAI).

Keywords: adaptability; socio-ecological system; flood risk; municipality; Nysa Kłodzka sub-basin

---

1. Introduction

The increasing occurrence of hazardous natural phenomena that have severe consequences, such as floods [1–3], is forcing the development of management tools that help minimize the adverse consequences of such phenomena. These tools may be helpful when making decisions in conditions of uncertainty resulting from the randomness of this hazardous phenomena [4].

Currently, the key role in the analysis of the adaptation of SESs to climate change, including the increasing frequency of natural phenomena such as floods [5], is the assessment of resilience [1,6,7]. Therefore, the flood hazard assessment of a socio-ecological system (Figure 1) is based on the assumption that every system is characterized by the scale of exposure of its elements to floods (as presented on flood hazard maps, etc.,) [8] and its vulnerability to this phenomenon [6,7,9–13]. Floods as a factor that disturbs the functioning of an SES may display diverse intensity at a given probability [2,4,14]. The exposure refers to all elements of an SES that are vulnerable to floods and located in the flood hazard area [4,15]. The sensitivity involves the characteristics (features) of an SES's elements exposed to flood hazard, which make floods a cause of various losses [16,17]. Both the exposure and the sensitivity of the elements of the system at risk of floods affect the scale of potential loss. The vulnerability is a result of the volume of potential flood losses and its resilience influencing the reduction of these losses [17]. In turn, a system's resilience to flood hazard is shaped by its durability, as well as its capability to adapt and to transform (this happens when adverse consequences and the scale of a cataclysm force a system to recreate its basic functions, i.e., to adapt) [18–21].

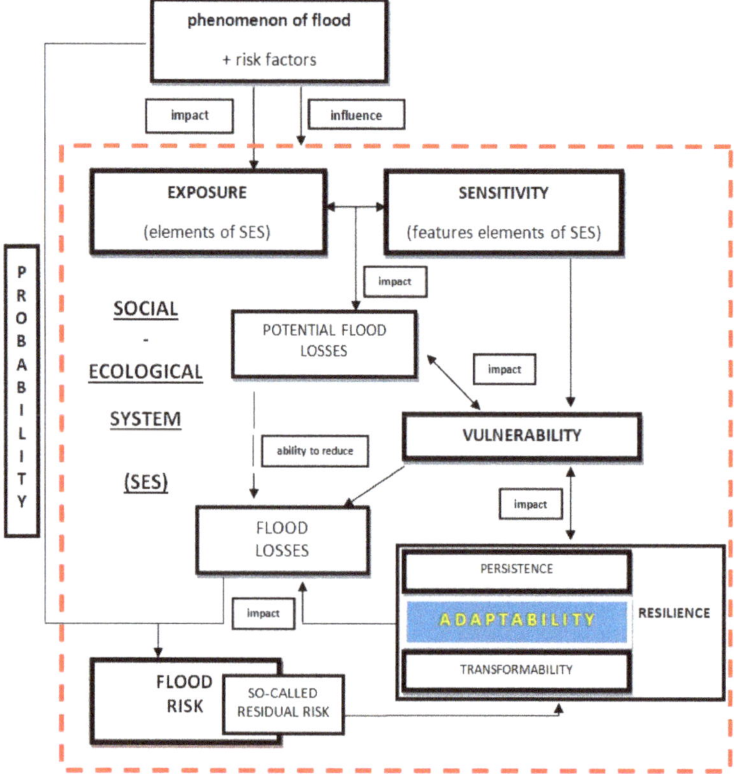

Figure 1. Algorithm for assessing the flood risk as a social-ecological system at risk of floods, indicating the role of a system's adaptability in risk reduction. Source: [19].

In Poland, the major natural hazard that may take the form of a cataclysm is a river flood [2,17,19,22]. As indicated by Pniewski and his team, the river flood hazard increases over years, and it is therefore necessary to develop a flood risk management system [2]. Although the last severe floods in Poland took place in 1997, 2001 and 2010, the fundamental flood protection infrastructure has not been restored yet in many regions of Poland, despite strong efforts [17,23]. One can therefore accept the notion that the resilience of the regions of Poland threatened by floods was not sufficient enough to accommodate enormous amounts of flood waters, because the adverse consequences for society and particular sectors of the economy were felt long after the flood waters subsided. This observation questions the state of the adaptability of Polish local government units to flood hazard [17,19].

In Poland, the provisions of the so-called 'flood directive' (i.e., Directive 2007/60/EC) [24] have been implemented since 2010 by the conducting of the Initial Flood Risk Assessment (IFRA), the preparation and public distribution of Maps of Flood Hazard (MFH) and Maps of Flood Risk (MFR) [8] and the introduction of Flood Risk Management by the Polish Parliament (FRMP) [17,24,25]. These documents and the current ongoing work to update them indicate that the national policy on flood risk management, understood as the sequence of decisions and actions (or lack of them) in the field of flood protection, is systematically being formed in Poland [1,23]. The update of IFRA, MFH, MFR and FRMP takes place in six-year planning cycles. Currently the IFRA verification and the MFH and MFR update is in process in Poland. This impacts the identification of areas of new river flood hazard that were not subject to hydraulic modelling in the first planning cycle to designate flood hazard zones [8].

The policy of flood risk management in Poland is implemented on every territorial level and meets the prerogatives assigned to the relevant authorities on the level of state, voivodeship, county and municipality [1,17,23], as well as water management regions [26]. Although the state and regional policy for shaping flood risk is usually carried out within a catchment, it is the municipality, as the basic local government unit in Poland, that is aware of flood risk level in its own territory and has to take up appropriate actions in order to protect itself against an identified threat [12,17,27]. The municipality is undoubtedly the subject (beneficiary) of the state and regional policy that shapes flood risk, but it is also the creator of the local policy on flood risk in its own territory, especially by way of adaptive actions, which are to make it immune to current and expected flood hazards. Disastrous floods cause serious disturbances in the functioning of a municipality as a SES [1,12,17,22]. The system loses its effectiveness and efficiency in terms of achieving its goals. Thus, the municipality-as-SES is supposed to build adaptive capital based on its own resources and shape its capability to adapt to identified and expected disturbances [19,27].

The aim of this article is to present a management tool which enables effective management in municipalities, understood as socio-ecological systems (SES), by way of the process of shaping municipalities' adaptability to flood hazard. The process of a municipality's adaptation to flood hazard may be improved by comparing it to other subjects threatened with flood. Such comparisons may be possible due to the introduction of 15 non-observable statistical features of a municipality-SES which determine its adaptability to flood hazard [19].

The conducted analyzes allowed the research questions to be answered:

Which of the 15 selected statistical features of the municipality-SES that influenced the assessment of adaptability:

1. distinguished the municipalities with the highest and the lowest adaptability assessment?
2. were characterized with the largest or the smallest variability?
3. adopted values distinguishing a municipality that can be a role model for other municipalities.

2. Features Which Determine the Adaptability of a Municipality-as-SES to Flood Hazard

It is necessary to highlight that the adaptation of units, such as municipalities and entire water regions, to flood hazard was not analyzed in the FRMP developed in Poland [17]. This means that the Polish Legislator assumed (wrongly, in the authors' opinion) that their adaptability level to flood hazard is the same across the entire area of the country [19]. Yet, it is hard to accept that all municipalities in Poland have comparable resources which they can use to limit the adverse consequences caused by floods, as well as during rebuilding the damage after a cataclysm subsides [16].

When assessing the adaptability level of Polish municipalities, it is necessary to note that:

- The municipality is a socio-ecological system (SES), because it is two interdependent systems, a social system and an ecological system, whose functioning is conditioned by numerous connections [17,19,28]. The aim of such an SES is social development, including the durability of the ecological system. The implementation of this aim can be achieved by shaping the attitudes and social needs and satisfying them, as well as by providing the opportunities to implement ecosystem services [19,29].
- The adaptability of a municipality-as-SES threatened with floods is a (current) capability of the system to limit the adverse consequences of floods. It results from the adaptive potential of the system, that is the quality and quantity of the resources possessed by the system, which are useful for conducting actions in the event of a cataclysm and after it subsides. It also results from the adaptive capability of the system, that is its capability to activate these resources (adaptive potential) [19,29].

- The current adaptive capability and adaptive potential of a municipality-as-SES, and in consequence its adaptability, can be described as the result of a set of SES's attributes/non-observable statistical features [19,29,30]. These features stimulate or break the process of shaping a municipality's capability to keep its effectiveness and efficiency in an event of disturbance (the phenomenon of flood).

Therefore, the catalog of 15 features of an SES have been proposed, and ordered into four categories [19,30]. On the one hand, these features (Figure 2) are a set of determinants which define the effectiveness of the ongoing SES adaptation to floods occurring in the past and expected in the future. These features also constitute a set of information on an SES, which are useful for assessing its current adaptability to flood hazard. Selected statistical features of a municipality-as-SES describe its adaptive potential (the quantity and quality of the resources) and adaptive capability (the capability to activate these resources) [19].

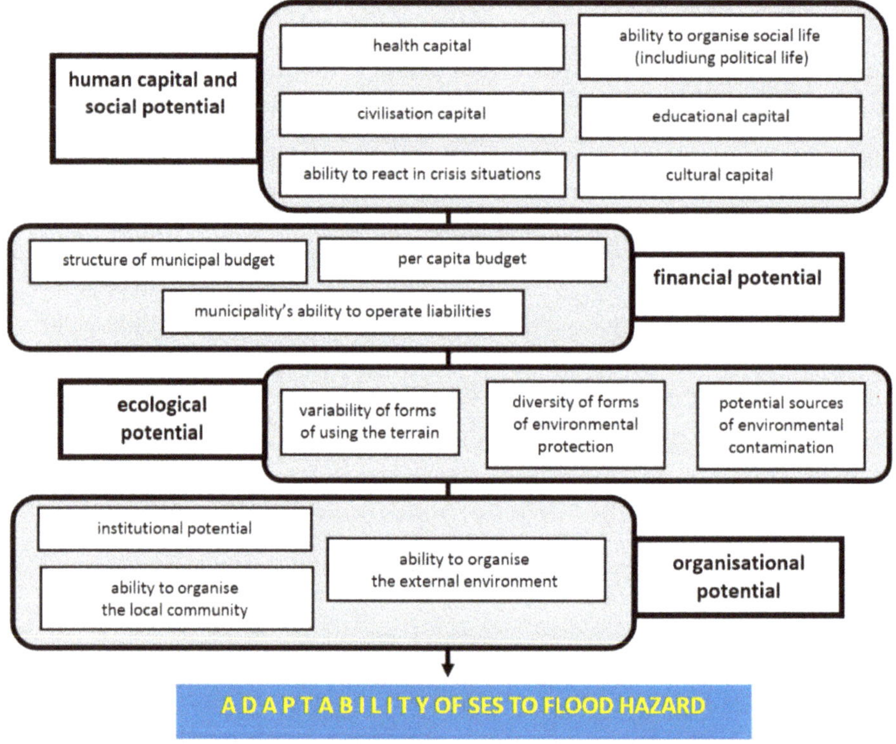

Figure 2. Features affecting the adaptability of a municipality as a socio-ecological system (SES) [source: own work].

Analysis with regard to the method of defining these authorial features has become the subject of papers [19,30,31].

3. Materials and Methods

3.1. Research Area

The Flood Risk Management Plans (FRMPs) have identified problem areas in Poland that are characterized by high flood risk. They require urgent intervention and are called the "hot spots".

One such region in Poland encompasses areas located along the course of the Nysa Kłodzka river. The river has a length of over 182 km and is the left bank tributary of the central Oder river. It passes through the following geographical areas: Kłodzka Valley, Sudetic Foothills and Silesian Lowland. In the region of the Kłodzko Valley, flood risk is very high.

In the area of the conducted studies, not only were the 7 municipalities located in the abovementioned hot spots included, but altogether 18 municipalities located in the Nysa Klodzka catchment (from the spring in the Śnieżnik Massif down to its mouth in the valley of Lewin Brzeski). Within the borders of these municipalities, areas of special flood hazard were identified and Maps of Flood Risk (MRP) and Maps of Flood Hazard (MFH) were developed. The research area included 7 rural municipalities: Kłodzko, Kamieniec Ząbkowicki, Łambinowice, Skoroszyce, Popielów, Olszanka and Skarbmierz; 10 rural-urban municipalities: Międzylesie, Bystrzyca Kłodzka, Bardo, Ziębice, Paczków, Otmuchów, Nysa, Niemodlin, Grodków and Lewin Brzeski, as well as one urban municipality—Kłodzko. The abovementioned municipalities became the subject of the study. The object of the study included the adaptability of these municipalities to flood hazard and the territorial spacing of the municipalities' level of adaptability to flood hazard [29,30]. The location of the research area is presented in Figure 3. In turn, spatial distribution of the integrated level of flood risk of this research area is presented in Figure 4.

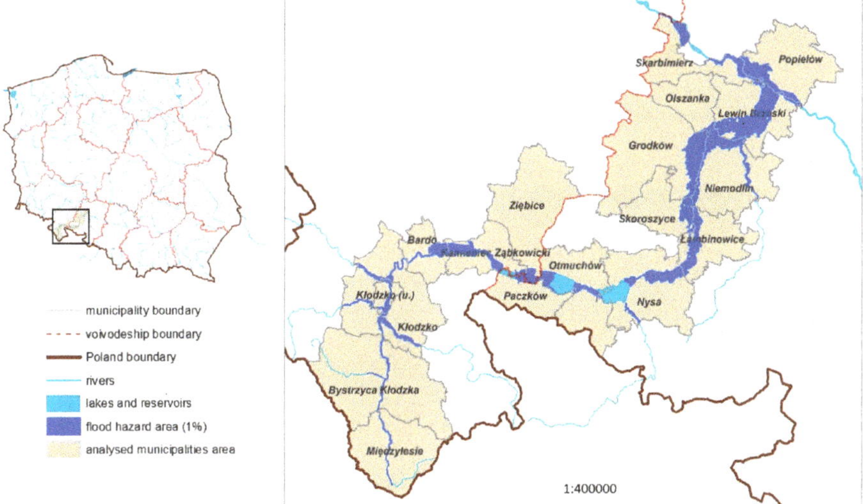

Figure 3. Location of the research area—municipalities along the course of the Nysa Kłodzka river [source: own work].

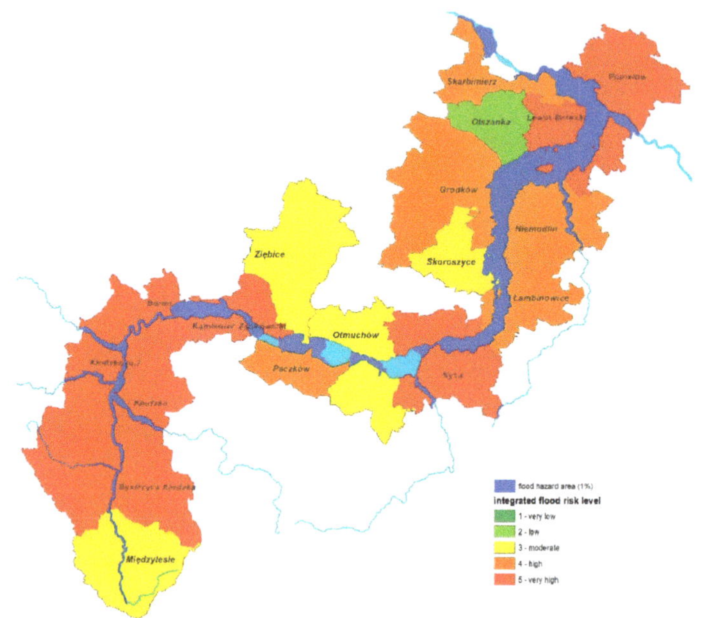

Figure 4. Municipalities of the sub-basin of the Nysa Kłodzka along with the assigned level of integrated flood risk set out in FRMP [source: own work].

It is worth underlining that most of the analyzed research area, i.e., Kłodzko Valley along with Kłodzko city, is a beneficiary of the currently implemented project "The Oder Flood Protection Project" within component 2: Kłodzko Valley Flood Protection. It was considered in this project that in order to build stable and sustainable social-economic development of selected areas of the Oder basin, it is necessary to strengthen flood protection [32,33].

The analysis of flood hazard and municipalities' adaptability in terms of the catchment appears to be the best solution [19,30,34]. In particular, flood protective investments implemented in Kłodzko Valley will shape flood risk of the municipalities located in the entire area of the Nysa Kłodzka catchment, and these municipalities will be able to plan their own adaptive actions towards flood hazard. Such a methodological approach also has practical justification as the results of the performed analyses and the developed research tool will support local decision-makers in setting strategic goals for adaptation to flood hazard. It will allow them to indicate these municipalities' features which influence the assessment of their adaptability to flood hazard, as such assessment deviates from the leaders' standpoint in this field and indicates areas requiring corrective actions [30].

3.2. Methods

It was considered while conducting the studies dedicated to developing the method of a municipality's adaptability assessment to flood hazard in Poland [1,17,19,29,30], that the adaptability of the municipality understood as a socio-ecological system (SES) is the result of many features of this system. These features characterize both the resources of the social subsystem (its capability to activate these resources in order to reduce adverse consequences of a flood) and the resources of the ecological subsystem with the capability of reducing the adverse consequences of a flood. In total, 15 features of the municipality have been distinguished (including 12 features of the social subsystem and 3 features of the ecological subsystem) [19,30], which form non-observable measurable statistical features. Therefore, it became necessary to create research procedures allowing for the collection of empirical observations which can be used to assess the municipality's adaptability to flood hazard [29].

In order to describe these 15 theoretical non-measurable statistical features of a municipality-as-SES, which determine its adaptability to flood hazard, a public domain search was carried out in terms of finding such observable variables that would comprehensively allow the diagnosis of these features. These variables were sought in the resources of the Central Statistical Office, including the Statistical Vademecum of the Local Government, and the Local Data Bank. Data obtained from the Supreme Chamber of Nurses and Midwives, the Supreme Medical Chamber and County Sanitary and Epidemiological Stations were also used. Due to these actions, data on 83 observable variables were obtained in the first phase of empirical/quantitative data collection.

During the process of the assignment of the obtained observable variables to particular latent statistical features, the problems of shortage or lack of such variables necessary to describe all characteristics of the municipality-as-SES, which determine its adaptability to flood hazard, were highlighted. Hidden statistical features, for which unsatisfactory numbers of observable variables were collected, were: the features belonging to category K1—Human capital and social potential: D4—the ability to respond to crises and D5—civilization capital; features from the category K2—financial potential: D7—adequacy of the municipality's budget structure to the hazard and D9—adequacy of the municipality's budget per capita and features belonging to the category of features K4—organizational potential i.e.,: D14—the ability to organize local community and D15—the ability to organize the external environment. In order to supplement the list of observable variables for the abovementioned hidden statistical features (7 were defined), a questionnaire was created, adopting the thesis that quantitative research is an effective source of obtaining data with regard to the new observable variables which describe the characteristics of the municipality determining its adaptability to flood hazard [29,30].

Using statistical data from all the mentioned sources, a matrix of averaged values of diagnostic variables was created and became the base for output data for the preliminary assessment of the municipality-as-SES' adaptability to flood hazard. The results of this preliminary assessment of adaptability of municipalities in the Nysa Kłodzka sub-basin are presented in this article [25].

It is necessary to emphasize that at the previous and current stage of studies [29], the authors did not decide on the significance/importance of both non-measurable statistical features describing the SES and the measurable observable variables forming these features, in view of their influence on the assessment of the SES's current adaptability to flood hazard. The same significance was assumed for each of 15 hidden statistical features in the built-up synthetic adaptability index (SAI).

Within the conducted studies, a comparison—so-called benchmarking—was made, which took into account 18 municipalities of the Nysa Kłodzka sub-basin. The benchmark was based on the individual features that influence the adaptability of an SES to flood hazard. The method of developing partial synthetic indicators of the adaptability in relation to the preliminary analyses [25] was modified in order to make the obtained rankings play the role of a management tool. The authors wanted to create a mechanism which could be applied in the process of the assessment of adaptability to flood hazard in relation to particular non-measurable statistical features, which are determinants of adaptability at the same time. The management tool is intended to be a useful form which allows the indication that these latent statistical features (determinants) of particular municipalities-as-SES which, in comparison to other units of local government, require improvement due to "unfavorably standing out".

The research method adopted by the authors is the multidimensional comparative analysis (MCA) linear ordering method [34–37] in its classic and non-standard approach, which is based on a synthetic variable. The MCA methods form a group of statistical methods used for studying complex phenomena, which are directly non-measurable and which characterize identified objects subject to analysis [38]. These objects should form a relatively homogeneous set. During the research, the objects subject to analysis are the 18 municipalities-as-SES located in the Nysa Kłodzka sub-basin. They caught the researchers' attention due to one of their features, of complex nature—the adaptability to flood hazard. The adaptability of a municipality-as-SES cannot be measured directly. Therefore, 15 authorial non-observable statistical features were proposed, describing the municipality-as-SES' adaptability to flood hazard. For this reason, the authors' interest focused on measurable observable variables, characterizing specific non-measurable statistical features of the municipality-as-SES. The observable (diagnostic) variables were also treated as equivalent (Appendix A; Appendix B: Figures A1 and A2). In this sense, the authors of this article use the terms: the feature of the municipality describing its adaptability (non-observable statistical feature); the observable variable (measurable statistical feature, indicator) and the category (category of the municipality's features; category of non-observable statistical features) which refer to the municipality as a social-ecological system threatened by flood (Figure 5).

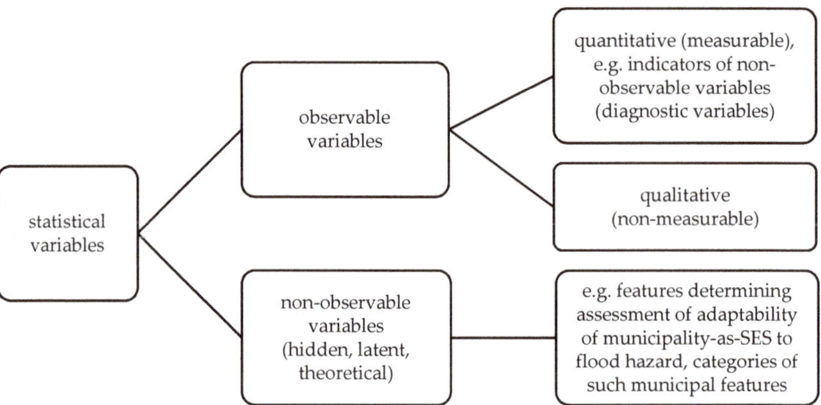

Figure 5. Types of statistical variables allowing the assessment of a municipality-as-SES' adaptability to flood hazard [source: own work].

The preliminary comparative assessment of municipalities' adaptability to flood hazard [29] allowed the identification of municipalities characterized by much greater adaptability to flood hazard than other municipalities. Thus, a general view of a municipality's adaptability level was obtained for the period 2010–2016. The collective classification of the considered municipalities became the source of this knowledge. It was created on the basis of the set of indicators of specially developed synthetic variable A the so-called SAI. In this way, a measurer was obtained, for which rising values reflect a higher assessment of adaptability of the given municipality. To build an SAI, the diagnostic variables (indicators) were used. The set of statistical data for the years 2010–2016 was obtained using the abovementioned sources. This is how the set of potential diagnostic variables X_j, $j = 1, \ldots, 110$ containing 110 elements was created, and these variables and the values of these variables were obtained in particular years. The statistical material was preliminarily developed using the complete review method. Then, using arithmetic average, the diagnostic variables were averaged. The structure of the final set of the diagnostic variables was two-stage and was based on the selection of potential indicators. The criteria of selection are mainly of a statistical character. The diagnostic variables were researched for the variability of their observed values, as well as for the mutual linear correlation of indexes, excluding the cases of apparent correlation. As a result of these analyses, out of 110 diagnostic variables 70 were selected and these were characterized by 14 statistical features (out of 15 subject to the analysis) of the municipalities-as-SES located in the Nysa Kłodzka sub-basin, which determine municipalities' adaptability to flood hazard. The ability to react to crisis situations (D4) is the only statistical feature of the municipality-as-SES which was excluded from analyses, because it was diagnosed by variables which absorbed homogeneous results in the population of the municipality in the Nysa Kłodzka sub-basin.

Next, the observable variables (diagnostic variables) were divided into stimulants and destimulants, which results, among other things, from the specifics of the adopted research method—MCA. The authors concluded that in the case of the problem in question, the nominants do not occur. The subsequent stage of work was the normalization of the values of the diagnostic variables by the method of zeroed unitarization, which has a lot of desirable features during the ongoing research [39–43]. Performing data normalization allows the comparison of the values of variables, including the diagnostic variables. Within the normalization and transforming destimulants into stimulants, the diagnostic variables were brought down to a mutual form or mutual character. At the end, the values a_i of the synthetic variables were indicated as the sum of the normalized values of the diagnostic variables. The municipality achieves a higher position in the classification as a greater value is achieved by the synthetic measure—that is, value a_i corresponding to a given municipality (see Appendix B: Figures A1 and A2).

The partial classifications were used in order to benchmark 18 municipalities-as-SES of the Nysa Kłodzka sub-basin with regard to individual non-measurable statistical features determining the SES's adaptability to flood hazard. Thus, the conducted research was based on performing partial qualifications (obtaining partial classifications) of the municipalities in question, where the criterion of classification was the assessment of each theoretical statistical feature affecting the adaptability of particular municipalities to flood hazard. At the same time, the assessment of these features was performed on the basis of a set of values of the specially created synthetic variables. In order to build particular synthetic variables, diagnostic variables describing a given statistical feature were used, i.e., the determinant of the municipality-as-SES affecting its adaptability.

For each of the 14 determinants D_k, where $k \in \{1, 2, 3, 5, 6, \ldots, 15\}$, which are the non-measurable features of the municipalities-as-SES affecting its adaptability (the adaptability determinants), the following actions were performed:

- normalized values of the diagnostic variables characterizing D_k were added, obtaining the sum of the form $d_{k,i} = \sum_j z_{ij}$, where z_{ij}—normalized value j—variable for the i municipality,

- for each k, where $k \in \{1, 2, 3, 5, 6, \ldots, 15\}$, normalization of the data $d_{k,1}, \ldots, d_{k,18}$ was performed by the method of zeroed unitarization, obtaining the data $\widetilde{d}_{k,1}, \ldots, \widetilde{d}_{k,18}$ (see Appendix B: Figures A3 and A4).

Then, taking as a basis the partial qualifications of the studied municipalities with regard to the non-measurable statistical features, the analyzed municipalities were classified again with regard to the adaptability determinants. Thus, the obtained classification of the adaptability level of the studied municipalities turned out to be slightly different from the one described in [25], because:

- the values $a_{D,i}$ of the statistical variable A_D were determined as the sum of the normalized values of the data $\widetilde{d}_{k,1}, \ldots, \widetilde{d}_{k,18}$, i.e., $a_{D,i} = \sum_k \widetilde{d}_{k,i}$, $i = 1, \ldots, 18$;

- for each i, where $i = 1, 2, \ldots, 18$, the normalization of the data $a_{D,1}, \ldots, a_{D,18}$ was performed by the method of zeroed unitarization, obtaining the values of the variable $\widetilde{A}_D$ (Figure 6).

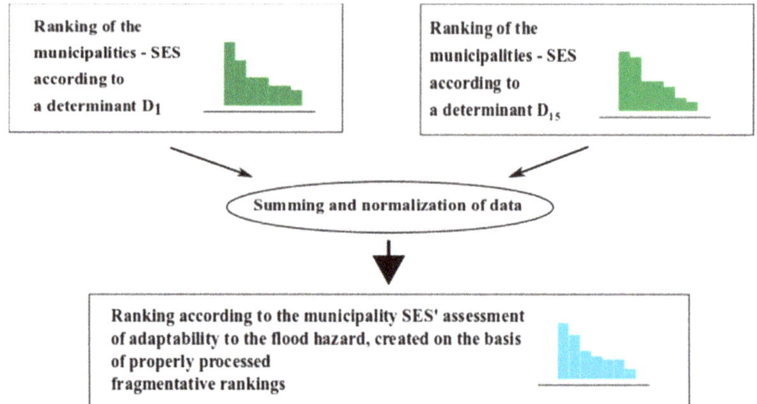

Figure 6. The procedure of creating a new ranking according to their adaptability to flood hazard [source: own work].

A general municipality classification of the studied area, with regard to the non-measurable statistical feature determining municipalities' adaptability to flood hazard, was created on the basis of a set of values of the variable $\widetilde{A}_D$ for each of the 14 features analyzed in the article. This classification is built with the assumption that the determinants have equal influence on the adaptability assessment. The higher a position a municipality achieves in the classification, the higher a value achieves the value of the variable $\widetilde{A}_D$ corresponding to it. These results are summarized in Figure 7.

It is necessary to emphasize that the obtained result in the form of the amount of measurable observable (diagnostic) variables, which describe the non-measurable statistical features (determinants) of a municipality-as-SES, is specific for the analyzed research area. Although the latent statistical feature (determinant) D4 (The ability to react in hazardous situations) did not maintain its form after the analyses of the representation as the observable (diagnostic) variables, it does not indicate that the feature is of marginal character.

It is highly probable that while conducting the analyses for the municipalities-as-SES located in another research area, the stake of selected variables and, in consequence, those described by these features (determinants), would be different. The features of a municipality, presented in the article, that determine the level of adaptability, can be applied within other areas of research, because they are features of universal character.

Although the observable variables as well as the non-observable features (adaptability determinants), ordered into four categories of features (determinants) of SES adaptability, may be considered universal because they result from the adopted concept of the municipality-as-SES, their usefulness in the description of an analyzed municipal population depends on the specifics of the analyzed SES and the time period of the conducted research as well as the observed values of individual statistical features.

4. Results

The collective results of the comparative analysis of the municipalities of the Nysa Kłodzka sub-basin are presented in Table 1. With regard to this catchment, the comparison criterion was determined by the values of the synthetic measures of 15 non-observable statistical features in relation to the characteristics of the municipalities, determining their adaptability to flood risk. Then, in Figure 8, a demonstrative drawing is presented which shows the spatial distribution of a given non-measurable statistical feature. These maps are to highlight the change of value of particular positions of the municipalities from the research area (features (F) 1–15; except for F4—insignificant statistical feature in an analyzed population of municipalities).

Finally, Figure 9 shows the spatial distribution of positions taken by the municipalities in the synthetic classification of the adaptability, based on the summary of partial classifications.

The ranking of municipalities according to the level of adaptability (synthetic adaptability index SAI form, Figure 7) is crucial for this work because it shows that each population (of municipalities) has its leader. It is also some kind of a clue for the rest of the municipalities as to how much their level of adaptability differs according to the leader's. It also suggests in which areas those municipalities need to improve. There is no adaptability threshold. Each population has its leader, thought this is not to say the leader is an ideal example.

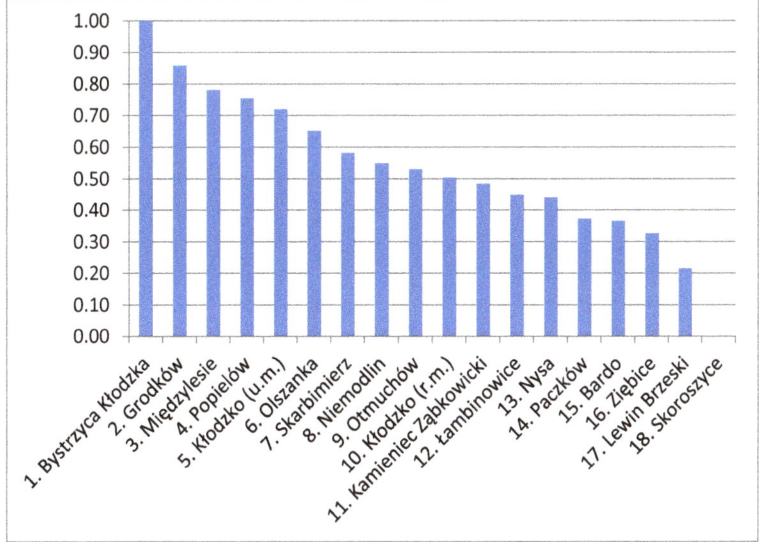

Figure 7. The classification of the adaptability of the municipalities-as-SES in the Nysa Kłodzka sub-basin to flood hazard, with the assumption of the homogeneous influence of non-measurable statistical features to the synthetic index of the adaptability [source: own work].

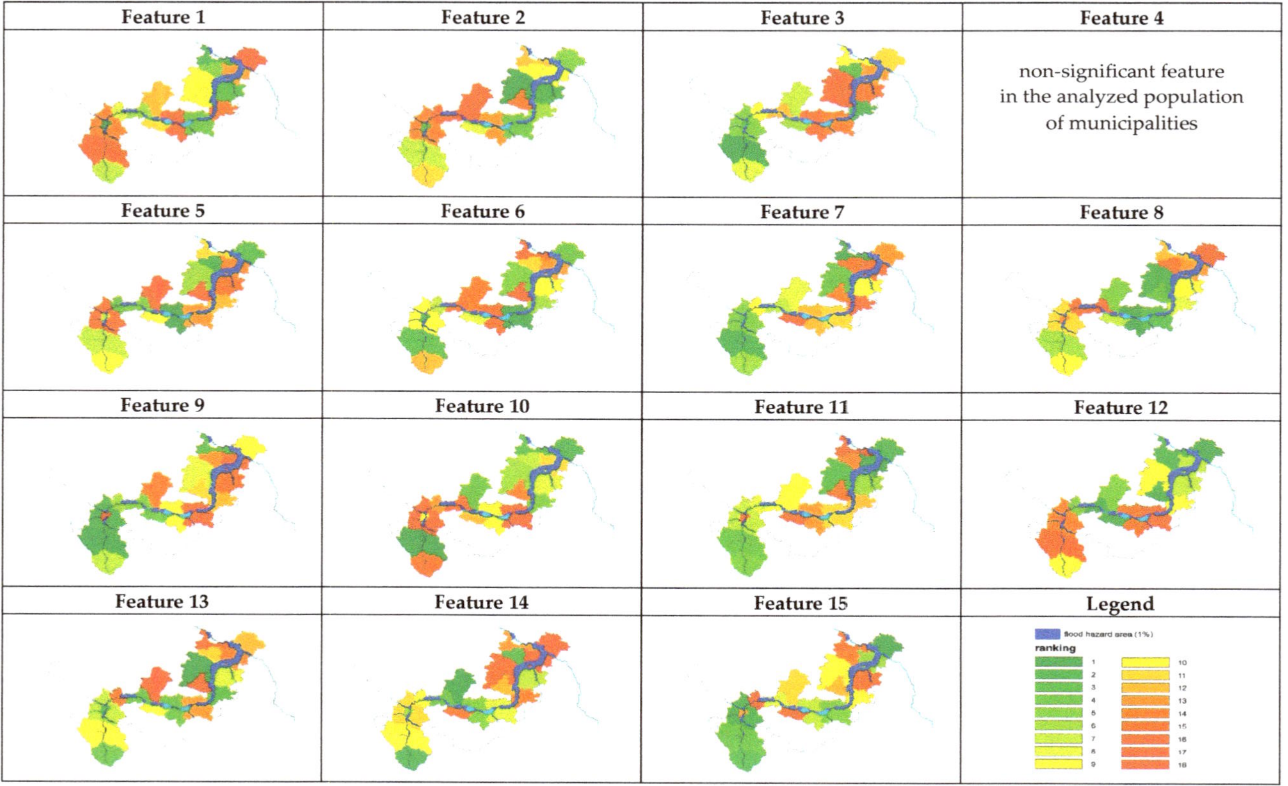

Figure 8. Demonstrative distribution of positions of the municipalities-as-SES according to non-measurable statistical features determining their adaptability to flood hazard [source: own work].

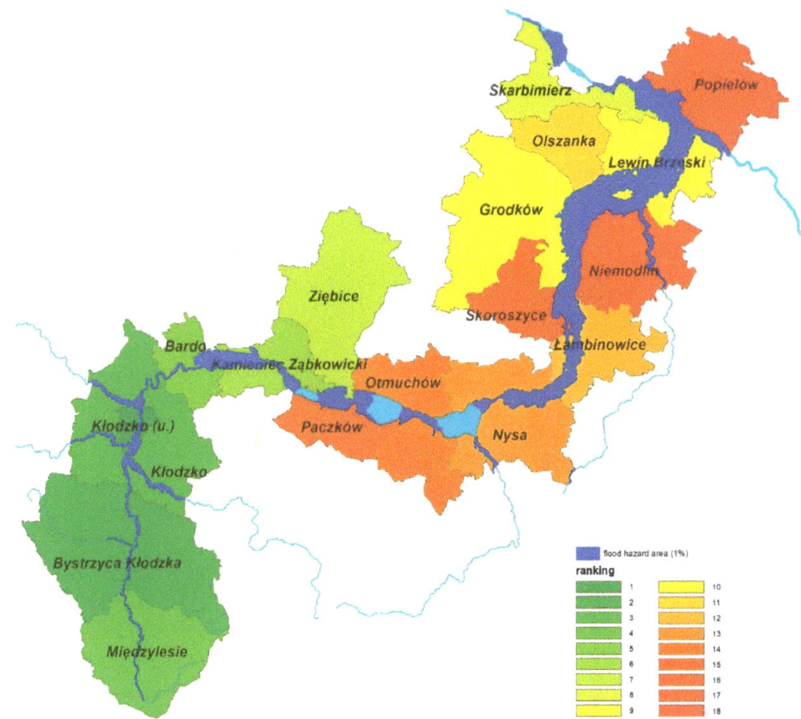

Figure 9. The synthetic classification of the adaptability to flood hazard in the municipalities-as-SES of the Nysa Kłodzka sub-basin [source: own work].

Table 1. Benchmarking of the municipalities-as-SES according to the assessment of statistical features of municipalities determining their adaptability to flood hazard.

No. of Features:	F1	F2	F3	F4	F5	F6	F7	F8	F9	F10	F11	F12	F13	F14	F15
name of non-measurable statistical features of a municipality-SES determining it's adaptability	Health capital	Educational capital	Civilisational capital	Ability to identify crisis situations	Ability to organize social life …	Cultural potential	Relevance of …	Municipality's ability to operate …	Municipal budget per capita	Variability of forms of development	Diversity of forms of environmental …	Risk of environmental …	Institutional potential	Ability to organize local …	Ability to organize …
No.	municipality							Position of the municipality's in the classification							
1	Międzylesie	7	11	8		9	12	6	10	7	15	5	11	4	2	3
2	Bystrzyca Kłodzka	16	7	1		7	3	2	6	1	1	4	16	10	10	4
3	Kłodzko (r.m.)	14	14	5		16	10	5	11	2	17	7	14	7	11	2
4	Kłodzko (u.m.)	1	3	3		8	2	3	14	18	10	16	17	2	6	13
5	Bardo	8	16	9		4	7	10	16	6	13	8	7	15	9	18
6	Kamieniec Ząbkowicki	6	18	12	non-significant feature in the analyzed population of municipalities	5	15	7	17	5	16	9	2	5	4	8
7	Ziębice	12	17	7		17	14	8	5	14	4	10	8	18	1	11
8	Paczków	11	8	6		10	8	17	7	4	12	17	3	9	15	17
9	Otmuchów	17	13	16		1	16	12	1	9	9	13	15	6	5	7
10	Nysa	4	4	14		13	1	11	3	17	18	11	18	13	8	5
11	Łambinowice	15	6	2		12	6	18	8	12	5	12	12	3	14	10
12	Skoroszyce	9	15	18		18	18	14	4	11	14	15	4	17	12	12
13	Niemodlin	3	2	15		15	9	9	9	15	8	6	9	8	7	16
14	Popielów	18	5	11		3	4	13	18	10	2	2	1	12	17	1
15	Grodków	10	1	17		6	5	4	2	8	6	3	13	1	16	9
16	Olszanka	5	10	4		2	11	15	15	13	7	14	5	11	3	14
17	Lewin Brzeski	13	9	13		14	13	16	13	16	11	1	10	14	18	6
18	Skarbimierz	2	12	10		11	17	1	12	3	3	18	6	16	13	15

The conducted comparative analysis of the municipalities-as-SES in the Nysa Kłodzka sub-basin according to the features determining municipalities' adaptability to flood hazard has allowed the following conclusions:

- The rural-urban municipality Bystrzyca Kłodzka three times scored first place in the classification of municipalities where the municipalities were compared according to their human capital (feature D3), their budget per citizen (feature D9), as well as diverse forms of land development (feature D11). Bystrzyca Kłodzka also obtained high scores in the classifications of municipalities, according to the municipal budget structure providing financing of adaptation actions (2nd place), and in the classification of municipalities according to cultural potential (3rd place). This municipality was classified first in the classification of municipal adaptability, however, in the case of classifications according to health capital and the threat of environmental contamination, its position was 16th. Therefore, it is obvious that the leader among the municipalities in the Nysa Kłodzka sub-basin may also use the experience of partner municipalities which are ahead of Bystrzyca Kłodzka in many other classifications when it comes to the process of shaping adaptability to flood hazard.
- The rural municipality Skoroszyce was classified last three times in the classification of the municipalities, where the municipalities were compared according to their civilization capital (feature D3), the ability to organize social life, including political life (feature D5), as well as cultural potential (feature D6). This municipality was also in far-reaching positions according to institutional potential (17th place) and the diverse form of land development (15th place). Skoroszyce, in the synthetic classification of adaptability to flood hazard, took the last (18th) position among all the analyzed municipalities. However, it has to be highlighted that there were two areas in which this municipality scored 4th place—the municipality's ability to service its obligations (feature D8) and the threat of environmental contamination by objects of environmental flood risk (feature D12).
- The biggest dispersions of values for the analyzed non-measurable statistical features of the municipalities-as-SES to flood hazard were noticed for the features: D11—diverse forms of nature protection and D14—the ability to organize the external environment.
- The smallest dispersions of the values for the analyzed non-measurable statistical features of the municipalities-SES to flood hazard were noticed for the features: D7—the structure of the municipal budget and D8—the municipality's ability to service its obligations (the municipality's financial condition).
- It is worth emphasizing that the smallest changes of values for the non-measurable statistical features of the municipalities-as-SES were noticed in the areas for which the information regarding the diagnostic variables were aggregated on the county level; this is mainly reflected by feature D12—the threat of environmental contamination from the objects of environmental flood risk.

Figure 8 shows the classification of municipalities' adaptability in the Nysa Kłodzka sub-basin to flood hazard created with regard to the synthetic adaptability index (SAI). On the basis of the conducted comparative analysis of the municipal adaptability to flood hazard (Figures 8 and 9), we can conclude that due to frequent historical floods, the municipalities located in the upper run of the Nysa Kłodzka river have already developed the ability to build their resilience to floods and this is reflected by their positions in the classification of municipalities: Bystrzyca Kłodzka (1), Międzylesie (3) and the urban municipality of Kłodzko (5).

5. Conclusions

It is necessary to highlight that the proposed research tool in the form of a synthetic adaptability index (SAI) used to assess the municipality-as-SES's adaptability to flood hazard is the first holistic conception of such a type, aggregating many types of municipality-as-SES's activity (human, capital and social potential, financial potential, ecological potential and organizational potential). Due to the fact that the presented studies are the first of their type in Poland, which undertake the subject of the assessment of the adaptability of Polish municipalities to flood hazard, they can be considered both pioneering studies and pilot studies. The obtained results now become the basis for verifying the content of the presented research tool in the aspect of its substantive construction (an adoption of certain—and no other—features, selection of these features, their quantity, giving equivalent weight). It is necessary to emphasize that in the Polish literature on the subject there have been studies regarding selected aspects of natural hazards or floods, however, they referred to one defined component in the field of conceptual flood risk assessment (Figure 1). These studies were performed in relation to social vulnerability [10,12,44], susceptibility [18] or SESs' adaptability to weather phenomena [27].

At this stage of the study, it was recognized that each of the features used to assess the adaptability (one of 15 components building SAI) is equal, regardless of the number of indicators included in it (observable variables). In the subsequent stage of study, it will be necessary to consider giving specific weight to particular statistical features.

Authorial non-measurable statistical features, though covering the entire spectrum of an SES's functioning, may be distinctive for a specific research population in a given country. Hence, social, economic or institutional conditions in other countries may raise barriers when trying to compare one SES with another, and thus, create limitations in the development of a universal tool for assessing the adaptability of different local government units which are all, however, at risk of flooding.

A similar problem may arise when building the indicators (observable variables) and using information from the research tool in the form of a survey questionnaire. It is necessary to emphasize that the presented non-observable statistical features were built on the basis of partial indicators sourced from public domain data and also from surveys. Data from the survey questionnaire may not only be subjective because of the person who fills in the required information; these data also lack information on archive data, their fragmentation, etc., (more on the construction of the research tool in the form of a survey questionnaire performed on the population of municipalities of the Nysa Kłodzka sub-basin in Poland is discussed in [30]).

Authors from other countries indicate that there are many research approaches in resilience assessment [29,45,46] (this article considers the adaptability as part of this body), and the conclusion that comes from the analysis of this part of the literature on the subject is that it is recommended that a combination of many of research approaches is used [47].

Author Contributions: Conceptualization, G.D., A.M. and A.T.; Data curation, G.D. and A.M.; Formal analysis, A.M.; Investigation, G.D.; Methodology, G.D., A.M. and A.T.; Project administration, G.D. and A.T.; Supervision, A.T.; Validation, G.D., A.M. and A.T.; Visualization, M.B. and A.M.; Writing—original draft, G.D., A.M. and A.T.; Writing—review & editing, G.D. All authors have read and agreed to the published version of the manuscript.

Funding: This research received no external funding.

Acknowledgments: The conducted studies constitute the repercussion of the project *Analysis of factor determining resilience of municipalities as social-ecological systems in the Nysa Kłodzka sub-basin*. This project received an award from the Wprost Weekly and the POLPHARMA Foundation within the remit of the competition *"Young & Talented 2016: My idea for Poland"*, the winner of which was Grzegorz Dumieński. The authors also want to thank the reviewers of the originally submitted paper, whose remarks and suggestions are reflected in the current version.

Conflicts of Interest: The authors declare no conflict of interest.

Appendix A

Construction of the set of diagnostic variables was two-step and was based on the selection of potential indicators. The criteria of choice are mainly of a statistical nature. Potential diagnostic variables removed from the set include:

1. ones that are not very diverse, i.e., those for which the absolute value of the classic coefficient of variation is less than 10%, as well as those of which the vast majority of realization is the same number;

2. ones that are too strongly correlated, i.e., with similar information potential, by eliminating one variable from each pair of too closely related indicators; however, cases of apparent correlation were not taken into account.

In the correlation analysis the critical value r* of Pearson's correlation coefficient was used [29]. This value was determined with the formula:

$$r^* = \sqrt{\frac{t^2}{t^2 + n - 2}} \tag{A1}$$

where:

n—number of data, in this case $n = 18$,

t—value of statistics read from the tables of t-Student for the level of significance α and for $n - 2$ degrees of ease, wherein it $\alpha = 0.05$ was adopted. As a result of the calculation, we get $r^* = 0.47$.

During the study, too strongly or otherwise significantly correlated variables were considered those for which the absolute value of Pearson's correlation coefficient are greater than 0.47.

The set of diagnostic variables were divided into stimulants and destimulants.

To compare the values of diagnostic variables, they should be normalized. In the subject literature, a number of methods of normalization are described [29].

The authors of this article insisted on the chosen normalization to meet the most possible of the requirements most often given to this type of methods.

These are the following postulates:

- depriving the titles in which the features are expressed;
- bringing the order of variable sizes to a state of comparability;
- the equality of the length of variability intervals of values of all normalized features (constancy of the range) and equality of the lower and upper limits of their variability interval, in particular the interval [0, 1];
- the ability to normalize the features of positive and negative values, or only the negative;
- the ability to normalize the features of the value that equals zero;
- not-negativeness of the value of the normalized features;
- the existence of simple formulas—within a given normalization formula—unifying the nature of the variables.

The normalization of the value of diagnostic variables was applied with the use of reset unitarization described with the formulas [29]:

- formula for stimulants: $z_{ij} = \dfrac{x_{ij} - \min\limits_{i}\{x_{ij}\}}{\max\limits_{i}\{x_{ij}\} - \min\limits_{i}\{x_{ij}\}}$

- formula for destimulants: $z_{ij} = \dfrac{\max\limits_{i}\{x_{ij}\} - x_{ij}}{\max\limits_{i}\{x_{ij}\} - \min\limits_{i}\{x_{ij}\}}$

where:

z_{ij}—normalized value j—variable for i—municipality.

Appendix B

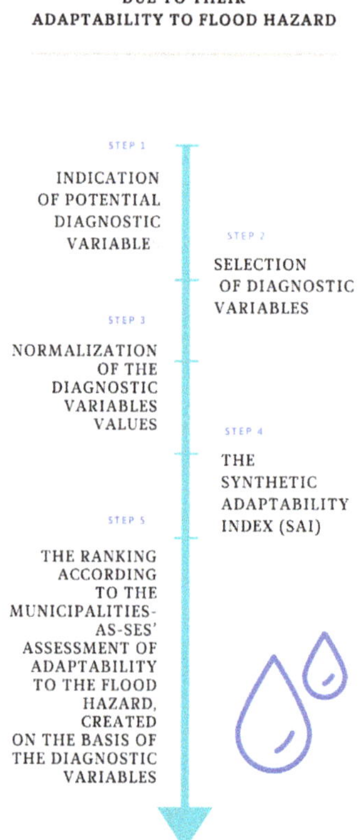

Figure A1. The procedure of creating a ranking of municipalities according to their adaptability to flood hazard [source: own work].

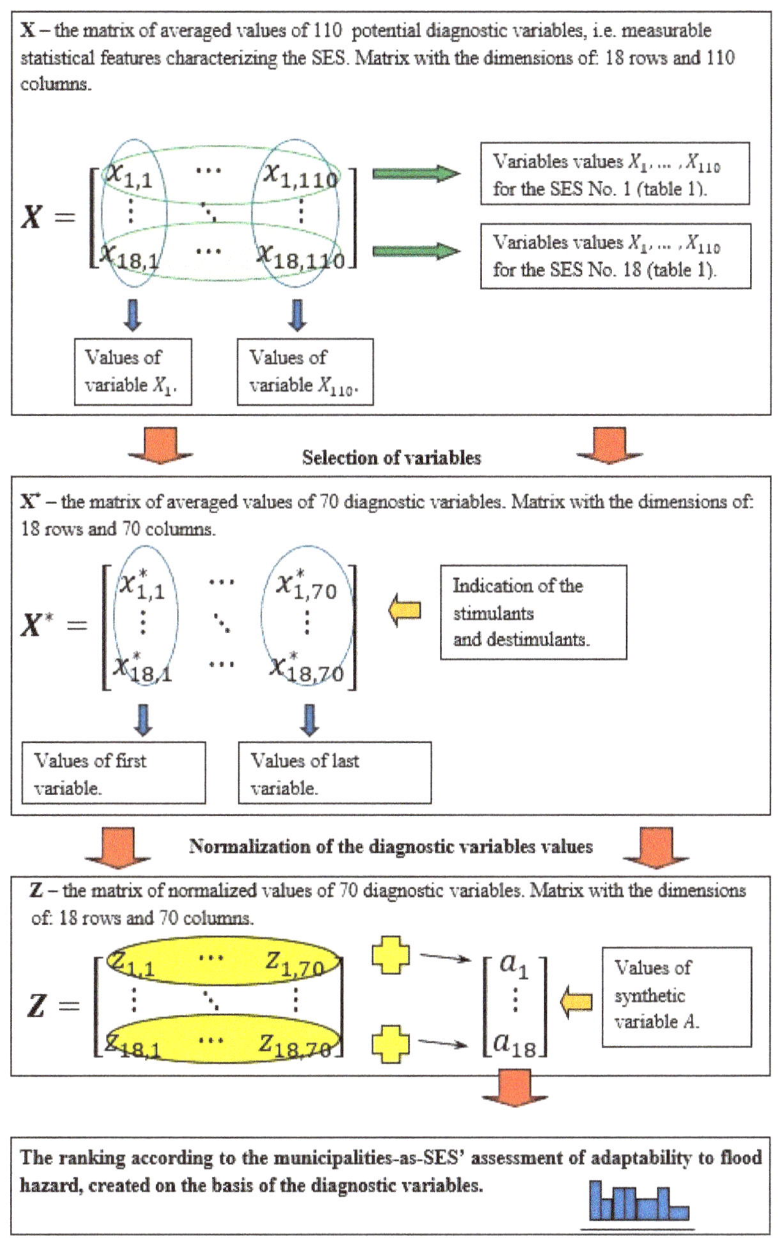

Figure A2. The procedure of making the ranking of municipalities due to their adaptability to flood hazard. [source: own work].

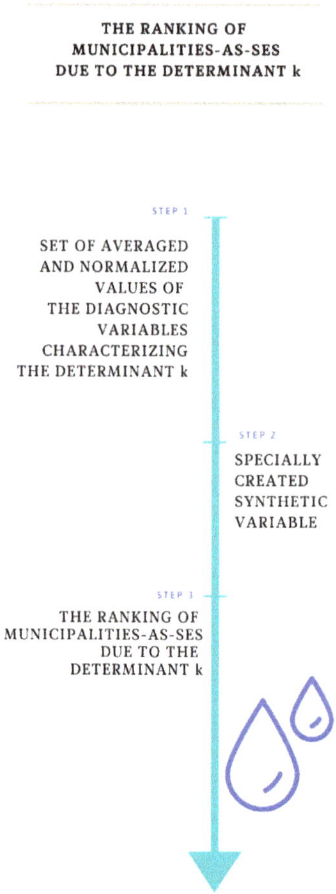

Figure A3. Procedure for creating a ranking of a municipality-as-SES's adaptability to flood hazard according to a random non-observable statistical feature of the municipality-as-SES determining its adaptability to flood hazard [source: own work].

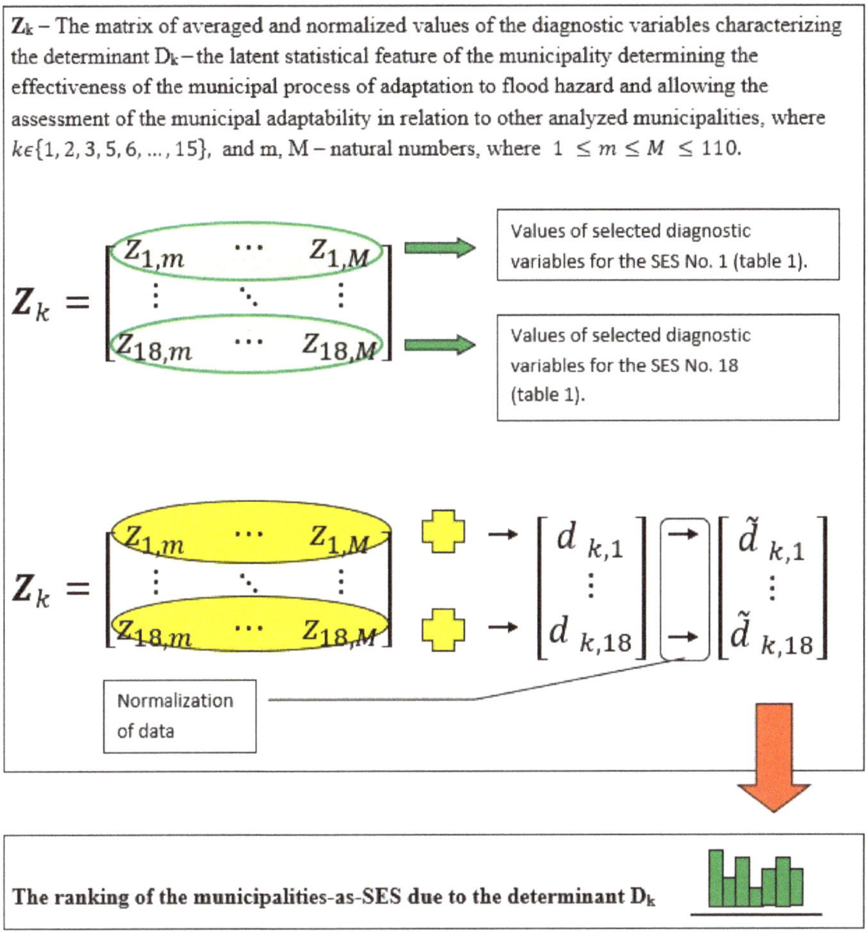

$\mathbf{Z_k}$ – The matrix of averaged and normalized values of the diagnostic variables characterizing the determinant D_k – the latent statistical feature of the municipality determining the effectiveness of the municipal process of adaptation to flood hazard and allowing the assessment of the municipal adaptability in relation to other analyzed municipalities, where $k \epsilon \{1, 2, 3, 5, 6, \dots, 15\}$, and m, M – natural numbers, where $1 \leq m \leq M \leq 110$.

$$Z_k = \begin{bmatrix} Z_{1,m} & \cdots & Z_{1,M} \\ \vdots & \ddots & \vdots \\ Z_{18,m} & \cdots & Z_{18,M} \end{bmatrix}$$

Values of selected diagnostic variables for the SES No. 1 (table 1).

Values of selected diagnostic variables for the SES No. 18 (table 1).

$$Z_k = \begin{bmatrix} Z_{1,m} & \cdots & Z_{1,M} \\ \vdots & \ddots & \vdots \\ Z_{18,m} & \cdots & Z_{18,M} \end{bmatrix} \rightarrow \begin{bmatrix} d_{k,1} \\ \vdots \\ d_{k,18} \end{bmatrix} \rightarrow \begin{bmatrix} \tilde{d}_{k,1} \\ \vdots \\ \tilde{d}_{k,18} \end{bmatrix}$$

Normalization of data

The ranking of the municipalities-as-SES due to the determinant D_k

Figure A4. The procedure of creating the ranking of adaptability of a municipality-as-SES to flood hazard in relation to a random non-measurable statistical feature of the municipality-as-SES determining its adaptability to the flood hazard [source: own work].

References

1. Dumieński, G.; Lisowska, A.; Tiukało, A. The Measurement of the Adaptive Capacity of the Social-Ecological System towards Flood Hazard. In *Proceedings of the 3rd Disaster Risk Reduction Conference, Warsaw, Poland, 12–13 October 2017*; Rucińska, D., Porczek, M., Moran, S., Eds.; Faculty of Geography and Regional Studies, University of Warsaw: Warsaw, Poland, 2017; p. 41.
2. Piniewski, M.; Marcinkowski, P.; Kundzewicz, Z.W. Trend detection in river flow indices in Poland. *Acta Geophys.* **2017**, *66*, 357–360. [CrossRef]
3. Burszta-Adamiak, E.; Licznar, P.; Zaleski, J. Criteria for identifying maximum rainfall determined by the peaks-over-threshold (POT) method under the Polish Atlas of Rainfall Intensities (PANDa) project. *Meteorol. Hydrol. Water Manag.* **2019**, *2*, 3–13. [CrossRef]
4. Szalińska, W.; Tokarczyk, T. Drought hazard assessment in the process of drought risk management. *Acta Sci. Pol. Form. Circumiectus* **2018**, *17*, 217–229.

5. Hu, M.; Zhang, X.; Li, Y.; Yang, H.; Tanka, K. Flood mitigation performance of low impact development technologies under different storms for retrofitting an urbanized area. *J. Clean. Prod.* **2019**, *222*, 373–380. [CrossRef]

6. Laurien, F.; Hochrainer-Stigler, S.; Keating, A.; Campbell, K.; Mechler, R.; Czajkowski, J. A typology of community flood resilience. *Reg. Environ. Chang.* **2020**, *20*, 1–14. [CrossRef]

7. Gallopin, G.C. Linkages between vulnerability, resilience, and adaptive capacity. *Glob. Environ. Chang.* **2006**, *16*, 293–303. [CrossRef]

8. Mapy Zagrożenia i Ryzyka Powodziowego. Available online: http://mapy.isok.gov.pl (accessed on 10 January 2020).

9. IPCC. *Climate Change 2014: Impacts, Adaptation, and Vulnerability*; Intergovernmental Panel of Climate Change: Cambridge, UK, 2014; pp. 1757–1775.

10. Działek, J.; Biernacki, W. Wrażliwość społeczna na klęski żywiołowe—Ujęcie teoretyczne i praktyka badawcza. *Prace Geogr.* **2014**, *55*, 27–41.

11. Birkmann, J. Risk and vulnerability indicators at different scales: Applicability, usefulness and policy implication. *Environ. Hazards* **2007**, *7*, 20–31. [CrossRef]

12. Działek, J.; Biernacki, W.; Konieczny, R.; Fiedeń, Ł.; Franczak, P.; Grzeszna, K.; Listwan-Franczak, K. *Zanim Nadejdzie Powódź. Wpływ Wyobrażeń Przestrzennych, Wrażliwości Społecznej na Klęski żywiołowe oraz Komunikowania Ryzyka na Przygotowanie Społeczności Lokalnych do Powodzi*; Instytut Geografii i Gospodarki Przestrzennej, Uniwersytet Jagielloński: Kraków, Poland, 2017.

13. Kerstin, F.; Schneiderbauer, S.; Bubeck, P.; Kienberg, S.; Buth, M.; Zebisch, M.; Kahlenborn, W. *The Vulnerability Sourcebook: Concept and Guidelines for Standardized Vulnerability Assessment*; Deutsche Gesellschaft fur Internationale Zusammenarbeit (GIZ): Bonn, Germany; Echborn, Germany, 2014; p. 24.

14. Lyu, J.; Mo, S.; Luo, P.; Zhou, M.; Shen, B.; Nover, D. A quantitative assessment of hydrological responses to climate change and human activities at spatiotemporal within a typical catchment on the Loess Plateau, China. *Quater. Inter.* **2019**, *527*, 1–11. [CrossRef]

15. Huo, A.; Yang, L.; Peng, J.; Cheng, Y.; Cheng, J. Spatial characteristics of the rainfall included landslide in the Chinese Loess Plateau. *Hum. Ecol. Risk Assess. Inter. J.* **2020**. [CrossRef]

16. Noy, I.; Yonson, R. Economic Vulnerability and Resilience to Natural Hazards: A Survey of Concepts and Measurements. *Sustainability* **2018**, *10*, 2850. [CrossRef]

17. Dumieński, G.; Tiukało, A. Ocena podatności systemu społeczno-ekologicznego zagrożonego powodzią. *Pr. Studia Geogr.* **2016**, *4*, 7–23.

18. Rucińska, D. Kwantyfikacja podatności na zagrożenia naturalne. Przegląd metod. *Pr. Studia Geogr.* **2015**, *57*, 43–53.

19. Dumieński, G.; Lisowska, A.; Tiukało, A. Adaptability of a socio-ecological system. The case study of a Polish municipality at risk of flood. *Pr. Geogr.* **2019**, *151*, 25–48.

20. Holling, C. Resilience and Stability of Ecological Systems. *Annu. Rev. Ecol. Syst.* **1973**, *4*, 1–23. [CrossRef]

21. UNSDR. *UNISDR Terminology on Disaster Risk Reduction*; United Nations International Strategy for Disaster Reduction: Geneva, Switezerland, 2009.

22. RCB. *Powódź. W Obliczu Zagrożenia*; Rządowe Centrum Bezpieczeństwa: Warszawa, Poland, 2013.

23. Kundzewicz, Z.W.; Lugeri, N.; Dankers, R.; Hirabayashi, Y.; Doll, P.; Pińskwar, P.; Dysarz, T.; Hochrainer, S.; Matczak, P. Assessing river flood risk and adaptation in Europe—Review of projections for the future. *Miti. Adap. Strat. Glob. Chan.* **2010**, *15*, 641–656. [CrossRef]

24. Dyrektywa 2007/60/WE Parlamentu Europejskiego i Rady z dn. 23 października 2007 r. w sprawie oceny ryzyka powodziowego i zarządzania nim, L 288/27. Available online: hhttps://www.kzgw.gov.pl/files/dyrektywa-powodziowa/tekst_Dyrektywy_Powodziowej_PL.pdf (accessed on 8 April 2020).

25. Aguiar, C.F.; Bentz, J.; Silva, J.M.N.; Fonseca, A.L.; Swart, R.; Duarte-Santos, F.; Penha-Lopes, G. Adaptation to climate change at local level in Europe: An overview. *Environ. Sci. Policy* **2018**, *86*, 38–63. [CrossRef]

26. Dyrektywa 2000/60/WE Parlamentu Europejskiego i Rady z dnia 23 października 2000 r. ustanawiająca ramy wspólnotowego działania w dziedzinie polityki wodnej (Dz. U. UE L z dnia 22 grudnia 2000 r). Available online: http://geoportal.pgi.gov.pl/css/powiaty/prawo (accessed on 31 March 2020).

27. Choryński, A. Adaptacja Wielkopolskich Gmin do Ekstremalnych Zdarzeń Pogodowych w Świetle Koncepcji Elastyczności (Resilience). Ph.D. Thesis, Wydział Socjologii, UAM, Poznań, Poland, 2019.

28. Folke, C. Traditional knowledge in social-ecological system. *Ecol. Soc.* **2004**, *9*, 8. [CrossRef]

29. Dumieński, G.; Mruklik, A.; Tiukało, A.; Lisowska, A. Preliminary research on adaptability of municipalities in the sub-basin of Nysa Kłodzka using multidimensional comparative analysis. *ITM Web Conf.* **2018**, *23*, 00008. [CrossRef]

30. Dumieński, G.; Lisowska, A.; Tiukało, A. Badania ilościowe źródłem danych o adaptacyjności gmin zagrożonych powodzią w zlewni Nysy Kłodzkiej. *Studia Reg. i Lokalne* **2020**, in review process.

31. Bedryj, M.; Dumieński, G.; Tiukało, A. Potential threat to Polish lakes and reservoirs from contamination by objects of environmental flood risk. *Limn. Rev.* **2018**, *18*, 137–147. [CrossRef]

32. Biuro Koordynacji Projektu Ochrony Przeciwpowodziowej Dorzecza Odry i Wisły. Available online: http://odrapcu.com.pl (accessed on 5 January 2020).

33. SWECO Poland. Available online: http://sweco.pl (accessed on 15 February 2020).

34. Measuring Vulnerability to Promote Disaster-Resilient Societies: Conceptual Frameworks and Definitions. In *Measuring Vulnerability to Natural Hazards: Toward Disaster Resilient Societies*; Birkmann, J., Ed.; United Nations University: New York, NY, USA, 2006.

35. Hellwig, Z. Wielowymiarowa analiza porównawcza i jej zastosowanie w badaniach wielocechowych obiektów gospodarczych. In *Metody i Modele Ekonomiczno-Matematyczne w Doskonaleniu Zarządzania Gospodarką Socjalistyczną*; Welfe, A., Ed.; PWE: Warsaw, Poland, 1981; p. 57.

36. Rencher, A.C. *Methods of Multivariate Analysis*; John Wiley & Sons: New York, NY, USA, 2002.

37. Giri, N.C. *Multivariate Statistical Analysis*; Marcel Dekker: New York, NY, USA, 2004.

38. Yang, J.; Zhang, H.; Ren, C.; Nan, Z.; Wei, X.; Li, C. A Cross-Reconstruction Method for Step-Changed Runoff Series to Implement Frequency Analysis under Changing Environment. *Int. J. Environ. Res. Public Health* **2019**, *16*, 4345. [CrossRef] [PubMed]

39. Panek, T. *Statystyczne Metody Wielowymiarowej Analizy Porównawczej*; Oficyna Wydawnicza SGH: Warszawa, Poland, 2009.

40. Panek, T.; Zwierzchowski, J. *Statystyczne Metody Wielowymiarowej Analizy Porównawczej, Teoria Zastosowania*; Oficyna Wydawnicza SGH: Warszawa, Poland, 2013.

41. Kukuła, K. Metoda unitaryzacji zerowanej na tle wybranych metod normowania cech diagnostycznych. *Acta Sci. Acad. Ostroviensis* **1999**, *4*, 5–31.

42. Dębkowska, K.; Jarocka, K. The impact of the methods of the data normalization on the result of linear ordering 2013. *Acta Univer. Lodziensis. Folia Oeconomika* **2013**, *286*, 181–188.

43. Jarocka, M. Wybór formuły normalizacyjnej w analizie porównawczej obiektów wielocechowych. *Econ. Manag.* **2015**, *1*, 113–126.

44. Walczykiewicz, T. *Metodyka Kwantyfikacji Wrażliwości Jako Jednego z Czynników Wpływających na Ryzyko Powodziowe*; Zasób własny IMGW-PIB: Warszawa, Poland, 2014.

45. Scherzer, S.; Lujala, P.; Rod, J.K. A community resilience index for Norway: An adaptation of the Baseline Resilience Indicators for Communities (BRIC). *Int. J. Disaster Risk Reduct.* **2019**, *36*, 101107. [CrossRef]

46. Cai, H.; Lam, N.S.N.; Qiang, Y.; Zou, L.; Correll, R.M.; Mihunov, V. A synthesis of disaster resilience measurement methods and indices. *Int. J. Disaster Risk Reduct.* **2018**, *31*, 844–855. [CrossRef]

47. Sharifi, A. A critical review of selected tools for assessing community resilience. *Ecol. Indic.* **2016**, *69*, 629–647. [CrossRef]

Article

Spatial Heterogeneous of Ecological Vulnerability in Arid and Semi-Arid Area: A Case of the Ningxia Hui Autonomous Region, China

Rong Li [1], Rui Han [1], Qianru Yu [1], Shuang Qi [2] and Luo Guo [1,*]

[1] College of the Life and Environmental Science, Minzu University of China, Beijing 100081, China; 19301382@muc.edu.cn (R.L.); 18301234@muc.edu.cn (R.H.); 18301244@muc.edu.cn (Q.Y.)

[2] Department of Geography, National University of Singapore; Singapore 117570, Singapore; e0457619@u.nus.edu

* Correspondence: guoluo@muc.edu.cn

Received: 25 April 2020; Accepted: 26 May 2020; Published: 28 May 2020

Abstract: Ecological vulnerability, as an important evaluation method reflecting regional ecological status and the degree of stability, is the key content in global change and sustainable development. Most studies mainly focus on changes of ecological vulnerability concerning the temporal trend, but rarely take arid and semi-arid areas into consideration to explore the spatial heterogeneity of the ecological vulnerability index (EVI) there. In this study, we selected the Ningxia Hui Autonomous Region on the Loess Plateau of China, a typical arid and semi-arid area, as a case to investigate the spatial heterogeneity of the EVI every five years, from 1990 to 2015. Based on remote sensing data, meteorological data, and economic statistical data, this study first evaluated the temporal-spatial change of ecological vulnerability in the study area by Geo-information Tupu. Further, we explored the spatial heterogeneity of the ecological vulnerability using Getis-Ord Gi*. Results show that: (1) the regions with high ecological vulnerability are mainly concentrated in the north of the study area, which has high levels of economic growth, while the regions with low ecological vulnerability are mainly distributed in the relatively poor regions in the south of the study area. (2) From 1990 to 2015, ecological vulnerability showed an increasing trend in the study area. Additionally, there is significant transformation between different grades of the EVI, where the area of transformation between a slight vulnerability level and a light vulnerability level accounts for 41.56% of the transformation area. (3) Hot-spot areas of the EVI are mainly concentrated in the north of the study area, and cold-spot areas are mainly concentrated in the center and south of the study area. Spatial heterogeneity of ecological vulnerability is significant in the central and southern areas but insignificant in the north of the study area. (4) The grassland area is the main driving factor of the change in ecological vulnerability, which is also affected by both arid and semi-arid climates and ecological projects. This study can provide theoretical references for sustainable development to present feasible suggestions on protection measures and management modes in arid and semi-arid areas.

Keywords: ecological vulnerability; geospatial analysis; sustainable development

---

1. Introduction

Ecological vulnerability, which refers to the self-recovery ability of ecosystems when they suffer disturbances at a specific temporal scale [1], has been an important concept for reflecting the deviation degree from the original ecological condition under the external interferences [2,3]. The assessment of ecological vulnerability, therefore, is critical for research on ecological change [4,5]. With the rapid development of society, however, the growth of anthropogenic activities has significantly aggravated ecological vulnerability on a global scale [6,7]. The sustainable development of arid and semi-arid regions is related to assessing ecological vulnerability because arid and semi-arid areas are especially sensitive to variable environmental changes and human influences.

The assessment of ecological vulnerability is essential for ensuring and managing eco-environmental stability. Ecological vulnerability can be assessed by the analytic hierarchy process (AHP) [1,8,9], principal component analysis (PCA) [10,11], fuzzy comprehensive evaluation [12], entropy weight analysis [13,14], spatial principal component analysis (SPCA), and other techniques. Studies have shown that the SPCA method, which is based on principal component analysis (PCA) and spatial feature extraction, has advantages in ecological vulnerability assessment [1,15,16]. One advantage of this method is that the SPCA not only adds spatial constraints to the traditional PCA but also considers the spatial dependence in data sets. This method has been widely used to describe the characteristics of changes in ecological vulnerability in arid and semi-arid areas because of its advantages in being good for map and comparison services [17,18].

In terms of the recognition of ecosystem-humanity interactions, human activities are considered to have a serious impact on regional ecological vulnerability [19,20]. For example, previous research shows that population and economic growth lead to increased conditions of ecological vulnerability, especially in close proximity to urban agglomerations [21–23]. Other studies imply that ecological vulnerability is negatively correlated with urbanization [14]. Although these studies explain the relationship between ecological vulnerability and human activities, some gaps in knowledge remain. Most studies have focused on the establishment of indicator systems of ecological vulnerability [1] or the spatial and temporal changes in ecological vulnerability [4,24] but they have paid little attention to the spatial heterogeneity of ecological vulnerability [22]. Currently, ecological vulnerability assessment focuses on unstable ecosystems, such as urban areas [22,25], specific habitats [26], or areas with a poor ecological environment [19,27,28]. These ecological vulnerability analyses fail to explore the local spatial differentiation of ecological vulnerability. Getis-Ord Gi* is a spatial statistical method, which can describe and visualize the spatial distribution, discover local spatial correlation patterns, identify heterogeneous units, and suggest spatial states [29,30]. Compared with the traditional method of identifying process-related regions, Getis-Ord Gi* has shown a certain effectiveness in exploring the spatial heterogeneity of ecological vulnerability, which lays a foundation for the study of the impact of human interference on ecological vulnerability.

This study takes the Ningxia Hui Autonomous Region, at the intersection of the Loess Plateau, the Mongolian Plateau, and the Qinghai-Tibet Plateau in China, as a case to examine ecological vulnerability in arid and semi-arid regions. Over the past few decades, Ningxia has undergone rapid urbanization, large-scale agricultural modernization, and continuous ecological conservation projects. The poor ecological conditions and the various human influences make the study area an ideal region to study ecological vulnerability. The objectives of this study are to: (1) analyze changes of ecological vulnerability in Ningxia, (2) visualize the spatial-temporal transformation of ecological vulnerability in Ningxia based on Geo-information Tupu, (3) explore the spatial heterogeneity of ecological vulnerability based on Getis-Ord Gi*, and (4) detect the driving factors of ecological vulnerability by SPCA. This study provides references for sustainable development in arid and semi-arid areas.

2. Materials and Methods

2.1. Study Area

The Ningxia Hui Autonomous Region (Ningxia) is located in the upper and middle reaches of the Yellow River in the northwest of China (Figure 1). The study area is situated at 35°14′–39°23′ N and 104°17′–107°39′ E in an arid and semi-arid region, which belongs to the temperate continental climate. The annual average temperature is about 5–9C. The annual precipitation is 150–600 mm, and the average annual surface evaporation is 1250 mm. The study area consists primarily of plains covered with grassland and farmland, which have provided the main agricultural function of the study area since the 16th century. In 2015, the study area included 5 cities and 22 counties [31].

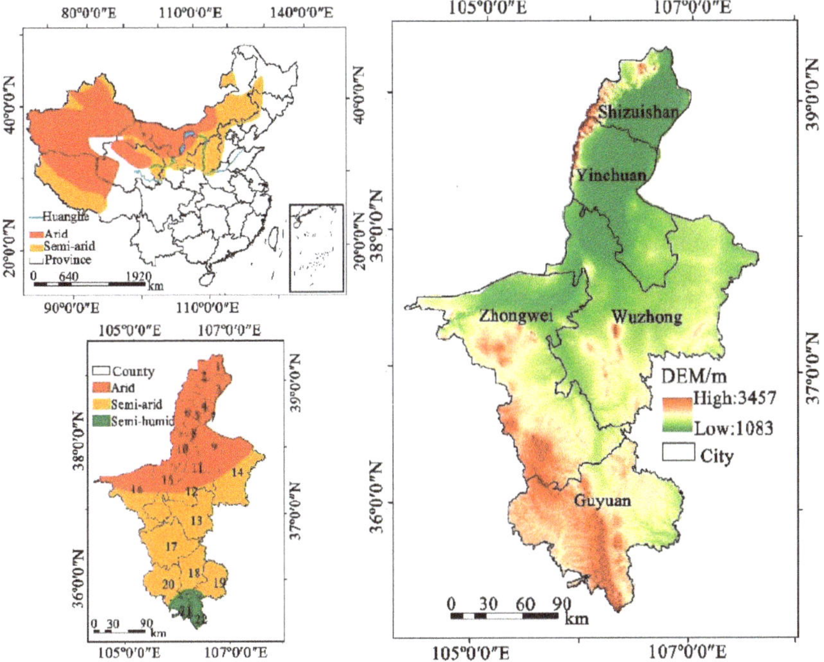

Figure 1. Location of the study area. (Numbers refer to counties: 1—Huinong district, 2—Dawukou district, 3—Pingluo county, 4—Helan county, 5—Jinfeng district, 6—Xixia district, 7—Xingqing district, 8—Yongning county, 9—Lingwu, 10—Qingtongxia, 11—Litong district, 12—HongSiBao district, 13—Tongxin county, 14—Yanchi county, 15—Zhongning county, 16—Shapotou district, 17—Haiyuan county, 18—Yuanzhou district, 19—Pengyang county, 20—Xiji county, 21—Longde county, and 22—Jingyuan county).

2.2. Data Collection

In this study, land use/land cover (LULC) data from 1990, 1995, 2000, 2005, 2010, and 2015 was provided by the Data Center for Resources and Environmental Sciences, Chinese Academy of Sciences (RESDC). The population density (POP) data was also from RESDC. The gross domestic product (GDP) data came from National Earth System Science Data Sharing Infrastructure, National Science and Technology Infrastructure of China. All spatial resolution was 1 km.

2.3. Methods

This study used the Data Management Tool module to generate cells by ArcGIS 10.4 and to perform all calculations for internal landscape units. This study adopted the sampling grids of 1×1 km with a total 66,400 cells for the six periods in this study area, which used ArcGIS 10.4, Python, and GeoDa 1.1 for further analyses.

2.3.1. Assessment of Ecological Vulnerability

Given the existing international evaluation principles and standards [1,2], the comprehensive evaluation system of ecological vulnerability was established combining the ecological conditions of the study area. The ecological vulnerability index is based on both natural and social factors. The natural factors include the digital elevation model (DEM), hours of sunshine, average annual precipitation, average annual temperature, normalized difference vegetation index (NDVI), soil erosion, and degree of land use. Social factors include the gross domestic product (GDP), agricultural output, industrial output, population density, and grassland area. All indicators are standardized. SPCA is employed to determine the weight of each factors. SPCA is used to add spatial features on the basis of PCA, and its calculation principle is consistent with PCA [1,15,16].

PCA is a statistical analysis method that transforms multiple variables into a few principal components (composite indicators) through dimensionality reduction [32–34]. In this study, PCA was used to make a linear combination of 12 standardized indexes to make them become new comprehensive indexes. The correlation coefficient matrix was solved to obtain the eigenvectors, thus obtaining 12 principal component results. The number of principal components was determined by the standard of a cumulative contribution rate ≥ 85%. From this w got our final principal component result. The calculation formula was as follows:

$$R = \frac{Z^T Z}{n} \tag{1}$$

$$|R - \lambda I| = 0 \tag{2}$$

$$CCR = \frac{\sum_{j=1}^{m} \lambda_j}{\sum_{j=1}^{n} \lambda_j} \geq 0.85 \tag{3}$$

$$P = Z \cdot W \tag{4}$$

where R was the correlation coefficient matrix, Z was the standardized value of each selected index, n was the number of indexes, λ was the eigenvalues of the R correlation coefficient matrix, I was the identity matrix, CCR was the cumulative contribution rate, m was the number of principal components that were determined, P was the matrix containing values of every considered principal component, and W was m number of eigenvectors with the largest eigenvalues selected to form the dimensional matrix. The SPCA was obtained by calculating PCA in ArcGIS 10.4. The SPCA results are shown in Table 1.

Table 1. The results of spatial principal component analysis.

	Principal Component	1990a	1995a	2000a	2005a	2010a	2015a
Eigenvalue/%	I	0.069	0.073	0.066	0.067	0.066	0.066
	II	0.036	0.039	0.038	0.042	0.041	0.044
	III	0.021	0.019	0.024	0.029	0.026	0.026
	IV	0.015	0.013	0.015	0.016	0.016	0.017
Contribution/%	I	42.315	44.965	40.530	37.930	38.657	37.612
	II	22.256	24.309	23.420	23.765	23.650	24.640
	III	13.157	11.545	14.908	16.288	15.106	14.890
	IV	9.453	7.748	9.526	9.006	9.108	9.543
Cumulative contribution/%	I	42.315	44.965	40.530	37.930	38.657	37.612
	II	64.571	69.274	63.950	61.694	62.306	62.252
	III	77.729	80.819	78.857	77.983	77.412	77.143
	IV	87.182	88.566	88.384	86.988	86.520	86.686

Based on spatial principal component analysis (SPCA), the ecological vulnerability index (EVI) was the sum of the weighted principal components [2].

$$EVI = \sum_{i=1}^{m} r_m P_m \tag{5}$$

$$r_i = \frac{n_i}{\sum_i^m n_i} \tag{6}$$

where *EVI* was the ecological vulnerability index, *r* was the contribution ratio, *P* is the principal component, *m* was the number of principal components, r_i was the contribution ratio of principal component *i*, and n_i was the eigenvalue of principal component *i*.

2.3.2. Gradient Classification of Ecological Vulnerability

In this paper, natural breaks classification (NBC) was used to classify the EVI to reflect different degrees of ecological vulnerability. NBC is a method of analyzing the statistical distribution of attribute space, which maximizes the difference between classes. In this study, the assessment of ecological vulnerability was divided into five grades [4,24], namely slight vulnerability: <0.1444, light vulnerability: 0.1444–0.2480, medium vulnerability: 0.2480–3982, heavy vulnerability: 0.3982–0.6282, and extreme vulnerability: >0.6282.

2.3.3. Geo-Information Tupu Change Analysis

Geo-information Tupu is an effective way to study the process integration of ecological vulnerability. Based on the ArcGIS tool, EVI process change is revealed by Geo-information Tupu [35]. The specific operational formula was [36]

$$T = Y_1 \times 10^{n-1} + Y_2 \times 10^{n-2} + \ldots + Y_n \times 10^{n-n} \tag{7}$$

where T was the Tupu unit code of the Tupu mode characteristics within the representation research stage, Y_n was the ecological vulnerability Tupu unit code of representation in a certain year, and n was the number of the ecological vulnerability grade.

2.3.4. Cold-Hotspot Study Change Analysis

In this study, the Getis-Ord Gi* index was used to analyze the high/low spatial aggregation degree of EVI changes, that is, the spatial distribution of cold/hot spots. The calculation formula was

$$Gi^* = \frac{\sum_{j=1}^{n} w_{ij}x_j - \overline{X}\sum_{j=1}^{n} w_{ij}}{s\sqrt{\left[n\sum_{j=1}^{n} w_{ij}^2 - \left(\sum_{j=1}^{n} w_{ij}\right)^2\right]/(n-1)}} \tag{8}$$

$$\overline{X} = \frac{1}{n}\sum_{j=1}^{n} x_j \tag{9}$$

$$\sqrt{\left(\frac{1}{n}\sum_{j=1}^{n} x_j^2 - \overline{X}^2\right)} \tag{10}$$

where G_i^* was the output statistical Z-score, x_j was the EVI change of space unit *j*, and w_{ij} was the spatial weight between adjacent space units *i* and *j*.

3. Results

3.1. Spatial and Temporal Characteristics of Ecological Vulnerability

The results of the ecological vulnerability index (EVI) in Ningxia were as follows:

1. $EVI_{1990} = 0.4852A_1 + 0.2553A_2 + 0.1509A_3 + 0.1084A_4;$
2. $EVI_{1995} = 0.5077B_1 + 0.2745B_2 + 0.1303B_3 + 0.0875B_4;$
3. $EVI_{2000} = 0.4586C_1 + 0.2650C_2 + 0.1687C_3 + 0.1078C_4;$
4. $EVI_{2005} = 0.4360D_1 + 0.2732D_2 + 0.1872D_3 + 0.1035D_4;$
5. $EVI_{2010} = 0.4468F_1 + 0.2733F_2 + 0.1746F_3 + 0.1053F_4;$

6. $EVI_{2015} = 0.4339P_1 + 0.2842P_2 + 0.1718P_3 + 0.1101P_4;$

where EVI_i were the ecological vulnerability index in year i; A1~A4, B1~B4, C1~C4, D1~D4, F1~F4, and P1~P4 were the principal components in 1990, 1995, 2000, 2005, 2010, and 2015, respectively.

Figure 2 shows the area percentages of different grades of ecological vulnerability. It can be seen that the sum of the area percentage (SSL) of slight vulnerability and light vulnerability of the whole area was ranked as $SSL_{2010} > SSL_{2005} > SSL_{2000} > SSL_{2015} > SSL_{1990} > SSL_{1995}$. In 2005, the whole degree of ecological vulnerability was at the lowest level, with the area of slight vulnerability accounting for 51% of the study area.

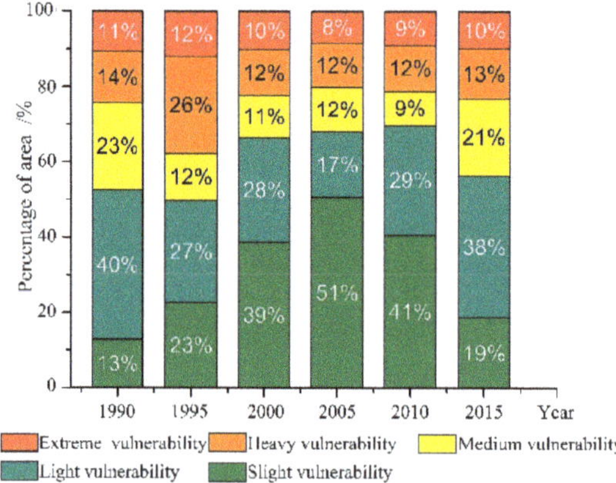

Figure 2. Changes of ecological vulnerability in Ningxia.

The spatial change of ecological vulnerability is shown in Figure 3. The spatial statistical model was used for data visualization and shows the spatial and temporal variation of ecological vulnerability in Ningxia more clearly and intuitively. The EVI showed a decreasing trend from north to south. The grade of ecological vulnerability in the north of Ningxia (Shizuishan City and Yinchuan City) from 1990 to 2015 was mainly extreme vulnerability, and the area of extreme vulnerability initially increased but decreased later. From 1990 to 2015, the grade of ecological vulnerability in the west of Ningxia (Zhongwei City) mainly consisted of medium vulnerability and heavy vulnerability. The areal proportions of medium vulnerability presented a trend of "increased-decreased-increased", while the area of heavy vulnerability initially increased but decreased later. The grade of ecological vulnerability in the east of Ningxia (Wuzhong City) was light vulnerability, and the area of light vulnerability initially increased but decreased in 2015. The grade of ecological vulnerability in the south of Ningxia (Guyuan City) mainly consisted of slight vulnerability and light vulnerability, and the area of slight vulnerability showed initially increased but decreased in the latter period.

3.2. Transformation of Ecological Vulnerability

Tupu is a spatial model that visualizes the transformation between different grades of ecological vulnerability at each patch. Tupu analysis can more clearly show the dynamic change of ecological vulnerability in Ningxia from 1990 to 2015 (Figure 4). In the past 25 years, the main changes of ecological vulnerability in the study area were that the area of light vulnerability decreased and the area of slight vulnerability increased. First, from 1990 to 2015, the transformation area of ecological vulnerability level accounted for 37.95% of the study area. There were 20 types of transformation, among which the main type was the transformation from light vulnerability to slight vulnerability, accounting for 10.37% the study area. Secondly, the area of slight vulnerability transforming into light vulnerability accounted for 5.39% of the study area.

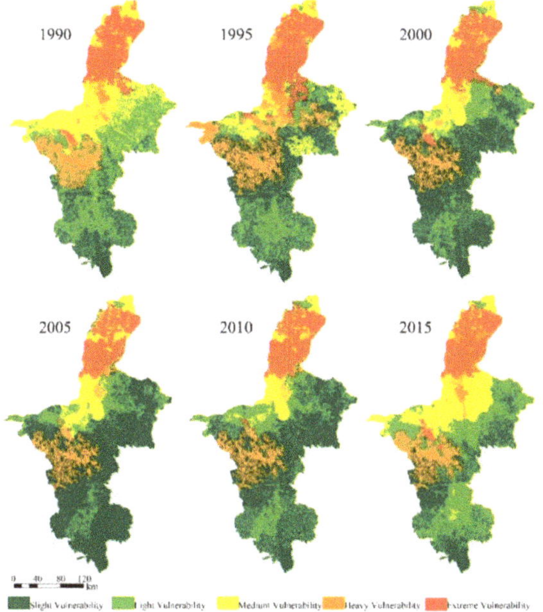

Figure 3. Spatial distribution of ecological vulnerability in the study area.

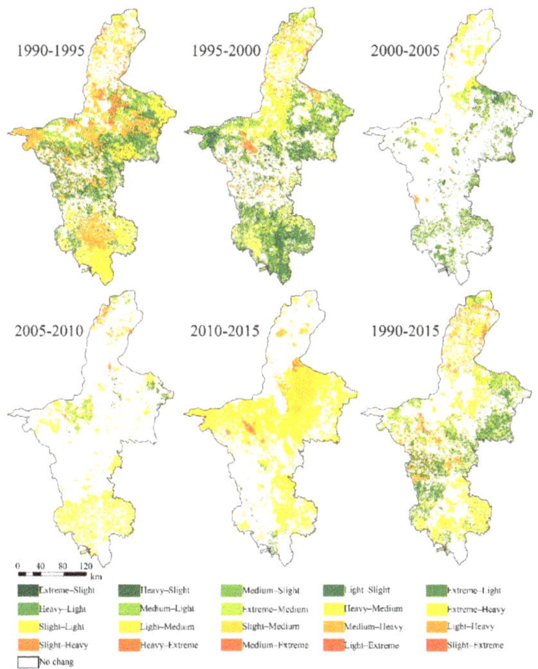

Figure 4. Tupu of study area ecological vulnerability.

There were 20 types of Tupu changes of ecological vulnerability, and the characteristics of Tupu changes were different in six periods. The transformation areas of ecological vulnerability accounted for 58.12%, 52.26%, 24.93%, 19.97%, 17.13%, and 38.43%, respectively, of the study area for the years of 1990, 1995, 2000, 2005, 2010, and 2015. The main change of ecological vulnerability was from light vulnerability to slight vulnerability, and the change was significant in the period of 1990–1995, followed by 1995–2000. Such changes mainly occurred in the center of the study area in 1990–1995 and in the center and south of the study area in 1995–2000. The transformation of ecological vulnerability was less from 2005 to 2010, when the main type was transformation from slight vulnerability to light vulnerability in the center of the study area.

3.3. Cold-Hotspot Analysis of EVI

Cold-hotspot is a spatial model used to display spatial aggregation calculated by Getis-Ord Gi*. The calculation of the EVI of Ningxia from 1990 to 2015 based on the cold-hotspot can intuitively see spatial aggregation of similar values. At the same time, by comparing the spatial distribution of cold spots and hot spots in different years, we can get the differences in the spatial change of EVI on the time scale (Figure 5). On the whole, the hot-spot areas were clustered in the north and the cold-spot areas were distributed in the south of the study area. In terms of the temporal trend, areas of hot spots and cold spots of the EVI all decreased from 1990 to 2005, but the trend was reversed from 2005 to 2015. This trend hints that north-south heterogeneity of the EVI was mitigated in the former period (1990–2005), while the changes of ecological vulnerability since 2005 are noteworthy.

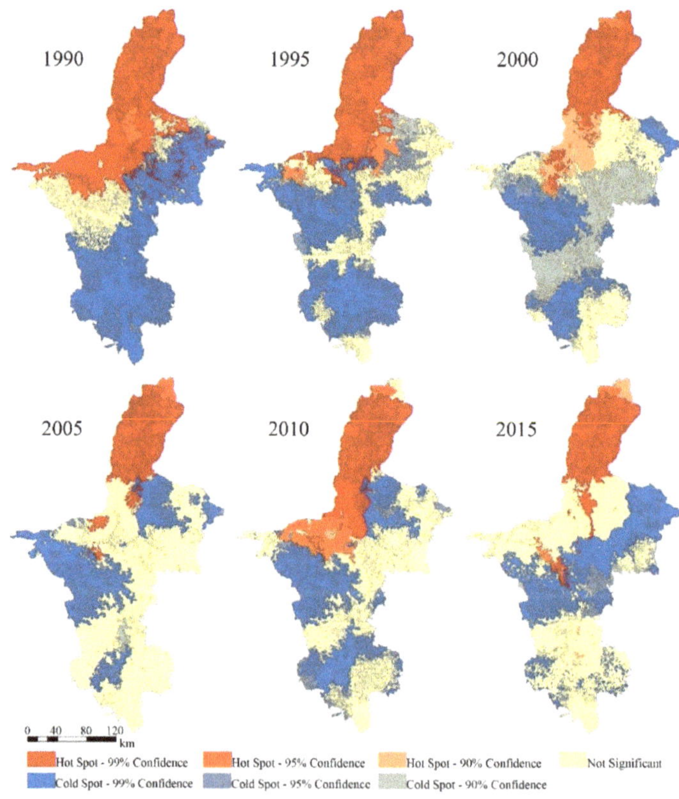

Figure 5. Getis-Ord Gi* scores of ecosystem vulnerability values.

In terms of spatial patterns, planners need to pay more attention to the high ecological vulnerability in the north and increasing ecological vulnerability in the south. In the north, the ecological conditions in Shizuishan City and Yinchuan City were always at a disadvantage from 1990 to 2015 (hot spots of the EVI). In the south, the cold spots of Guyuan City showed a decreasing trend, which means an increasing ecological sensitivity. Fortunately, environmental conditions in the west and east of the study area seemed to become better during the study period. In the west, hot spots of the EVI in Zhongwei City decreased, implying that the ecological vulnerability weakened. In the east, Wuzhong City was at a low level of ecological vulnerability in 2015, showing that the environmental condition was in a good state, although it seemed to suffer from 1990 to 2010.

3.4. Driving Factors Analysis

Principal component vectors were employed to detect the driving factors of ecological vulnerability change, as shown in Figure 6.

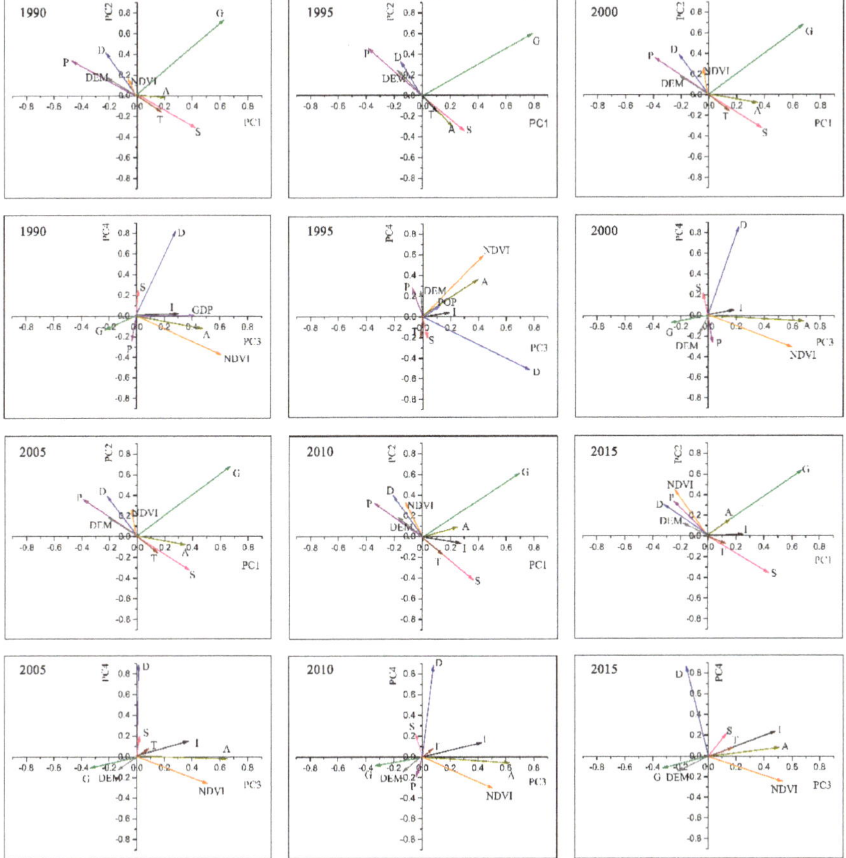

Figure 6. Correlation analysis diagram of principal components and factors. DEM—Digital Elevation Model, P—Annual Average Precipitation, T—Annual Average Temperature, S—Hours of Sunshine, NDVI—Normalized Difference Vegetation Index, SE—Soil Erosion, D—Degree of Land Use, A—Agricultural Output, I—Industrial Output, POP—Population Density, GDP—Gross Domestic Product, and G—Grassland Area.

In 1990, principal components I, II, III, and IV all reflected socio-economic background factors, while principal components I and II also reflected meteorological factors. The contribution rate of principal component I was 42.32%. This factor was highly correlated with the grassland area, hours of sunshine, and GDP, with the correlation coefficients of 62.87%, 41.53%, and 16.86%, respectively. The contribution rate of principal component II was 22.26%. This factor was highly correlated with the grassland area, degree of land use, and annual average precipitation, with the correlation coefficients of 73.23%, 42.13%, and 33.08%, respectively. The contribution rate of principal component III was 13.16%. The correlation coefficient between this factor and the NDVI, agricultural output, and GDP was high at 60.68%, 47.69%, and 41.14%, respectively. The contribution rate of principal component IV was 9.45%, and the correlation coefficient between this factor and the degree of land use was higher (82.53%). The GDP, agricultural output, and land use are commonly used indicators to describe factors of social and economic background. Therefore, principal components I, II, III, and IV have received a more comprehensive response in human society and economy. It can be seen that human social and economic factors were the main driving force of ecological vulnerability in Ningxia. At the same time, the hours of sunshine and annual average precipitation are the main indicators used to describe meteorological factors. Principal component I and principal component II were highly correlated with the hours of sunshine and annual average precipitation, respectively. It can be seen that meteorological factors were external forces affecting the ecological vulnerability of Ningxia.

In 1995, principal components II, III, and IV all reflected socio-economic background factors. The contribution rate of principal component I was 44.97%, which was mainly correlated with the grassland area (79.42%) and hours of sunshine (29.48%). The contribution rate of principal component II was 24.31%, which was mainly correlated with the grassland area (59.80%), agricultural output (46.22%), and degree of land use (33.40%). The contribution rate of principal component III was 11.55%, which was mainly correlated with the degree of land use (76.91%), NDVI (43.57%), and agricultural output (39.83%). The contribution rate of principal component IV was 7.75%, which was mainly correlated with the NDVI (59.75%) and agricultural output (36.47%). Agricultural output and degree of land use are common indicators to describe factors of social and economy background. Therefore, principal components II, III, and IV reflect that human social and economic factors are the main driving force of ecological vulnerability. The hours of sunshine are often the meteorological factor, so the meteorological factor was the external force that affected the ecological vulnerability of Ningxia.

In 2000, principal components I, II, III, and IV all reflected socio-economic background factors, while principal components I and II also reflected meteorological factors. The contribution rate of principal component I was 40.53%, which was mainly correlated with the grassland area, hours of sunshine, and agricultural output (67.19%, 37.58%, and 34.90%, respectively). The contribution rate of principal component II was 23.42%, which was mainly correlated with the grassland area, degree of land use, and annual average precipitation (68.32%, 39.90%, and 36.68%, respectively). The contribution rate of principal component III was 14.91%, which was mainly correlated with the agricultural output and NDVI (68.26% and 60.02%, respectively). The contribution rate of principal component IV was 9.53%, which was mainly correlated with the degree of land use (86.01%).

In 2005, principal components II, II, and IV all reflected socio-economic background factors. The contribution rate of principal component I was 37.93%, which was mainly correlated with the grassland area and hours of sunshine (66.34% and 42.09%, respectively). The contribution rate of principal component II was 23.77%, which was mainly correlated with the grassland area, degree of land use, and NDVI (66.34%, 36.48%, and 35.34%, respectively). The contribution rate of principal componen III was 16.29%, which was mainly correlated with the agricultural output and NDVI (65.65% and 50.88%, respectively). The contribution rate of principal component IV was 9.01%, which was mainly correlated with the degree of land use (89.27%).

In 2010, principal components II, II, and IV all reflect socio-economic background factors. The contribution rate of principal component I was 38.66%, which was mainly correlated with the grassland area and hours of sunshine (70.17% and 36.40%, respectively). The contribution rate of principal component II was 23.65%, which was mainly correlated with the grassland area, degree of land use, and NDVI, which were 61.74%, 40.41%, and 33.48%, respectively. The contribution rate of principal componen III was 15.11%, which was mainly correlated with the agricultural output and NDVI (62.33% and 51.04%, respectively). The contribution rate of principal component IV was 9.11%, which was mainly correlated with the degree of land use (87.83%).

In 2015, principal components III and IV reflected the socio-economic background factors. The contribution rate of principal component I was 37.61%, which was mainly correlated with the grassland area and hours of sunshine (67.27% and 42.88%, respectively). The contribution rate of principal component II was 24.64%, which was mainly correlated with the grassland area and NDVI (65.00% and 45.41%, respectively). The contribution rate of principal component III was 14.89%, which was mainly correlated with the NDVI and agricultural output (53.00% and 50.47%, respectively). The contribution rate of principal component IV was 9.54%, which was mainly correlated with degree of land use (88.07%).

In Figure 6, in principal component I (PC1), there was a positive correlation between the EVI and grassland area and hours of sunshine from 1990 to 2015. In principal component II (PC2), the EVI was highly correlated with the grassland area and degree of land use from 1990 to 2015. From 2005 to 2015, the EVI was also highly relevant to the agricultural output and average annual precipitation. In principal component III (PC3), there was a high correlation between the EVI and NDVI and agricultural output from 1990 to 2015, as well as a high correlation with the GDP in 1990 and 2015, degree of land use in 1995 and 2000, and industrial output, respectively, in 2005 and 2010. In principal component IV (PC4), from 1990 to 2015, the degree of land use had more than 80% correlation with the EVI.

We can see that principal component I, principal component II, principal component III, and principal component IV reflect the social and economic factors' impact on the ecological conditions. At the same time, principal component I and principal component II reflect the influence of climatic factors, such as the hours of sunshine and average annual precipitation, on the conditions of ecological vulnerability.

4. Discussion

4.1. Spatial-Temporal Characteristics of Ecological Vulnerability

This study visualized the temporal and spatial changes of ecological vulnerability in Ningxia by using the Tupu and Getis-Ord Gi* spatial analysis models. Tupu clearly shows the transformation of grades of ecological vulnerability in Ningxia. It can be seen from the Tupu analysis that the transformation of the ecological vulnerability grade showed obvious differences in space-time. At the same time, the Getis-Ord Gi* spatial model was used to gather similar ecological vulnerability indexes in spatial statistics, and the results were divided into cold spots, hot spots, and not significant spots.

In the north of Ningxia, the EVI was stable from 1990 to 2015, when the main grade of ecological vulnerability was extreme vulnerability. The north of the study area was also the hot-spot region of the EVI during the study period. The north of Ningxia is mainly constituted of Shizuishan City and Yinchuan City. These two cities feature high-density populations and economic development, which explains why ecological vulnerability was at a high level.

In the west of Ningxia, the EVI was unstable in the study period, with the main grades of ecological vulnerability medium vulnerability and heavy vulnerability. The west of the study area, which is crossed by the Yellow River of China, constitutes Zhongwei City. The region is an important base of grain suppliers, aquatic products, and vegetable facilities in China. The stable economic structure is also the main factor that causes the stabilization of the EVI in Zhongwei City.

In the east of Ningxia, the EVI decreased in the former period but increased later. Based on the analysis spatial heterogeneity, the EVI is not stable in the east of Ningxia. Wuzhong City is the main district in the east of Ningxia. Figure 2 shows that the spatial heterogeneity of the EVI in Wuzhong is significant. As a node city in the New Silk Road Economic Belt of China, Wuzhong City is active economically, which has a great impact on the local ecological conditions.

In the south of Ningxia, the EVI increased but reversed the trend later. Cold spots are mainly distributed in the south of study area. In addition, and the EVI changed insignificant in this region. Guyuan City in the south of Ningxia has a poor preference in economic and population density, so ecological conditions of this region have rarely been influenced by local human activities.

From the perspective of arid and semi-arid climate zones, the ecological vulnerability in the north of Ningxia, which belongs to an arid region, is much worse than that in central and southern Ningxia, which belongs to a semi-arid region. At the same time, the spatial heterogeneity of the EVI in arid regions is insignificant, while in semi-arid regions it is significant.

4.2. The Impact of Ecological Projects on Ecological Vulnerability

Grassland is one of the main driving factors of ecological vulnerability in the study area. The change of the grassland area is influenced by the policy. It can be seen that policies can indirectly affect ecological vulnerability. Due to the ecological characteristics of each patch being different, the implementation of policy is different in space. The Grain for Green Program (GGP), China's ecosystem restoration project, aims to transform farmland, on steep hillsides or in areas of severe desertification, into forestland or grassland in order to increase vegetation coverage and reduce water loss and soil erosion. GGP has a great influence on ecological vulnerability. Ningxia started the GGP in 1999 and the first phase ended in 2006. Therefore, the degree of ecological vulnerability was low in 2005, when the area proportion of slight vulnerability was 51%; after 2005, with a decrease in the intensity of the GGP, the proportion of area with slight vulnerability decreased to 41% by 2010 and sharply decreased to 19% in 2015. It can be seen that GGP has a great impact on the ecological vulnerability of Ningxia. The implementation of GGP in the early stages indeed increased the grassland area in Ningxia, while the economic input was reduced in the latter stages. Additionally, the work of grassland management and grassland maintenance was not very good [37], so the area proportion of slight vulnerability reduced.

Another ecological project that affects grassland area is the Prohibited Grazing Policy (PGP). The PGP is an important ecological service project launched by the Chinese government in 2003 to restore degraded grassland in severely degraded grasslands and ecologically vulnerable areas. Many studies have found that the implementation of the PGP plays an important role in ecological protection, and it has achieved remarkable ecological benefits [25,38,39]. Yanchi County, located in the east of the study area, is a study hot spot of the PGP. According to the results of this study, the ecological vulnerability of Yanchi County in 2005 and 2010 was better than that of other years. The grade of ecological vulnerability was mainly slight vulnerability and light vulnerability in 2005 and 2010, and the degree of ecological vulnerability increased in 2015. The results were similar to other studies [40,41]. It can be seen that the PGP has had a great impact on Yanchi County's environment conditions.

4.3. Suggestions

Ningxia is located in the transition zone between the Loess Plateau and the Inner Mongolia Plateau. The Maowusu Sandy Land is in the east and the Tengger Desert is in the west in the north of the study area. Therefore, the ecological situation in the north of the study area is weak. It is easy to cause desertification in the north of the study area if it is not well managed. Desertification is highlighted in Goal 15 of the UN Sustainable Development Goals (SDGs). In this study, based on the study results of ecological vulnerability in Ningxia, we provide reasonable suggestions for sustainable development. The following section gives detailed suggestions on ecological vulnerability protection.

First, the SDGs emphasize the balanced development of society and ecosystems. This study showed that the north of Ningxia, which belongs to an arid region, is mainly at extreme vulnerability in the grades of ecological vulnerability. Meanwhile, the north of Ningxia is mainly the hot spot area of the ENI. The Yellow River is throughout the north of Ningxia, making it have the highest level of economic development in Ningxia. Additionally, it is the region with the fastest development of urbanization in Ningxia, which leads to changes of land use. Therefore, it is suggested that the government should carry out urban planning reasonably and reduce changes of land use as far as possible, which means the social activities of encroachment on to grassland, forest, and other natural ecosystems should be prohibited in this region. In this way, the unreasonable occupation of the natural ecosystem by social development can be effectively alleviated, and their coordinated development can be guaranteed to a certain extent, so as to promote the sustainable development of society and the ecosystem. Second, the SDGs also focus on desertification. According to the analysis results of the driving factors of ecological vulnerability, the grassland area is the main influencing factor of ecological vulnerability in Ningxia. Meanwhile, changes in grassland areas is closely related to ecological projects. Therefore, it is suggested that the construction of ecological projects should be strengthened in the future, and the management of ecological projects should have strengthened management in the latter stages, so as to ensure long-term sustainable development. In this way, the conditions of ecological vulnerability can be effectively protected, and environmental problems such as desertification can be effectively slowed down in the study area, so as to promote sustainable development. At the same time, the government should increase publicity efforts to raise public awareness of environmental protection, and provide convenient communication platforms for the public, such as Internet platforms, to improve public participation. Finally, according to the spatial heterogeneity of the EVI in Ningxia, we suggest that different levels of government should strengthen inter-regional cooperation and promote the establishment of a unified and synchronized platform for information sharing and monitoring, which could make the ecological condition develop in a better direction in Ningxia.

5. Conclusions

In this study, spatial distribution of ecological vulnerability of the Ningxia Hui Autonomous Region in an arid and semi-arid region was studied by using Tupu and Getis-Ord Gi*. Tupu can be used to understand the transformation of the grade of ecological vulnerability in Ningxia. Getis-Ord Gi* can be used to intuitively understand the spatial aggregation state of the EVI in Ningxia. The results of Tupu and cold-hotspot show that the ecological vulnerability of Ningxia has obvious heterogeneity in space. The results showed that conditions of ecological vulnerability were unstable from 1990 to 2015 in the study area, with the degree of ecological vulnerability mainly decreasing from 1990 to 2005 and increasing from 2005 to 2015. During the study period, the degree of ecological vulnerability gradually increased from the south to the north. The hot-spot areas of the EVI were mainly concentrated in the north of Ningxia, while the cold-spots areas of the EVI were scattered in the central and southern areas of Ningxia. At the same time, the results showed that the grassland area was the main driving factor of ecological vulnerability of Ningxia from 1990 to 2015. Through the discussion, it was found that the change in the grassland area is obviously related to the implementation of ecological projects.

The spatial distribution of the EVI was shaped by interaction between human activities and environmental factors. In this study, we found that the ecological vulnerability was greatly affected by ecological projects in Ningxia. Therefore, the government should continue to implement relevant ecological projects in the future and limit construction activities that could cause an increase in the EVI in order to achieve the sustainable development goals in arid and semi-arid areas.

Author Contributions: Conceptualization, R.L., R.H., and L.G.; methodology, R.L.; R.H., and Q.Y.; software, R.L. and L.G.; formal analysis, R.L. and R.H.; investigation, R.L. and S.Q.; resources, R.L. and L.G.; data curation, L.G.; writing—original draft preparation, R.L.; writing—review and editing, R.H. and L.G.; project administration, L.G.; funding acquisition, L.G. All authors have read and agreed to the published version of the manuscript.

Funding: This research study was supported by the key research project of the Chinese Ministry of Science and Technology (2017YFC0505601) and is an innovation team project of the Chinese Nationalities Affairs Commission, grant number 10301-0190040129.

Acknowledgments: We are grateful for the comments of the anonymous reviewers, which greatly improved the quality of this paper.

Conflicts of Interest: No conflict of interest exits in this manuscript, and the manuscript has been approved by all authors for publication. The authors declare that the work described was original research that has not been published previously and is not under consideration for publication elsewhere, in whole or in part.

References

1. Hou, K.; Li, X.; Zhang, J. GIS Analysis of Changes in Ecological Vulnerability Using a SPCA Model in the Loess Plateau of Northern Shaanxi, China. *Int. J. Environ. Res. Public Health* **2015**, *12*, 4292–4305. [CrossRef] [PubMed]
2. Kang, H.; Tao, W.; Chang, Y.; Zhang, Y.; Xuxiang, L.; Chen, P. A feasible method for the division of ecological vulnerability and its driving forces in Southern Shaanxi. *J. Clean. Prod.* **2018**, *205*, 619–628. [CrossRef]
3. De Lange, H.J.; Sala, S.; Vighi, M.; Faber, J.H. Ecological vulnerability in risk assessment da review and perspectives. *Sci. Total Environ.* **2010**, *408*, 3871–3879. [CrossRef] [PubMed]
4. Hou, K.; Li, X.; Wang, J.; Zhang, J. Evaluating Ecological Vulnerability Using the GIS and Analytic Hierarchy Process (AHP) Method in Yan'an, China. *Pol. J. Environ. Stud.* **2016**, *25*, 599–605. [CrossRef]
5. Xiaodan, W.; Xianghao, Z.; Pan, G. A GIS-based decision support system for regional eco-security assessment and its application on the Tibetan Plateau. *J. Environ. Manag.* **2010**, *91*, 1981–1990. [CrossRef] [PubMed]
6. Cutter, S.L.; Boruff, B.J.; Shirley, W.L. Social vulnerability to environmental hazards. *Soc. Sci. Q.* **2003**, *84*, 242–261. [CrossRef]
7. Mi, Z.; Wei, Y.M.; Wang, B.; Meng, J.; Liu, Z.; Shan, Y.; Liu, J.; Guan, D. Socio-economic impact assessment of China's CO_2 emissions peak prior to 2030. *J. Clean. Prod.* **2017**, *142*, 2227–2236. [CrossRef]
8. Songa, G.; Chena, Y.; Tianb, M.; Lvb, S.H.; Zhanga, S.H.; Liua, S. The ecological vulnerability evaluation in southwestern mountain region of China based on GIS and AHP method. *Procedia Environ. Sci.* **2010**, *2*, 465–475. [CrossRef]
9. Ren, X.; Dong, Z.; Hu, G.; Zhang, D.; Li, Q. A GIS-Based Assessment of Vulnerability to Aeolian Desertification in the Source Areas of the Yangtze and Yellow Rivers. *Remote. Sens.* **2016**, *8*, 626. [CrossRef]
10. Abson, D.J.; Dougill, A.J.; Stringer, L.C. Using principal component analysis for information-rich socio-ecological vulnerability mapping in Southern Africa. *Appl. Geogr.* **2012**, *35*, 515–524. [CrossRef]
11. Inostroza, L.; Palme, M.; de la Barrera, F. A Heat Vulnerability Index: Spatial Patterns of Exposure, Sensitivity and Adaptive Capacity for Santiago de Chile. *PLoS ONE* **2016**, *11*, e0162464. [CrossRef] [PubMed]
12. Li, L.; Jiang, K.Y.; Liu, F.P.; Dou, J.X. Fuzzy comprehensive evaluation and application for vulnerability of complex geological disaster in southwest mountainous area in China. *Electron. J. Geotech. Eng.* **2014**, *19*, 6929–6937.
13. Zhao, J.; Ji, G.; Tian, Y.; Chen, Y.; Wang, Z. Environmental vulnerability assessment for mainland China based on entropy method. *Ecol. Indicat.* **2018**, *91*, 410–422. [CrossRef]
14. He, G.; Bao, K.; Wang, W.; Zhu, Y.; Li, S.; Jin, L. Assessment of ecological vulnerability of resource-based cities based on entropy-set pair analysis. *Environ. Technol.* **2019**, 1–11. [CrossRef]
15. Takane, S. Effect of domain selection for compact representation of spatial variation of head-related transfer function in all directions based on Spatial Principal Components Analysis. *Appl. Acoust.* **2016**, *101*, 64–77. [CrossRef]
16. Xue, L.; Wang, J.; Zhang, L.; Wei, G.; Zhu, B. Spatiotemporal analysis of ecological vulnerability and management in the Tarim River Basin, China. *Sci. Total Environ.* **2019**, *649*, 876–888. [CrossRef]
17. Xi, M.; Kong, F.; Li, Y.; Yu, X. Analysis on characteristics of soil salinization in coastal wetlands of Jiaozhou Bay. *Bull. Soil Water Conserv.* **2016**, *36*, 288–292.
18. Wei, W.; Shi, S.; Zhang, X.; Zhou, L.; Xie, B.; Zhou, J.; Li, C. Regional-scale assessment of environmental vulnerability in an arid inland basin. *Ecol. Indic.* **2020**, *109*, 105792. [CrossRef]
19. Yu, X.; Li, Y.; Xi, M.; Kong, F.; Pang, M.; Yu, Z. Ecological vulnerability analysis of Beidagang National Park, China. *Front. Earth Sci.* **2019**, *13*, 385–397. [CrossRef]
20. Kurniawan, F.; Adrianto, L.; Bengen, D.G.; Prasetyo, L.B. Vulnerability assessment of small islands to tourism: The case of the Marine Tourism Park of the Gili Matra Islands, Indonesia. *Glob. Ecol. Conserv.* **2016**, *6*, 308–326. [CrossRef]

21. Mansur, A.V.; Brondízio, E.S.; Roy, S.; Hetrick, S.; Vogt, N.D.; Newton, A. An assessment of urban vulnerability in the Amazon Delta and Estuary: A multi-criterion index of flood exposure, socio-economic conditions and infrastructure. *Sustain. Sci.* **2016**, *11*, 625–643. [CrossRef]
22. Peng, B.; Huang, Q.; Elahi, E.; Wei, G. Ecological Environment Vulnerability and Driving Force of Yangtze River Urban Agglomeration. *Sustainability* **2019**, *11*, 6623. [CrossRef]
23. He, L.; Shen, J.; Zhang, Y. Ecological vulnerability assessment for ecological conservation and environmental management. *J. Environ. Manag.* **2018**, *206*, 1115–1125. [CrossRef] [PubMed]
24. Ding, Q.; Shi, X.; Zhuang, D.; Wang, Y. Temporal and Spatial Distributions of Ecological Vulnerability under the Influence of Natural and Anthropogenic Factors in an Eco-Province under Construction in China. *Sustainability* **2018**, *10*, 3087. [CrossRef]
25. Wang, W.; Zhou, L.; Yang, G.; Sun, Y.; Chen, Y. Prohibited Grazing Policy Satisfaction and Life Satisfaction in Rural Northwest China—A Case Study in Yanchi County, Ningxia Hui Autonomous Region. *Int. J. Environ. Res. Public Health* **2019**, *16*, 4374. [CrossRef]
26. Zhang, X.; Wang, L.; Fu, X.; Li, H.; Xu, C. Ecological vulnerability assessment based on PSSR in Yellow River Delta. *J. Clean. Prod.* **2017**, *167*, 1106–1111. [CrossRef]
27. Kan, A.K.; Li, G.Q.; Yang, X.; Zeng, Y.L.; Tesren, L.; He, J. Ecological vulnerability analysis of Tibetan towns with tourism-based economy: A case study of the Bayi District. *J. Mt. Sci.* **2018**, *15*, 1101–1114. [CrossRef]
28. Satyanarayana, B.; van der Stocken, T.; Rans, G.; Kodikara, K.A.S.; Ronsmans, G.; PulukkuttigeJayatissa, L.; Husain, M.; Koedam, N.; Dahdouh-Guebas, F. Island-wide coastal vulnerability assessment of Sri Lanka reveals that sand dunes, planted trees and natural vegetation may play a role as potential barriers against ocean surges. *Glob. Ecol. Conserv.* **2017**, *12*, 144–157. [CrossRef]
29. Anselin, L. Local indicators of spatial association-LISA. *Geogr. Anal.* **1995**, *27*, 93–115. [CrossRef]
30. Han, R.; Feng, C.-C.; Xu, N.; Guo, L. Spatial heterogeneous relationship between ecosystem services and human disturbances: A case study in Chuandong, China. *Sci. Total Environ.* **2020**, 137818. [CrossRef]
31. Statistical Bureau of Ningxia. *Ningxia Statistical Yearbook*; China Statistics Press: Beijing, China, 2016.
32. Leśniewska-Napierała, K.; Nalej, M.; Napierała, T. The Impact of EU Grants Absorption on Land Cover Changes—The Case of Poland. *Remote. Sens.* **2019**, *11*, 2359. [CrossRef]
33. Dunteman, G.H. *Principal Component Analysis*; Sage Publications: Newburry Park, CA, USA, 1989.
34. Jolliffe, I.T. *Principal Component Analysis*; Springer: New York, NY, USA, 2002.
35. Ye, Q.H.; Liu, G.H.; Lu, Z.; Gong, Z.H.; Marco. Research of TUPU on land use/land cover change based on GIS. *Prog. Geogr.* **2002**, *21*, 349–357.
36. Wang, J.L.; Shao, J.A.; Li, Y.B. Geospectrum based analysis of crop and forest land use change in the recent 20 years in the three gorges reservoir area. *J. Nat. Resour.* **2015**, *30*, 235–247.
37. Yan, W. Reflections on ecological environment construction in Ningxia. *Chin. For. Econ.* **2017**, *2*, 79–80, 97. [CrossRef]
38. Chen, Y.; Zhou, L. Farmers' Perception of the Decade-Long Grazing Ban Policy in Northern China: A Case Study of Yanchi County. *Sustainability* **2016**, *8*, 1113. [CrossRef]
39. Hou, C.; Zhou, L.; Wen, Y.; Chen, Y. Farmers' adaptability to the policy of ecological protection in China—A case study in Yanchi County, China. *Soc. Sci. J.* **2017**, *55*, 404–412. [CrossRef]
40. Chen, Y.; Wang, T.; Zhou, L.; Liu, N.; Huang, S. Effect of prohibiting grazing policy in northern China: A case study of Yanchi County. *Environ. Earth Sci.* **2013**, *72*, 67–77. [CrossRef]
41. Wei, X.; Zhou, L.; Yang, G.; Wang, Y.; Chen, Y. Assessing the Effects of Desertification Control Projects from the Farmers' Perspective: A Case Study of Yanchi County, Northern China. *Int. J. Environ. Res. Public Health* **2020**, *17*, 983. [CrossRef]

Review

Shifting Perspectives in Assessing Socio-Environmental Vulnerability

Jonathan W. Long [1,*] **and E. Ashley Steel** [2,3]

1   USDA Forest Service, PSW Research Station, Davis, CA 95618, USA
2   USDA Forest Service, PNW Research Station, Seattle, WA 98103, USA
3   Present: Forestry Department, Food and Agricultural Organization (FAO) of the United Nations, Viale delle Terme di Caracalla, 00153 Rome, Italy; ashley.steel@fao.org
*   Correspondence: jonathan.w.long@usda.gov; Tel.: +01-530-759-1744

Received: 17 February 2020; Accepted: 24 March 2020; Published: 26 March 2020

Abstract: Governments and institutions across the globe are conducting vulnerability assessments and developing adaptation plans to confront rapidly changing climatic conditions. Interrelated priorities, including the conservation of biodiversity, ecological restoration, sustainable development, and social justice often underlie these efforts. We collaborated with colleagues in an effort to help guide vulnerability assessment and adaptation (VAA) generally in Southeast Asia and specifically in the watershed of the Sirindhorn International Environmental Park (SIEP) in Phetchaburi Province, Thailand. Reflecting upon our experiences and a review of recent VAA literature, we examine a series of seven questions that help to frame the socio-ecological context for VAAs. We then propose a three-dimensional framework for understanding common orientations of VAAs and how they appear to be shifting and broadening over time, particularly in the USA. For example, key leaders in the SIEP project emphasized social development and community-based approaches over more ecology-centric approaches; this orientation was consistent with other examples from SE Asia. In contrast, many efforts for US national forests have evaluated vulnerability based on projected shifts in vegetation and have promoted adaptation options based upon ecological restoration. Illustrating a third, highly integrated approach, many VAAs prepared by indigenous tribes in the USA have emphasized restoring historical ecological conditions within a broader context of promoting cultural traditions, social justice, and adaptive capacity. We conclude with lessons learned and suggestions for advancing integrated approaches.

Keywords: vulnerability and adaptation assessment; climate change; ecological restoration; watershed management; sustainable development; community-based assessment; indigenous peoples

---

1. Introduction

Entities around the globe are increasingly engaged in assessing vulnerability and developing plans to adapt to climate change. International efforts were catalyzed at the 21st Conference of the Parties to the United Nations Framework Convention on Climate Change (COP21) by the adoption of an agreement for all participating countries to develop climate change adaptation plans [1]. Such planning efforts are now connected to broader sustainability initiatives, such as the UN Decade on Ecosystem Restoration (2021–2030), for which the United Nations General Assembly has set a goal of massively scaling up the restoration of degraded and destroyed ecosystems to address climate change and promote sustainable ecosystems. A first step in identifying appropriate actions and policies in the face of climate change is often to conduct a vulnerability and adaptation assessment (VAA). In fact, this is frequently a required step in accessing climate preparedness funds [2]. Vulnerability assessments identify future risks induced by climate change, identify key vulnerable resources, and provide a sound basis for designing adaptation strategies [3]. A recent trend among vulnerability assessments has been to expand considerations to include social, non-climatic determinants of vulnerability, such as adaptive capacity, and to shift from simply estimating expected damages to identifying opportunities to reduce damages [4]. Another thrust has been to recognize the concerns of local communities, particularly underserved and highly vulnerable communities, including indigenous peoples [5]. Both COP21 and a related international agreement, the Convention on Biological Diversity, have explicitly called for the consideration and integration of indigenous communities into their strategies [6]. Alongside these trends there has been a shift from qualitative descriptions of vulnerability to more quantified evaluations of risk, including the probability of occurrence of hazards, potential exposure to hazards, and the severity of potential impacts [7].

1.1. Background

We were invited to share the perspectives and ideas used by the US Forest Service (USFS) in developing VAAs with a team of government agencies, universities, and non-governmental organizations (NGOs) in central Thailand that was initiating one of the first watershed-scale VAA efforts in Southeast Asia. Over the course of three in-person visits and extensive remote collaboration, we both contributed to the completion of the project and learned about collaborative, large-scale VAA initiatives. In this paper, we consider the key issues and lessons that emerged from that experience in the context of scientific literature describing the principles and approaches of successful VAA efforts. We begin by describing the framework of the particular project in Thailand and then review guidance from the USFS as well as literature regarding VAAs and ecological restoration. We then present seven questions that emerged as points of tension during the project, and we reframe those issues in light of recent literature and published examples of VAAs from national forests and tribes in the USA. Based upon that review and our engagement in the cross-cultural project as a case study, we suggest ways to shift toward more integrated VAAs.

1.2. SIEP Project Case Study: Watershed-Scale Climate Vulnerability Assessment and Adaptation Planning in Thailand

The Watershed-Based Adaptation to Climate Change project was a regional collaborative initiative financed by the National Research Council of Thailand (NRCT), Royal Thai Government, and US Agency for International Development (USAID) with technical support from the USFS. The initiative aimed to create a model for watershed-scale planning through a VAA for Sirindhorn International Environment Park (SIEP) in Cha-am District, Phetchaburi Province, Thailand (hereafter, the SIEP project) (Figure 1). The park has been developed to educate visitors about energy, natural resources, and the environment in the coastal region of Thailand, with a special emphasis on the conservation of mangrove habitats. It is located in one of the driest parts of Thailand, a region important for agriculture, including lemons and pineapple, and tourism, including golf courses and hotels. The SIEP project included the participation of research faculty from two Thai universities, park staff, representatives of several ministries, and leadership by the Sustainable Development Foundation (SDF), a well-established non-governmental organization in the region.

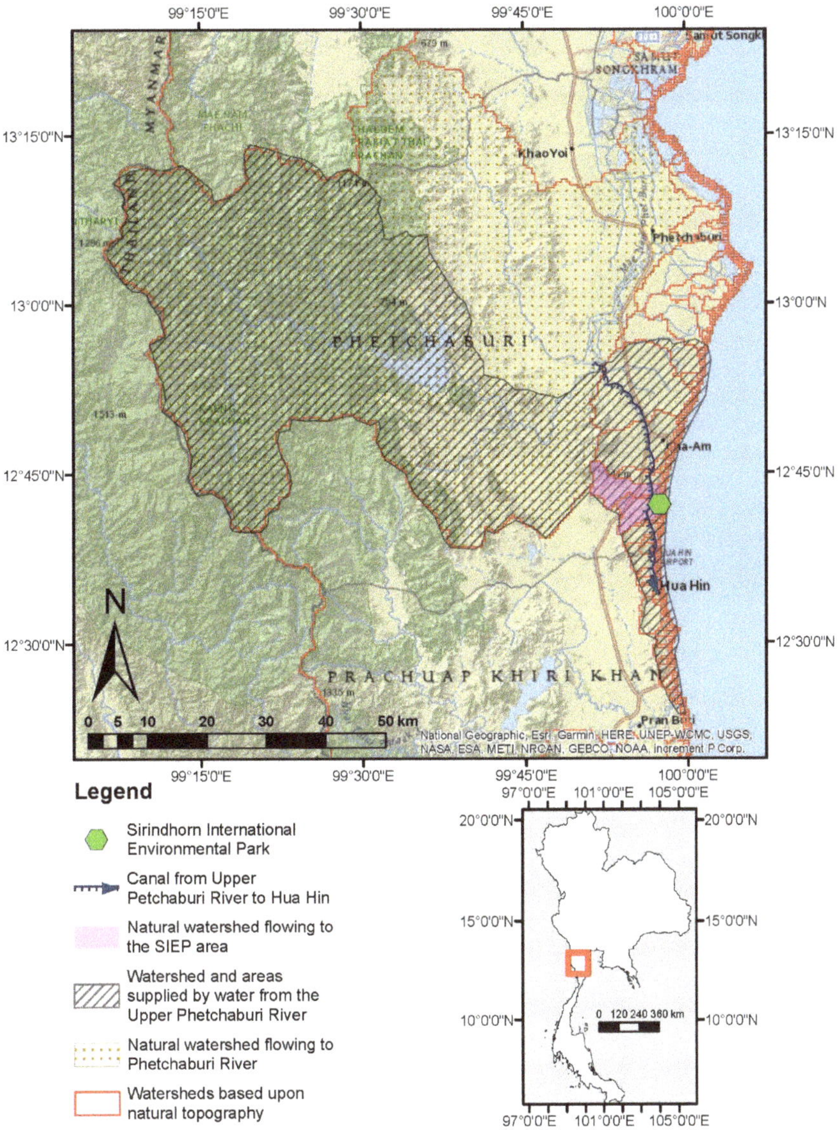

Figure 1. The natural watershed of the Sirindhorn International Environmental Park (SIEP) (light purple) is much smaller than the watershed of the Upper Phetchaburi River (brown stippling) that supplies water to coastal areas, including the area around the park (gray hatching).

The project included climate downscaling work across the entire Phetchaburi River Basin (Figure 1) as well as on-the-ground data collection in communities selected by the SDF to represent the main economic sectors in the watershed. In the upper watershed, the SDF focused on a Karen community that was experiencing conflict with the government regarding land rights and agricultural activities, as are other indigenous people in mountainous forested areas [8]. In the central watershed, the SDF identified four villages that each relied on a particular cash crop and two communities struggling with urban expansion and water supply management. In the lower watershed, the SDF focused on three communities suffering from floods and droughts as well as a community where many livelihoods depended on salt farming. Their methodology explicitly considered both climate and non-climate factors that contribute to vulnerability [9].

2. Review of Vulnerability Literature

2.1. Guidance for VAAs in the US

USFS reports and related journal articles from the past dozen years reveal how ecological and socio-economic issues have been addressed in VAA guidance over time. In 2008, the USFS released a strategic framework for responding to climate change [10], which was followed by a more detailed "roadmap" [11]. Additionally in 2008, the US government published a "preliminary review of adaptation options for climate-sensitive ecosystems and resources"; the authors of the chapter on national forests authored a related journal article the following year [12]. Over the next few years, the USFS issued guidance reports for VAAs including a guidebook for developing adaptation options with case study examples in 2010 [3], a guide to adaptation tools and resources references in 2012 [13], and a framework with pilot assessments for watershed vulnerability in 2013 [14].

The early guidance and examples of VAAs on national forests focused heavily on the ecological dimensions of vulnerability. Although they also mentioned ecosystem services, they often considered socio-economic factors in terms of being added stressors or constraints. For example, Spies et al. [15] noted how "social license", policies, and laws might constitute barriers to adaptation. Vulnerability was commonly evaluated in terms of biophysical values, including wildlife, forests, fisheries, and infrastructure. For example, three pilot examples of watershed vulnerability assessments from national forests in the Pacific Northwest evaluated only the impacts to aquatic species listed as rare or threatened by government agencies as well as to infrastructure including roads, campgrounds, and water diversions [14]. Most national forests in the USA are situated within inland mountain regions; consequently, most examples of detailed vulnerability analyses from the USFS are from such areas [3]. These rural areas have relatively low human population densities and few permanently populated areas. Early VAAs may have largely ignored the social vulnerability of local communities in order to focus on the quantification of biophysical risks. Many of these early VAAs stand in contrast to examples from Southeast (SE) Asia that focused on densely populated urban areas [16].

USFS guidance for VAAs has recently shifted toward a greater focus on socio-economic dimensions of vulnerability through the lenses of ecosystem services and household adaptive capacity. In 2013, Fischer et al. [17] published an article calling for greater attention to social vulnerability in assessments for national forests. In 2016, the USFS updated the adaptation guidance published four years earlier to include urban forests [18]. Most recently, the USFS issued guidance for integrating socio-economic concerns into vulnerability frameworks [19] and protocols for evaluating socio-economic vulnerability [20].

The impact of climate change on indigenous communities has also been an emerging concern in VAA literature pertaining to US national forests. Many of the USFS guidance documents do not specifically draw attention to indigenous peoples, although one of the recent reports [20] mentions some values that are particularly important to such communities, including traditional foods and medicinal herbs. To address this gap, the USFS recently published a framework for understanding climate impacts on indigenous peoples that highlights impacts on tribal social systems [21]. That guidance notes that classifications such as "natural," "cultural", and "economic" reflect colonial systems of thinking that impose false divisions between fundamentally interconnected systems, as highlighted in other tribally directed research [22]. A group led by a team of indigenous people in the North Central USA recently published an adaptation menu that responded to earlier USFS guidance by featuring words, processes, and values that they thought would be more productive and respectful for work involving indigenous communities [23]. American Indian tribes have been broadening the scope of VAAs in the USA—a recent review [24] found that VAAs written by or for tribes in the Pacific Northwest USA emphasized how climate change would impact their distinctive cultural values.

Nearly half of the plans examined in that review specifically discussed impacts to public health, tribal economies, and cultural well-being, including traditional knowledge and cultural practices.

2.2. An Evolving Foundational Concept: Ecological Restoration

Ecological restoration has emerged as a common framework for addressing multiple natural resource management challenges, including the promotion of local livelihoods, sustainable development, conservation of biodiversity, mitigation of climate change, and adaptation to disturbances including climate change [25]. Proponents of restoration suggest that these efforts can be broadly relevant when customized to particular landscape and social contexts [26]. For example, the mission statement for the UN Decade of Landscape Restoration suggests that achievement of sustainable development goals depends on addressing ecological degradation. The concept of ecological restoration has had a wide and deep influence on conservation efforts in the United States and within the USFS in particular. The USFS roadmap invoked a future-looking form of ecological restoration as a central pillar to its efforts: "the agency is responding to climate change through adaptive restoration—by restoring the functions and processes characteristic of healthy ecosystems, whether or not those systems are within the historical range of variation" [11]. Additionally, restoration has been incorporated in the new forest planning rule to guide the revision of land management plans across the USA.

The ecological restoration framework is not rigidly prescriptive, but instead emphasizes principles, indicators, and best practices. Ecological restoration plans in the Pacific Northwest of the USA have often set targets based upon conditions prior to colonization by Europeans that are characterized as the "historical" or "natural range of variation" [27]. These restoration efforts may focus on evaluating and restoring native species diversity as well as natural processes that reset succession, such as fires and floods. While restoration has focused on removing artificial constraints on those natural disturbances, adaptation efforts have often focused on building resistance and resilience to uncharacteristically severe disturbances [12].

Linkages between ecological restoration and adaptation to climate change are common in USFS VAA literature. For example, Peterson et al. [3] described one climate change adaptation strategy of "forestalling change" as "intensifying management to return a site to a prior condition of that ecosystem in the face of climate change", which mirrors ecological restoration. Through interviews for his dissertation research, Timberlake [28] found that national forestland managers were largely focused on the goal of ecological restoration in pursuing climate change adaptation.

Ecological restoration appears less common, however, in VAA examples from SE Asia. For instance, a recently published guide to VAAs in the Greater Mekong subregion [2] did not include the terms "restore" or "restoration". Similarly, in a comparative study of socio-ecological system resilience in the United States and SE Asia, MacQuarrie [29] included several references to salmon and river restoration in his case study of the Columbia River Basin (US) but only one to forest restoration in Cambodia. An example for high mountain areas in Asia produced by the World Wildlife Fund did propose ecosystem restoration as one solution to reduce climate vulnerability [30]. Because there are so many diverse examples of VAAs in both regions, it is less important to demonstrate consistent differences between them than it is to consider how particular orientations can influence project outcomes.

3. Findings from the Review and Case Study

3.1. Important Framing Questions Considered in the SIEP Project

Through our efforts to bring ideas and frameworks from the USFS to the SIEP project, we identified several framing questions for consideration in structuring the scope and scale of any large-scale or watershed-scale VAA.

3.1.1. What Is the Appropriate Management Unit?

Approaches applied to US national forests and in Southeast Asia have both emphasized the importance of using watersheds as analysis and management units. Guidance for conducting climate change vulnerability assessments in the USFS references a framework for assessing the watershed vulnerability at relevant scales, typically with hydrologic units that are 4000 to 16,000 ha [14]. This approach has been widely used in USFS lands that typically constitute headwater forests. The USFS has examined the national vulnerability of water supplies to climate change [31]. As a simplifying assumption, it did not consider the adaptive capacity of the water supply systems to respond to changes in precipitation; yet, extensive reservoir and irrigation systems do allow for storage and redistribution of water, representing adaptive capacity to manage changes in precipitation.

The SIEP project applied a "ridge to reef" approach that considered watershed areas in a similar size range. During our project, however, we confronted a challenge in simply identifying the boundaries of the watershed for the park because trans-basin diversions (Figure 2) had transformed the natural "ecological" watershed (based upon topography and historical conditions) into a much larger "water supply" watershed (Figure 1). Specifically, the park is located within the Bangtranoi watershed (64 km^2) with hydrologic connections to the neighboring Huaisai watershed (20 km^2) [32]. Although the Thai Royal Forest Department manages the forested mountain areas that have been recognized as priority water quality management zones, concerns over water supply were pronounced and not only related to headwater conditions. Both the agriculture and tourism sectors depend on water supplies from outside of the study-area basin. Water distribution plans were designed to balance human needs while leaving some minimal in-stream flows. In the project, the emphasis on consumptive use drew attention away from the forested highlands of the interior and toward the agricultural and urbanized areas along the coast.

Figure 2. Water conveyance structure delivering water from storage reservoirs outside the SIEP watershed to downstream agriculture, cities, and tourism-related development. Photo by Jonathan Long.

3.1.2. Is Downscaled Climate Data Necessary?

Because projections from global climate models are generally coarse, regional and local projects may rely on downscaling [33] to understand how climate change is likely to affect particular watersheds and landscapes [34]. Decisionmakers may prize this customized scientific information for gauging the susceptibility of highly built environments to natural disturbance. For example, downscaled climate data can play a role in predicting future flood flows and the vulnerability of bridges and other infrastructure. However, downscaled climate data may give the impression of being more precise and accurate that it truly is. It is not necessarily useful [35] and, in particular, it may be a distraction when managing wildland systems, which have evolved under a wide range of natural variation and which, therefore, may be more tolerant of climatic fluctuations than constructed systems. Additionally, thresholds for natural systems are less likely to be known, so precise estimates of environmental change may not be of high value.

Organizers of the SIEP project had a strong desire to create and utilize downscaled climate change data. Previously published research by Shrestha [36] included climate projections on an approximately 22×22 km^2 scale, to evaluate conditions across nine hydrological response units in Thailand; SIEP is included in the Eastern Gulf unit. Their findings suggest an increase in water availability during both the wet and dry seasons; however, they noted that the models did not perform as well in the coastal regions as in other regions. To address such concerns, the SIEP project commissioned downscaled climate projections at the 1 km scale.

While developing such finely downscaled climate projections can increase the reported precision of predictions and bolster confidence among decision makers, the fine scale of the predictions may also inflate expectations that the projections are accurate. Because there can be considerable uncertainty associated with localized forecasts, USFS guidance on VAAs [14] urges staff on individual national forests to proceed with VAAs even without finely downscaled climate projections. Often existing information, such as historical information on climate change and observations in extreme years, can be used for framing climate change in terms that are clearly understood and appreciated [37]. In fact, the ranges and directions of projected change may be more important than specific estimates for future temperature and runoff [14].

3.1.3. Is Ecological Restoration or Socio-Economic Development the Priority?

Many VAAs undertaken in Asia emphasize socio-economic development and immediate livelihood needs [35]. This focus contrasts with ecologically oriented VAA examples from the USFS. Even nature-oriented institutions such as the World Wildlife Fund have generated vulnerability assessment reports that explicitly evaluate impacts on ecosystem services and community livelihoods [30]. In the SIEP project, scientists and managers did highlight historical degradation including the loss of forest cover, soil erosion and reduced infiltration, sedimentation, the loss of mangrove habitats, and losses of important wildlife species including axis deer (*Axis porcinus*). Local watershed managers and park exhibits described conditions prior to the death of King Rama VI in 1926 as being a reference, comparable to the way the USFS often describes reference conditions prior to "Euro-American settlement". However, during the SIEP project, we observed that managers were focused on increasing forest and ground cover more than on restoring historical composition and structure. For example, activities described under forest restoration included planting vetiver grass (*Chrysopogon zizanioides*, native to India) and terracing landscapes to reduce soil erosion (Figure 3). Such actions, while bearing similarities to early soil restoration efforts on US national forests [38], are not aimed at restoring pre-existing conditions but rather at moving the ecosystem forward toward greater productivity. Managers may have favored such reparative actions as being more practical than attempting to recreate historical conditions. Some VAAs have explicitly blended restoration and economic development goals. One such VAA from the Philippines [39] concluded that an "ideal program" for reducing the vulnerability of farmers should combine reforestation to improve denuded watershed conditions with infrastructure enhancements to help farmers cope with floods and water shortages.

Figure 3. Erosion control measures including rock check dams and plantings of vetiver grass. Photo by Jonathan Long.

3.1.4. Are Roads an Ecological Liability or a Resource for Reducing Social Vulnerability?

Roads represent a significant boundary element in social and ecological systems, illustrating the need for an integrated perspective that weighs a variety of ecological and social values. In many US examples, VAAs have suggested closing and even obliterating roads because of concerns over their impacts on watershed condition [40]. These are most often in watersheds where limiting human impact is a goal. On the other hand, tribes have emphasized the need for access via roads to maintain traditional practices [41]. In SE Asia, VAAs have also recognized the importance of roads as adaptive capacity [42], because road access is essential for getting goods to market and for receiving services during emergencies.

3.1.5. What Are the Costs and Benefits of Dams and Diversions?

The negative impacts of dams and diversions have been highlighted in VAAs in the USA, while the benefits of such hydrologic infrastructure for ameliorating the impacts of floods and droughts are recognized in watershed management plans for the Phetchaburi River in Thailand. Adaptation plans in the USA and Europe emphasize reducing the impacts of dams and diversions, leaving "room for the river" [43] and otherwise promoting the use of soft adaptable green bank stabilization measures rather than streambank armoring and channelization [44] (Figure 4). VAAs for the Mekong River, where millions of people depend on fishing for their livelihoods and diet, have similarly emphasized these ecological principles [2]. In comparing system resilience in the USA and SE Asia, MacQuarrie [29] concluded that the construction of large dams in the Pacific Northwest USA had reduced ecological resilience, but that those impacts were counterbalanced by increased social resilience that stemmed from a focus on ecological restoration as well as on the legal and political power of tribes, fishers (commercial and recreational), and other stakeholders. In contrast, he concluded that social systems in SE Asia were less socially resilient, but that they had not experienced as much loss in ecological resilience. However, ecosystems in SE Asia, like the Mekong, are also experiencing declines in large migratory fish species (comparable to salmon and sturgeon in the Mekong, are also experiencing declines in large migratory fish species (comparable to salmon and sturgeon in the Columbia Basin) due to the construction of large dams [45]. Trans-boundary institutions have developed in the region to reduce the ecological impacts of dams, representing another source of adaptive capacity [46].

Figure 4. Channelization on a river system in the SIEP watershed. Photo by Jonathan Long.

3.1.6. Is the Focus on Sustaining Forests or Sustaining Forest-Dependent People?

In the SIEP case study, the use of forest lands by local communities, both for subsistence agriculture (Figure 5) and for the gathering of forest products, emerged as an important issue. Such issues also occur on federal lands in the United States, and they are a particularly significant issue for Native Americans [41,47]; however, neither have featured prominently in climate change vulnerability assessments. Both national parks and national forests in the United States have a long history of driving out and excluding indigenous peoples [48,49]. NGO staff leading the community-based assessment for the SIEP project described the adoption across SE Asia of what they considered to be a "Western model" of evicting local peoples from conservation lands such as national parks (see [50] for more on that history). Similarly, a VAA for the Philippines [39] expressed concern that adaptation strategies focusing on forest restoration and protection might not address the needs of communities, noting that forest protection rules could adversely affect farmers without official land tenure.

Figure 5. Crops planted within patches cleared from forests in the upper watershed by local communities. Photo by Jonathan Long.

3.1.7. Is Fire to Be Restored or Excluded?

Tensions in applying an ecological restoration framework are particularly evident in debates over whether human-caused burns should be part of a management regime to restore reference conditions. In tropical parts of the USA and affiliated islands, suppression of wildfires is a predominant concern and human-set fires are often regarded as unnatural and harmful [51]. A similar view appears to hold in many parts of SE Asia, although some researchers and community organizations maintain that rotational agriculture using fire to clear forest patches is a long-standing practice, going far back in human history. The dry dipterocarp forest found in the SIEP project area has likely had a long history of fire influence; however, the importance of fire in altering its succession to evergreen forest remains a topic of some uncertainty [52]. In the national park at the headwaters of the SIEP project area, the impacts of Karen people using fire to clear forests were a high concern; yet, community members asserted that they were continuing a long-standing tradition of swidden agriculture. In the Western USA, management of fire regimes has long been a dominant concern in watershed management and the adaptation to climate change, but the focus has recently shifted toward restoring more frequent and lower intensity fire that mimics cultural burning by indigenous peoples [41]. Some tribes in the USA have also sought to restore traditional agroforestry practices, such as cultural burning, to national forests encompassing ancestral lands [41] as part of a management plan to promote resilience to future climate change [53,54].

3.2. General Themes in VAAs Represented by Three Axes

By considering the key questions above that emerged in our case study as well as the trends in the reports and journal articles discussed above, we identified three major axes shaping the orientation of VAAs (Figure 6). First is the extent of focus on the ecological versus socio-economic dimensions of systems. Second is the level of reliance on technical "experts" versus broader community participation as sources of information. Third is the time frame for evaluating departure from desired conditions. The three axes in the figure are in theory independent, but as a practical matter, assessments have tended to cluster between top-down and ecologically focused versus bottom-up and socially focused.

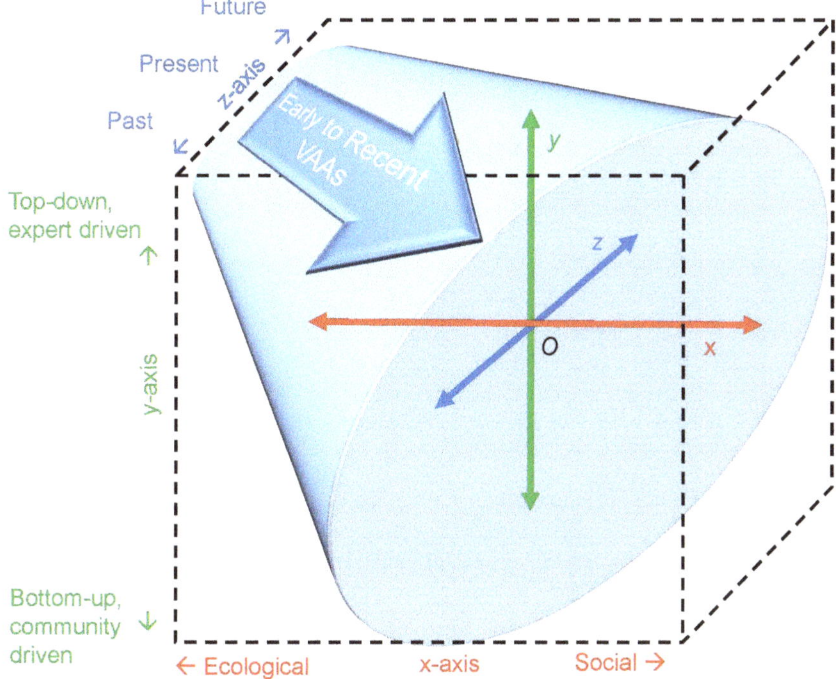

Figure 6. The scope of vulnerability and adaptation assessments. The x-axis represents the social-ecological orientation. The y-axis represents bottom-up versus top-down processes, and the z-axis represents the temporal perspective including past, present, and future reference points.

3.2.1. Social-Ecological Considerations

The first axis describes the relative emphasis on social versus ecological issues and outcomes. Many of the differences between VAA along this axis relate to broader tensions between social development and conservation.

Early reports and associated articles written on VAA issues for national forests in the USA recognized both ecological and socio-economic systems; however, they focused heavily on the ecological aspects, such as how the distribution of dominant species or the associated plant communities may shift. Social concerns have been incorporated into such frameworks by focusing on species that are harvested for commercial gain or subsistence, or which are "charismatic" [55]. Social dimensions in early VAA efforts also included the consideration of ecosystem services [10], such as impacts to water supply and recreation [14] as well as the consideration of how to bolster internal agency adaptive capacity and reduce external social and economic barriers to adaptation [3,56].

The language used in many VAAs indicates their predominantly ecological orientation. For example, Timberlake and Schultz [57] noted that "In (US) Forest Service documents, the agency frequently uses resilience in the sense of ecological resilience; it sometimes describes social-ecological resilience but to date has rarely worked this type of resilience into planning and management activities." In another article, they noted that considerations of ecosystem services were often limited to qualitative descriptions [58]. An ecological focus is often reflected in terms such as "landscape" to characterize areas. In contrast, the perspective of tribes or other indigenous communities is often important in viewing such areas instead as "homelands" that they depend upon to sustain their livelihoods and identities.

In Southeast Asia, many regional VAAs have emphasized poverty reduction and the associated socio-economic concerns. For example, the USAID Mekong Adaptation and Resilience to Climate Change (ARCC) project [2] considered impacts to livestock, fisheries, agriculture, natural systems, and health and rural infrastructure. It focused its assessment on six principal crops and applied downscaled global climate projections to evaluate impacts to livelihoods. Impacts to "natural systems" were considered through the lenses of non-timber forest products and wild relatives of key crops. This perspective is consistent with the idea that many forested landscapes in the Mekong Basin are used for agroforestry. Other regional assessments have also integrated social and ecological evaluations [42].

3.2.2. Information Sources and Process

The second dimension (the y-axis in Figure 6) focuses on the sources of information used to prepare the assessment. A "bottom-up" approach focuses on local knowledge. It may rely upon local observations of climate trends and qualitative information, often obtained through interviews with community members or participatory processes. A "top-down" approach depends more heavily on detailed scientific data, published studies, downscaled climate projections, and expert opinion. For example, experts often review literature and suggest indicators that they think are useful for measuring vulnerability [59].

Bottom-up approaches have become a priority for many communities that have been frustrated when government agencies develop plans first and seek buy-in from local communities second. Communities often prefer to have priorities and plans emerge from local concerns and to gain agency buy-in as a second step. Green et al. [60] stressed that the latter, "bottom-up", approaches may better develop practical activities to reduce vulnerability; they cautioned that decision makers may become stuck in an information gathering and research phase if they prioritize having quantitative justification for particular policy options. Eriksen et al. [61] also emphasized the need to pursue adaptation actions that promote both social justice and environmental integrity by integrating local knowledge into adaptation responses and by recognizing the importance of non-climate stressors [15]. Non-profit organizations leading VAA efforts in other parts of Asia have emphasized the need to address the non-climate dimensions of vulnerability through community engagement at local scales [35].

In the USA, individual national forests have typically completed VAAs on their own, although one plan was developed jointly between the Olympic National Forest and Olympic National Park [40]. Early VAAs by the USFS placed considerable emphasis on the technical expertise of internal agency staff, as well as external scientists from universities and non-profit conservation organizations. These efforts may not be strictly characterized as "top-down", because they were often built on information from stakeholder workshops with participation from universities, NGOs, local governments, and other interests. Yet, lists of participants suggest that staff and scientists with biophysical expertise often predominated. Timberlake [28] similarly noted that the USFS has emphasized a "coproduction" model involving scientists and managers. One of the planning guides suggested that broader community considerations could be addressed by previewing VAAs with stakeholders who may be affected [24]. The top-down approach may also be triggered by short planning horizons; plans are often created relatively rapidly, taking between eight months and one year to complete [14].

In contrast, VAAs in SE Asia frequently emphasize multi-agency approaches, which may take several years to complete. For example, the Mekong ARCC project worked over several years assessing conditions across many communities. The SIEP project involved multiple consultations with stakeholders mostly composed of local community members. The consultations were of various sizes and locations and took place over two years. Several recent examples of VAAs have highlighted more participatory processes that seek to integrate top-down and bottom-up approaches [62,63].

3.2.3. Temporal References

The third axis describes the temporal orientation of the assessments, which typically compare conditions during two different time periods. For example, restoration plans commonly compare present conditions to those during a past reference period. Using a past reference period can help address the pervasive effects of colonization, which are a major concern of indigenous groups [64]. As Helen Fillmore, a tribal member and climate change researcher recently explained, "The degradation that they [our elders] couldn't fathom is already our current condition" [65]. To address these concerns, researchers have attempted to estimate historical changes in indicators that are socially and ecologically important, such as the availability of traditionally harvested wild foods [22].

Some researchers contend that historical conditions may no longer be appropriate or realistic in the face of changing climate [3]. In that vein, early USFS guidance suggested that VAAs be more forward-looking by considering simulations of future climate, vegetation, and species movements [3]. Indeed, early examples of VAAs in the USA often assessed ecological vulnerability based upon such modeled climate projections and judgments of their adaptive capacity by professional scientists. Many of these assessments implicitly set a present-day reference by comparing current conditions to future conditions that are projected using models of climate change.

Many socially oriented VAAs are oriented toward current conditions, rather than forecasts, to evaluate vulnerability. As a practical matter, trends in key social indicators are often challenging to predict. For example, fewer than half of health-oriented VAAs internationally have estimated the future health impacts of climate change; that limitation reflects a common lack of data and analysis regarding relationships between climate drivers and health outcomes on which to build predictive models [66]. Other socially oriented VAAs have focused on vulnerability to widely predicted climatic impacts, particularly for low-lying coastal regions, such as increases in temperature, increased risk of intense storms and flooding, and sea-level rise [67–69].

Some researchers contend that while modeling to predict future conditions is valuable, it is important to focus on present-day vulnerabilities informed by community input [35]. For example, O'Brien et al. [70] argued for an emphasis on human security and present difficulties in coping with external stressors, which they characterized as "context vulnerability". They contrasted that approach with the assessment of "outcome vulnerability" that may develop in the future. Williams et al. [71] further emphasized this point by distinguishing between impacts that have already been "realized" versus "potential" impacts that might occur. Some economists have suggested that focusing on current demand for services, rather than their potential supply, is important for aligning policy with sustainable development goals [72].

3.3. Lessons Learned for Achieving Better Integrated VAAs

Over the past dozen years, there has been a trend toward greater social-ecological integration of VAAs in the USA. Here we present lessons learned that may reinforce this shift toward more integrated perspectives in advancing the sustainability of coupled human-natural systems.

3.3.1. Humans are Part of Most Natural Systems

Frameworks in North America have broadened to consider the influence of humans in the evolution of natural ecosystems. For example, the use of fire by indigenous peoples, which has occurred for thousands of years in North America, is increasingly regarded as part of the natural disturbance regime under which many ecosystems in the Western United States have evolved [73,74]. Accordingly, particular ecological communities, including groves of large California black oaks (*Quercus kelloggii*) and Oregon white oaks (*Quercus garryana*), camas grounds (*Camassia* spp.), and rock-walled "clam gardens" are recognized as having been produced or shaped by indigenous practices and are now targets for ecological restoration [41,75,76]. A recent indigenous-led publication supplemented USFS guidance on VAAs by emphasizing more holistic perspectives in adaptation planning rather than worldviews and language that tend to separate humans from the natural world [23].

3.3.2. The Ecological Restoration Framework Can Be Broadened to Include Social Considerations

The longstanding ecological restoration framework remains a powerful and appealing tool for evaluating conditions and setting management directions; however, research has recommended greater focus on social indicators and consideration of local community perceptions when evaluating baselines for ecological restoration [77]. Furthermore, ecological restoration has the potential to contribute to injustices if strategies do not account for social impacts. Accordingly, many planners, even some within the restoration field, have encouraged greater attention to social objectives [78].Ecological restoration can be more closely allied with environmental justice and indigenous concerns than has been commonly recognized, as revealed in the ecocultural restoration movement being advanced by many tribes [79]. This movement emphasizes the important of access to traditional foods and medicines for community health, which aligns with calls for VAAs to evaluate the impacts of climate change on the public health and food security of vulnerable populations [66,80].

3.3.3. Consideration of Social Adaptive Capacity, Including Governance Institutions, May Increase the Relevance of VAAs

Socially oriented VAAs consider how communities can adapt to the future climate given institutions, policies, property rights, and other constraints that humans are able to change [81]. Promoting adaptive social capacity may align with many ecological restoration goals, particularly for communities that have been impacted by declines in the productivity of ecosystems. However, that focus may also bring a wider range of possible targets under management consideration than would be considered under a purely ecological restoration perspective, which may become narrowly fixated on past ecological conditions. Governance institutions, which are an important component of adaptive capacity, are especially important for the management of harvested systems or resource allocations [82,83]. The ecological restoration framework in the USA can also be used to align shared ecological interests of non-governmental entities including watershed councils, prescribed fire councils, and other groups. An increasing responsiveness of institutions to the interests of indigenous communities, with support or pressure from federal law, has been important for promoting system resilience, especially in the Pacific Northwest of the USA [29]. Tribes in the USA have pursued institutional relationships that facilitate more cooperative management, such as comanagement agreements and the establishment of special demonstration areas [41,75]. These developments have propelled important ecological restoration efforts, such as the removal of large dams that have inhibited recovery of salmonids.

3.3.4. Community Engagement in the Development of Indicators May Increase Their Effectiveness

Evaluations of socio-ecological resilience may be greatly influenced by the selection of indicators. Because of that, MacQuarrie [29] noted that adaptive capacity could be increased by engaging local stakeholders in creating and studying a more complete inventory of indicators. During the SIEP project, we discussed a variety of climate indicators or "facets" that would translate climate change projections into locally meaningful terms, such as the duration of floods and length of rainy periods that impact specific crops. For example, the local community wanted to create an indicator describing the maximum number of consecutive days without rain in the growing season. This indicator contained information critical to anticipating climate effects on particular crops and to making decisions about what to plant in the future. While emphasizing values that are important to community stakeholders, these facets can be built on recommended principles of scientific reliability [59]. Work with tribes in the US has similarly emphasized biophysical metrics to quantify the usability and availability of key forest product species, including the abundance of mature hardwood trees that produce culturally important foods [84,85]. Assessments will be more likely to result in social change where indicators have been developed in collaboration with local communities [86].

3.3.5. Integrated Science and Monitoring Contribute to Effective Management in the Long Term

Monitoring efficient, ecologically important, and socially meaningful indicators can help ensure that adaptation efforts are effective in reducing vulnerability. However, the outcome of a particular adaptation approach may not be evident for years after it is put into practice, and long-term, retrospective evaluations of plan implementation are rare. Most countries have not completed a cycle of adaptive management by evaluating the sufficiency of baseline assessments and monitoring the success of adaptation efforts [66]. An openness to experimentation may be particularly important for advancing the adaptation science knowledge base, particularly when experiments are designed with consideration of governance institutions and with the engagement of local communities [87]. Support for conducting experiments may depend on alignment with social and cultural values. For example, attitudes that regard humans as part of nature, as held by many Native Americans, have been associated with public support for interventions to conserve Alaska yellow cedar under a warming climate, even within protected forest areas [88].

3.3.6. Integration of Top-Down and Bottom-Up Approaches Is Necessary to Solve Complex Problems

Many recent examples of VAAs aim to integrate perspectives from professional scientists and local communities. The value of such integration is that broader knowledge is brought to bear on complex problems. The USAID Mekong ARCC framework [2] pursued this goal by integrating a "community-based vulnerability assessment" with a "science-based vulnerability assessment," to build a community-driven adaptation plan, which was subjected to expert review. A recent VAA in Thailand illustrated such an approach, as researchers compared community views that precipitation was becoming more variable with projections of increased precipitation overall [62]. Another effort in the Mekong Basin found that participatory approaches helped to bridge divides over adaptation priorities between investments in infrastructure versus changes in land-use policies [63]. In their review of USFS VAAs, Timberlake and Schultz [58] suggested that a greater focus on social–ecological linkages and key ecosystem services could be achieved by involving stakeholders in participatory processes that integrate qualitative case studies and top-down quantitative assessments. While bottom-up approaches are important for ensuring community engagement, top-down support can help to design and implement more rigorous experimental efforts that can be replicated and upscaled [87]. As part of such a mixed-methods integrative approach, local qualitative inquiries can help stakeholders prioritize quantitative investigations on broader scales [89].

4. Conclusions

Tension between top-down approaches, which are steeped in the quantitative analysis of biophysical impacts of climate change, and bottom-up approaches, which emphasize participatory and qualitative examination of social capacity and development opportunities, will always be present in VAAs. However, the past decade of VAA applications in the USA has shifted from predominantly "ecologically focused" VAAs toward more holistic ones. This trend reflects a broader recognition that planners and managers need to move beyond problematic distinctions between "ecological" and "social" dimensions when promoting sustainable systems. Our summary risks overgeneralizing differences between VAA approaches applied in US national forest contexts and ones that have been commonly applied in SE Asia. The seeming divergence likely reflects multiple factors, including the different priorities of federal land management agencies such as the USFS, compared to non-governmental socio-economic development institutions that have had prominent roles in VAAs in SE Asia. Moreover, VAAs in both regions are increasingly assessing joint ecological and social impacts and identifying opportunities to increase adaptive capacity through institutional change. Assessing vulnerability and developing effective adaptation plans will require a skillful combination of perspectives, including relatively standardized, quantitative measures of future risk along with more locally customized assessments of community vulnerability and adaptive capacity.

Author Contributions: Conceptualization, writing, and editing: J.W.L. and E.A.S.; literature review and figures: J.W.L. All authors have read and agreed to the published version of the manuscript.

Funding: This research received funding support from the USDA Forest Service Pacific Northwest Research Station and Pacific Southwest Research Station based upon a partnership with the US Agency for International Development.

Acknowledgments: We thank Monthip Sriratana Tabucanon, Ravadee Prasertcharoensuk, Alex Smajgl, Jonathan Shott, Jerasorn Santisirisomboon, and Robert Dobias for sharing insights regarding vulnerability assessments in Southeast Asia. We also thank Kathy Lynn and Haley Case-Scott for sharing findings from their compilation and review of vulnerability assessment and adaptation plans developed by tribes in the USA. We thank Geoffrey Blate for his expertise and his contributions to illuminating insights from the SIEP project as well as for providing feedback on a draft of this manuscript.

Conflicts of Interest: The authors declare no conflict of interest.

References

1. United Nations Framework Convention on Climate Change. *Adoption of the Paris Agreement FCCC/CP/2015/L.9/Rev.1*; United Nations Framework Convention on Climate Change: Paris, France, 2015.
2. Smajgl, A.; Salamanca, A.; Steel, A.; Blate, G.; Long, J.; Hartman, P.; Manuamom, O.; Sawhney, P.; Friend, R.; Gustafson, S.; et al. *Watershed Vulnerability and Adaptation Assessments in the Greater Mekong Subregion: Guidelines for Climate Change Practitioners*; Greater Mekong Subregion (GMS) Roundtable for Climate Adaptation: Bangkok, Thailand, 2018.
3. Peterson, D.L.; Millar, C.I.; Joyce, L.A.; Furniss, M.; Halofsky, J.E.; Neilson, R.; Morelli, T.L. *Responding to Climate Change in National Forests: A Guidebook for Developing Adaptation Options*; USDA Forest Service, Pacific Northwest Research Station: Portland, OR, USA, 2011.
4. Füssel, H.-M.; Klein, R.J. Climate change vulnerability assessments: An evolution of conceptual thinking. *Clim. Chang.* **2006**, *75*, 301–329. [CrossRef]
5. David-Chavez, D.M.; Gavin, M.C. A global assessment of Indigenous community engagement in climate research. *Environ. Res. Lett.* **2018**, *13*, 123005. [CrossRef]
6. Amiott, J. Investigating the Convention on Biological Diversity's protections for traditional knowledge. *Mo. Environ. Law Policy Rev.* **2003**, *11*, 3–37.
7. Connelly, A.; Carter, J.; Handley, J.F.; Hincks, S. Enhancing the practical utility of risk assessments in climate change adaptation. *Sustainability* **2018**, *10*, 1399. [CrossRef]
8. Hares, M. Forest conflict in Thailand: Northern minorities in focus. *Environ. Manag.* **2008**, *43*, 381–395. [CrossRef] [PubMed]
9. Shott, J. Vulnerability and Capacity Assessment (VCA): An Integrated Approach for Local Stakeholders. Powerpoint; Asia Pacific Adaption Network. Available online: http://www.asiapacificadapt.net/adaptationforum2012/sites/default/files/SDF%20VCA%20Synthensized%20Ver06%20Adapt%20Forum.pdf (accessed on 3 February 2020).
10. Dillard, D.; Rose, C.; Conard, S.; MacCleery, D.; Ford, L.; Conant, K.; Cundiff, A.; Trapani, J. *Forest Service Strategic Framework for Responding to Climate Change*; USDA Forest Service: Washington, DC, USA, 2008.
11. Tkacz, B.; Brown, H.; Daniels, A.; Acheson, A.; Cleland, D.; Fay, F.; Johnson, R.; Kujawa, G.; Roper, B.; Stritch, L. *National Roadmap for Responding to Climate Change*; USDA Forest Service: Washington, DC, USA, 2010.
12. Joyce, L.A.; Blate, G.M.; McNulty, S.G.; Millar, C.I.; Moser, S.; Neilson, R.P.; Peterson, D.L. Managing for multiple resources under climate change: National forests. *Environ. Manag.* **2009**, *44*, 1022–1032. [CrossRef]
13. Swanston, C.; Maria, E.J. *Forest Adaptation Resources: Climate Change Tools and Approaches for Land Managers*; Gen. Tech. Rep. NRS-GTR-87; USDA Forest Service, Northern Research Station: Newtown Square, PA, USA, 2012.
14. Furniss, M.; Roby, K.B.; Cenderelli, D.; Chatel, J.; Clifton, C.F.; Clingenpeel, A.; Hays, P.E.; Higgins, D.; Hodges, K.; Howe, C.; et al. *Assessing the Vulnerability of Watersheds to Climate Change: Results of National forest Watershed Vulnerability Pilot Assessments*; Gen. Tech. Rep. PNW-GTR-884; USDA Forest Service, Pacific Northwest Research Station: Portland, OR, USA, 2013.

15. Spies, T.A.; Giesen, T.W.; Swanson, F.J.; Franklin, J.F.; Lach, D.; Johnson, K.N. Climate change adaptation strategies for federal forests of the Pacific Northwest, USA: Ecological, policy, and socio-economic perspectives. *Landsc. Ecol.* **2010**, *25*, 1185–1199. [CrossRef]

16. MacClune, K.; Opitz-Stapleton, S. *Building Urban Resilience to Climate Change*; ISET-International: Boulder, CO, USA, 2012.

17. Fischer, A.P.; Paveglio, T.; Brenkert-Smith, H.; Carroll, M.; Murphy, D. Assessing social vulnerability to climate change in human communities near public forests and grasslands: A framework for resource managers and planners. *J. For.* **2013**, *111*, 357–365. [CrossRef]

18. Swanston, C.W.; Janowiak, M.K.; Brandt, L.A.; Butler, P.R.; Handler, S.D.; Shannon, P.D.; Lewis, A.D.; Hall, K.R.; Fahey, R.T.; Scott, L.; et al. *Forest Adaptation Resources: Climate Change Tools and Approaches for Land Managers*; Gen. Tech. Rep. NRS-GTR-87–2; USDA Forest Service, Northern Research Station: Newtown Square, PA, USA, 2016.

19. Hand, M.S.; Eichman, H.; Triepke, F.J.; Jaworski, D. *Socioeconomic Vulnerability to Ecological Changes to National Forests and Grasslands in the Southwest*; Gen. Tech. Rep. RMRS-GTR-383; USDA Forest Service, Rocky Mountain Research Station: Fort Collins, CO, USA, 2018.

20. Armatas, C.A.; Borrie, W.T.; Watson, A.E. *Protocol for Social Vulnerability Assessment to Support National Forest Planning and Management: A Technical Manual for Engaging the Public to Understand Ecosystem Service Tradeoffs and Drivers of Change*; Gen. Tech. Rep. RMRS-GTR-396; USDA Forest Service, Rocky Mountain Research Station: Fort Collins, CO, USA, 2019.

21. Norton-Smith, K.; Lynn, K.; Chief, K.; Cozzetto, K.; Donatuto, J.; Redsteer, M.H.; Kruger, L.E.; Maldonado, J.; Viles, C.; Whyte, K.P. *Climate Change and Indigenous Peoples: A Synthesis of Current Impacts and Experiences*; Gen. Tech. Rep. PNW-GTR-944; USDA Forest Service, Pacific Northwest Research Station: Portland, OR, USA, 2016.

22. Norgaard, K.M. *Salmon & Acorns Feed Our People: Colonialism, Nature, & Social Action*; Rutgers University Press: New Brunswick, NJ, USA, 2019.

23. Tribal Adaptation Menu Team. *Dibaginjigaadeg Anishinaabe Ezhitwaad: A Tribal Climate Adaptation Menu*; Great Lakes Indian Fish and Wildlife Commission: Odanah, WI, USA, 2019.

24. Case-Scott, H.; Lynn, K. Review of tribal climate change plans. In *Understanding Effects of Climate Change on Forest Resources for Tribal Communities in the Pacific Northwest and Northern California*; Lake, F.K., Long, J.W., Norgaard, K.M., Lynn, K., Case-Scott, H., Eds.; (unpublished, manuscript in preparation).

25. Reinecke, S.; Blum, M. Discourses across scales on forest landscape restoration. *Sustainability* **2018**, *10*, 613. [CrossRef]

26. Biswas, S.; Mallik, A.U.; Choudhury, J.K.; Nishat, A. A unified framework for the restoration of Southeast Asian mangroves—Bridging ecology, society and economics. *Wetl. Ecol. Manag.* **2008**, *17*, 365–383. [CrossRef]

27. Keane, R.; Hessburg, P.F.; Landres, P.B.; Swanson, F.J. The use of historical range and variability (HRV) in landscape management. *For. Ecol. Manag.* **2009**, *258*, 1025–1037. [CrossRef]

28. Timberlake, T.J. Climate Change Adaptation on Public Lands: Policy, Vulnerability Assessments, and Resilience in the US Forest Service. Ph.D. Thesis, Colorado State University, Fort Collins, CO, USA, 2019.

29. MacQuarrie, P.R. Resilience of Large River Basins: Applying Social-Ecological Systems Theory, Conflict Management, and Collaboration on the Mekong and Columbia Basins. Ph.D. Thesis, Oregon State University, Corvallis, OR, USA, 2012.

30. Smith, T. *Climate Vulnerability in Asia's High Mountains*; Technical Report to USAID; World Wildlife Fund: Washington, DC, USA, 2014.

31. Foti, R.; Ramirez, J.A.; Brown, T.C. *Vulnerability of US Water Supply to Shortage: A Technical Document Supporting the Forest Service 2010 RPA Assessment*; USDA Forest Service, Rocky Mountain Research Station: Fort Collins, CO, USA, 2012.

32. Boonyawat, S.; Kheereemangkla, Y.; Puk-ngam, S.; Thueksathit, S.; Tongdeenok, T.; Kaewjampa, N.; Havanond, S.; Sittiyanpaiboon, A. *Integrated Watershed Ecosystem Management: A Case of Forest and Water Resources in Bangtranoi and Huaisai Watershed, Cha-am District, Phetchaburi Province*; International Conference on Climate Change: Phetchaburi Province, Thailand, 2016.

33. Campbell, J.D.; Taylor, M.A.; Stephenson, T.S.; Watson, R.A.; Whyte, F.S. Future climate of the Caribbean from a regional climate model. *Int. J. Clim.* **2010**, *31*, 1866–1878. [CrossRef]

34. Cayan, D.R.; Maurer, E.; Dettinger, M.D.; Tyree, M.; Hayhoe, K. Climate change scenarios for the California region. *Clim. Chang.* **2008**, *87*, 21–42. [CrossRef]

35. Opitz-Stapleton, S.; MacClune, K. Chapter 11 Scientific and social uncertainties in climate change: The Hindu Kush-Himalaya in regional perspective. In *Climate Change Modeling for Local Adaptation in the Hindu Kush-Himalayan Region*; Lamadrid, A., Kelman, I., Eds.; Emerald Group Publishing Limited: Bingley, UK, 2012.

36. Shrestha, S. Assessment of water availability under climate change scenarios in Thailand. In *Climate Change Impacts and Adaptation in Water Resources and Water Use Sectors*; Shrestha, S., Ed.; Springer International Publishing: New York, NY, USA, 2014.

37. Steel, E.A.; Marsha, A.; Fullerton, A.H.; Olden, J.D.; Larkin, N.K.; Lee, S.-Y.; Ferguson, A. Thermal landscapes in a changing climate: Biological implications of water temperature patterns in an extreme year. *Can. J. Fish. Aquat. Sci.* **2019**, *76*, 1740–1756. [CrossRef]

38. Hall, M. *Earth Repair: A Transatlantic History of Environmental Restoration*; University of Virginia Press: Charlottesville, VA, USA, 2005.

39. Lasco, R.D.; Cruz, R.; Pulhin, J.; Pulhin, F. *Assessing Climate Change Impacts, Vulnerability and Adaptation: The Case of Pantabangan-Carranglan Watershed*; World Agroforestry Centre-Philippines: Laguna, Philippines, 2010.

40. Halofsky, J.E.; Peterson, D.L.; O'Halloran, K.A.; Hoffman, C.H. *Adapting to Climate Change at Olympic National Forest and Olympic National Park*; Gen. Tech. Rep. PNW-GTR-844; USDA Forest Service, Pacific Northwest Research Station: Portland, OR, USA, 2011.

41. Long, J.W.; Lake, F.K. Escaping social-ecological traps through tribal stewardship on national forest lands in the Pacific Northwest, United States of America. *Ecol. Soc.* **2018**, *23*, 1–14. [CrossRef]

42. Yusuf, A.A.; Francisco, H. Climate change vulnerability mapping for Southeast Asia. In *Economy and Environment Program for Southeast*; Economy and Environment Program for Southeast Asia (EEPSEA): Singapore, 2009.

43. Kondolf, G.M. The Espace de Liberté and restoration of fluvial process: When can the river restore itself and when must we intervene. In *River Conservation and Management*; John Wiley & Sons, Ltd.: Chichester, UK, 2012; pp. 223–241.

44. Chornesky, E.A.; Ackerly, D.D.; Beier, P.; Davis, F.W.; Flint, L.E.; Lawler, J.J.; Moyle, P.B.; Moritz, M.A.; Scoonover, M.; Byrd, K. Adapting California's ecosystems to a changing climate. *Bioscience* **2015**, *65*, 247–262. [CrossRef]

45. Hogan, Z.; Na-Nakorn, U.; Kong, H. Threatened fishes of the world: *Pangasius sanitwongsei* Smith 1931 (Siluriformes: Pangasiidae). *Environ. Biol. Fishes* **2008**, *84*, 305–306. [CrossRef]

46. Kittikhoun, A.; Staubli, D.M. Water diplomacy and conflict management in the Mekong: From rivalries to cooperation. *J. Hydrol.* **2018**, *567*, 654–667. [CrossRef]

47. Flood, J.P.; McAvoy, L.H. Use of national forests by Salish? Kootenai tribal members: Traditional recreation and a legacy of cultural values. *Leisure/Loisir* **2007**, *31*, 191–216. [CrossRef]

48. Keller, R.H.; Turek, M.F. *American Indians and National Parks*; University of Arizona Press: Tucson, AZ, USA, 1999.

49. Catton, T. *American Indians and National Forests*; University of Arizona Press: Tucson, AZ, USA, 2016.

50. McElwee, P.D. Displacement and relocation redux: Stories from Southeast Asia. *Conserv. Soc.* **2006**, *4*, 396–403.

51. Jennings, L.N.; Douglas, J.; Treasure, E.; González, G. *Climate Change Effects in El Yunque National Forest, Puerto Rico, and the Caribbean Region*; Gen. Tech. Rep. SRS-GTR-193; USDA Forest Service, Southern Research Station: Asheville, NC, USA, 2014.

52. Stott, P.A.; Goldammer, J.G.; Werner, W.L. The role of fire in the tropical lowland deciduous forests of Asia. In *Ecological Studies*; Springer Science and Business Media LLC: Berlin/Heidelberg, Germany, 1990; Volume 84, pp. 32–44.

53. Lake, F.K.; Long, J.W. *Fire and Tribal Cultural Resources*; Gen. Tech. Rep. PSW-GTR-247; USDA Forest Service, Pacific Southwest Research Station: Albany, CA, USA, 2014.

54. Lynn, K.; Daigle, J.; Hoffman, J.; Lake, F.; Michelle, N.; Ranco, D.; Viles, C.; Voggesser, G.; Williams, P. The impacts of climate change on tribal traditional foods. *Clim. Chang.* **2013**, *120*, 545–556. [CrossRef]

55. Kershner, J.; Samhouri, J.F.; James, C.A.; Levin, P.S. Selecting indicator portfolios for marine species and food webs: A Puget Sound case study. *PLoS ONE* **2011**, *6*, e25248. [CrossRef] [PubMed]

56. Spies, T.A.; White, E.M.; Kline, J.D.; Fischer, A.P.; Ager, A.A.; Bailey, J.D.; Bolte, J.; Koch, J.; Platt, E.; Olsen, C.S.; et al. Examining fire-prone forest landscapes as coupled human and natural systems. *Ecol. Soc.* **2014**, *19*, 9. [CrossRef]

57. Timberlake, T.J.; Schultz, C.A. Policy, practice, and partnerships for climate change adaptation on US national forests. *Clim. Chang.* **2017**, *144*, 257–269. [CrossRef]

58. Timberlake, T.J.; Schultz, C.A. Climate change vulnerability assessment for forest management: The case of the U.S. forest service. *Forests* **2019**, *10*, 1030. [CrossRef]

59. James, C.A.; Kershner, J.; Samhouri, J.; O'Neill, S.; Levin, P.S.; Kershner, J. A methodology for evaluating and ranking water quantity indicators in support of ecosystem-based management. *Environ. Manag.* **2012**, *49*, 703–719. [CrossRef]

60. Green, D.; Niall, S.; Morrison, J. Bridging the gap between theory and practice in climate change vulnerability assessments for remote Indigenous communities in northern Australia. *Local Environ.* **2012**, *17*, 295–315. [CrossRef]

61. Eriksen, S.E.H.; Aldunce, P.; Bahinipati, C.S.; Martins, R.; Molefe, J.I.; Nhemachena, C.; O'Brien, K.; Olorunfemi, F.; Park, J.; Sygna, L.; et al. When not every response to climate change is a good one: Identifying principles for sustainable adaptation. *Clim. Dev.* **2011**, *3*, 7–20. [CrossRef]

62. Gustafson, S.; Cadena, A.J.; Ngo, C.C.; Kawash, A.; Saenghkaew, I.; Hartman, P. Merging science into community adaptation planning processes: A cross-site comparison of four distinct areas of the Lower Mekong Basin. *Clim. Chang.* **2017**, *149*, 91–106. [CrossRef]

63. Smajgl, A.; Ward, J. Evaluating participatory research: Framework, methods and implementation results. *J. Environ. Manag.* **2015**, *157*, 311–319. [CrossRef]

64. Whyte, K. Indigenous climate change studies: Indigenizing futures, decolonizing the anthropocene. *Engl. Lang. Notes* **2017**, *55*, 153–162. [CrossRef]

65. Fillmore, H.M. TahoeLand Live! *National Public Radio TahoeLand Podcast.* Romero, E.D., Ed.; Available online: http://www.capradio.org/news/tahoeland/2019/10/10/tahoeland-live/ (accessed on 18 October 2019).

66. Berry, P.; Enright, P.M.; Shumake-Guillemot, J.; Prats, E.V.; Campbell-Lendrum, D. Assessing health vulnerabilities and adaptation to climate change: A review of international progress. *Int. J. Environ. Res. Public Health* **2018**, *15*, 2626. [CrossRef] [PubMed]

67. Avelino, J.E.; Crichton, R.; Valenzuela, V.P.B.; Odara, M.G.N.; Padilla, M.A.T.; Kiet, N.; Danh, T.N.; Van, P.C.; Bao, H.D.; Linh, M.T.Y.; et al. Survey tool for rapid assessment of socio-economic vulnerability of fishing communities in Vietnam to climate change. *Geosciences* **2018**, *8*, 452. [CrossRef]

68. Amuzu, J.; Jallow, B.P.; Kabo-Bah, A.T.; Yaffa, S. The climate change vulnerability and risk management matrix for the coastal zone of the Gambia. *Hydrology* **2018**, *5*, 14. [CrossRef]

69. Younus, A.F. An assessment of vulnerability and adaptation to cyclones through impact assessment guidelines: A bottom-up case study from Bangladesh coast. *Nat. Hazards* **2017**, *89*, 1437–1459. [CrossRef]

70. O'Brien, K.; Eriksen, S.; Nygaard, L.P.; Schjolden, A.N.E. Why different interpretations of vulnerability matter in climate change discourses. *Clim. Policy* **2007**, *7*, 73–88. [CrossRef]

71. Williams, S.E.; Shoo, L.P.; Isaac, J.L.; Hoffmann, A.A.; Langham, G. Towards an integrated framework for assessing the vulnerability of species to climate change. *PLoS Biol.* **2008**, *6*, 2621–2626. [CrossRef]

72. Aziz, T.; Van Cappellen, P. Comparative valuation of potential and realized ecosystem services in Southern Ontario, Canada. *Environ. Sci. Policy* **2019**, *100*, 105–112. [CrossRef]

73. Taylor, A.H.; Trouet, V.; Skinner, C.N.; Stephens, S. Socioecological transitions trigger fire regime shifts and modulate fire-climate interactions in the Sierra Nevada, USA, 1600–2015 CE. *Proc. Natl. Acad. Sci. USA* **2016**, *113*, 13684–13689. [CrossRef]

74. Walsh, M.K.; Duke, H.J.; Haydon, K.C. Toward a better understanding of climate and human impacts on late Holocene fire regimes in the Pacific Northwest, USA. *Prog. Phys. Geogr.* **2018**, *42*, 478–512. [CrossRef]

75. Gomes, T.C. Novel ecosystems in the restoration of cultural landscapes of Tl'chés, West Chatham Island, British Columbia, Canada. *Ecol. Process.* **2013**, *2*, 15. [CrossRef]

76. Augustine, S.; Dearden, P. Changing paradigms in marine and coastal conservation: A case study of clam gardens in the Southern Gulf Islands, Canada. *Can. Geogr. Géogr. Can.* **2014**, *58*, 305–314. [CrossRef]

77. Guerrero-Gatica, M.; Aliste, E.; Simonetti, J.A. Shifting gears for the use of the shifting baseline syndrome in ecological restoration. *Sustainability* **2019**, *11*, 1458. [CrossRef]

78. Egan, D. Ecological restoration and sustainable development. *Ecol. Restor.* **2003**, *21*, 161–162. [CrossRef]

79. Tomblin, D.C. The ecological restoration movement. *Organ. Environ.* **2009**, *22*, 185–207. [CrossRef]

80. Nagy, G.J.; Filho, W.L.; Azeiteiro, U.M.; Heimfarth, J.; Verocai, J.E.; Li, C. An assessment of the relationships between extreme weather events, vulnerability, and the impacts on human wellbeing in Latin America. *Int. J. Environ. Res. Public Health* **2018**, *15*, 1802. [CrossRef]

81. Kelly, P.M.; Adger, W.N. Theory and practice in assessing vulnerability to climate change and facilitating adaptation. *Clim. Chang.* **2000**, *47*, 325–352. [CrossRef]

82. Whitney, C.; Bennett, N.J.; Ban, N.C.; Allison, E.; Armitage, D.; Blythe, J.; Burt, J.; Cheung, W.W.L.; Finkbeiner, E.M.; Kaplan-Hallam, M.; et al. Adaptive capacity: From assessment to action in coastal social-ecological systems. *Ecol. Soc.* **2017**, *22*. [CrossRef]

83. Creed, I.F.; Van Noordwijk, M. *Forest and Water on a Changing Planet: Vulnerability, Adaptation and Governance Opportunities*; A Global Assessment Report; International Union of Forest Research Organizations: Vienna, Austria, 2018.

84. Marks-Block, T.; Lake, F.K.; Curran, L.M. Effects of understory fire management treatments on California Hazelnut, an ecocultural resource of the Karuk and Yurok Indians in the Pacific Northwest. *For. Ecol. Manag.* **2019**, *450*, 117517. [CrossRef]

85. Long, J.W.; Gray, A.N.; Lake, F.K. Recent trends in large hardwoods in the Pacific Northwest, USA. *Forests* **2018**, *9*, 651. [CrossRef]

86. Lemos, M.C.; Kirchhoff, C.J.; Ramprasad, V. Narrowing the climate information usability gap. *Nat. Clim. Chang.* **2012**, *2*, 789–794. [CrossRef]

87. Hildén, M.; Jordan, A.; Huitema, D. Special issue on experimentation for climate change solutions editorial: The search for climate change and sustainability solutions—The promise and the pitfalls of experimentation. *J. Clean. Prod.* **2017**, *169*, 1–7. [CrossRef]

88. Oakes, L.E.; Hennon, P.E.; Ardoin, N.M.; D'Amore, D.V.; Ferguson, A.J.; Steel, E.A.; Wittwer, D.T.; Lambin, E.F. Conservation in a social-ecological system experiencing climate-induced tree mortality. *Biol. Conserv.* **2015**, *192*, 276–285. [CrossRef]

89. Kmoch, L.; Pagella, T.; Palm, M.; Sinclair, F. Using local agroecological knowledge in climate change adaptation: A study of tree-based options in Northern Morocco. *Sustainability* **2018**, *10*, 3719. [CrossRef]

Article

From Single-Use Community Facilities Support to Integrated Sustainable Development: The Aims of Inter-Municipal Cooperation in Poland, 1990–2018

Marek Furmankiewicz [1,*] and Adrian Campbell [2]

[1]   Department of Spatial Economy, The Faculty of Environmental Engineering and Geodesy,
      Wroclaw University of Environmental and Life Sciences, Grunwaldzka 55, 50-357 Wrocław, Poland
[2]   International Development Department, School of Government and Society, University of Birmingham,
      Birmingham 15 2TT, UK; a.campbell@bham.ac.uk
*   Correspondence: marek.furmankiewicz@upwr.edu.pl

Received: 4 October 2019; Accepted: 21 October 2019; Published: 23 October 2019

Abstract: The paper explores and compares the aims of the three most common legal forms of inter-municipal cooperation in Poland (engaging rural, urban-rural and urban municipalities) during the years 1990–2018: Mono-sectoral Special Purpose Unions, Municipal Associations and cross-sectoral Local Action Groups. Content analysis was applied and development priorities from the statutes and strategies were studied. The main form of territorial association evolved from, initially, mono-functional bodies concerned mainly with local infrastructural investment and managed solely by a group of local authorities, to a devolved type, consisting of multi-purpose associations managed with the participation of economic and third sector representatives. This was the result of the European Union policy of promoting territorial governance and integrated development in functional regions, this being considered as part of the process of Europeanisation. However, these successive forms of municipal cooperation do not appear to have replaced the pre-existing forms, but they have introduced additional modes of governance of local resources. The findings show that the most "integrated" and "sustainable" management of local resources is observed mainly in cross-sectoral partnerships, like Local Action Groups, but not so often in mono-sectoral municipal unions and associations led solely by local government and focused more on hard infrastructure and municipal facilities. However, given the specialisation shown by each of the three types of association, it is likely that the full range of development possibilities in the areas concerned can only be realised if all three forms of cooperation are present. The analysis confirms the positive role of local economic and social sector participation in shaping sustainable development. The findings also indicate the utility of the concept of cross-sectoral territorial partnerships in post-socialist and post-authoritarian countries lacking traditions of grassroots or participative development.

Keywords: inter-municipal cooperation; cross-sectoral partnerships; place-based and integrated development; sustainable management; integrated planning index; Poland

---

1. Introduction

In recent decades, best practice in public policy and administration has been declared to be undergoing a gravitational shift from top-down state intervention to bottom-up participation and co-management with collaboration across agency, state and non-government organisation (third sector) boundaries, a trend associated with the growing popularity of the territorial governance concept [1–3]. It has been seen to support decentralisation (empowering local government) and devolution (empowering non-governmental stakeholders in public sector decision making). This may be seen as facilitating a "neo-endogenous" or "integrated place based" model of development, in contrast to a top-down, exogenous and sectoral model focusing on single issues such as transport or agriculture and managed solely by government [4–6]. The place-based, neo-endogenous model implies central government support for regional and local bottom-up initiatives for development and well-being, facing one particular territory (natural or economic region, agglomeration, etc.), often requiring cooperation between stakeholders from fragmented administrative areas [7–9]. Integrated planning in this context is understood to involve drawing together local social, economic and ecological issues, rather than focusing only on, for instance, technical infrastructure [10]. It is often closely related to "sustainable" development, which involves an additional concern with long-term retention of local natural resources [11–13]. In the countries where these processes first emerged, they were rooted in pre-existing democratic traditions, whereas in the European post-socialist countries such principles have largely developed since the beginning of the 21st century where they were disseminated, mainly with the support of European Union (EU) policy and funding as a means of achieving social, economic and environmental regeneration [14–16].

Cooperation between democratically elected local governments in the management of local resources, know-how exchange and common interest representation has a long tradition in Europe and takes many domestic and cross-border forms (see, for example, [17–19]). These range from simple networks and contractual agreements to dedicated structures, which may be legal entities in their own right, for politically-sensitive or capital-intensive tasks [20–22]. However, from the 1980s onwards, a new, broader model of territorial governance and development based on cross-sectoral partnership between representatives of local governments, non-governmental organisations, private firms and state institutions began to take the place of traditional mono-sectoral or hierarchical approaches, the latter being increasingly seen as remote from local concerns, as well as being weak on sustainability [23].

Integrated and sustainable development are frequently cited as aims of EU programs and projects related to territorial cooperation in functional areas, not least in Poland [7,24]. However, the question arises as to how far these fashionable slogans, proclaimed to facilitate access to European Union funds, are actually reflected in the objectives and tasks of the territorial organisations created by these funds, with the participation of local authorities. Is the shift from focusing on communal infrastructure to more sustainable and integrated management of local resources, taking into consideration environmental, business, cultural and social issues, perceptible in the priorities espoused by collaborative municipal organisations? To answer this question we will assess the evolution of municipal cooperation objectives, based on the Polish case and its three most common forms of territorial cooperation: Special Purpose (municipal) Unions, Municipal Associations and "special" (cross-sectoral) association (Local Action Groups) and their respective aims, as listed in their statutes or strategies. We seek to evaluate how the change of policy context was taken into account in the reforms and policy changes implemented. Local government cooperation and cross-sectoral partnerships are often considered and discussed separately in the literature [18,25,26]. However, they are both forms of municipal cooperation, only differing in the degree of local community participation in decision-making. In this paper we compare both kinds of territorial inter-municipal cooperation, thereby filling what appears to be a gap in the current literature.

The following section gives a brief overview of the emergence of the place-based, integrated and sustainable development concept, which is strongly connected to the concepts of neo-endogenous development and territorial (collaborative) governance. Next, the legislative, institutional and economic basis of municipal cooperation in Poland in the years 1990–2018 is analysed in terms of the main aims of municipal unions and associations. We found that the evolution from mono-functional authority-led to multi-purpose partnership-based co-operation did not occur gradually but was directly initiated by administrative reforms or by EU policy and, moreover, that the three forms of cooperation typically differ in their development priorities.

2. Conceptual Background

Regarding economic development strategies, two basic models have been identified: exo- and endogenous development [27]. The first emphasizes top-down management using external technology, capital and human resources, often organised in separate, sectoral agencies under central government control (e.g., agriculture, industry, culture). The second concept, the endogenous (or grass-roots) approach, encourages bottom-up development factors, trusting in the sustainable use of local resources for socio-economic development by cooperating residents [28,29].

The exogenous model of sectoral development policies was common in democratic countries in the mid-20th century and dominated in the socialist (or "communist") countries of Central and Eastern Europe to such an extent that independent local and regional self-government did not exist, in Poland for example, until these were re-introduced in 1990 [30–32]. However, in the Western liberal democracies, in the 1980s, that model of development policy began to be perceived as ineffective and even detrimental, leading to regional disparity, environmental damage and the degradation of nature [33–35]. As a result, a neo-endogenous approach, based on territorial cross-sectoral cooperation of local stakeholders (i.e., on territorial governance processes) began to gain favour [4,36]. The model of territorial governance as a horizontal, cross-sectoral regional network has tended to be most pronounced in the Anglo-Saxon countries and has remained popular through much of Western Europe where traditions of co-operative governance are strong, not least because the state is sufficiently powerful to be confident of maintaining the initiative in partnerships [26,37,38]. In developing countries the notion of collaborative governance has been widely promoted as a means of legitimising and strengthening the role of civil society in the management of public goods in the absence of sufficient local state capacity or for increasing state accountability [39,40]. Collaborative territorial governance may be seen as an instrument not only for the production of public goods and management of local, limited resources, but also for changing the nature and logic of state-society relations in a given country [41]. It was disseminated throughout most European post-socialist countries from the early 21st century, mainly after their accession to the European Union [42,43]. The term "endogenous development" is considered as more or less synonymous with the "grassroots", "bottom-up" or "participatory" approach to management of local resources and is closely related to "sustainable development" in policy practice [44,45].

The advantages attributed to local stakeholders' territorial cooperation encouraged EU policy makers to promote neo-endogenous and place-based integrated development, in which such governance processes have an important role, and have similar features and are considered more sustainable than the top-down exogenous approach [23,46,47]. These models seek to construct a hybrid of top-down and bottom-up development by incorporating top-down support for bottom-up territorial governance in functional areas [9]. According to Ray [4], neo-endogenous development has a strong emphasis on local resources, capacities, perspectives, needs and knowledge but it also involves hierarchical management of incentives and the replacement of the principle of "administrative territory" by a functional principle, with inter-sectoral co-operation through regional networks and partnerships and increased self-governing responsibilities at regional or local levels [48]. In other words, these models assume the mobilisation of local stakeholders, their knowledge, demands and concerns by instruments and territorial organisational forms determined by central government, for example, via financial grants with territorial partnership creation as conditionality. The approach is seen to further the European Union's objectives of smart, inclusive and sustainable growth [45].

The term "place-based development" was popularised initially by the Organisation for Economic Co-operation and Development (OECD) report [23] and the Barca report to the European Union [49] as a means of social inclusion and participatory and fair process in development policy. It implied territorially sensitive, multi-sectoral approaches, involving multilevel governance and co-ordination of all relevant levels, policies and stakeholders [50,51]. The approach informed the creation of area-based development partnerships in functional regions, managed by diverse local stakeholders, for instance EU Local Action Groups in rural and urban areas within the framework of Community-Led Local Development [52–54], Integrated Territorial Investment Partnerships in metropolitan areas [15,55], "Urban Action Europe" (URBACT) Local Support Groups in cities [56] and others. Place-based development, according to Zaucha and Świątek [7], is a long-term development strategy aimed at reducing persistent inefficiency (underutilisation of the full potential) and inequality (the share of people below a given standard of well-being and/or extent of interpersonal disparities) in specific places. It can be attained through the production of bundles of integrated, place-tailored public goods and services, designed and implemented by eliciting and aggregating local preferences and knowledge through participatory political institutions, and by establishing linkages with other places. It can be promoted from outside by a system of multilevel governance in which grants, subject to conditionality on both objectives and institutions, are transferred from higher to lower levels of government [57]. Both place-based and neo-endogenous development emphasise the need to take into simultaneous consideration the social, economic and environmental needs of the locality, through a cooperative network of different local and external stakeholders, an approach which is termed as "sustainable" or "integrated" [45,55,58].

This strong emphasis on stimulating cooperation in functional regions, which often cross administrative boundaries, means that various forms of inter-communal networks become a common feature of place-based or neo-endogenous development initiatives. The main aims of local government are to fulfil local residents' needs and the sustainable well-being of their territory [59,60]. However, if local self-governments are focused on activities inside their territory, why should they cooperate? There are two main reasons; the need to deal with problems that go beyond the boundaries of the territorial unit and the opportunity to obtain additional benefits resulting from group work [61–63]. Collaboration can facilitate economies of scale, the attainment of "critical mass" in specialised activities, synergy, policy learning and better quality of municipal services [64–66]. Inter-municipal, territorial cooperation helps cope with problems that are not contained within the local administrative boundary [67,68]. According to Bennett [69], many administrative structures are "under-bounded": the functional activity space crosses over many small communal or local government boundaries (i.e., lakes, rivers, public transport areas). Local communities' cooperation is also very important in sustainable management of territorially limited resources [70]. As a result, local area-based cooperation is common, especially given that, in a purely territorial system, each community interacts much more with neighbouring units than with those who are far away [71]. In planning practice, a territory can be delimited by a grouping of dominant functions, not only by administrative division, creating a functional region [10,72]. The negative effects of local self-government fragmentation into a large number of small municipalities may encourage local authorities to pool the management of local resources, co-operation being the only practical alternative to amalgamation as a solution for this problem, especially if the many types of functional areas that overlap each other are taken into account [73].

In traditional democratic governmental systems, local authorities have carried out their responsibilities through their own sectoral cooperation, which is a common historical phenomenon given that they are typically part of the same functional local economy [71]. The empirical data shows that local authorities in many developing countries focus on infrastructural and economic issues [72,74]. However, the needs and ideas of local inhabitants may go beyond visible infrastructure investment. As a result, the newer concepts of development and governance support also cross-sectoral cooperation, gathering public as well as private stakeholders (both business and NGOs) to take into consideration their different needs. Area-based partnership in functional regions is seen to create better mutual understanding, added value, increased budgets and a reduction in duplication of provision [41,75].

In the context of EU policy, partnerships were expected to enable co-ordination of individual actions and information exchange leading to more sustainable and integrated development, not only economic growth [24,45,76]. The reality has not always conformed to this idealistic template. The wider literature testifies to the existence of partnership governance problems: Interest group conflicts, state domination, weak accountability, possible clientelism, exclusion of the weakest local stakeholders' needs—especially in post-socialist countries [77–79] and Mediterranean countries [80], but, to a lesser extent, also in the older western democracies [81]. However these constraints are not the main subject of this paper.

How far is the dissemination of sustainable and integrated development policies, based on territorial cooperation in functional regions supported by external programmes, actually reflected in the goals and priorities of inter-municipal organisations in, for example, Poland? Did the changing development preferences of inter-municipal cooperation over the last three decades lead to observable differences between their mono-sectoral and cross-sectoral forms?

The case of Poland, which started deep transformation in 1990 after more than 40 years of a top-down, centralist socialist system, may be instructive for countries that are preparing or implementing reforms after a period of centralist, populist or authoritarian rule, not only in Central and Eastern Europe, characterised by similar historical-cultural conditions, but also further afield, wherever top-down exogenous development policies have been applied. The results of our analysis may also be useful for policy makers and managers of development programs in the European Union. This is especially important because various programs and projects are already operating to disseminate methods of territorial partnership governance or place-based development at the local and sub-regional level in newer democracies, for example, in Georgia [82]. Our results are also an argument not to consider mono-sectoral cooperation of local governments and cross-sectoral territorial partnerships with the participation of a minimum of two municipalities as separate phenomena, which until now was the norm in the literature discussing forms of inter-municipal cooperation. It also supports the thesis that wide cooperation of local communities in the management of local resources (participative governance) leads to more sustainable development.

3. Materials and Methods

We analysed three of the most common forms of domestic municipal cooperation in Poland: Special Purpose Unions (SPUs), Municipal Associations (MAs) and Local Action Groups (LAGs), which have the legal status of special associations and are a kind of cross-sectoral partnership. In the paper, we use the name "municipality" for all three kinds of local government in Poland; rural commune, rural-urban and urban municipality (*gmina: wiejska, miejsko-wiejska* and *miejska*). In the research, we omitted the commercial partnerships and foundations created by municipalities to maintain selected community services, local tourist organisations (some created without local government) and individual agreements of municipalities (which are not registered at national level). According to Kołsut [83], they constitute less than 18% of the instances of local self-government cooperation, so we can focus on formal associations and unions. We also omitted associations for international cooperation (i.e., Euroregional associations, town-twinning) and focused only on in-country co-operation. We also did not consider Integrated Territorial Investment Partnerships in urban agglomerations, because they have various legal forms and organisation. We analysed only organisations with first-level local governments' involvement, so inter-county (*powiat*) and inter-regional (*województwo*, voivodeship) cooperation were not covered by our comparative study.

We used inductive methods of content analysis from "social artefacts" (manuscripts) used initially in social sciences [84,85]. We analysed whether a given subject was taken into consideration in priorities and formal tasks listed in the statutes or (if these existed and were available) the main strategies of the analysed organisations. The subjects were categorised in terms of eight main types:

- Common spatial or strategic planning,
- Common interest representation or promotion (both tourist and investment promotion, lobbying of higher public authorities),
- Social or human capital development (social inclusion, education and training, support for local voluntary organisations and cultural affairs),
- Enterprise (entrepreneurship) or employment development (local product development, local services development, enhancing employment),
- Tourism or recreation (small tourist and sport infrastructure, support for agro-tourism),

- Social services or infrastructure (refurbishment of public buildings such as social welfare establishments, voluntary fire brigade buildings, pavements, playgrounds, neighbourhood schools and kindergartens, public transport services),
- Technical infrastructure or roads (water supply systems, gas and electricity networks, internet networks, car parks and roads),
- Environmental protection or ecological infrastructure (sewage systems, sewage treatment plants, thermo-modernization of public buildings, ecological waste management, nature conservation).

This methodology has a potential weakness in that the authors' subjectively assign freely formulated aims into strictly defined categories, and these might not show the real level of cooperation. However, it is considered acceptable to study development policy preferences from strategic or programming documents [86–88]. The LAGs concerned were each obliged to prepare a common strategy, so this was taken into consideration in Figures 4 and 5 in the results section, even when "common planning" was not listed among the aims. We consciously did not analyse the intensity of cooperation (which in SPUs can be much stronger and based on infrastructure construction and sustaining, than in MAs, focusing on small soft projects) but only the development preferences listed in the goals and tasks. We considered only LAGs with a minimum of two municipalities (communes) as members.

Research periods have been distinguished on the basis of the analysis of the year of establishment of organisations that existed in 2018 (Figure 1) and the year of new development policy implementation (reforms, joining the EU, EU programming periods, etc.). We have no comparable data about all liquidated (unregistered before 2018) units.

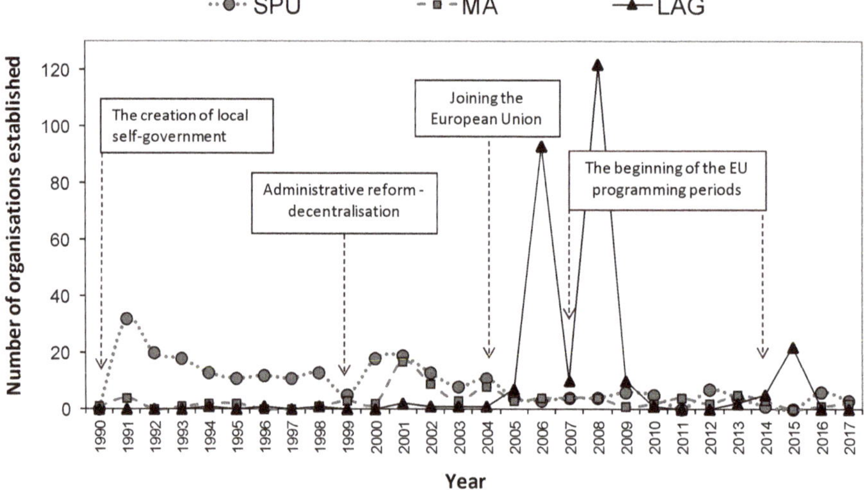

Figure 1. The number of analysed organisations established in a given year (only those existing in 2018 were analysed). SPU—Special Purpose Unions; MA—Municipal Associations; LAG—Local Action Groups. There is a visibly the growing number of establishments over some years (this is analysed in Section 4). Source: The Authors' own research.

Data was gathered from the registry of Special Purpose Unions run by the Polish Ministry of Internal Affairs and Administration, valid on 30.06.2018 (downloaded in January 2019), the list of Municipal Associations from the National Court Register (*Krajowy Rejestr Sądowy*, own query October-December 2018) and from the territorial partnership strategies published obligatory by the Local Action Groups, valid on 31.12.2011 and 31.12.2018 (downloaded in 2012 and 2019). These registers of different forms of cooperation contain similar data on the year in which cooperation was established and the associations' main aims or tasks, listed in their statutes or strategies. The last year fully considered for all researched forms of cooperation is 2017; however, most of the data are valid also for 2018 (for instance, there were no changes in the development priorities of studied organisations).

Additionally, we devised a simple integrated planning index (IPI, Tables 1 and 2 in Section 4) showing the average percentage of organisations choosing a given subject (the index is lowest when all organisations list only one subject, and highest where all organisations take into consideration all subjects in goals or tasks). The high IPI is considered as a feature of integrated and sustainable development.

4. Results

Local self-government, based on the rural commune, rural-urban and urban municipality (called *gmina*) in Poland was established according to the Local Government Act of 8 March 1990 and by subsequent legislation. A crucial role in installing both democracy and local cooperation was played also by the Law on Associations. This enabled both individuals and legal entities, including territorial self-government, to organize formal associations to deal with issues that concerned them. This widespread type of collaborative organisation may be termed municipal associations (MAs, *stowarzyszenie gmin*). The first MAs were initially the country-wide ones such as The Union of Polish Metropolitan Towns (1990), The Association of Polish Cities (1991), The Association of Rural Gminas of the Republic of Poland (1993) and others. Territorial associations of municipalities typically appeared later. The law did not permit them to provide basic communal services, so they focused on representation of common interests and on human capital building (Figure 2), typically education and trainings for clerks. There were only a little over a dozen MAs created across Poland before 1998.

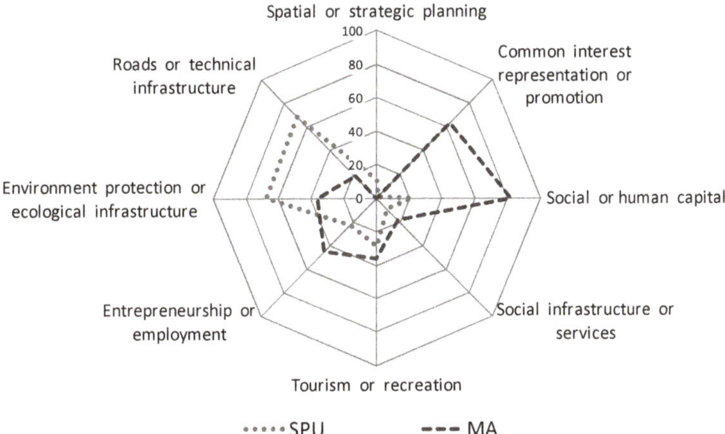

Figure 2. The percentage share of analysed organisations including a given subject in their priorities, by organisations created in the years 1990–1998: SPU—Special Purpose Unions; MA—Municipal Associations. The degree of specialisation in these forms of cooperation is evident. The Authors' own research.

Polish law also allowed individual municipalities to form Special Purpose Unions (*celowy związek międzygminny*). The SPUs can provide communal services on participants' territory (e.g., sewage treatment, public transportation, water supply) and have been the most numerous bodies of local territorial collaboration from the outset. During the years 1990–1998, between 13 and 47 SPUs were established annually to deal with public utilities, focusing on investment in hard technical and ecological infrastructure (Figure 2) like gas, water and sewage supply, waste dumping sites and roads, and some of them were even deregistered after investment completion. There were almost no cross-sectoral territorial partnerships involving non-state partners at that time. The literature gives only one example of a rural bottom-up, area-based partnership in four municipalities in south-east Poland [89].

From 1990 to the end of 1998, the only form of self-government was the municipal (*gmina*) level. The 49 voivodeships (*województwa*) were central government regional administration. On 1 January 1999, however, a three-tier system of government administration was introduced. The upper levels of self-government, named *powiat* (county, district), and *województwo* (voivodeship, region) were created [31,90]. These soon began to engage in various forms of cooperation.

After the reform, at the local level, the SPUs remained the main form of municipal territorial cooperation (from 5 to 23 SPUs were established yearly, 1999–2003), but the role of Municipal Associations (MAs) increased. At the end of 2004, according to researched sources, there were 247 SPUs and about 46 multi-purpose MAs. According to our data, in 2001 about 90% of inter-municipal networks were of a local (micro-regional) character, working most commonly on functional areas lower than the region, most frequently with several municipality members.

Most SPUs existing at that time had been created during the first half of the 1990s, while the MAs were established most frequently in the years 2000–2003. The SPUs have usually remained mono-functional, focusing on one or two targets (Figure 3), while MAs had three aims on average. According to our data, about 30% of MAs in 2001 had some development strategy. We can consider them as a little more related to the concept of integrated development, preferring realisation of long-term common targets in several domains. The main aims were similar as in the former analysed period, however, the SPUs reduced pressure on road construction and technical infrastructure, while MAs more often considered tourist development as one of the priorities for action.

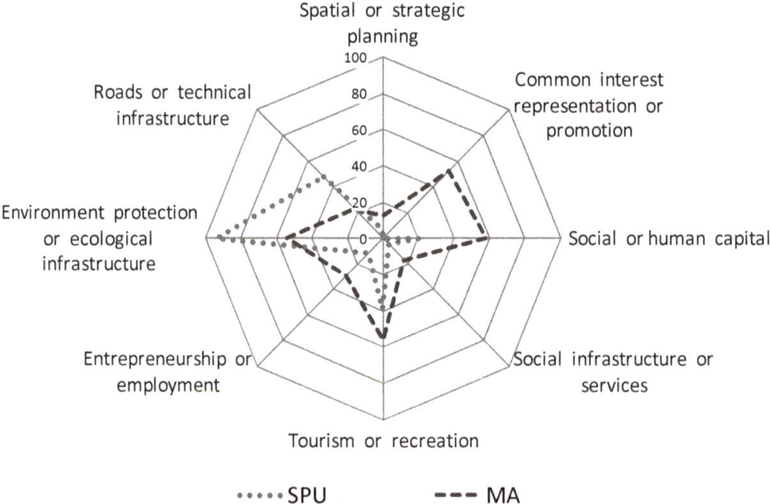

Figure 3. The percentage share of analysed organisations including a given subject in their priorities, by organisations created in years 1999–2004. SPU—Special Purpose Unions; MA—Municipal Associations. There were no formal LAGs in that period. The Authors' own research.

There were no formal LAGs in that period, however. A total of nearly 15–20 territorial partnership networks were established across the country, but only a few of these had legal status or formal structures such as a dedicated secretariat, so we cannot analyse their priorities from statutes or formal documents. Whilst the legal basis for municipal associations was straightforward, cross-sectoral partnership-based forms of cooperation faced a more difficult legal environment. The main problem was the absence of any legal framework of cooperation within which self-governments, voluntary organisations and business could participate on an equal basis, and there was often a lack of public government interest in co-operating with non-state actors in practice, especially if this meant transferring competences involving the management of public funds or having NGOs participating in decision making [91,92]. The Law on Associations allowed the creation of civil associations (in which only individual persons are full members) and, separately, municipal associations (in which only municipalities are members).

Poland's accession to the European Union on 1 May 2004 had major repercussions in terms of partnership creation. The EU programmes supported the creation of area-based, cross-sectoral rural development partnerships called LEADER Local Action Groups (LEADER is an acronym in French—*Liaison entre actions de développement de l'économie rurale*—meaning Links between actions for the development of the rural economy) and public municipal investment [93–95]. Initially, LAGs in Poland had to be a formal association, foundation or union of associations, on a compact territory inhabited by 10,000–100,000 persons, with a decision-making body in which the public sector could have no more than a 50 per cent share (LEADER + Pilot Programme [96]). In 2007, consequent to European Union demands and financial support, a new law enabled the creation of special associations with both individuals and different legal types of organisation as possible members in rural areas inhabited by 10,000-150,000 persons. This led to a total of 336 LAGs being registered across the country to participate in the EU Rural Development Programme 2007–2013 (initially 338 LAGs were created, but two did not eventually join the UE programme), which became the main form of co-operation with municipal engagement in rural areas; the most popular across the country. Most of them (324 LAGs) took the form of inter-municipal cooperation, with participation from 2 to 23 local governments in a cohesive functional region. They were obliged to create local development strategies. According to our study, LAGs in 2011 were focused on social or human capital, social infrastructure and services, entrepreneurship or employment and tourism or recreation (Figure 4). In comparison, SPUs established in this period focused continuously on environment protection and ecological infrastructure, while MAs focused on common interest representation and promotion. Comparing preferred goals of analysed forms of inter-municipal co-operation, we can see their specialisation in some issues.

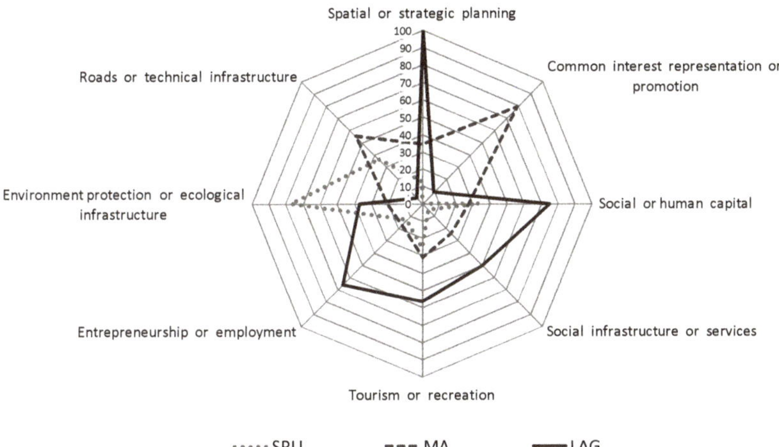

Figure 4. The percentage share of analysed organisations including a given subject in their priorities in units created in the years 2005–2013: SPU—Special Purpose Unions; MA—Municipal Associations; LAG—Local Actions Groups (data for 324 inter-municipal LAGs from 2011). The Authors' own research.

At the beginning of the EU 2014–2020 programming period, new rules and policies appeared. Collaboration in urban functional areas started to be a significant beneficiary of EU Integrated Territorial Investments (ITI) funding policy, pushing local governments to create unions or to sign agreements in 16 areas of voivodeships (regional) capitals (for detail see [15,55]). The main aim was to improve the quality of life in metropolitan functional regions consisting of urban and rural municipalities and impacted by urban sprawl [97–99]. The ITI Partnerships have various legal forms, most commonly municipal association and inter-municipal agreements, so they were not analysed separately in our research (a comparative study requires the separation of the forms studied). They focus on the development of public transport (including cycling paths), educational programmes for local inhabitants and thermo-modernisation of public buildings. They did not have strong cross-sectoral features, compared to LAGs.

In the case of LAGs, the central government new rules demanded cross-sectoral cooperation in a minimum of two municipalities (except urban LAGs) with 30,000–150,000 inhabitants (for more detailed rules see [53]). As a result, about 30 small LAGs had to be dissolved, and their member municipalities created new ones or joined larger neighbouring organisations. Less than 10 new SPUs and six MAs were created in the years 2014–2018.

In the middle of 2018, there were, in Poland, 322 LAGs (279 rural development, 35 piscatorial/fishing and eight urban in one municipality), 251 SPUs (243 inter-municipal and eight municipal-county) and nearly 92 MAs, so the cross-sectoral partnership remained the most common form of co-operation with municipal engagement.

Compared to the SPUs and MAs, the EU rules on inclusion of participants from business and the community sector in LAGs led to an increased emphasis on support for local business, social and human capital and social infrastructure (Figure 5, data from all organisations existing in 2018). However, the figure shows the extent of specialisation; SPUs deal mainly with technical and ecological infrastructure, whereas LAGs focus on social issues, local business, tourism and recreation. The LAGs are also more multi-purpose in character, because typically five to six types of aims were considered to be important for local integrated sustainable management, hence the integrated planning index (IPI) being higher for LAGs than for SPUs or MAs, even without a common "spatial or strategic planning" category, which is most commonly not stated in priorities but is obligatory for LAGs (Table 1).

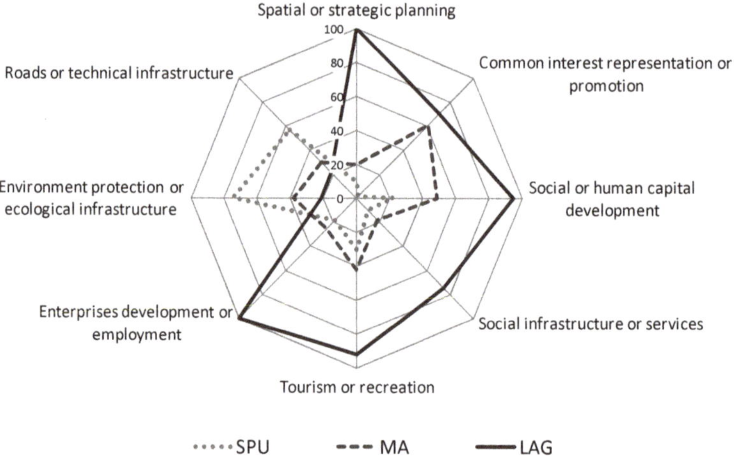

Figure 5. The percentage share of analysed organisations counting given subject in their priorities, by all analysed organisations existing in 2018: SPU—Special Purpose Unions; MA—Municipal Associations; LAG—Community-Led Local Development Local Actions Groups (data for 314 inter-municipal LAGs from 2018). The Authors' own research.

Table 1. The integrated planning index (IPI) in three main forms of associations engaging local municipalities in Poland (2018). SPU—Special Purpose Unions; MA—Municipal Associations; LAG—Local Action Groups.

SPU [1]	MA [2]	LAG [3]	Comments
27.7	35.4	71.9	With "spatial or strategic planning" category
30.3	37.6	59.6	Without "spatial or strategic planning" category

Source: Aims or main tasks listed in: ([1]) register of municipal SPUs and municipal-county SPUs (N = 321); ([2]) in statutes of MAs (N = 92); ([3]) in inter-municipal LAGs strategies (N = 314). The Authors' own research (2018).

Table 2. The integrated planning index (IPI) in organisations established in given research periods in three main forms of cooperation engaging local municipalities in Poland. SPU—Special Purpose Unions; MA—Municipal associations; LAG—Local Action Groups.

Period	SPU	MA	LAG	Mean—All Forms
1990–1998	28.7	37.5	No LAGs	33.1
1999–2004	27.8	37.8	No LAGs	32.8
2005–2013	25.3	36.2	49.7*	37.1
2014–2018	17.5	32.1	71.9**	40.5

*data for 324 inter-municipal LAGs existing in 2011; **data for 314 inter-municipal LAGs existing in 2018. Source: The Authors' own research (2019).

Analysis of IPI in the years 1990–2018 (Table 2) shows that the mean index increased from 33.1 to 40.5. The SPUs remain specialised in technical and ecological infrastructure (and it tends to increase, because IPI decreased from 28.7 to 17.5) and the MAs in common interest representation, tourist promotion and social and human capital development (with the most stable IPI; however, it decreased from 37.8 to 32.1). The LAGs take into consideration a much wider scope of works (IPI 71,9 in 2018), with special attention to business and social issues. Most likely, the LAGs took over part of the activities in the field of social issues and small business, and SPUs focused even more on large infrastructure investments, waste and transport management, which cannot be implemented by LAGs. The analysed data shows that the LAGs preferences are the closest to the concept of integrated and sustainable development, however, only simultaneous use of the analysed three forms of cooperation ensures that all analysed categories are covered by actions (visible on Figure 5).

5. Discussion

Our analyses have shown that mono-sectoral forms of inter-municipal cooperation in Poland differ from cross-sectoral, participative ones in terms of their local development priorities. The thesis was confirmed that participative territorial governance leads to a more balanced development, which takes account of various local social needs, without focusing only on infrastructure investments. This is partly the result of the requirements of the programs that control the creation of partnerships, but partly it can also be a bottom-up process, in line with Ostrom's theories [70]. However, since there are no similar comparative studies from other countries, it is difficult for us to assess whether this is a typical worldwide or a characteristic phenomenon mainly for Poland. Inter-Municipal Cooperation (IMC) analyses available in the literature suggest that former Eastern Bloc countries may have similar experiences, while countries with long democratic traditions and continually existing local government may slightly differ in the priorities of similar forms of territorial cooperation of local municipalities [18].

In this paper, the three main phases in the evolution of territorial cooperation in Poland have been identified, which can be categorised in terms of Bennett's stages of partnership cooperation development [61]:

- In the first phase after communism, the fragmentation of local self-governments created networks that were focused on fulfilling immediate physiological needs in underdeveloped areas (water and gas supply, waste management, etc.) and, accordingly, it was primarily sectoral, Special Purpose

Unions that were created (Bennet's "early network"). The priorities were related to the main municipal functions. The extension of market relations made municipalities group together for economies of scale in management of resources, especially where natural environmental areas and economic links crossed local administrative boundaries. There were some features of bottom-up territorial development, however, the voluntary and business sectors were not included widely in the process of strategic planning and decision making. The government, even when decentralised, did not want to lose power. In that period, local government tended to regard Non-Governmental Organisations (NGOs) as rivals and a potential threat to their own influence and, at best, as not to be trusted collaborators [100]. This is perhaps not surprising as NGOs and local governments have often been on opposing sides in spatial planning conflicts [101,102]. Democratisation after 1990 gave local communities in Poland new freedoms and responsibilities, but modes of behaviour associated with the socialist system did not disappear and tended to co-exist alongside the new organisation of the state, as noted by Regulska [103] and Grosse [92]. Nonetheless, voluntary organisations did develop strong relationships with local authorities so that by the end of the 1990s local authorities had emerged as the most common supporter of NGOs [100]. However, support did not mean wide engagement in the management of municipal resources.

- When, through national administrative reform measures, new levels of self-governance were created in 1999—the districts (*powiat*) and regional councils (*województwo*)—local municipalities were mobilized to establish the regional and local associations to enhance their capacity to lobby regional and central government. The new bodies engaged more with business, tourism and social and human capital than had their predecessors. However, they remained mono-sectoral, municipal organisations with a relatively narrow list of aims and can be considered also as examples of Bennet's "early network".

- The last phase began after accession to the European Union, with minor changes at the beginning of EU programming periods. The creation of new forms of territorial cooperation with local municipality engagement were clearly stipulated by the EU policies and funds (Bennett's phase of "developing trust" between collaborators, with external agency support). The new forms were created mainly at the beginning of the support programmes, which provided a powerful incentive for their organisation. Such arrangements had scarcely been established prior to funding being available, so cannot be considered as a genuine national self-organisation process, but rather as policy transfer of the concept of place-based development. It can be considered also as "Europeanisation"; in this case, a conscious process of territorial development policy transfer to post-socialist countries led by European Union policymakers [43,104]. Funds and programmes originating in Western Europe played a significant role in mobilising local stakeholders to create partnerships and provide technical assistance, funding and investment possibilities. This process was common across all new EU member post-socialist states in Central and Eastern Europe [78,105,106].

This evolution of cooperation can be considered as a "learning process" of the partnership formed management of local resources as stated by Bennett [61]. He compares different levels of effectiveness of partnerships (on the vertical axis) with the time it takes to develop effective working relationships (on the horizontal axis). After the initial stage of isolation, the following stage is development of expertise and experience (early networks for specific projects, in our case SPUs and MAs). Often in later stages, a broad range of agents interested in the development of the local economy may be linked together in effective commitment. At that stage, an external force such as a development agency or other institution (e.g., an external support programme; in our case, EU programmes) may play a central role, mobilising for cooperation, organising resources, providing technical consulting, etc. This can be referred to as the stage of trust development. A facilitator in the form of a neutral third party or a most active participant who serves as the initiator of the collaboration is critical in the formation and success of co-operative efforts [71].

We can deal with the most advanced stage of territorial partnership development (Bennet's "mature" stage) when it will be capable of functioning even without external financial support, based on local resources. However, it is difficult to say how many LAGs would survive after the end of support. In Pawłowska's survey, between 518 members from 26 LAGs in the Sub-Carpathian region in Poland, only 32% of respondents were sure that the LAGs can survive after closing EU financial support [107]. Such studies, assessing the extent to which Europeanisation processes are persistent, may be carried out after Britain's exit from the European Union. However, we must take into consideration that public-private partnerships for local development had a longer tradition in Anglo-Saxon countries, like the United Kingdom, than EU LEADER type programmes supporting rural development partnerships [26].

In many of the LAG strategies analysed, especially in 2011, the objectives were formulated as the names of the main budget lines of the support program. This suggests that the authors of the strategy first adapted to the program requirements to obtain funds more easily than to real local needs, which were then "tailored" to the requirements of the support program. According to some authors, this is a deliberate operation in place-based and neo-endogenous development concepts [7,48]. It assumes "steering" local initiatives with a "reward" in the form of financial resources, which can be largely associated with the theory of rational choice [108]. There are no restrictions or orders to act, as in authoritarian systems, but funds are obtained only for actions designated by central authorities in the support program procedures. Therefore, it cannot be considered as a classic grass-roots development, where local resources are used by residents completely according to their needs and priorities [29]. The mass establishment of LAGs in all EU countries partly confirms the effectiveness of such motivation of local elites for territorial cooperation [52].

In the current partnership model in Poland and other post-socialist countries, local authorities have consistently played an important role, although sometimes with features of tokenism [109–111]. However, in the management of common resources, there was a significant increase in the formal involvement of business and the voluntary sector. For local post-socialist communities, this stage can be considered as the learning process of new models of the integrated development and territorial governance [61,112]. However, also in developed countries, partnership relations are considered as a positive "adaptive co-management" based on self-learning social networks [113]. In this stage, the role of cultural and social issues increased in territorial strategies, with a significant movement towards the sustainable development idea [45].

The research findings show increased attention to social and local small business affairs in the new form of cross-sectoral collaboration, whereas the initial sectoral, inter-municipal cooperation had been focused mainly on technical infrastructure and public utilities and had been weakly integrated and was low on sustainability. These findings are similar to those of LAGs studies in EU post-socialist member states [12,114]. The mono-sectoral inter-municipal cooperation in this region are most often focused on the development and modernization of infrastructure necessary for directly carrying out their tasks, for instance in the Czech Republic, Slovenia and Romania [62,72,115]. However, according to Potkański [74], such an approach is short-sighted because, despite improving the quality of infrastructure and the momentary satisfaction of voters, it does not ensure long-term economic development, creating jobs or income sources for the local population. The LAGs helped to change the ordering of local priorities, paying greater attention to the role of social and human capital and small local business in local development. Partnership governance used in sustainable and integrated (place-based or neo-endogenous) development promotes the understanding that community development encompasses more than economic development. Wealth and technical infrastructure is an important element of quality of life, but it is only one element, and community development requires that quality of life issues are also addressed [24,45,116]. Bringing different social groups to a consensus is unlikely to be accomplished through a narrowly economic focus or infrastructure investments, as in SPUs.

As technical infrastructural issues became less urgent, after a dozen years of investment, issues such as social and human capital building, socio-cultural infrastructure and small business have become a major focus for area-based cross-sectoral cooperation. The LAGs have pursued a wider range of aims per organisation, so they claim with justification to have a more comprehensive (integrated or sustainable) approach than have mono-functional SPUs and MAs. Such observations are in line with Ostrom's concepts [70]. She found that for any group to manage common resources (e.g., aquifers, judicial systems, pastures) for optimal sustainable production there should be established cooperation relations with a small set of design principles, addressing how to cope with free-riding, solving commitment problems, arranging for the supply of new institutions and monitoring individual compliance with that set of rules. Ostrom found that a group can successfully manage local resources when they have defined their territory clearly, the rules of use of collective goods are well matched to local needs and conditions and when most individuals affected by these rules can participate in modifying the rules, which are respected by external authorities. The group should have a common system of cooperation for monitoring members' behaviour (and the community members themselves should undertake this monitoring, using a graduated system of sanctions for users who break the rules) and the community members should have access to low-cost conflict resolution mechanisms. LAGs, although their features have been partly imposed by the requirements of the support program, partly confirm this approach. Despite the problems noted regarding power relations and participation of non-public entities, they clearly enable meeting a greater range of local communities' needs and are characterised by more balanced and integrated local development planning than in the case of mono-sectoral cooperation of public authorities. However, it is probably more typical for post-socialist new EU member states, because in the established democracies, IMC deals with social and cultural issues to a greater extent [117–119]. For example, in Switzerland, municipal cooperation focused in the 1990s on school issues and health care, followed by sewage system and waste disposal [20]. However, the differences could be results of formal competences and public administration organisation, not only of local government.

These analyses also indicate that in Poland changes in methods of co-management of local resources have not been of a gradual and evolutionary character, as in many countries with long local democratic traditions [120,121], but rather have a "stepping" character. Each phase of territorial co-operation in Poland involving municipalities was initiated by radical legislative changes (creation of self-government in 1990, administrative reform implemented in 1999, and joining the European Union in 2004), which created the impulse towards other forms of co-operation development. This shows a degree of inertia in administration systems, such that only a strong external impulse can initiate significant changes. Hence, the entry of post-socialist countries into the European Union was for them a very important factor that accelerated their transformation. Despite the inclination of some populist governments of these countries to limit democratic principles [122], or to take control on cross-sectoral partnerships [111], it is possible that some of the positive changes will take root in time sufficiently to survive possible future reductions in financial support for participatory forms of inter-municipal cooperation [123].

6. Conclusions

In this paper, we analysed the evolution of development priorities in three main forms of domestic municipal cooperation in Poland: mono-sectoral Special Purpose Unions and Municipal Association, and in cross-sectoral Local Action Groups, which are also common in other European countries. The main findings can be summarised in six groups:

- First, the process of area-based partnership creation in post-socialist EU member states was a conscious process of place-based or neo-endogenous development policy transfer, considered as an element of the process of Europeanisation. The rapid development of territorial partnerships is self-evidently the result of special EU financial support policy (supporting the theory of rational choice [108]) and it is therefore difficult to predict the long-term durability of the sponsored organisations if this support is reduced or terminated;

- Second, the evolution of inter-municipal cooperation in transitional post-socialist Poland has more of a step character than the longer-term evolution associated with countries with longer experience of democratic local self-government. The main stages were initiated by radical legislative changes (creation of self-government in 1990, administrative reform implemented in 1999, and joining the European Union in 2004 with its consequences in terms of changed policies in their programming periods);
- Third, the territorial cross-sectoral partnerships are most frequently forms of inter-municipal cooperation that have more sustainable and integrated features of development planning than mono-sectoral forms led solely by government, and are taking more account of social and small business issues;
- Fourth, the three forms of cooperation present some features of specialisation—SPUs focus most commonly on "hard" infrastructure and waste management; MAs on "soft" actions like common municipal representation and promotion and public administration development; LAGs on social and human capital building, small business development, tourism and recreation;
- Finally, the new forms of cooperation do not replace the old ones, but can and do exist simultaneously. However, the greater potential of area-based cross-sectoral partnerships (such as LAGs) has become increasingly evident.

According to the literature, cross-sectoral territorial partnerships can be effective in the implementation of integrated strategies for sustainable management of local resources in developing countries [124–126]. Even if scholars observe problems of power-relations in bottom-up governance [111,127], this learning process of cross-sectoral collaboration in functional areas could be important for more participative and sustainable planning in countries with weak democratic traditions. In addition, we recommend that we should not limit the overviews of inter-municipal cooperation to the mono-sectoral cooperation of local authorities, but also take into account inter-municipal, cross-sectoral partnerships, in which local authorities make decisions together with representatives of the social and economic sector.

Author Contributions: Conceptualization, M.F.; methodology, M.F.; software, M.F.; validation, M.F.; investigation, M.F.; formal analysis, M.F.; resources, M.F.; data curation, M.F.; writing—original draft, M.F. and A.C.; writing—review and editing, M.F. and A.C.; visualization, M.F.; supervision, M.F.; project administration, M.F.; funding acquisition, M.F.; overall share in works: M.F, 70%; A.C., 30%.

Funding: We are grateful to the Ministry of Science and Higher Education (MNiSW, Poland) for supporting open access publishing in the framework of the Statutory Funds of Department of Spatial Economy at the Wrocław University of Environmental and Life Sciences.

Conflicts of Interest: The authors declare no conflict of interest. The funders had no role in the design of the study; in the collection, analyses, or interpretation of data; in the writing of the manuscript, or in the decision to publish the results.

References

1. *Adaptative Co-menagement: Collaboration, Learning and Multi-Level Governance*; Armitage, D.F.; Berkes, F.; Doubleday, N. (Eds.) University of British Columbia: Vancouver, Canada, 2007.
2. Simard, J.-F.; Chiasson, G. Introduction: Territorial governance: A new take on development. *Can. J. Reg. Sci.* **2008**, *31*, 471–485.
3. Stead, D. The Rise of Territorial Governance in European Policy. *Eur. Plan. Stud.* **2014**, *22*, 1368–1383. [CrossRef]
4. Ray, C. Neo-endogenous rural development in the EU. In *Handbook of Rural Studies*; Cloke, P., Marsden, T., Mooney, P., Eds.; Sage: London, UK, 2006; pp. 278–291.
5. Canete, A.J.; Navarro, F.; Cejudo, E. Territorially unequal rural development: The cases of the LEADER Initiative and the PRODER Programme in Andalusia (Spain). *Eur. Plan. Stud.* **2018**, *26*, 726–744. [CrossRef]

6. Scott, M. Delivering Integrated Rural Development: Insights from Northern Ireland. *Eur. Plan. Stud.* **2002**, *10*, 1013–1025. [CrossRef]

7. Zaucha, J.; Świątek, D. *Place-Based Territorially Sensitive and Integrated Development*; Ministry of Regional Development: Warsaw, Poland, 2013.

8. George, C.; Reed, M.G. Operationalising just sustainability: Towards a model for place-based governance. *Local Environ.* **2015**, *22*, 1105–1123. [CrossRef]

9. Bosworth, G.; Annibal, I.; Carroll, T.; Price, L.; Sellick, J.; Shepherd, J. Empowering Local Action through Neo-Endogenous Development; The Case of LEADER in England. *Sociol. Rural.* **2016**, *56*, 427–449. [CrossRef]

10. Markowski, T. Territorial dimensions of integrated development policy—Expectations and challenges concerning planning and institutional systems. *Stud. Reg.* **2013**, *35*, 51–64.

11. Therese, B.; Sandström, C. Public-private partnerships in a Swedish rural context—A policy tool for the authorities to achieve sustainable rural development? *J. Rural Stud.* **2016**, *49*, 58–68.

12. Kis, K.; Gal, J.; Veha, A. Effectiveness, efficiency and sustainability in local rural development partnerships. *APSTRACT Appl. Stud. Agribus. Commer.* **2012**, *6*, 31–38. [CrossRef]

13. Gorlach, K.; Adamski, T.; Klekotko, M. Poland: Designing Nature and Resource Management Strategies. In *Rural Sustainable Development in the Knowledge Society*; Bruckmeier, K., Tovey, H., Eds.; Ashgate: Surrey, UK, 2009; pp. 187–202.

14. Macken-Walsh, Á.; Curtin, C. Governance and Rural Development: The Case of the Rural Partnership Programme (RPP) in Post-Socialist Lithuania. *Sociol. Rural.* **2013**, *53*, 246–264. [CrossRef]

15. Kaczmarek, T.; Kociuba, D. Models of governance in the urban functional areas: Policy lessons from the implementation of integrated territorial investments (ITIs) in Poland. *Quaest. Geogr.* **2017**, *36*, 47–64. [CrossRef]

16. Doitchinova, J.; Stoyanova, Z. Activation of local communities for development of rural areas. *Econ. Agric.* **2014**, *61*, 661–675. [CrossRef]

17. *Inter-Municipal Cooperation in Europe*; Hulst, R.; van Montfort, A. (Eds.) Springer: Dordrecht, The Netherlands, 2007.

18. *Inter-Municipal Cooperation in Europe: Institutions and Governance*; Teles, F.; Swianiewicz, P. (Eds.) Palgrave Macmilian: Cham, Switzerland, 2018.

19. Dołzbłasz, S.; Raczyk, A. Different Borders-Different Cooperation? Transborder Cooperation in Poland. *Geogr. Rev.* **2015**, *105*, 360–376. [CrossRef]

20. Steiner, R. The causes, spread and effects of intermunicipal cooperation and municipal mergers in Switzerland. *Public Manag. Rev.* **2003**, *5*, 551–571. [CrossRef]

21. Hulst, R.; van Montfort, A. Institutional features of inter-municipal cooperation: Cooperative arrangements and their national contexts. *Public Policy Adm.* **2012**, *27*, 121–144. [CrossRef]

22. Namyślak, B. Cooperation and Forming Networks of Creative Cities: Polish Experiences. *Eur. Plan. Stud.* **2014**, *22*, 2411–2427. [CrossRef]

23. OECD. *The Future of Rural Policy. From Sectoral to Place-Based Policies in Rural Areas*; Organisation for Economic Co-Operation and Development: Paris, France, 2003.

24. Shortall, S.; Shucksmith, M. Integrated rural development: Issues arising from the Scottish experience. *Eur. Plan. Stud.* **1998**, *6*, 73–89. [CrossRef]

25. Swianiewicz, P.; Gendźwiłł, A.; Krukowska, J.; Lackowska, M.; Picej, A. *Współpraca międzygminna w Polsce*; Wydawnictwo Naukowe Scholar: Warszawa, Poland, 2016.

26. *Local Partnerships for Rural Development. The European Experience*; Moseley, M.J. (Ed.) CABI Publishing: Wallingford Oxon, UK, 2003.

27. Lowe, P.; Murdoch, J.; Ward, N. Beyond endogenous and exogenous models: Networks in rural development. In *Beyond Modernisation: The Impact of Endogenous Rural Development*; van der Ploeg, J.D., van Dijk, G., Eds.; Van Gorcum: Assen, The Netherlands, 1995; pp. 87–105.

28. Slee, B. Theoretical aspects of the study of endogenous development. In *Born From within: Practice and Perspectives of Endogenous Rural Development*; van der Ploeg, J.D., Long, A., Eds.; Van Gorcum: Assen, The Netherlands, 1994; pp. 184–194.

29. Willis, K. *Theories and Practices of Development*; Routledge: Oxon, NY, USA, 2005.

30. Bański, J. Dilemmas for Regional Development in the Concepts Seeking to Develop Poland's Spatial Structure. *Reg. Stud.* **2010**, *44*, 535–549. [CrossRef]

31. Regulski, J. *Local Government Reform in Poland: An Insider's Story*; Open Society and Local Government and Public Service Reform Initiative: Budapest, Hungary, 2003; p. 263.

32. Coulson, A.; Campbell, A. Local government in central and Eastern Europe: Introduction. *Local Gov. Stud.* **2006**, *32*, 539–541. [CrossRef]

33. Bowler, I. Endogenous agricultural development in Western Europe. *Tijdschrift Economische Sociale Geografie* **1999**, *90*, 260–271. [CrossRef]

34. Dutkowski, M.; Parysek, J.J. Going green: Sustainable development as a model of socio-economic development in European post-communist countries. *Eur. Plan. Stud.* **1994**, *2*, 419–434.

35. Świąder, M. The implementation of the concept of environmental carrying capacity into spatial management of cities: A review. *Manag. Environ. Qual.* **2018**, *29*, 1059–1074. [CrossRef]

36. Barke, M.; Newton, M. The EU LEADER Initiative and Endogenous Rural Development: The Application of the Programme in Two Rural Areas of Andalusia, Southern Spain. *J. Rural Stud.* **1997**, *13*, 319–341. [CrossRef]

37. Rhodes, R.A.W. The New Governance: Governing without Government. *Polit. Stud.* **1996**, *44*, 652–667. [CrossRef]

38. Peters, B.G.; Pierre, J. Governance Without Government? Rethinking Public Administration. *J. Public Adm. Res. Theory* **1998**, *8*, 223–243. [CrossRef]

39. Abers, R. From clientelism to cooperation: Local government, participatory policy, and civic organizing in Porto Alegre, Brazil. *Polit. Soc.* **1998**, *26*, 511–538. [CrossRef]

40. Ackerman, J. Co-Governance for Accountability: Beyond "Exit" and "Voice". *World Dev.* **2004**, *32*, 447–463. [CrossRef]

41. Thuesen, A.A.; Nielsen, N.C. A territorial perspective on EU's LEADER approach in Denmark: The added value of community-led local development of rural and coastal areas in a multi-level governance settings. *Eur. Countrys.* **2014**, *6*, 307–326. [CrossRef]

42. Chevalier, P.; Mačiulytė, J.; Razafimahefa, L.; Dedeire, M. The LEADER programme as a model of institutional transfer: Learning from its local implementation in France and Lithuania. *Eur. Countrys.* **2017**, *2*, 317–341. [CrossRef]

43. Dąbrowski, M. Shallow or deep Europeanisation? The uneven impact of EU cohesion policy on the regional and local authorities in Poland. *Environ. Plan. C Gov. Policy* **2012**, *30*, 730–745. [CrossRef]

44. Kanak, S.; Iiguni, Y. The role of social capital in endogenous development. *J. Rural Probl.* **2005**, *41*, 180–184. [CrossRef]

45. Dargan, L.; Shucksmith, M. LEADER and Innovation. *Soc. Rural.* **2008**, *48*, 274–291. [CrossRef]

46. Petrick, M. Reversing the rural race to the bottom: An evolutionary model of neo-endogenous rural development. *Eu. Rev. Agric. Econ.* **2013**, *40*, 707–735. [CrossRef]

47. Wellbrock, W.; Roep, D.; Mahon, M.; Kairyte, E.; Nienaber, B.; Dolores, M.; García, D.; Kriszan, M.; Farrell, M. Arranging public support to unfold collaborative modes of governance in rural areas. *J. Rural Stud.* **2013**, *32*, 420–429. [CrossRef]

48. Böcher, M. Regional Governance and Rural Development in Germany: The Implementation of LEADER+. *Soc. Rural.* **2008**, *48*, 372–388. [CrossRef]

49. Barca, F. *An Agenda for a Reformed Cohesion Policy. A Place-Based Approach to Meeting European Union Challenges and Expectations*; Economics and Econometrics Research Institute: Brussels, Belgium, 2009.

50. Mendez, C. The post-2013 reform of EU cohesion policy and the place-based narrative. *J. Eur. Public Policy* **2013**, *20*, 639–659. [CrossRef]

51. Churski, P.; Kociuba, D.; Ochojski, A.; Polko, A. Towards Policy—Place-Based Policy and Smart Specialisation. In *Measuring Regional Specialisation*; Kopczewska, K., Churski, P., Ochojski, A., Polko, A., Eds.; Springer: Berlin, Germany, 2017; pp. 267–380.

52. Konečný, O. The Leader Approach Across the European Union: One Method of Rural Development, Many Forms of Implementation. *Eur. Countrys.* **2019**, *11*, 1–16. [CrossRef]

53. Aldorfai, G.; Czabadai, L.; Topa, Z. An innovative methodology for supporting the CLLD. *Pol. J. Manag. Stud.* **2016**, *13*, 7–17. [CrossRef]

54. Servillo, L.; De Bruijn, M. From LEADER to CLLD: The Adoption of the New Fund Opportunities and of Their Local Development Options. *Eur. Struct. Invest. Funds J.* **2018**, *6*, 223–233.

55. Kociuba, D. Implementation of Integrated Territorial Investments in Poland—Rationale, Results, and Recommendations. *Quaest. Geogr.* **2018**, *37*, 81–98. [CrossRef]

56. Zajda, K.; Kołomycew, A.; Sykała, Ł.; Janas, K. *Leader and Community-Led Development Approach. Polish Experiences*; Wydawnictwo Uniwersytetu Łódzkiego: Łódź, Poland, 2017.
57. Faludi, A. Place is a no-man's land. *Geographia Polonica* **2015**, *88*, 5–20. [CrossRef]
58. Masot, A.N.; Alonso, G.C.; Moreno, L.M.C. Principal Component Analysis of the LEADER Approach (2007–2013) in South Western Europe (Extremadura and Alentejo). *Sustainability* **2019**, *11*, 4034. [CrossRef]
59. Hełdak, M.; Raszka, B. Evaluation of the Local Spatial Policy in Poland with Regard to Sustainable Development. *Pol. J. Environ. Stud.* **2013**, *22*, 395–402.
60. Stacherzak, A.; Hełdak, M.; Raszka, B. Planning Documents and Sustainable Development of a Commune in Poland. *Wit Trans. Ecol. Environ.* **2012**, *162*, 23–34.
61. Bennett, R.J. *Local Government in Post-Socialist Cities*; Open Society Institute, Local Government and Public Service Reform Initiative: Budapest, Hungary, 1997.
62. Săgeată, R. Inter-communal cooperation and regional development: The case of Romania. *Quaest. Geogr.* **2012**, *31*, 95–106. [CrossRef]
63. Kozlova, O.A.; Makarova, M.N. Inter-Municipal Cooperation as an Institution of Strategic Development of Territories. *Econ. Soc. Chang. Facts Trends Forecast* **2018**, *11*, 132–144.
64. Nunn, S.; Rosentraub, M.S. Dimension of interjurisdictional cooperation. *J. Am. Plan. Assoc.* **1997**, *63*, 205–219. [CrossRef]
65. Ruszkai, C.; Kovács, T. The Community Initiative LEADER I and the implementation and results of the Hungarian Pilot LEADER programme in rural development. *Bull. Geogr. Soc. Econ. Ser.* **2012**, *19*, 87–97. [CrossRef]
66. Bel, G.; Fageda, X.; Mur, M. Why Do Municipalities Cooperate to Provide Local Public Services? An Empirical Analysis. *Local Gov. Stud.* **2013**, *39*, 435–454. [CrossRef]
67. Szewrański, S.; Kazak, J.; Szkaradkiewicz, M.; Sasik, J. Flood risk factors in suburban area in the context of climate change adaptation policies—Case study of Wroclaw, Poland. *J. Ecol. Eng.* **2015**, *16*, 13–18. [CrossRef]
68. Szewrański, S.; Świąder, M.; Kazak, J.; Tokarczyk-Dorociak, K.; Van Hoof, J. Socio-Environmental Vulnerability Mapping for Environmental and Flood Resilience Assessment: The Case of Ageing and Poverty in the City of Wroclaw, Poland. *Integr. Environ. Assesment Manag.* **2018**, *14*, 592–597. [CrossRef]
69. Bennett, R.J. Administrative Systems and Economic Spaces. *Reg. Stud.* **1997**, *31*, 323–336. [CrossRef]
70. Ostrom, E. *Governing the Commons. The Evolutions of Institutions for Collective Action*; Cambridge University Press: Cambridge, UK, 1990.
71. Lackey, S.B.; Freshwater, D.; Rupasingha, A. Factors Influencing Local Government Cooperation in Rural Areas: Evidence from the Tennessee Valley. *Econ. Dev. Q.* **2002**, *16*, 138–154. [CrossRef]
72. Rus, P.; Nared, J.; Bojnec, Š. Forms, areas, and spatial characteristics of intermunicipal cooperation in the Ljubljana Urban Region. *Acta Geogr. Slov.* **2018**, *58*, 47–61. [CrossRef]
73. Swianiewicz, P. If territorial fragmentation is a problem, is amalgamation a solution? An East European perspective. *Local Gov. Stud.* **2010**, *36*, 183–203. [CrossRef]
74. *Współpraca jednostek samorządu terytorialnego narzędziem wsparcia polskiej polityki rozwoju*; Potkański, T. (Ed.) Związek Miast Polskich: Poznań, Poland, 2016.
75. Hutchinson, J. The Practice of Partnership in Local Economic Development. *Local Gov. Stud.* **1994**, *20*, 335–344. [CrossRef]
76. Bristow, G. Structure, Strategy and Space: Issues of Progressing Integrated Rural Development in Wales. *Eur. Urban Reg. Stud.* **2000**, *7*, 19–33. [CrossRef]
77. Zajda, K. Problems of functioning of Polish local action groups from the perspective of the social capital concept. *Eastern Eur. Countrys.* **2014**, *20*, 73–97. [CrossRef]
78. Bumbalová, M.; Takáč, I.; Tvrdoňová, J.; Valach, M. Are stakeholders in Slovakia ready for community-led local development? Case study findings. *Eur. Countrys.* **2016**, *2*, 160–174. [CrossRef]
79. Furmankiewicz, M. Współrządzenie czy ukryta dominacja sektora publicznego? Koncepcja governance w praktyce Lokalnych Grup Działania LEADER. *Studia Regionalne Lokalne* **2013**, *1*, 71–89.
80. Esparcia, J.; Escribano, J.; Serrano, J.J. From development to power relations and territorial governance: Increasing the leadership role of LEADER Local Action Groups in Spain. *J. Rural Stud.* **2015**, *42*, 29–42. [CrossRef]
81. Goodwin, M. The Governance of Rural Areas: Some Emerging Research Issues and Agendas. *J. Rural Stud.* **1998**, *14*, 5–12. [CrossRef]

82. *The LEADER Method. Transferring Experience of the Visegrad Group Countries to Georgia*; SykałaŁDej, M.; Wolski, O. (Eds.) Institute of Urban Development: Kraków, Poland, 2015.

83. Kołsut, B. *Zinstytucjonalizowane Sieci Współdziałania Międzygminnego*; Bogucki Wydawnictwo Naukowe: Poznań, Poland, 2015.

84. Prasal, D.B. Content Analysis. A method of Social Science Research. In *Research Methods for Social Work*; Lal Das, D.K., Bhaskaran, V., Eds.; Rawat: New Delhi, India, 2008; pp. 173–193.

85. Babbie, E. *The Basics of Social Research*, 5th ed.; Cengage Learning: Wadsworth, UH, USA, 2011.

86. Lošťák, M.; Hudečková, H. Preliminary impacts of the LEADER+ approach in the Czech Republic. *Agric. Econ. Zemědělská Ekonomika* **2010**, *56*, 249–265. [CrossRef]

87. Svobodová, H. Do the Czech Local Action Groups Respect the LEADER Method? *Acta Univ. Agric. Silvicult. Mendelianae Brunensis* **2015**, *63*, 1769–1777. [CrossRef]

88. Chen, C.; Li, D.; Man, C. Toward Sustainable Development? A Bibliometric Analysis of PPP-Related Policies in China between 1980 and 2017. *Sustainability* **2019**, *11*, 142. [CrossRef]

89. Furmankiewicz, M.; Thompson, N.; Zielińska, M. Area-based partnerships in rural Poland: The post-accession experience. *J. Rural Stud.* **2010**, *26*, 52–62. [CrossRef]

90. Meurs, M.; Kochut, R. Local government performance in rural Poland: The roles of local government characteristics and inherited conditions. *Landbauforschung Appl. Agric. Forest. Res.* **2014**, *64*, 163–178.

91. Dąbrowski, M. EU cohesion policy, horizontal partnership and the patterns of sub-national governance: Insights from Central and Eastern Europe. *Eur. Urban Reg. Stud.* **2014**, *21*, 364–383. [CrossRef]

92. Grosse, T.G. Social dialogue during Enlargement: The case of Poland and Estonia. *Acta Politica* **2010**, *45*, 112–135. [CrossRef]

93. Chmieliński, P. On Community Involevement in Rural Development—A Case of Leader Programme in Poland. *Econ. Soc.* **2011**, *4*, 120–128. [CrossRef]

94. Błąd, M.; Kamiński, R. Social capital enhancement in the Polish countryside: Experiences from the implementation of LEADER-type programmes. In *Development in the Enlarged European Union*; Zawalińska, K., Ed.; Institute of Rural and Agricultural Development, Polish Academy of Sciences: Warszawa, Poland, 2005; pp. 235–247.

95. Kachniarz, M.; Szewranski, S.; Kazak, J. The Use of European Funds in Polish and Czech Municipalities. A Study of the Lower Silesia Voivodship and Hradec Kralove Region. *IOP Conf. Ser. Mater. Sci. Eng.* **2019**, *471*, 1–8. [CrossRef]

96. Furmankiewicz, M.; Knieć, W.; Atterton, J. Rural governance in the new EU member states: The experience of the Polish LEADER+ Pilot Programme (2004–2008). In *Governance in Transition*; Buček, J., Ryder, A., Eds.; Springer Geography: Dordrecht, The Netherlands, 2015; pp. 133–153.

97. Kaczmarek, T. Collaboration between local governments in metropolitan areas—Determinants, current situation and perspectives for the future. In *Towards Urban-Rural Partnerships in Poland: Preconditions and Potential*; Dej, M., Janas, K., Wolski, O., Eds.; Instytut Rozwoju Miast: Kraków, Poland, 2014; pp. 41–55.

98. Przybyła, K.; Kulczyk-Dynowska, A.; Kachniarz, M. Quality of Life in the Regional Capitals of Poland. *J. Econ. Issues* **2014**, *48*, 181–195. [CrossRef]

99. Tokarczyk-Dorociak, K.; Kazak, J.; Szewrański, S. The Impact of a Large City on Land Use in Suburban Area—The Case of Wroclaw (Poland). *J. Ecol. Eng.* **2018**, *19*, 89–98. [CrossRef]

100. Grochowski, M.; Regulska, J. New Partnership and Collaboration: Local Government and Its Supporting Institutions—The Case of Poland. In *Towards a New Concept of Local Self-Government? Recent Local Government Legislation in Comparative Perspective*; Amnå, E., Montin, S., Eds.; Fagbokforlaget: Bergen, Norway, 2000; pp. 73–100.

101. Kurek, W.; Faracik, R.; Mika, M. Ecological conflicts in Poland. *GeoJournal* **2001**, *54*, 507–516. [CrossRef]

102. Furmankiewicz, M.; Potocki, J.; Kazak, J. Land-Use Conflicts in the Sudetes, Poland. *IOP Conf. Ser. Mater. Sci. Eng.* **2019**, *471*, 1–10. [CrossRef]

103. Regulska, J. Governance or Self-governance in Poland? Benefits and Threats 20 Years Later. *Int. J. Polit. Cult. Soc.* **2009**, *22*, 537–556. [CrossRef]

104. Kołsut, B. Inter-Municipal Cooperation in Waste Management: The Case of Poland. *Quaest. Geogr.* **2016**, *35*, 91–104. [CrossRef]

105. Katona-Kovács, J.; High, C.; Nemes, G. Importance of Animation Actions in the Operation of Hungarian Local Action Groups. *Eur. Countrys.* **2011**, *3*, 227–240. [CrossRef]

106. Doitchinova, J.; Miteva, A.; Stoyanova, Z. The process of creating local action groups in Bulgaria—Problems and prospects. *Sci. Ann. Alexandru Ioan Cuza Univ. Iaşi Econ. Sci.* **2012**, *59*, 183–208. [CrossRef]

107. Pawłowska, A. Territorial Partnerships in Rural Regions—Neo-Institutional Perspective. *Pol. Sociol. Rev.* **2017**, *1*, 95–108.

108. Furmankiewicz, M. LEADER+ territorial governance in Poland: Successes and failures as a rational choice effect. *Tijdschrift Economische Sociale Geografie* **2012**, *103*, 261–275. [CrossRef]

109. Marquardt, D.; Möllers, J.; Buchenrieder, G. Social Networks and Rural Development: LEADER in Romania. *Soc. Rural.* **2012**, *52*, 398–431. [CrossRef]

110. Zajda, K. *New Forms of Social Capital of Rural Areas. A Case Study of Selected Polish Local Action Groups*; LAP Lambert Academic Publishing: Saarbrücken, Germany, 2014.

111. Furmankiewicz, M.; Macken-Walsh, Á. Government within governance? Polish rural development partnerships through the lens of functional representation. *J. Rural Stud.* **2016**, *46*, 12–22. [CrossRef]

112. Adamski, T.; Gorlach, K. Neo-Endogenous Development and the Revalidation of Local Knowledge. *Pol. Sociol. Rev.* **2007**, *160*, 481–497.

113. Pappalardo, G.; Sisto, R.; Pecorino, B. Is the Partnership Governance Able to Promote Endogenous Rural Development? A Preliminary Assessment Under the Adaptive Co-Management Approach. *Eur. Countrys.* **2018**, *10*, 543–565. [CrossRef]

114. Naydenov, N.; Bogdanova, Z. Strategic & organizational problems of "LEADER" approach application as a tool for sustainable rural development in Bulgaria. In *Values and Challenges in Designing the European Rural Structures—Research Network Experience*; Voicilas, D.M., Tudor, M., Eds.; European Rural Development Network: Warsaw, Poland, 2007; pp. 53–61.

115. Lysek, J.; Šaradín, P. Mapping the Success: Inter-municipal Cooperation in Two Czech Micro-regions. In *Inter-Municipal Cooperation in Europe: Institutions and Governance*; Teles, F., Swianiewicz, P., Eds.; Palgrave Macmilian: Cham, Switzerland, 2018; pp. 315–326.

116. Freshwater, D.; Thurston, L.; Ehrensaft, P. Characteristics of successful community development partnership strategies. In *The Structure, Theory and Practice of Partnership in Rural Development*; Rounds, R.C., Ed.; Agriculure and Rural Restructuring Group, Rural Development Institute: Brandon, MB, Canada, 1993; pp. 3–23.

117. Swianiewicz, P.; Teles, F. Inter-municipal Cooperation Diversity, Evolution and Future Research Agenda. In *Inter-Municipal Cooperation in Europe: Institutions and Governance*; Teles, F., Swianiewicz, P., Eds.; Palgrave Macmilian: Cham, Switzerland, 2018; pp. 335–350.

118. Klok, P.-J.; Boogers, M.; Denters, B.; Sanders, I.T. Inter-municipal Cooperation in the Netherlands. In *Inter-Municipal Cooperation in Europe: Institutions and Governance*; Teles, F., Swianiewicz, P., Eds.; Palgrave Macmilian: Cham, Switzerland, 2018; pp. 157–171.

119. Eythórsson, G.T. Bigger and Stronger Together: How Icelandic Municipalities Solve their Lack of Capacity and Scale Economy. In *Inter-Municipal Cooperation in Europe: Institutions and Governance*; Teles, F., Swianiewicz, P., Eds.; Palgrave Macmilian: Cham, Switzerland, 2018; pp. 209–224.

120. Hertzog, R. Inter-municipal Cooperation in France: A Continuous Reform, New Trends. In *Inter-Municipal Cooperation in Europe: Institutions and Governance*; Teles, F., Swianiewicz, P., Eds.; Palgrave Macmilian: Cham, Switzerland, 2018; pp. 133–155.

121. Steiner, R.; Kaiser, C. Inter-municipal Cooperation in Switzerland. In *Inter-Municipal Cooperation in Europe: Institutions and Governance*; Teles, F., Swianiewicz, P., Eds.; Palgrave Macmilian: Cham, Switzerland, 2018; pp. 173–187.

122. Levitsky, S.; Ziblatt, D. *How Democracies Die*; Penguin Books Ltd.: London, UK, 2018.

123. Kovách, I.; Kučerová, E. The Project Class in Central Europe: The Czech and Hungarian Cases. *Soc. Rural.* **2006**, *46*, 3–21. [CrossRef]

124. Petrick, M.; Gramzow, A. Harnessing Communities, Markets and the State for Public Goods Provision: Evidence from Post-Socialist Rural Poland. *World Dev.* **2012**, *40*, 2342–2354. [CrossRef]

125. Vrabková, I.; Pavel, Š. The Technical Efficiency of Local Action Groups: A Czech Republic Case Study. *Acta Univ. Agric. Silvicult. Mendelianae Brunensis* **2017**, *65*, 1065–1074. [CrossRef]

126. Chmieliński, P.; Faccilongo, N.; Fiore, M.; La Sala, P. Design and implementation of the Local Development Strategy: A case study of Polish and Italian Local Action Groups in 2007–2013. *Stud. Agric. Econ.* **2018**, *120*, 25–31. [CrossRef]
127. Antonio Navarro, F.; Woods, M.; Cejudo, E. The LEADER Initiative has been a Victim of Its Own Success. The Decline of the Bottom-Up Approach in Rural Development Programmes. The Cases of Wales and Andalusia. *Sociol. Rural.* **2016**, *56*, 270–288. [CrossRef]

Article

Sustainable Tourism Development in Protected Areas of Rivers and Water Sources: A Case Study of Jiuqu Stream in China

Chin-Hsien Hsu [1], Hsiao-Hsien Lin [1,*] and Shangwun Jhang [2]

[1]   Department of Recreation and Sports Management, National Chin-Yi University of Technology, Taichung 41170, Taiwan; hsu6292000@yahoo.com.tw
[2]   Division of Neurosurgery, Department of Surgery, Changhua Christian Hospital, Changhua 50006, Taiwan; 133393@cch.org.tw
*   Correspondence: chrishome12001@yahoo.com.tw

Received: 2 June 2020; Accepted: 23 June 2020; Published: 29 June 2020

Abstract: This paper discusses the status quo of tourism and policy development regarding the Jiuqu Stream in China from different stakeholder perspectives. By combining field investigations, questionnaires, and statistical examinations of collected data, 812 samples were analyzed using multivariate analysis. The results indicate that increased visibility, employment opportunities, and real estate values in the scenic areas along the river will attract residents to return for future development, while public safety and conservation policies, featured architecture and tourism signage planning, increased cost of living, and waste and pollution will cause disincentives. Visitors will be attracted by the natural and ecological features, transportation planning, unique local culture, and events. Recreational facilities and architectural planning, merchandise lacking characteristics, tourist consumer expenditure, smoke and pollution from motor vehicles, and how it feels to interact with residents will influence the desire to visit the place. Development of an area should consider the different needs of every stakeholder in terms of recreational facilities, local infrastructure, expenditure and income, public safety and health, waste disposal, ecology and environmental conservation, tourism, and the quality of life.

Keywords: Jiuqu stream; tourism impact; impact perception; stakeholders

1. Introduction

The Jiuqu Stream is located among peaks and valleys on Mount Wuyi in the Fujian Province, China, and it is 60 km long. The crisscrossing surrounding peaks and boulders shaped the stream into the nine bends from which its name is derived (*Jiuqu* means "nine bends" in Chinese), as shown in Figure 1. Rich in natural resources, history, and culture, the Jiuqu Stream is a World Heritage site [1]. In addition to the booming local tea industry, enterprises are encouraged to invest in setting up hotels. Taking advantage of the Jiuqu River's water resources, bamboo rafting experiences can be set up from Seongchon Township to explore the surrounding natural ecological resources. In addition, tourism can be promoted by combining the ancient features of Taoist monuments, such as the Wuyi Temple, with the art and cultural performances (Impression Dahongpao). According to 2018 statistics, the stream attracted 15,146,900 tourists and created approximately 43.5 billion USD in business opportunities [2], demonstrating the enthusiastic development of local tourism.

Figure 1. Location and characteristics of the Jiuqu Stream in China.

Decision-making affects the direction and efficacy of tourism development [3]. Local development can promote economic improvement that increases living standards for residents and the health of the surrounding environment, but it can have negative outcomes [4]. These changes can be explored from economic, social, and environmental perspectives [5].

Exploring changes from the economic perspective involves examining cost of living, industrial construction, and village development [6] to understand concerns regarding employment, wages, consumerism, construction, industries, facilities, prices, premiums, health, culture and creativity, recreation, community feedback, and strategy coordination [6–8].

Social investigations involve tourism facilities, community development, living atmosphere, and cultural public safety [6] to understand community recognition, service and activity quality, political participation, tourism organizational planning, cultural and architectural traits, public safety enforcement, community facilities, and public interactions [9–13].

The environmental perspective involves tourism and recreational facilities and natural ecology to understand concerns of public transportation, parking and recreational areas, tourists' environmental literacy, waste volume, forest lands and ecological habitats, automotive exhaust, water supply, and air quality.

Decision-making—in terms of the economy, society, and environment—can be investigated jointly by using policy announcements, public sentiment, economic development, online marketing, medical facilities, industry distribution, public safety, maintenance of historical sites, community assistance and development, tourism resource chains, public facility management, environmental campaigns, industrial traits, ecological conservation, environmental education, personnel training, travel planning, development monitoring, and traffic management [6,14].

After activities conclude [4,15,16], the resulting changes from impacts and effects can be collected from resident and tourist experiences [6–14]. Residents' perspectives offer insight into changes in the area [6–10], whereas tourists' perspectives can reveal shortcomings in development [11–13]; simultaneous investigations of both residents' and tourists' perspectives can generate detailed understanding of shortcomings [6,12–14]. These investigations can help decision-makers resolve challenges and achieve balanced sustainable development. Therefore, the researchers collected and analyzed the perceptions of residents and tourists on development near Jiuqu Stream and identified shortcomings in the developmental status quo.

The Earth's ecological environment provides an abundance of natural resources, sufficient to fulfill the needs of all life on the planet [13]. To meet their needs for survival, human beings exploit natural resources to varying degrees in order to acquire them [17]. Although the goals and expectations are different for different individuals, for human beings, acquiring natural, ecological, social, humanistic, and economic resources to improve individual living standards is a universal goal [18]. This is also the case with tourism behavior and development.

The phenomenon of tourism is a result of the development of human technology and civilization, which has increased the efficiency of work and transportation, generating leisure time. The local tourism resources are used as an appeal to attract tourists to visit and spend money in the hope of gaining relaxation and satisfying their psychological needs [7,17,19–23]. The residents expect tourism development to stimulate local economic development, increase job opportunities, and improve the quality of life [4,24–26]. With tourism development as the main axis, they both interact with each other in the same region, in different positions, at different times, in different spaces, and with different resources, expecting to improve the psychological needs and status of individuals [27,28]. Although differences in needs and perceptions exist between the two stances [17,27], there have been many research reports on the effectiveness of tourism development from the perspective of residents [21–23]. However, there are quite a few researchers who investigate the defects of development through tourists [26,27]. However, not many researchers have explored the issue from the perspectives of both residents and tourists, and there is also a lack of studies on the development of river or stream tourism. Therefore, the investigators believe that sequentially understanding the feelings of residents and tourists towards the current situation of the development area and then analyzing the differences between them in terms of the local development, in order to obtain common or different viewpoints, can help to obtain more appropriate improvement suggestions [13,28]. Suggestions are then proposed based on this study's findings to provide a reference for relevant agencies and to improve the sustainable development goals for Jiuqu Stream.

In order to properly address this study, the following research questions have been raised:

Research Objective 1: What impact do residents feel on the current economic, social, and environmental development?
Research Objective 2: What impact do tourists feel on the current economic, social, and environmental development?
Research Objective 3: What impact did the two groups have on the current economic, social, and environmental development?

2. Methods and Instruments

2.1. Study Framework and Hypotheses

This study aimed to understand and compare the perceptions of residents and tourists on tourism strategies and developmental effects on Jiuqu Stream. Future developmental trends and suggestions for Jiuqu Stream are also proposed.

The perceptions of residents, tourists, and both groups on tourism and strategic development for Jiuqu Stream were collected. Resident and tourist opinions were obtained from the data to address developmental challenges and enable facilities to meet residents' and tourists' expectations. This study was developed [16] using questionnaire tools referenced from relevant literature [3–15], case studies, the researchers' experiences, common understandings between residents and tourists [6,22,24,26,27], multiple research methods, data collection [29], and data comparison and testing by performing induction, organization, and analysis [30]. Collecting accurate and reasonable information can enable the revision of strategic and development planning for the Jiuqu Stream. Figure 2 presents the research framework based on the study goal and theoretical review.

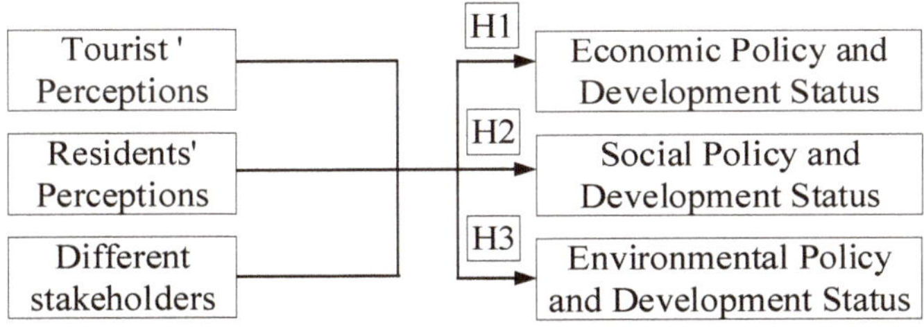

Figure 2. Study framework.

On the basis of this research framework, this study developed the following three hypotheses: (1) Residents have a Tourism and Policy Development Status of consistent awareness; (2) tourists have a Policy Development Status of consistent awareness; (3) different stakeholders have the same correct understanding of the status of cognitive tourism development.

2.2. Study Procedure and Instruments

The research used mixed research methods. The survey outline was formulated by referring to research theory and literature [1–16]. Adopted the concepts of existing theories, explained the research results, and sought the opinions of experts and scholars. We compiled a questionnaire on the current situation of tourism development, which is divided into 40 questions: Economy (15), society (15), and environment (10), as shown in Table 1.

Table 1. Initial questionnaire issue preparation.

		Residents	Tourists
Economic	Cost of living	Increase job opportunities Increase income Increase the cost of living	Multiple job opportunities Consumption increase Increase consumption costs
	Industry construction	Increase sightseeing facilities Increased tourism industry Combination of industrial characteristics Increased leisure opportunities Have preferential measures	Increase sightseeing facilities Increased tourism industry Sufficient industrial specialty commodities Increased leisure opportunities Have preferential measures
	Village development	Increased public construction Enterprise feedback Land and price increases Medical and health improvement Government communication platform Tourism and conservation policy Development of creative products Increase job opportunities	Adequate public construction Industry Promotion Rent and price increases Adequate medical and health level Sufficient tourism complaint platform Adequate tourism and conservation policies Diversified creative products Multiple job opportunities
Social	Company building	Increase visibility Increase service quality Increase activity quality Increase community development and willingness to participate Tourism indicators increase Increase the choice of tourist facilities Youth's willingness to return home	Increased visibility Service quality improvement Event quality improvement Popular community development and willingness to participate Adequate tourism index Multiple choices of tourist facilities Youth return to employment increases
	Life	Educational vocational training opportunities Monuments and cultural preservation Community environment commercialization Tourist friendly Good interaction between tourists and residents	Diversified educational vocational training opportunities Monuments and cultural preservation Community environment commercialization Travellers' friendly spending attitude Good interaction between tourists and residents
	Cultural security	Increase the security staff Increase willingness to purchase Increase visibility	Adequate police staff Increase in willingness to travel again Increased visibility
Environmental	Tourist rest facilities	Water pollution River pollution Vegetation change Biological habitat change Steam locomotive, oil, and smoke pollution Noise and garbage pollution Alien species threat	Water pollution River pollution Vegetation is scarce Biomass changes Steam locomotive pollution Increased noise and garbage Increased alien species
	Natural ecosystems	Tourist transportation Parking space Tourism environment destruction Water pollution	Convenient transportation Ample parking The tourist environment is damaged Water pollution

The questionnaire was designed using a five-point Likert's scale, with 1 representing "strongly disagree" and 5 representing "strongly agree". Fifty questionnaires were distributed for pre-testing. When Kaiser-Meyer-Olkin (KMO)> 0.06 and the p-value in the Bartlett test is less than 0.01 ($p < 0.01$), this indicates that the scale is suitable for continued factor analysis [31]. Then, the coefficient α is greater than 0.60. Tests show that the questionnaire has good reliability [32].

Economy (15) had a KMO > 0.627, with a Bartlett approximate χ^2 value of 274.688, a degree of freedom (df) of 120, and a significance of 0.000 ($p < 0.001$), and was therefore suitable for factor analysis. The explainable variations of the scale were 10.339%, 3.418%, and 3.223%, and the total explainable variation was 16.98%. After factor analysis and considering the understanding of the actual current situation of economic development, all were retained. Three aspects were named: Cost of living (3), industrial construction (5), and village development (7). They contained a total of 15 questions, and the three scales were 0.603, 0.601, and 0.600, respectively.

Society (15) had a KMO > 0. 688, with a Bartlett approximate χ^2 value of 413.731, a df of 105, and a significance of 0.000 ($p < 0.001$) and was therefore suitable for factor analysis. The explainable variations of the scale were 15.559%, 4.050%, and 3.65%, and the total explainable variation was 23.259%. After factor analysis and considering the intention of understanding the actual current situation of economic development, all were retained. Three aspects were named: Community building (6), living atmosphere (5), and cultural public safety (4). They contained a total of 15 questions, and the three scales were 0.644, 0.656, and 0.707, respectively.

Environment (110) had a KMO > 0. 914, with a Bartlett approximate χ^2 value of 1280.57, a df of 55, and significance of 0.000 ($p < 0.001$), and was therefore suitable for factor analysis. The explainable variations of the scale were 47.496% and 7.834%, and the total explainable variation was 55.331%. After factor analysis, all were retained. Two aspects were named: Tourism and recreational facilities (4) and natural ecology (6). They contained a total of 10 questions, and the scales were 0.859 and 0.855.

Based on the results of the above analysis, it can be seen that the questions in the tourism and policy development awareness questionnaire designed by this study are all reliable. Afterwards, we analyzed the survey's reliability with statistical verification and analyzed the results with descriptive analyses and t-tests, as shown in Table 2.

Table 2. Constructs involved in the questionnaire on perceptions of the impact of residents and tourism on the Jiuqu Stream.

Construct	Subfacet	Issues of Perceptions about the Impact of Tourism	Cronbach's α
Economic	Cost of living	Employment opportunities, income, expenditures	0.60–0.62
	Industrial construction	Tourism facilities, tourism industries, Commodities combining local characteristics and products, leisure opportunities, tourism premiums	0.60–0.61
	Village development	Facility maintenance, development feedback, real-estate prices, medical care, community communications, political participation, cultural and creative products	0.59–0.63
Social	Community building	Recognition, quality of services and activities, community engagement, sufficient signage, travel and recreational alternatives, organizational capacity	0.63–0.66
	Living atmosphere	Young people returning, job-training opportunities, cultural preservation, architecture, tourist attitudes	0.63–0.71
	Cultural public safety	Public interaction, public safety enforcement, Re-tourism intentions or land purchase	0.58–0.65
Environmental	Tourism and recreational facilities	Waste disposal, transportation, parking and recreational areas, environmental damage by tourists	0.85–0.89
	Natural ecology	Water resources, lakes, flora, biological habitats, automotive exhaust, noise pollution, external ecological threats	0.85–0.86

Next, interviews were used to supplement the sample information. With the consent of the respondents, six participants were interviewed, including local tourism practitioners, residents, and scholars with travel experience with regard to the Jiugqu River using the semi-structured design and open interviews to obtain their opinions on the analysis results presented in the survey, as shown in Table 3.

Table 3. Background and interview topics.

Gender	Length of Stay/Seniority	Occupation	Gender	Length of Stay/Seniority	Occupation
male	30	Residents	male	12	prof
female	28	Residents	male	6	prof
male	25	Residents	female	15	tourist guide

interview topics Description

1. What do you think is the impact of the tourism development of the Jiuquxi Scenic Area on the local economic development? Please explain the reasons in detail.
2. How do you think the tourism development of the Jiuquxi Scenic affects the local social and cultural development? Please elaborate.
3. What do you think is the impact of the tourism development of the Jiuquxi Scenic Area on the development of the local natural environment? Please explain the reasons in detail.
4. What do you think is the impact of tourism development in the Jiuquxi Scenic Area on local tourism policies? Please explain in detail why.

After the participants verified the accuracy of the recorded content, the researchers integrated the information of the questionnaire, analyzed the results, and completed the research paper through the processes of induction, organization, and analysis [10,29]. Finally, a multivariate validation analysis method was adopted to combine the information obtained from different research subjects, theories, and methods to validate multiple data from multiple perspectives and to compare the results of different studies [29,30] in order to acquire accurate knowledge and implications. Ultimately, we hope to explore the current state of the Jiuqu River's tourism development from the perspectives of both residents and visitors, and to offer suggestions for improvement based on the views of both.

2.3. Study Scope and Limitations

Jiuqu Stream is located between the peaks and valleys of Wuyi Mountain, Fujian Province, China, and is 60 kilometers long. The criss-crossing peaks and boulders shape the river into nine bends; the surrounding natural ecology is diverse, the mountain landscape is diverse, and it has become the main tourist area for major tourists. The surrounding Xing-cun Town and Gong-guan Village are the closest towns to the Jiuquxi Scenic Area. The area also takes Jiuqu Stream and surrounding tourist attractions as tourist destinations to improve local development.

The study sample was collected from October 2019 and completed in February 2020. Initially, questionnaires were collected on site. Due to cost, manpower, and material considerations, as well as restricted working hours of residents and visitors' willingness to be interviewed, the survey was conducted using random sampling. Interviews were conducted right after seeking the consent of the participants. However, due to the outbreak of COVID-19, the sampling was changed to an online questionnaire platform, and since it was not easy to confirm the respondents, snowball sampling was used instead. Summing up the above explanations, it is unlikely that more comprehensive information can be obtained due to the limitations of the sample background. If this results in any discrepancy in the study, it will be taken into consideration for the further study.

3. Analysis of Results

3.1. Descriptive Characteristics of the Sample

Based on the analysis of the 812 samples, the Jiuqu Stream is considered a well-known tourist area with rich natural ecology and numerous historic sites. The area is popular for light tourist activities (42.4%) and among women (57.6%). Average consumer expenditure per visit was 420–700 USD (36.5%). Most tourists (68%) arranged two trips to visit local scenery per year (35.5%), as shown in Table 4.

Table 4. Descriptive characteristics of the participants.

Identity			
Identity	**Percentage**	**Age**	**Percentage**
Residents	32%	Under 20	5.4%
Tourists	68%	21–30	33.5%
Gender	**Percentage**	31–40	17.2%
Male	42.4%	41–50	13.3%
Female	57.6%	51–60	11.3%
		Over 61	19.2%
Spending on Trips and Number of Tours			
Annual trips	**Percentage**	**Travel Consumption (USD)**	**Percentage**
1	19.2%	140 下	8.9%
2	35.5%	140–420	30.5%
3	27.1%	420–700	36.5%
4	14.8%	700–1120	11.8%
5	3.4%	1120–1700	9.9%
		1700 上	2.2%
Tourist purpose			
Tourism	42.4%	Administrative affairs	9.9%
Shopping consumption	14.8%	Fire safety	3.0%
Work	13.8%	Academic research	2.5%
Leisure and sport	13.8%		

3.2. Perceptions of Economic Strategies and Developmental Efficacy

Respondents responded based on the content of the questionnaire. Statistical verification analysis was used. A score of 1 indicates strong disagreement, and a score of 5 indicates a strong agreement. Because residents, tourists, and other stakeholders had inconsistent perceptions of economic strategies and development efficacy, Hypothesis 1 did not hold, as shown in Table 5.

Table 5. Residents' and tourists' perceptions of current economic development.

	Facets	Highest	M	Lowest	M
	Price of people	Job opportunity	4.26	Cost of expenditure	3.86
Residents	Industry construction	Tourism industry	4.06	Sightseeing discount	3.96
	Village development	Land price	4.40	Protection policy	3.96
	Price of people	Income	3.94	Cost	3.88
Tourist	Industry construction	Leisure opportunities	3.78	Building	3.71
	Village development	Health and medical facilities	3.84	Feedback measures	3.63

The government uses the natural environment, cultural monuments, and agricultural products to develop tourism and corresponding industries that enhance real-estate value and create employment and business opportunities. However, tourist facilities and resources are primarily targeted towards tourists; residents are few and scattered, and transportation in mountainous areas is inconvenient.

Residents lacked access to transportation, resource pipelines, or tourism premiums, and protection policies could not be implemented. As a result, residents perceived improvements in employment opportunities, the tourism industry, and real-estate prices, but they believed that efficacy in expenditures and costs, tourism premiums, and protection policies were lacking.

Rich natural resources as well as rafting and performance activities expand travel options. In addition, improvements to medical care also increase tourists' willingness to visit a destination. However, remote locations, insufficient tourist facilities, and crude town facilities affect tourists' willingness to consume. Therefore, tourists perceived improvements in income, medical care, and leisure opportunities, but they felt that efficacy in expenditures, costs, and development feedback were lacking.

As a result, stakeholder perceptions of employment opportunities, salary income, tourist facilities, real-estate prices, and medical services differed, causing divergences ($p < 0.001$), as shown in Table 6.

Table 6. Recognition of economic development status by persons with different rights.

Issue		Resident		Passenger		T	p-Value
		M	SD	M	SD		
Price of people	Job	4.26	0.565	3.91	0.369	5.075	0.000 *
	Income	4.16	0.650	3.94	0.367	2.967	0.000 *
	Cost	3.86	0.535	3.88	0.443	−0.294	0.082
Industry construction	Building	3.98	0.553	3.71	0.561	3.012	0.002 *
	Business	4.06	0.620	3.72	0.531	3.778	0.371
	Spot color combination	4.04	0.699	3.75	0.544	3.090	0.467
	Fallow machine	4.00	0.535	3.78	0.490	2.723	0.050
	Grace	3.96	0.638	3.75	0.541	2.258	0.602
Village development	Installation protection	4.02	0.589	3.75	0.541	2.977	0.108
	Circulation	3.98	0.685	3.63	0.549	3.701	0.484
	Land and kitchen	4.40	0.639	3.83	0.535	6.221	0.002 *
	Hygiene	4.26	0.694	3.84	0.488	4.689	0.000 *
	Mizonori cooperation	4.16	0.584	3.77	0.506	4.533	0.649
	Policy	4.08	0.601	3.69	0.579	4.137	0.041
	Protective measures	3.96	0.638	3.77	0.519	2.105	0.986
	Creative products	4.14	0.606	3.75	0.517	4.416	0.772

$* = p < 0.001$.

Tourist and medical facilities were sufficient to meet tourist needs and increase their willingness to consume. These facilities indirectly provided employment opportunities to area residents. However, jobs in the tourism industry have long work hours and low pay, and residents live far from these workplaces, thereby increasing their cost of living. Figure 3 demonstrates the perceptions of social strategies and developmental efficacy.

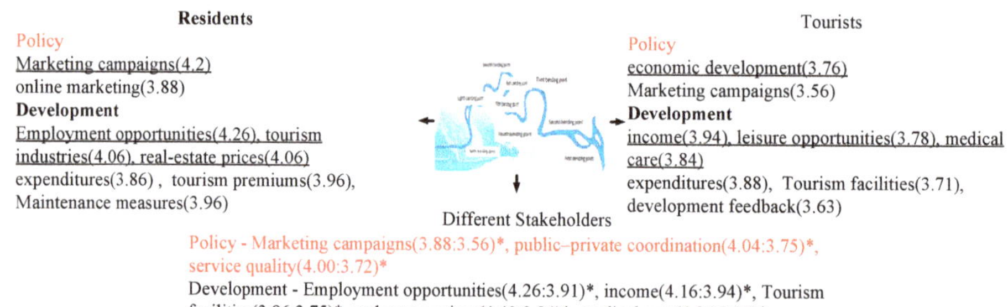

Figure 3. Perceptions of economic strategies and developmental efficacy. *: $p < 0.01$ (Residents: Tourists); attractive, repulsive force.

3.3. Perceptions of Social Strategies and Developmental Efficacy

Respondents responded based on the content of the questionnaire. Statistical verification analysis was used. A score of 1 indicates strong disagreement, and a score of 5 indicates strong agreement.

Because residents, tourists, and other stakeholders had inconsistent perceptions of social strategies and developmental efficacy, Hypothesis 2 did not hold, as shown in Table 7.

Table 7. Residents' and tourists' perceptions of social development.

	Figure	Highest	M	Lowest	M
Residents	Company building	Popularity	4.28	Indicator satisfaction	3.94
	Life	Cultural preservation	4.48	Building style	1.96
	Cultural security	Replay	4.16	Public security management	3.82
Tourist	Company building	Activity quality	3.86	Community involvement	3.66
	Life	Vocational training opportunities	3.90	Building style	2.70
	Cultural security	Public interaction	3.75	Re-travel	2.51

Government relocation of residents can protect the environment by preserving culture, creating business opportunities by attracting tourists, and increasing residents' willingness to invest. However, the preservation of monuments and public safety is jeopardized in large and remote scenic areas, which have insufficient signage, buildings, and marketing traits and small and scattered residential populations, leading to shortages in staffing to maintain the scenic area. As a result, residents perceived increased recognition, cultural preservation, and willingness to revisit and acquire property, but they felt that efficacy in architecture, signages, and public safety enforcement were lacking.

The government's integration of local cultures and industries attracted corporate investments and created high-quality tourist environments. However, the small and scattered residential population caused staffing shortages for maintaining the large scenic area, and government control of decision-making has caused low public involvement in policy. As a result, tourists perceived increases activity quality, job-training opportunities, and public interaction, but they felt that efficacy in architecture, community participation, and willingness to revisit and acquire property were lacking.

As a result, stakeholders perceived signages and public safety planning, job-training opportunities, architecture, and revisiting and property acquisitions differently, leading to divergences ($p < 0.001$), as shown in Table 8.

Table 8. Recognition of social development status by different stakeholders.

Issue		Resident		Passenger		t	p-Value
		M	SD	M	SD		
Company building	Popularity	4.28	0.536	3.79	0.509	5.826	0.237
	Service quality	4.00	0.535	3.77	0.493	2.788	0.024
	Activity quality	4.00	0.639	3.86	0.505	1.632	0.434
	Community involvement	3.98	0.622	3.66	0.552	3.445	0.033
	Adequate indicators	3.94	0.818	3.82	0.488	1.220	0.000 *
	Recreational choices	4.04	0.570	3.73	0.529	3.582	0.053
	Increased organization	4.18	0.629	3.76	0.500	4.844	0.130
Life	Youth return home	4.42	0.538	3.82	0.527	6.911	0.027
	Vocational training opportunities	4.42	0.538	3.90	0.455	6.669	0.000 *
	Cultural preservation	4.48	0.544	3.80	0.574	7.321	0.185
	Building style	1.96	0.402	2.70	0.918	−0.516	0.000 *
	Tourist attitude	3.94	0.424	3.86	0.501	0.981	0.055
Cultural security	Public interaction	3.88	0.480	3.75	0.507	1.655	0.020
	Public security management	3.82	0.482	3.37	0.523	5.433	0.002 *
	Re-travel	4.16	0.738	2.51	0.954	11.182	0.006 *

*: $p < 0.01$.

Abundant natural resources and historic buildings primarily attract tourists and create business opportunities. However, signage and facilities were for shuttle services and tourist centers because tourists are unfamiliar with the geography of the scenic area, thereby increasing the risk to visitor safety because of the area's size. Figure 4 demonstrates the perceptions of social strategies and developmental efficacy.

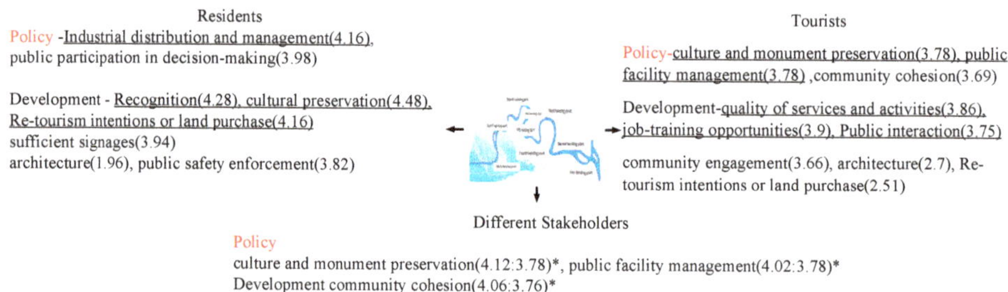

Figure 4. Perceptions of social strategies and developmental efficacy. *: $p < 0.01$; residents; tourists; attractive, repulsive forces.

3.4. Perceptions of Environmental Strategies and Developmental Efficacy

Respondents responded based on the content of the questionnaire. Statistical verification analysis was used. A score of 1 indicates strong disagreement, and a score of 5 indicates strong agreement.

Residents, tourists, and different stakeholders had inconsistent perceptions of environmental strategies and developmental efficacy; therefore, Hypothesis 3 did not hold, as shown in Table 9.

Table 9. Residents' and tourists' perceptions of the current situation of environmental development.

	Facets	Highest	M	Lowest	M
Residents	Tourist rest facilities	Tourist transportation	4.36	Garbage placement	1.88
	Natural ecosystems	Steam locomotive fume	2.30	Alien species threat	1.94
Tourists	Tourist rest facilities	Tourist transportation	3.75	Garbage placement	2.46
	Natural ecosystems	Steam locomotive fume	2.60	Biological habitat	2.34

Comprehensive bus services in the scenic area help residents commute. However, large numbers of tourists rapidly increase automotive transportation and human-made waste, and the small local population has insufficient cleaning personnel, thereby increasing the amount of waste in the scenic area. As a result, residents perceived adequate ecological and tourist transportation planning, but they felt apprehension regarding waste disposal and automotive exhaust problems.

Comprehensive ecological management and transportation planning can protect the environment and reduce transportation time, thereby increasing willingness to travel. However, because of the scenic area's large size, transportation time increased, and the few residents and staff cannot easily maintain facilities and the environment or manage waste. As a result, tourists perceived comprehensive ecological protections and tourism transportation planning as well as problems of waste disposal and automotive exhaust.

As a result, different stakeholders had divergent views on waste disposal, water sources, lakes, flora, biological habitats, noise pollution, and external ecological threats ($p < 0.001$), as shown in Table 10.

Table 10. Recognition of environmental development status by different stakeholders.

Issue		Resident		Tourist		t	*p*-Value
		M	SD	M	SD		
Tourist rest facilities	Water source	2.28	0.536	2.53	0.866	1.917	0.000 *
	River	2.06	0.240	2.42	0.809	3.141	0.000 *
	vegetation	2.00	0.202	2.37	0.784	3.260	0.000 *
	Biological habitat	1.96	0.198	2.34	0.736	3.602	0.000 *
	Steam locomotive fumes	2.30	0.678	2.60	0.781	2.443	0.021
	Noise pollution	2.00	0.350	2.52	0.795	4.502	0.000 *
	Alien species threat	1.94	0.373	2.57	0.923	4.685	0.000 *
Natural ecosystems	Garbage placement	1.88	0.521	2.46	0.770	5.001	0.000 *
	Tourist transportation	4.36	0.631	3.75	0.553	6.513	0.037
	Parking space	3.80	0.700	3.54	0.550	2.746	0.190
	Tourist environmental destruction	3.40	1.030	3.15	1.018	1.501	0.672

*: $p < 0.01$.

The local area is mountainous and has dense forests, clear water, and a beautiful natural environment. However, the large increase in tourists and vehicles coupled with tourists' boisterousness and environmental illiteracy have caused environmental destruction. Figure 5 demonstrates the perceptions of environmental strategies and developmental efficacy.

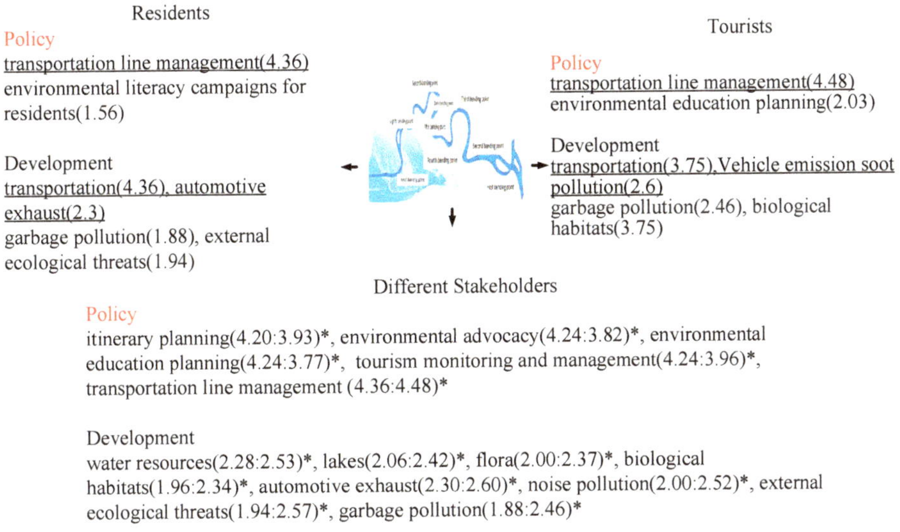

Figure 5. Perceptions of environmental strategies and developmental efficacy. *: $p < 0.01$; residents; tourists; attractive, repulsive forces.

4. Conclusions and Recommendations

4.1. Conclusions

Every stakeholder has different demands in terms of recreational facilities, local infrastructure, expenditure and income, public safety and health, waste disposal, and ecological and environmental conservation, as well as tourism and the quality of life. Residents' willingness to return to their hometowns for development depends on the visibility, employment opportunities, and the growth of real estate. Public safety and protection policies, insurance expenses, construction, inadequate signage, increased cost of living, and waste and pollution are the deterrents to residents returning for future development. Tourists are attracted by the natural and ecological features, well-planned transportation, and unique local culture and events, but are usually deterred by poor recreational facilities and architectural planning, merchandise lacking characteristics, high tourist consumer expenditure, smoke and pollution from motor vehicles, and unpleasant encounters with locals.

4.2. Recommendations

1. Community development and local facilities should be improved while valuing public sentiment in mutually beneficial situations for the government and residents.
2. Environmental sanitation requires improvement by defining construction and cultural traits and resolving waste and pollution problems.
3. Tourism management, marketing, and service quality require improvement, and travel itineraries should be developed to enhance residents' employment skills; travel safety should be enhanced, and signage, waste, and noise pollution problems should be addressed. Water sources, lakes, and flora environments should be maintained to increase residents' and tourists' willingness to invest and travel in the area.
4. The development of ecotourism in other regions and countries warrants study from different perspectives.
5. Political, social, and industrial developments warrant investigation from different perspectives to collect comprehensive research data.

Author Contributions: Conceptualization, C.-H.H. and H.-H.L.; methodology, H.-H.L.; software, H.-H.L.; validation, C.-H.H. and H.-H.L. and S.J.; formal analysis, H.-H.L.; investigation, H.-H.L.; resources, H.-H.L.; data curation, H.-H.L.; writing—original draft preparation, H.-H.L.; writing—review and editing, H.-H.L.; visualization, C.-H.H.; supervision, C.-H.H.; project administration, C.C.H.; funding acquisition, S.J. All authors have read and agreed to the published version of the manuscript.

Funding: This research received no external funding.

Conflicts of Interest: The authors declare no conflict of interest

References

1. Ministry of Culture and Tourism of the People's Republic of China (2020). Overview of the 2018 Tourism Market. Available online: http://zwgk.mct (accessed on 31 March 2020).
2. NPTA.CN. Jiuqu Stream in Mt. Wuyi. 2019. Available online: http://www.npta.cn/spots/show_82.html (accessed on 3 March 2020).
3. Kingdon, J.W. *Agendas, Alternatives, and Public Policies*; Addison-Wesley: Boston, MA, USA, 1995.
4. Ap, J.; Crompton, J.L. Developing and testing a tourism impact scale. *J. Travel Res.* **1998**, *37*, 120–130. [CrossRef]
5. Johnson, J.D.; Snepenger, D.J.; Akis, S. Residents' perceptions of tourism development. *Ann. Tour. Res.* **1994**, *21*, 629–642. [CrossRef]
6. Lin, H.H. The Study on Tourism Policy, Current Development Status, and Impact Perception of Sun Moon Lake. Doctoral Dissertation, DaYeh University, Institute of Environmental Engineering Department, Changhua, China, 2019.
7. Lankford, S.V.; Howard, D.R. Developing a Tourism Impacts Attitude Scale. *Ann. Tour. Res.* **1994**, *21*, 121–139. [CrossRef]
8. Rothman, R. Residnets an dtransients: Community reaction to seasonal visitors. *J. Travel Res.* **1978**, *16*, 8–13. [CrossRef]
9. Husbands, W. Social status and perception of tourism in Zambia. *Ann. Torusim Res.* **1989**, *16*, 237–253. [CrossRef]

10. Gursoy, D.; Jurowski, C.; Uysal, M. Resident Attitudes: A structural modeling approach. *Ann. Tour. Res.* **2002**, *20*, 79–105. [CrossRef]

11. de Waal, F. Are we Smart Enough to Know How Smart Animals Are? Wilson Hall Auditorium: Harrisonburg, VA, USA, 2017.

12. Paulo, E.J.; Belmiro, V.P. Olfactory information from the path is relevant to the homing process of adult pigeons. *Behav. Ecol. Sociobiol.* **2017**, *72*, 5. [CrossRef]

13. Lin, H.; Lee, S.; Perng, Y. Investigation about the Impact of Tourism Development on a Water Conservation Area in Taiwan. *Sustainability* **2018**, *10*, 2328. [CrossRef]

14. Anderson, J.E. *Public Policy-Making*; CBS College Publishing: New York, NY, USA, 1990.

15. Jurowski, C.; Uysal, M.; Williams, D.R. A theoretical analysis of host community resident reactions to tourism. *J. Travel Res.* **1997**, *36*, 3–11. [CrossRef]

16. Strauss & Corbin. *Basics of Qualitative Research: Grounded Theory Procedures and Techniques*; Sage: Newbury Park, CA, USA, 1990.

17. Schreiber, R.L.; Diamond, A.W.; Peterson, R.T.; Cronkite, W. *Save the Birds*; Cambridge, University Press: Lanham, MD, USA, 1987; p. 384. ISBN 10: 0395511720.

18. Otto, J.E.; Ritchie, J.R.B. The service experience in tourism. *Tour. Manag.* **1996**, *17*, 165–174. [CrossRef]

19. Goffi, G.; Cladera, M.; Osti, L. Sun, Sand, and … Sustainability in Developing Countries from a Tourists' Perspective. The Case of Punta Cana. *Sustainability* **2020**, *12*, 4743. [CrossRef]

20. Zhang, A.; Zhong, L.; Xu, Y.; Wang, H.; Dang, L. Tourists' Perception of Haze Pollution and the Potential Impacts on Travel: Reshaping the Features of Tourism Seasonality in Beijing, China. *Sustainability* **2015**, *7*, 2397–2414. [CrossRef]

21. Tsai, K.-T.; Lin, T.-P.; Lin, Y.-H.; Tung, C.-H.; Chiu, Y.-T. The Carbon Impact of International Tourists to an Island Country. *Sustainability* **2018**, *10*, 1386. [CrossRef]

22. Butler, R.W. The concept of a tourist area cycle of evolution: Implications for management of resources. *Can. Geogr.* **1980**, *24*, 5–12. [CrossRef]

23. Campón-Cerro, A.M.; Folgado-Fernández, J.A.; Hernández-Mogollón, J.M. Rural Destination Development Based on Olive Oil Tourism: The Impact of Residents' Community Attachment and Quality of Life on Their Support for Tourism Development. *Sustainability* **2017**, *9*, 1624. [CrossRef]

24. Galeote, L.C.; Mestanza, J.G. Qualitative Impact Analysis of International Tourists and Residents' Perceptions of Málaga-Costa Del Sol Airport. *Sustainability* **2020**, *12*, 4725. [CrossRef]

25. Katsikari, C.; Hatzithomas, L.; Fotiadis, T.; Folinas, D. Push and Pull Travel Motivation: Segmentation of the Greek Market for Social Media Marketing in Tourism. *Sustainability* **2020**, *12*, 4770. [CrossRef]

26. Perdue, R.R.; Long, P.T.; Allen, L. Resident support for tourism developments. *Ann. Tour. Res.* **1990**, *14*, 420–429. [CrossRef]

27. Lin, Z.; Chen, Y.; Filieri, R. Resident-tourist value co-creation: The role of residents' perceived tourism impacts and life satisfaction. *Tour. Manag.* **2017**, *61*, 236–442. [CrossRef]

28. Riu, A.S.; Caldés, R.G.; Donada, J.T. Furthering Internal Border Area Studies: An Analysis of Dysfunctions and Cooperation Mechanisms in the Water and River Management of Catalonia, Aragon and the Valencian Community (Spain). *Sustainability* **2019**, *11*, 4499. [CrossRef]

29. Anselm, L. Strauss, Juliet, Corbin. In *Basics of Qualitative Research: Grounded Theory Procedures and Techniques*, 2nd ed.; Sage: Newbury Park, CA, USA, 1998.

30. Janesick, V.J. The Choreography of Qualitative Research Design: Minuets, Improvisations, and Crystallization. In *Handbook of Qualitative Research*; Denzin, N.K., Lincoln, Y.S., Eds.; Sage: Thousand Oak, CA, USA, 2000; pp. 379–399.

31. Ajzen, I. The Theory of Planned Behavior. *Organ. Behav. Human Decis. Process.* **1991**, *50*, 179–211. [CrossRef]

32. Devellis, R.F. *Scale Development: Theory and Applications*; Sage: Newbury Park, CA, USA, 1991.

 *sustainability*

Article

The Influence of Negative Political Environment on Sustainable Tourism: A Study of Aksu-Jabagly World Heritage Site, Kazakhstan

Imanaly Akbar [1,2]**, Zhaoping Yang** [1,]*****, Fang Han** [1] **and Gulnar Kanat** [1,2]

[1] State Key Laboratory of Desert and Oasis Ecology, Xinjiang Institute of Ecology and Geography, Chinese Academy of Sciences, Urumqi 830011, China; yimanaili_akebaier@yahoo.com (I.A.); hanfang@ms.xjb.ac.cn (F.H.); melody622@mails.ucas.ac.cn (G.K.)

[2] Xinjiang Institute of Ecology and Geography, University of Chinese Academy of Sciences, Beijing 100049, China

* Correspondence: yangzp@ms.xjb.ac.cn; Tel.: +86-991-788-5349

Received: 30 November 2019; Accepted: 17 December 2019; Published: 23 December 2019

Abstract: The political environment of a tourism destination is the most important element in planning, implementing, and controlling sustainable tourism development. The political environment refers to the coordination and cooperation among many participants to formulate and apply tourism policies. In our study the term political environment refers to political power, leadership, structures, mechanisms, and strategies, or policies for the implementation of sustainable tourism development. The main purpose of this article is to, through the example of Aksu-Jabagly natural heritage site in Kazakhstan, study how the negative political environment (NPE) of a tourism destination inhibits the implementation of sustainable tourism development in Kazakhstan. This study draws on in-depth interviews with local residents who are considered as one of the key stakeholders in the tourism industry. In our research, we conducted a questionnaire survey of 222 representative households from the neighboring village of Aksu-Jabagly, a natural world heritage site. Results show that because of negative political environments, the residents highly perceive the negative economic and environmental impacts of tourism development. Although the residents highly evaluated tourism's positive sociocultural impacts, its relevance to other indicators was relatively weak. The residents are dissatisfied with tourism development, and their participation level in tourism was low. The results also reveal that highly perceived negative economic and negative environmental impacts of tourism are the main cause of residents' dissatisfaction with tourism development and residents' lack of participation in tourism.

Keywords: political environment; sustainable tourism; impact; world heritage; Aksu-Jabagly nature reserve; Kazakhstan

---

1. Introduction

Political environment is the government actions which affect the operations of a company or business. These actions may be on the local, regional, national, or international levels. Business owners and managers pay close attention to the political environment to gauge how government actions will affect their company. Regarding the World Economic Forum's travel and tourism competitiveness index, it seems that some new models are emerging [1]. The "enabling policy environment for tourism" indicator shows that countries with more favorable environmental conditions are more likely to develop management plans for each World Heritage site (either existing or out of date), and it is more likely that there is no plan or claim prepared by those states with unfavorable political environments [2]. The statement of the political environment has scarcely been recognized in previous studies [3,4], therefore, it is not widely used in tourism research. The political environment (an older expression) overlaps with an emerging understanding of destination governance (a newer expression) [5,6]. Some authors have extended the concept of a political environment to further development of the three pillars of environmental sustainability, which has continued to generate more debates [4,7–9]. Dialogs on the management of tourist destinations have collected some elements of our perception of the political environment with a focus on how tourist destinations guide and govern the planning, implementation, and monitoring processes of tourism development [5,6]. Mihalič et al. (2016) elaborated on the issue by emphasizing the significance of the influence of the political environment and destination governance on sustainable tourism development. They addressed issues limited to understanding the importance of the political environment to sustainable tourism's implementation. Mihalič et al. (2016) argued that the political environment does not indicate political parties or systems (although both may be related to tourism development), but indicates political power, leadership, structure, mechanisms, and strategies, or policies as critical to the implementation of sustainable tourism development. In relation to the agreement with the three sustainable development environments (economic, environmental and sociocultural), the concept of the political environment has not been recognized with such force, and its designation as a missing element has not yet been achieved unanimously in the field of sustainable development [10].

The development of sustainable tourism destinations has attracted great attention from researchers over the years, especially the positive and negative tourism's influences on resources and destination communities [11]. Tourism can have a positive and negative impact on the community, but the development of tourism can also depend on how the locals of the destination feel about these effects [10]. As described by the social exchange theory, destination residents show their support for tourism development based on their satisfaction with the sustainable livedoid in the communities [12,13]. Destination resources are generally understood as economic, sociocultural, and environmental, which is similar to the so-called three-pillar sustainable tourism principle [14]. Apart from tourism's positive and negative impacts on the destination community, residents' perceptions of these impacts can affect sustainable tourism development. Tourism should properly consider its current and future economic, social, and environmental impacts to meet the needs of tourists, industry, environment, and the host community.

In the above context, it is understood that the destination resources are economic, sociocultural, and natural (or environmental, in a narrow sense), which is consistent with the concept the so-called three pillars of sustainable tourism [15]. However, researchers should distinguish the practical implementations of three pillars of sustainability in tourism [10]. Though it is difficult to perform sustainable tourism in practice [16], Mihalič et al. (2016) argue that this problem can be minimized if the concept of three pillars of sustainability is extended to include some "pushing forces" to ensure the effective implementation of sustainability in business and tourism destinations. Many authors have discussed other requirements for implementing sustainability, such as political support, power, critical mass, consensus, environmental education, awareness, and ethics [4,9]. Ritchie and Crouch (2003) showed that the debate on sustainable tourism must be extended with political sustainability. Lately, some of these "forces" have been debated under the theme of destination governance, which is interested in how tourism destinations guide and manage the implementation (and planning and control) process of sustainable tourism development [5].

However, many previous studies have shown that there is little interest in studying sustainable tourism development in terms of residents' attitudes towards tourism and the political aspects of tourism governance [17]. This article discusses this issue and contributes to understanding the constricting role of the negative political environment in implementing sustainable tourism.

1.1. Conceptual Model and Hypotheses

1.1.1. Residents' Perception of Tourism Impacts

Sustainability is often understood as the three-pillar concepts of economic, natural, and sociocultural environments. involves providing opportunities to promote economic growth, protect the location, and improve the quality of life of residents while increasing future opportunities through the development of tourism and the quality of the environment [18]. However, not all environments are subject to the same research and practical concerns [10]. Many previous studies focused only on economic or environmental pillars, which may not fully reflect community concerns [19–21]. In this case, the three pillars of sustainability concepts provide a well-structured framework for studying the positive and negative tourism's economic, environmental and sociocultural impacts.

1.1.2. Residents' Satisfaction with Tourism

In the past, when studying the satisfaction of residents, scholars divided the perception of tourism impact into two factors, such as positive tourism impact perception and negative tourism impact perception [22–24], or divided it into three factors like cost-awareness, material benefit perception, and spiritual benefit perception [25], and then the relationship between tourism impact and satisfaction could be comprehended. In the context of applying social exchange theory to residents' attitudes towards tourism, most studies involve the impacts of tourism and support for tourism, while some studies also include satisfaction with the quality of life in the tourist destinations or tourism development. Recent tourism studies indicate that tourism impacts the quality of life [26–28]. Moreover, previous studies on the impact of tourism on residents' well-being simulated the overall satisfaction of individuals with life, which stems from satisfaction with several areas of life [10]. Satisfaction with community, material, emotional, health, and safety are sources of general satisfaction with life [29].

1.1.3. Residents' Participation in Tourism at World Heritage Sites (WHS).

The variety of residents' perceptions of tourism development influences the level of residents' support and participation in tourism development [30]. Numerous studies have proven the importance of community involvement in heritage conservation and tourism development [31–34]. Local residents' involvement in WHS management can resolve conflicts between the economic and development benefits of the community and the need to preserve WHS destinations as valuable resources and can help clarify the concept of heritage among residents [33,35]. Several studies on heritage management have confirmed the importance of community participation in sustainable conservation programs [32,36]. Local residents' involvement in heritage management contributes to improving their quality of life, economic development of the local region, and sustainability of conservation programs [31,32,35].

Thus, with Aksu-Jabagly state nature reserve and the adjacent Jabagly village as a study area, this research examines the indirect impacts of the negative political environment of a tourism destination on local residents' lack of participation in tourism development through assessing the perceptions of the neighboring community from tourism in their hometown. Additionally, a number of determinants (residents' dissatisfaction with tourism and tourism's negative economic, negative environmental and positive sociocultural impacts) that influence this relationship should also be checked. The model proposed in this study assumes the relationship between the aforementioned indicators. Our structural model takes into consideration the indirect impacts of negative political environment on residents' lack of participation in tourism, which have been studied in the context of sustainable tourism development by very few scholars so far. Therefore, we incorporated both observations in our proposed model of the relationship between the three tourism pillars and residents' dissatisfaction with tourism development. Simultaneously, the following seven hypotheses (Figure 1) were developed and tested in the current study:

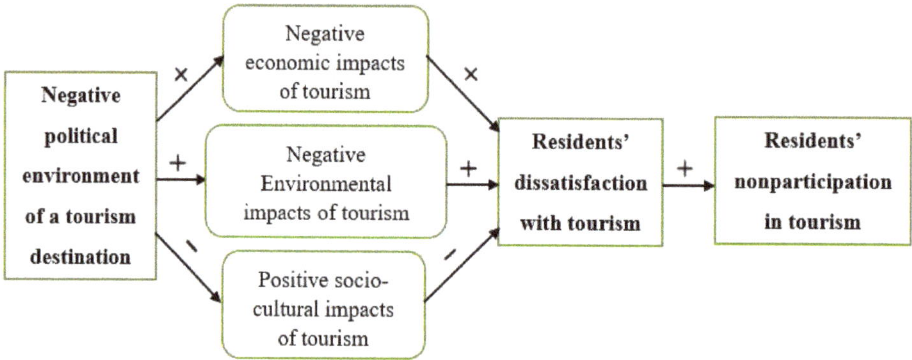

Figure 1. Theoretical model, direct paths.

Hypothesis 1. *The negative political environment has a direct positive effect on negative economic impacts of tourism;*

Hypothesis 2. *The negative political environment has a direct positive effect on negative environmental impacts of tourism;*

Hypothesis 3. *The negative political environment has a direct negative effect on positive sociocultural impacts of tourism;*

Hypothesis 4. *The negative economic impacts of tourism have a direct positive effect on residents' dissatisfaction with tourism development;*

Hypothesis 5. *The negative environmental impacts of tourism have a direct positive effect on residents' dissatisfaction with tourism development;*

Hypothesis 6. *The positive sociocultural impacts of tourism have a direct negative effect on residents' dissatisfaction with tourism development;*

Hypothesis 7. *Residents' dissatisfaction with tourism development has a direct positive effect on residents' not participation in tourism development;*

1.2. Aksu-Jabagly Natural World Heritage Site and Jabagly Village

Aksu-Jabagly Reserve is Kazakhstan's second natural world heritage site and it offers a stunningly diverse landscape from semi-deserts to snow-capped peaks. The Aksu-Jabagly, which was listed on UNESCO under the criteria of (vii) and (x) on 17 July 2016, is a unique wilderness experience where marmots, ibex, lynx, wolves, bears, argalis, and deer live [37]. The Aksu-Jabagly State Nature Reserve was established in 1926 and it is located in the north-west of Talas Alatau and the south of Karatau in the West Tien Shan. It is home to 48% of regional bird species, 72.5% of vertebrates, 221 out of 254 fungi species, 63 out of 80 moss species, 15 out of 64 vegetation types, and 114 out of 180 plant formations found in the West Tien Shan. Approximately 2500 insect species have been recorded in the reserve [38]. The wild tulips, the unique natural apples, and the snow leopards (which roam the high mountains of this area) in the Aksu-Jabagly reserve spread its name all over the world [39]. Aksu-Jabagly State Nature Reserve consists of three zones, it lies in Tulkibas district of Turkistan region and Jualy district of Jambyl region of the Republic of Qazaqstan. The main part of the nature reserve (N42 16 34, E70 40 27) has 131,704 ha property zone and 25,800 ha buffer zone. The other two zones are Karabastau paleontological area (N42 56 24, E69 54 54) and Aulie paleontological area (N42 54 18, E70 00 00) with only property zones, 100 and 130 ha, respectively [37]. The Aksu-Jabagly Biosphere Reserve is located in four districts of two oblasts in the most densely populated region of Kazakhstan (Figure 2), with a total population of about three million people. Approximately 150,000 people live in the transition area of the reserve. The main economic activities are agriculture, plant growing, and cattle breeding. In the last 10 years, ecological tourism has become highly popular in the reserve, mainly due to ornithological and botanical foreign tourism, and local recreational tourism [38].

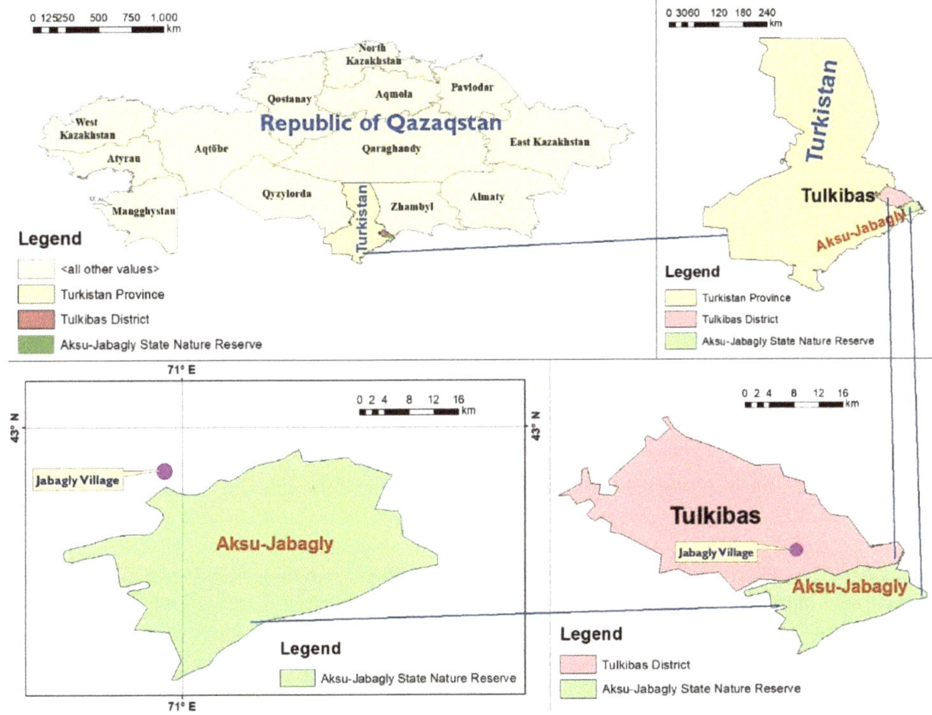

Figure 2. Aksu-Jabagly Natural World Heritage site and Jabagly village.

Jabagly village is an administrative unit of Tulkibas district. It includes the settlement of Jabagly, Abaiyl, and Russian Railway 115. The total population of the Jabagly village is 3048 people, including 2401 people of Jabagly settlement, 545 people of Abaiyl settlement, and 102 people of settlement Russian Railway 115. The center of the village is Jabagly settlement. And Jabagly settlement is 17 km southeast to the Turar Ryskulov town (former Vannovka), the administrative center of Tulkibas district. Jabagly settlement has a public transport connection with Turar Ryskulov town and Shymkent city (passport of Jabagly village, 2019). Lying adjacent to the West Tien Shan Mountains, Jabagly settlement is the gateway to Aksu-Jabagly State Natural Reserve (Figure 2). The main economic activities are agriculture, plant growing, and cattle breeding. The 59 km area of Tulkibas is located along Western Europe-Western China (WE-WC) Highway, and it provides convenience to travel to Jabagly village by car for visitors [39].

1.3. Tourism Development at Aksu-Jabagly Heritage Site

For a long time, tourism has been seen as an important means of achieving protection outcomes and as a potential source of negative impact. Decades of academic research and practical experience have indicated that the relationship between tourism and protected areas is complex. Part of the reason is that the economic priorities of tourism often conflict with each other, and the protection priorities of protected area stakeholders are also important [40]. Particularly, protection objectives may be compromised by the negative effects of visitors and commercial activities [41]. However, since the establishment of the earliest national parks, tourism and recreation have often become the main drivers of land protection [42]. Ecotourism, sustainable tourism, or "conscience tourism" is a less intrusive, more eco-friendly way to experience the world's unique natural and cultural treasures. Today, tourism faces more challenges than ever to protect and promote cultural heritage and the environment, while helping to reduce poverty by creating jobs around the world [2]. The implementation of the world heritage structure, especially in rural areas, has achieved a global impact because it has become a venue characterized by a global vision and traditional rural elements. Moreover, construction always increases local and regional development possibilities since conservation measures tend to stimulate tourism [43]. Natural World Heritage Sites (WHS) are widely recognized as the world's most important protected areas. Therefore, in order to develop tourism at a world heritage site it is necessary to consider its characteristics, for example, when developing tourism in ecologically sensitive protected areas the best strategy is to organize tourism activities at the buffer zone of the protected areas. In this respect, our research area, Aksu-Jabagly biodiversity conservation site, can be one of the best examples because, in accordance with the "Specially Protected Natural Territories" law of the Republic of Kazakhstan, areas that are not included in especially valuable ecological systems are allowed to organize ecological excursions under the control of authorities, as well as excursion paths and routes for regular tourism created by the licensed tourism sectors [38].

Without doubt, the most important indicator which shows the tourism development status of one tourist destination is the number of visitors and tourism revenue volume. Tulkibas district mayor Nurbol Turashbekov (2017) said "In 2016 more than 12 thousand tourists had visited Tulkibas district to see Aksu-Jabagly nature reserve and other places of interests, including 7% foreigners". Apparently, the aforementioned numbers are very small considering its high potential for tourism development. Below, we analyze some statistics which indicate domestic and foreign visitors to Aksu-Jabagly state nature reserve in the last 10 years.

We can easily see from Figure 3 that the number of total visitors and domestic tourists was higher in 2011 with 2890 and 2104 people, respectively. Additionally, in 2015, there were fewer visitors to the Aksu-Jabagly nature reserve, the total number of tourists decreased to 1471 people. The total and the domestic number of travelers has been increasing slowly in the last three years. As far as foreign visitors are concerned, there has been a fluctuation in the number. The year when there were fewer foreign tourists was 2013 with 666 people, while more foreign tourists visited the nature reserve than other times in 2017, the number reached 1098. It can be concluded from the above analysis that although there is a higher potential for planning tourism activities in Jabagly village, for instance, the quality of accommodation and convenience of accessibility are higher and even in line with international standards, the development of tourism in Aksu-Jabagly is still in the primary stage or even undeveloped. Therefore, we suppose one of the main factors which impede the development of tourism in our research area is the lack of favorable political environment for sustainable tourism development. Thus, the main content of this research is the impact of the negative political environment on implementing sustainable tourism development.

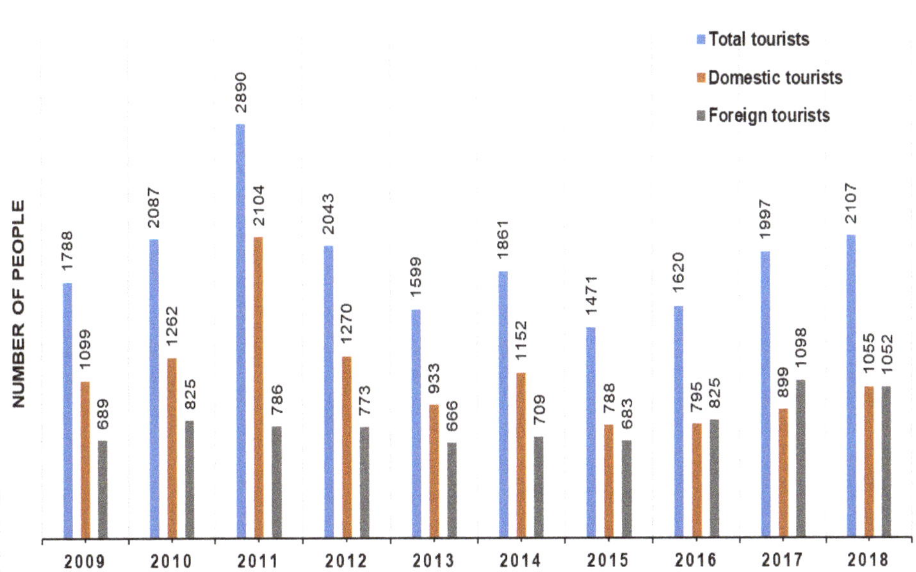

Figure 3. The number of tourists to Aksu-Jabagly heritage site from 2009 to 2018. Source of statistics: Tourism Department of Aksu-Jabagly State Nature Reserve Office.

2. Materials and Methods

2.1. Data Collection and Methodology

Data collection occurred over a 20-day period from 2 March to 22 March 2019. To develop our measurement instrument, we first created a list of sustainable tourism development indicators based on the literature review. Our list of questionnaires included over 30 indicators that were used in the subsequent survey. The self-administered questionnaire was distributed to Jabagly village's residents. To obtain more accurate opinions and perceptions, we conducted our survey by issuing the prepared questionnaire to each occupied household (328 households in total). Residents were asked to complete the questionnaire and return it to the municipality or the central school in the village (if they have a child who goes to high school there). Overall, 229 questionnaires were returned, and among them, 222 were qualified.

In order to develop our measurement tools, we first created a list of sustainable tourism development indicators based on a literature review. The list initially includes more than 100 indicators. These indicators were evaluated by relevant experts in the tourism fields. The expert group consists of scholars and local tourism industry representatives. Those experts reduced the initial list of indicators to 30, which were used for follow-up investigations.

The questionnaire for all relevant respondents was designed with three major sections. Section one was designed by ticking "√" on the corresponding option to acquire basic information about their gender, age, ethnic and the education level. Section two was designed with some multiple-choice questions indicating annual household income, current engaging industry, the number of tourism-engaged people in their family, and tourism income rate in their annual household income to understand local residents' economic situation and participation level in tourism generally. Section three evaluates respondents' perceptions of statements regarding negative political environment at the Aksu-Jabagly tourism destination, three pillars of sustainable tourism (including the negative economic, negative environmental, and positive sociocultural impacts of tourism), residents' dissatisfaction with tourism and residents' participation in tourism development. All indicators were designed as statements in section three to encourage respondents to rate on five-point Likert scale questions with 1 (completely disagree), 2 (disagree), 3 (neutral), 4 (agree) and 5 (completely agree).

The collected data were analyzed with principal component analysis (PCA) to reduce the number of variables in the model. When we removed the problematic items, we deleted some indicators, which have bigger or smaller variances of errors, in Amos output. The correlation matrix was then checked to reveal any possibly problematic variables. After this data-reduction procedure, only 18 (out of 30) indicators were used in further analysis (see Table 1). Thus, a six-factor model was then inputted into the confirmatory factor analysis (CFA) and, finally, structural equation modeling was used to establish the connections between the factors. Before the factor analysis, Cronbach's Alpha and KMO test were needed. When we tested When reliabilities between measurements, the Cronbach's Alpha coefficient of all measurement dimensions was checked and the "KMO test" was done. As shown in Table 1, in order to examine the standard intrinsic fitness level of the model, composite reliability of the latent variables and the average variance extracted value were calculated with the formula of $CR = \frac{(\sum L)^2}{(\sum L)^2 + (\sum e)}$, $AVE = \frac{\sum L^2}{n}$, ($e = 1 - L^2$ and L is completely standardized loading). Using IBM SPSS Amos 25.0 software, we conducted the confirmatory factor analysis (CFA). When we checked the overall model fit, the following equation model's fitness indices, such as CMIN/DF, NFI, TLI, CFI, RMSEA, PNFI, and PCFI, were checked. Finally, we tested the hypothesized relationships between the constructs, p value (indicating statistically significance) and critical ratio (CR) as substitute to t value and β (significant influence) were used.

Table 1. Descriptive statistics and measurement model results (n = 222).

Constructs and Indicators: $CR=\frac{(\sum L)^2}{(\sum L)^2+(\sum e)}$, $AVE=\frac{\sum L^2}{n}$, ($e = 1 - L^2$)	Mean	St. dev.	Responses in %			Model Results		
			Agree Rate * %	Neutral Rate %	Disagree Rate ** %	CSL (L)	C R	AVE
Negative political environment	3.470					L	0.984	0.939
PE_1 Tourism development is less supported by relative organizations.	3.37	1.226	62.2	6.8	31.0	0.963		
PE_2 The local area has fewer benefits from tourism development.	3.41	1.183	63.1	7.6	29.3	0.991		
PE_3 Local residents are rarely informed about tourism development there.	3.58	1.192	68.9	6.3	24.8	0.963		
PE_4 Tourism businesses are monopolized by a few politically powerful people.	3.52	1.172	67.1	5.4	27.5	0.958		
Negative economic impacts of tourism	3.393						0.927	0.811
EI_1 Tourism has increased the gap between the rich and poor in this village.	3.39	1.163	62.1	9.5	28.4	0.908		
EI_2 Local prices and the necessary cost of living for residents has increased.	3.38	1.162	61.7	9.0	29.3	0.977		
EI_5 Most of the local money is earned by outsiders.	3.41	1.173	63.1	8.1	28.8	0.808		
Negative environmental impacts	3.375						0.829	0.708
EI_3 A large influx of tourists has a great impact on the normal life of the flora and fauna.	3.38	1.181	65.4	7.2	27.4	0.860		
NEI_4 Development of tourism contributes to pollution (throw rubbish and make noise, etc.).	3.37	1.255	63.5	6.3	30.2	0.823		
Positive sociocultural impacts of tourism	3.527						0.979	0.940
ScI_2 Tourism provides an incentive for the preservation of local culture in Jabagly village.	3.59	1.096	72.1	5.8	22.1	0.965		
ScI_3 Tourism grows the cultural exchanges between tourists and residents.	3.48	1.062	63.1	13.5	23.4	0.966		
ScI_4 1 Infrastructure of this region has improved due to tourism development.	3.51	1.001	67.6	11.7	20.7	0.978		
Residents' dissatisfaction with tourism	3.275						0.986	0.960
Sat_2 I am dissatisfied with local's employment in the tourism industry here.	3.53	1.228	66.2	9.9	23.9	0.961		
Sat_3 I am dissatisfied with residents' involvement and influence in the planning and development of tourism in the Aksu-Jabagly heritage site.	3.39	1.143	61.2	11.7	27.1	0.990		
Sat_4 I am dissatisfied with the tourism generated benefits for ecological protection and regional development.	3.64	1.104	73.4	7.2	19.4	0.988		

Table 1. *Cont.*

Constructs and Indicators: $CR=\frac{(\sum L)^2}{(\sum L)^2+(\sum e)}$, $AVE=\frac{\sum L^2}{n}$, (e = 1 − L^2)	Mean	St. dev.	Responses in %			Model Results		
			Agree Rate * %	Neutral Rate %	Disagree Rate ** %	CSL (L)	C R	AVE
Residents' not participation in tourism	3.450						0.952	0.870
Par_1 I do not participate in decision making about tourism development.	3.69	1.268	68.9	11.3	19.8	0.892		
Par_2 I do not participate in planning works of tourism development.	3.58	1.297	66.2	5.4	28.4	0.991		
Par_4 I do not participate in the ecological protection works of this tourism destination.	3.08	1.037	31.1	45.0	23.9	0.912		

Notes: * Agreement rate: Completely agree + agree, and ** Disagreement rate: Completely disagree + disagree.

2.2. Sample Characteristics

Table 2 shows that the majority of respondents were male (66.67%), and most of them were Kazakhs (91.89%), while Russian and other ethnicities were only 4.50% and 3.60%, respectively. The highest proportion of respondents were aged 35–54 (51.80%), followed by 18–34 years old, accounting for 36.94% and the lowest proportion of respondents were the elder group, aged above 55 (11.26%). The proportion of people who have attended school or college (considered as middle-level education) was the largest (86.49%) and only 13.51% were those who have received higher education (including university or above). Table 2 also shows that nearly half of the respondents (50.45%) have an annual household income of 500,000–1 million KZ Tenge, followed by a family annual income of 1–1.5 million KZ Tenge, accounting for 34.68%. The populations with annual household income of below 500,000 and higher than 1.5 million KZ Tenge were the lowest and approximately the same proportion with 7.66% and 7.21%, respectively. As far as their current engaging industries are concerned, there were more residents (41.44%) engaged in animal husbandry, followed by farming (20.72%) and other industries (18.47%), the proportion of people engaged in business and tourism was the lowest and accounted for 10.81% and 8.56%, respectively, indicating that animal husbandry is the main industry and tourism is the least developed industry of community residents. Concerning the tourism income rate in household income, the population whose tourism income accounts for 0% of the annual household income was 91.44%, and other three tourism income ranges (of 1%–20%, 21%–60%, and 61%–100%) were 6.31%, 1.80%, and 0.45%, respectively. From the above statistical analysis, we can easily conclude that although they live adjacent to one of the most famous tourism destinations in Kazakhstan, residents of Jabagly village had a weak involvement in and they have nearly no tourism income (Table 2).

Table 2. Details of resident sample responses (*n* = 222).

Characteristics	Frequency	Percentage
Gender:		
Male	148	66.67
Female	74	33.33
Age (years):		
Young (18–34)	82	36.94
Middle age (35–54)	115	51.8
Elder (≥55)	25	11.26
Ethnicity:		
Kazakh	204	91.9
Russian	10	4.5
Other	8	3.6

Table 2. Cont.

Characteristics	Frequency	Percentage
Education:		
Middle (school or college)	192	86.49
High (university or above)	30	13.51
Annual household income: (KZ Tenge, 1$ = 375 tenge)		
Below 500,000	17	7.66
500,000–1 million	112	50.45
1–1.5 million	77	34.68
1.5 million and above	16	7.21
Current engaging industry:		
Tourism	19	8.56
Animal husbandry	92	41.44
Farming	46	20.72
Business	24	10.81
Other industry	41	18.47
Tourism income rate in your annual household income:		
0%	203	91.44
1%–20%	14	6.31
21%–60%	4	1.8
61%–100%	1	0.45

3. Results

3.1. Reliability and Validity Test

When testing reliability, Cronbach's Alpha is needed. Reliability analysis is used to evaluate the stability or reliability of the questionnaire. It examines the degree of consistency of the results obtained by repeated measurements of the same thing using questionnaires [44]. It is generally believed that when the reliability coefficient value reaches 0.8–0.9, the reliability of the scale is very good. When the reliability coefficient reaches 0.7–0.8, the scale has considerable reliability. Using the reliability analysis function in SPSS, the reliability test of the measurement items in the questionnaire scale was carried out, as a result, the Cronbach's Alpha coefficient of all measurement dimensions is greater than 0.8. It indicates that the reliability of all the scales is very good, and the scales have considerable reliability, and the reliability test is passed.

Validity refers to the degree of effectiveness of the measurement. It refers to the extent to which the measurement tool or means can accurately measure the things that need to be measured. The validity of the questionnaire is tested from two aspects: Content validity and structural validity. Content validity is mainly investigated by logic analysis. The structural validity of the questionnaire is usually measured by factor analysis [45]. Before the factor analysis, the KMO test is needed. When the KMO value is greater than 0.9, the effect is best, 0.7 or more is acceptable, and 0.5 or less is not suitable for factor analysis [44]. Using the factor analysis function in SPSS, the validity of all measurement items above were tested. The calculated KMO values of all items in this model are greater than 0.8, and $p <$ 0.001, reaching a very significant level, indicating that the scale is more effective.

The standard intrinsic fitness level of the model requires that the composite reliability of the latent variables is greater than 0.60, and the average variance extracted value is greater than 0.50 [46]. All of the composite reliabilities of the model in this study are greater than 0.8, and the average variance extracted values are between 0.708 and 0.960 (Table 1), it indicates that the model meets the criteria for fitness very well.

3.2. Confirmatory Factor Analysis

Based on the reliability and validity test, the model was tested for confirmatory factors using AMOS 25.0 software (IBM, New York, United States). The confirmatory factor analysis (CFA) includes three aspects: The basic fitness level of the model, the overall model fitness level, and the intrinsic fitness level of the model. The basic fitness level of the model for confirmatory factor analysis requires that the factor loadings (or completely standardized loading) must be between 0.5 and 0.95 [46]. The factor loads of all indicators in this model are above 0.5, all of them are between 0.8 and 1. This means that the basic fitness level of this model is very good.

The researchers recommended some indices to evaluate the overall model fit, including CMIN/DF (Chi-square/df), RMSEA, NFI [47], IFI, TLI, CFI, PNFI, PCFI, CN [48]. Among them, when the CMIN/DF value is between 1 and 3, the model has a simple adaptation degree. The standard of IFI value, TLI value and CFI value is above 0.9, and the standard of RMSEA value is lower than 0.05 (good fit) and less than 0.08 (suitable), PNFI and PCFI values are above 0.5, and CN should be greater than 200 [46]. When the variance for the whole model was checked, 12 variables (two from negative economic impacts of tourism, two from negative environmental impacts, three from positive sociocultural impacts of tourism, three from residents' dissatisfaction and two from the residents' not participation in tourism) were excluded from further analysis due to the higher p-value ($p > 0.05$) and 18 indicators remained in our proposed model. After deleting those 12 indicators, nearly all p-values were smaller than 0.05. After the modification, the indexes of the overall model fitness were: CMIN/DF = 2.699, NFI = 0.931, TLI = 0.941, CFI = 0.955, RMSEA = 0.062, PNFI = 0.710, PCFI = 0.729, CN > 200. From the results, it can be easily seen that the corrected model fits well.

Finally, structural equation modelling was undertaken to test the hypothesized relationships between the factors. The resulting structural model provides evidence for the proposed relationships between the constructs and their indicators. All measures tested above provide evidence of a good model fit.

SEM confirms the connections among the negative political environment, the perceived negative economic, negative environmental and positive sociocultural impacts of tourism, residents' dissatisfaction and residents' nonparticipation in tourism development. Not all constructs are relatively well explained by their predictors, as suggested by the explained variance, which ranges from 0.01 to 0.70. However, most of the path coefficients (6 out of 7) between the two constructs are still significant (Table 3).

Table 3. The path coefficients between the two constructs.

Constructs			C.R. (t)	*p* Value
Negative_Environmental_Impacts	<—	Negative__Political_Environment	16.207	***
Positive__Sociocultural_Impacts	<—	Negative__Political_Environment	−0.440	0.660
Negative__Economic_Impacts	<—	Negative__Political_Environment	2.573	0.010
Residents__Dissatisfaction	<—	Negative__Economic_Impacts	4.932	***
Residents__Dissatisfaction	<—	Negative_Environmental_Impacts	2.008	0.045
Residents__Dissatisfaction	<—	Positive__Sociocultural_Impacts	−2.033	0.042
Residents_Nonparticipation	<—	Residents__Dissatisfaction	6.075	***

Notes: *** Statistically significant at $p < 0.001$.

Further analysis of the structural part of the model reveals that the negative political environment has a significant positive effect on negative environmental impacts of tourism, negative economic influences of tourism have a significant positive effect on residents' dissatisfaction with tourism development and residents' dissatisfaction has a significant positive effect on residents' nonparticipation in tourism development ($\beta = 0.83$, t = 16.207, $p < 0.001$; $\beta = 0.32$, t = 4.932, $p < 0.001$ and $\beta = 0.39$, t = 6.075, $p < 0.001$, respectively), indicating a significant and strong positive relationship between negative political environment and negative environmental impacts of tourism, tourism's negative economic impacts and residents' dissatisfaction with tourism, residents' dissatisfaction and residents' not participation in tourism. it means that the higher perception of residents on the negative political environment, negative impacts of tourism and dissatisfaction of residents, the higher perception of residents on negative environmental impacts of tourism, residents' dissatisfaction with tourism and residents' not participation in tourism.

Similarly, the path coefficient between negative political environment and negative economic impacts of tourism is 0.18 (t = 2.573, $p < 0.05$) and the path coefficient between negative environmental impacts of tourism and residents' dissatisfaction with tourism is 0.14 (t = 2.573, $p < 0.05$). It indicates that the negative political environment has a positive significant influence on tourism's negative economic impacts, at the same time tourism's negative environmental impacts have a positive significant influence on residents' dissatisfaction with tourism. Therefore, H1, H2, H4, H5, and H7 were all proven.

There was no significant relationship between the negative political environment and positive sociocultural impacts of tourism but the negative relationship between positive sociocultural impacts of tourism and residents' dissatisfaction with tourism development was relatively significant. The negative political environment has a very weak negative influence on the positive sociocultural impacts of tourism ($\beta = -0.03$, t = -0.440, $p > 0.05$), and positive environmental impacts of tourism have a significant negative effect on residents' dissatisfaction ($\beta = -0.13$, t = -2.033, $p < 0.05$), so H6 was proven, but H3 was not proven, indicating that negative political environment is not a function of positive sociocultural impacts of tourism development at the Aksu-Jabagly tourism destination.

In the seven relationship hypotheses in the proposed model, six were true but one was not. H3 is not valid because the path analysis results are contrary to the proposed assumption (Figure 4).

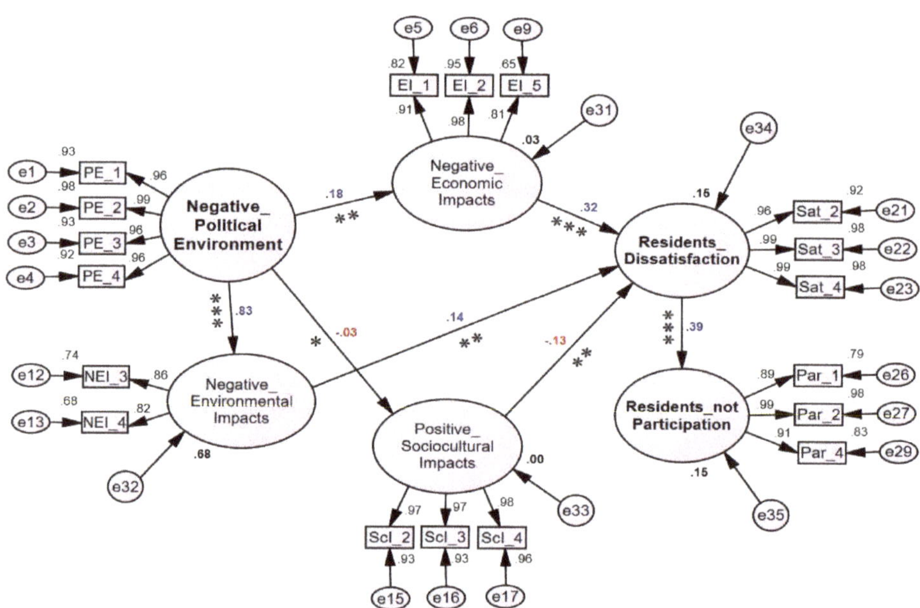

Figure 4. The model of residents' participation in tourism. *** Statistically significant at $p < 0.001$, ** Statistically significant at $p < 0.05$, and * Statistically insignificant.

4. Discussion

Our proposed model in this study was developed based on the six-factor model constructed by Mihalič et al. (2016), which includes factors such as the political environment, the three pillars of sustainable tourism development (economic, sociocultural and natural), residents' satisfaction and residents' support for tourism. In their model, Mihalič et al. (2016) explore the direct impacts of the political environment on three dimensions of sustainable tourism development, direct impacts of the three pillars of sustainable tourism development on residents' satisfaction with tourism, indirect impacts of the political environment on residents' satisfaction with tourism and direct impacts of residents' satisfaction with tourism on residents' support for tourism. And in our model, only seven direct connections were originally hypothesized. Instead of residents' support for tourism, we used residents' participation in tourism. Our study confirmed a total of 18 indicators that created a six-factor model in line with our theoretical model construct (Table 1). More specifically, these factors were the negative political environment of tourism destination, negative economic, positive sociocultural and negative environmental impacts of tourism, residents' dissatisfaction with tourism development, residents' nonparticipation in tourism. Each of these represents a self-standing construct in our model.

This paper contributes to the tourism knowledge base by integrating the dimensions of the political environment into sustainable models that could survey community participation in tourism. Therefore, our proposed model and discussion begin with the negative political environment which was measured by four indicators: Insufficient support from relative tourism organizations for locals, fewer benefits from tourism development for the local residents, less information about tourism development for local residents, and monopolization of tourism businesses by a few people in the village. Then, our model analyzed the direct impacts of the six factors.

It was concluded that the residents of Jabagly highly perceive the negative political environment because they agree with statements about describing negative political environment elements, the four indicators of negative political environments were evaluated with the mean value of 3.470. According to the respondents, "tourism development is less supported by relevant organizations" (mean = 3.37), "the local residents have fewer benefits from tourism development" (mean = 3.41), "local residents are rarely informed about tourism development" (mean = 3.58), and "tourism businesses are monopolized by a few politically powerful people" (mean = 3.52), which can be significantly improved. The negative political environment factors had a relatively high composite reliability (CR = 0.984), revealing the construct's high level of internal consistency. Furthermore, residents' dissatisfaction with tourism development also received higher mean values (3.275). Finally, it can be seen from the respondents' evaluation that residents not participating in tourism similarly received higher mean values (3.540). One can easily imagine that further improvement of the (at present) relatively negative political environment would result in even higher residents' satisfaction with tourism development and increase active participation of locals in the tourism industry.

With respect to the impact of Aksu-Jabagly's political environment, based on the other factors, seven direct connections were originally hypothesized. The first group of hypotheses assumes the role of the negative political environment on the three pillars of sustainable tourism (negative economic, negative environmental, and positive sociocultural impacts of tourism) and how they are perceived by the community (H1, H2, and H3). The results confirm that the impact of the negative political environment on negative economic (H1: $\beta = 0.18$, $p < 0.05$) and negative environmental (H2: $\beta = 0.83$, $p < 0.001$) impacts of tourism at the Aksu-Jabagly heritage site were significant and positive, while the impact of the negative political environment on positive sociocultural impacts of tourism (H3: $\beta = -0.03$, $p > 0.05$) was not statistically significant, indicating that a negative political environment impact increased the residents' evaluation of negative economic and environmental impacts of tourism. By forming different factors with CFA, this study affirmed that, in reality, the impacts of tourism on the destination could be divided into the sociocultural, natural, and economic impacts [49–51]. The respondents also gave a relatively high score to negative economic and environmental impacts of tourism development in their hometown. The three indicators of negative economic impacts of tourism were evaluated with the mean value of 3.393, meanwhile, the two indicators of negative environmental impacts of tourism were evaluated with the mean value of 3.375.

Figure 4 shows that negative economic and environmental impacts of tourism had a significant positive effect on residents' dissatisfaction with tourism (H4, $\beta = 0.32$, $p < 0.001$ and H5: $\beta = 0.14$, $p < 0.05$, respectively). It can be observed that residents largely agree that tourism caused negative economic and environmental impacts, such as the widened gap between the rich and poor (mean = 3.39), risen local prices, and the necessary cost of living for residents (mean = 3.38), the leakage of local money (mean = 3.41), the high numbers of tourists who disturb the normal life of the flora and fauna in the reserve (mean = 3.38), and the tourism-generated pollution (throwing rubbish and making noise, etc.) in the tourism destination (mean = 3.37). Based on residents' assessment in Jabagly village, they also gave relatively high scores to the effects of tourism's positive sociocultural impacts (with an average of 3.527) on residents' dissatisfaction with tourism, however, the higher perception had a weaker negative influence on residents' dissatisfaction with tourism (H6, β = −0.13, $p < 0.05$), more research is needed in this area.

Therefore, it is assumed that the more positive political environment for tourism development is seen to be benefiting the positive economic development and environmental protection in Aksu-Jabagly, given that more residents are satisfied with tourism and embrace tourism development in their communities.

The seventh hypothesis (H7), which proposed residents' dissatisfaction with tourism development positively affects residents' nonparticipation in tourism, was proved (H3: β = 0.39, $p < 0.001$). It was found that in Aksu-Jabagly natural heritage tourism destination, higher dissatisfaction of local residents resulted in residents' weak participation in tourism development in their village. The high rate of dissatisfaction and low participation is noteworthy. From the investigation, a conclusion can be drawn that although the direct reason for residents' low involvement in the tourism sector was due to the dissatisfaction of local community with tourism development, one of the most primary indirect reasons for passive participation in the tourism industry was the negative political environment in the Aksu-Jabagly tourism destination. If local residents believe that authorities and government officials are interested in hearing their voices and providing them with an opportunity to participate in the decision-making process, it will be a big encouragement for their participation. In the end, residents will participate in conservation programs and tourism development within the scope of what they believe the local government allows [52–54]. However, in underdeveloped and rural destinations, especially in developing countries, residents believe that the political structure of centralization and the tendency of local policymakers to evade power sharing will be detrimental to them [52,55,56]. Therefore, for rural residents in many developing countries, negative political environments, such as hiding preferential policies, unequal participation opportunities, and unequal benefit sharing, will limit their enthusiasm for participating in the tourism industry.

To sum up, the participation of local residents in the World Heritage tourism development in their hometown is one of the main prerequisites for sustainable tourism. If the political environment for implementing tourism development is beneficial for local residents, they will actively participate in the measures of protecting the world heritage sites within their communities. When implementing effective measures of sustainable tourism development, the local people play a very important role because they are more familiar with those antiquities and know well what it takes to protect and promote them.

5. Conclusions

This study highlights the importance of the political environment for sustainable tourism development. Residents' support for tourism may be affected by a well-developed political environment and destination governance [10]. Our results confirm that the negative political environment of a tourism destination can determine residents' negative assessment of the three pillars of sustainability (economic, environmental, and sociocultural) These negative assessments of the three pillars of sustainability can increase residents' dissatisfaction with the pace of tourism development. Therefore, residents' participation in tourism may be affected by the badly-developed political environment in the tourism destination.

Based on the above findings, the study also helps local communities and the government to realize the importance of the positive political environment of the tourist destination in developing sustainable tourism. Based on the identified connections and impacts, the Aksu-Jabagly community has the potential to increase residents' participation level in tourism development by improving the dimensions of the political environment. In this regard, in order to improve the current situation in Aksu-Jabagly world natural heritage tourism destination, the following measures are recommended: The relevant organizations should provide adequate support for tourism development, the tourism development generates more benefit to the development of the local area, relevant authorities provide local residents with comprehensive information about tourism development, opportunities of engaging in the tourism sector should be equally given everyone in the local area. Additionally, in order to achieve sustainable tourism, tourism development should recognize and encourage a higher level of local community satisfaction because local residents are one of the key stakeholders in tourism destinations. This requires a modification of the destination governance system to effectively develop and implement tourism policies based on the coordination and cooperation of all stakeholders. To ensure a higher level of coordination in the tourism industry itself, governance must overcome barriers of incoherent industries that do not adequately represent vulnerable interest groups [57], which are usually composed of local community residents. In short, in the case of Kazakhstan, reducing the influence of the negative political system and power structure on the tourism industry is one of the key ways to achieve sustainability in the most vulnerable heritage tourism destinations, specifically heritage sites like Aksu-Jabagly Biodiversity Reserve. Therefore, it is important to have a clear understanding of political issues, the interests of key political actors and how to mitigate personal interests in order to promote and maintain sustainable tourism development in this developing country.

Those aforementioned effective measures will increase local residents' satisfaction with tourism and active participation in tourism development, as a result, sustainable tourism development can be realized at this vulnerable biodiversity heritage site. The proposed model can also serve as a pioneer in further research to determine whether the model can be adapted and applied to other destinations to improve the political environment of a tourism destination and implementation of sustainable tourism development.

This study was not without its limitations that can affect the applicability of the results. This study surveyed a small sample size of local residents but did not investigate the perceptions of other stakeholder groups, such as tourists, government/local authorities, or tourism industry/the private sector. Therefore, a broad view of all tourism relevant stakeholders may not have been captured. In this regard, future research may be required to test the above-mentioned indicators, especially the poorly-developed political environment indicator. Furthermore, respondents can be selected from the various stakeholders in the tourism sector in further studies.

Author Contributions: Conceptualization: I.A. and Z.Y.; methodology, formal analysis; visualization and writing—original draft preparation: I.A.; investigation: I.A. and G.K.; supervision Z.Y.; project administration: F.H. All authors have read and agreed to the published version of the manuscript.

Funding: This research was sponsored by the projects of the National Key Research and Development Project of China with grant no. (2016YFC0503307), the National Natural Science Foundation of China under Grant no. (41971192) and the Chinese Government Scholarship under Grant number (80001).

Acknowledgments: Jumanov Smatulla Zhorauly, the deputy director of scientific research department of Aksu-Jabagly State Nature Reserve office in Jabagly village, provided great help during the field and social surveys. Special thanks Beknur Izenbaev (Ph.D from M. Auezov South Kazakhstan State University in Shymkent) and Ordenbek Mazbayev, Aday Seken and Gauhar Ospan (L.N. Gumilyov Eurasian National University in Nur-Sultan), who helped to design questionnaires for local residents.

References

1. Crotti, R.; Misrahi, T. *The Travel & Tourism Competitiveness Report 2017. Paving the Way for a More Sustainable and Inclusive Future*; World Economic Forum: Geneva, Switzerland, 2017.
2. Becken, S.; Wardle, C. *Tourism Planning in Natural World Heritage Sites*; Griffith Institute for Tourism Research Report No 13; Griffith University: Queensland, Australia, 2017.

3. Mihalič, T.; Kaspar, C. *Umweltökonomie im Tourismus*; Haupt: Bern, Switzerland, 1996.
4. Ritchie, J.B.; Crouch, G.I. *The Competitive Destination: A Sustainable Tourism Perspective*; CABI: Wallingford, UK, 2003.
5. Beritelli, P. *Tourist Destination Governance through Local Elites: Looking Beyond the Stakeholder Level*; University of St. Gallen: St. Gallen, Switzerland, 2011.
6. Bramwell, B.; Lane, B. Critical research on the governance of tourism and sustainability. *J. Sustain. Tour.* **2011**, *19*, 411–421. [CrossRef]
7. ECETAT, ECOTRANS; The European VISIT Initiative. *Tourism Eco Labelling in Europe—Moving the Markets towards Sustainability*; European Centre for Eco Agro Tourism: Amsterdam, The Netherlands, 2004.
8. Mihalič, T. Sodobna opredelitev razvoja in uspešnosti poslovanja. In *Oblikovanje Modela Merjenja Uspešnosti Poslovanja Hotelskih Podjetij*; RCEF: Ljubljana, Slovenia; ITEF: Ljubljana, Slovenia, 2009.
9. Mihalič, T.; Žabkar, V.; Knežević Cvelbar, L. A hotel sustainability business model: Evidence from Slovenia. *J. Sustain. Tour.* **2012**, *20*, 701–719. [CrossRef]
10. Mihalič, T.; Šegota, T.; Knežević Cvelbar, L.; Kuščer, K. The influence of the political environment and destination governance on sustainable tourism development: A study of Bled, Slovenia. *J. Sustain. Tour.* **2016**, *24*, 1489–1505. [CrossRef]
11. Kreag, G. *The Impacts of Tourism*; Minnesota Sea Grant: Duluth, MN, USA, 2001.
12. Dyer, P.; Gursoy, D.; Sharma, B.; Carter, J. Structural modeling of resident perceptions of tourism and associated development on the Sunshine Coast, Australia. *Tour. Manag.* **2007**, *28*, 409–422. [CrossRef]
13. Vargas-Sanchez, A.; Porras-Bueno, N.; De Los Ángeles Plaza-Mejía, M. Explaining residents' attitudes to tourism: Is a universal model possible? *Ann. Tour. Res.* **2011**, *38*, 460–480. [CrossRef]
14. UNWTO. *Making Tourism More Sustainable: A Guide for Policy Makers*; UNWTO: Madrid, Spain, 2005; pp. 11–12.
15. Hall, C.M. Policy learning and policy failure in sustainable tourism governance: From first-and second-order to third-order change? In *Tourism Governance*; Bramwell, B., Lane, B., Eds.; Routledge: London, UK, 2013; pp. 249–272.
16. Higgins-Desbiolles, F. The elusiveness of sustainability in tourism: The culture-ideology of consumerism and its implications. *Tour. Hosp. Res.* **2010**, *10*, 116–129. [CrossRef]
17. Jurowski, C.; Gursoy, D. Distance effects on residents' attitudes toward tourism. *Ann. Tour. Res.* **2004**, *31*, 296–312. [CrossRef]
18. Eagles, P.F.J.; McCool, S.F.; Haynes, C.D.; Phillips, A. *Sustainable Tourism in Protected Areas: Guidelines for Planning and Management*; Best Practice Protected Area Guidelines Series No. 8; IUCN: Gland, Switzerland, 2002.
19. Harrill, R. Residents' attitudes toward tourism development: A literature review with implications for tourism planning. *J. Plan. Lit.* **2004**, *18*, 251–266. [CrossRef]
20. Nunkoo, R.; Smith, S.L.; Ramkissoon, H. Residents' attitudes to tourism: A longitudinal study of 140 articles from 1984 to 2010. *J. Sustain. Tour.* **2013**, *21*, 5–25. [CrossRef]
21. Sharpley, R. Host perceptions of tourism: A review of the research. *Tour. Manag.* **2014**, *42*, 37–49. [CrossRef]
22. Ko, D.-W.; Stewart, W.P. A structural equation model of residents' attitudes for tourism development. *Tour. Manag.* **2002**, *23*, 521–530. [CrossRef]
23. Wang, X.; Zhen, F.; Wu, X.-G.; Zhang, H.; Liu, Z.-H. Driving factors of resident satisfaction with tourism development: A case study of Yangshuo in Guangxi Zhuang Autonomous Region. *Geogr. Res.* **2010**, *29*, 841–851.
24. Wang, S.; Xu, H. Influence of place-based senses of distinctiveness, continuity, self-esteem and self-efficacy on residents' attitudes toward tourism. *Tour. Manag.* **2015**, *47*, 241–250. [CrossRef]
25. Li, Y.; Zhang, J.; Liu, Z.; Wu, J.; Xiong, J. Structural relationship of residents' perception of tourism impacts: A case study in world natural heritage Mount Sanqingshan. *Prog. Geogr.* **2014**, *33*, 584–592.
26. Tokarchuk, O.; Gabriele, R.; Maurer, O. Tourism intensity impact on satisfaction with life of German residents. *Tour. Econ.* **2016**, *22*, 1315–1331. [CrossRef]
27. Kim, Y. *The Korean Wave: Korean Media Go Global*; Routledge: London, UK, 2013.
28. Woo, E.; Kim, H.; Uysal, M. Life satisfaction and support for tourism development. *Ann. Tour. Res.* **2015**, *50*, 84–97. [CrossRef]
29. Kim, K.; Uysal, M.; Sirgy, M.J. How does tourism in a community impact the quality of life of community residents? *Tour. Manag.* **2013**, *36*, 527–540. [CrossRef]

30. Easterling, D.S. The residents' perspective in tourism research: A review and synthesis. *J. Travel Tour. Mark.* **2005**, *17*, 45–62. [CrossRef]

31. Jaafar, M.; Noor, S.M.; Rasoolimanesh, S.M. Perception of young local residents toward sustainable conservation programmes: A case study of the Lenggong World Cultural Heritage Site. *Tour. Manag.* **2015**, *48*, 154–163. [CrossRef]

32. Nicholas, L.N.; Thapa, B.; Ko, Y.J. Residents' perspectives of a world heritage site: The Pitons Management Area, St. Lucia. *Ann. Tour. Res.* **2009**, *36*, 390–412. [CrossRef]

33. Su, M.M.; Wall, G. Community participation in tourism at a world heritage site: Mutianyu Great Wall, Beijing, China. *Int. J. Tour. Res.* **2014**, *16*, 146–156. [CrossRef]

34. Yung, E.H.; Chan, E.H. Evaluation for the conservation of historic buildings: Differences between the laymen, professionals and policy makers. *Facilities* **2013**, *31*, 542–564. [CrossRef]

35. Sirisrisak, T. Conservation of Bangkok old town. *Habitat Int.* **2009**, *33*, 405–411. [CrossRef]

36. Yung, E.H.; Chan, E.H. Problem issues of public participation in built-heritage conservation: Two controversial cases in Hong Kong. *Habitat Int.* **2011**, *35*, 457–466. [CrossRef]

37. Nomination Dossier: Western Tien-Shan. Proposal for Inscription on the UNESCO World Cultural and Natural Heritage List. Available online: https://whc.unesco.org (accessed on 16 December 2015).

38. Kazakhstan National Committee. *Aksu-Zhabagly: Biosphere Reserve*; Institute of Zoology: Almaty, Kazakhstan, 2014.

39. Rakhimova, Y.V.; Nam, G.A.; Yermekova, B.D.; Jetigenova, U.K.; Yessengulova, B.Z. Ecological and trophic differentiation of fungal diversity in Aksu-Zhabagly Nature Reserve (Kazakhstan). *Contemp. Probl. Ecol.* **2017**, *10*, 511–523. [CrossRef]

40. Wilson, E.; Nielsen, N.; Buultjens, J. From lessees to partners: Exploring tourism public–private partnerships within the New South Wales national parks and wildlife service. *J. Sustain. Tour.* **2009**, *17*, 269–285. [CrossRef]

41. Jamal, T.; Stronza, A. Collaboration theory and tourism practice in protected areas: Stakeholders, structuring and sustainability. *J. Sustain. Tour.* **2009**, *17*, 169–189. [CrossRef]

42. Liburd, J.J. Sustainable tourism and national park development in St. Lucia. In *Tourism and Social Identities: Global Frameworks and Local Realities*; Burns, P.M., Novelli, M., Eds.; Elsevier: Oxford, UK, 2006; pp. 155–174.

43. Butler, R.; Hall, C.M.; Jenkins, J.M. (Eds.) *Tourism and Recreation in Rural Areas*; John Wiley & Sons, Inc.: Hoboken, NJ, USA, 1998.

44. Qiang, D.; Jia, L. *SPSS Statistical Analysis from Entry to the Master*; Posts & Telecom Press: Beijing, China, 2009. (In Chinese)

45. Wang, C. Perception of the Impact of Residents of World Heritage Sites on Tourism Ecological Migration. Master Thesis, Hunan Normal University, Changsha, China, 2011. (In Chinese).

46. Wu, M. *Structural Equation Model: Operation and Application of AMOS*; Chongqing University Press: Chongqing, China, 2009. (In Chinese)

47. Doh, J.P.; Guay, T.R. Corporate social responsibility, public policy, and NGO activism in Europe and the United States: An institutional-stakeholder perspective. *J. Manag. Stud.* **2006**, *43*, 47–73. [CrossRef]

48. Nunkoo, R.; Ramkissoon, H. Developing a community support model for tourism. *Ann. Tour. Res.* **2011**, *38*, 964–988. [CrossRef]

49. Jurowski, C.; Uysal, M.; Williams, D.R. A theoretical analysis of host community resident reactions to tourism. *J. Travel Res.* **1997**, *36*, 3–11. [CrossRef]

50. Stylidis, D.; Terzidou, M. Tourism and the economic crisis in Kavala, Greece. *Ann. Tour. Res.* **2014**, *44*, 210–226. [CrossRef]

51. Yoon, Y.; Gursoy, D.; Chen, J.S. Validating a tourism development theory with structural equation modeling. *Tour. Manag.* **2001**, *22*, 363–372. [CrossRef]

52. Aas, C.; Ladkin, A.; Fletcher, J. Stakeholder collaboration and heritage management. *Ann. Tour. Res.* **2005**, *32*, 28–48. [CrossRef]

53. Tosun, C. Limits to community participation in the tourism development process in developing countries. *Tour. Manag.* **2000**, *21*, 613–633. [CrossRef]

54. Jepson, A.; Clarke, A.; Ragsdell, G. Investigating the application of the motivation-opportunity-ability model to reveal factors which facilitate or inhibit inclusive engagement within local community festivals. *Scand. J. Hosp. Tour.* **2014**, *14*, 331–348. [CrossRef]

55. Tosun, C. Expected nature of community participation in tourism development. *Tour. Manag.* **2006**, *27*, 493–504. [CrossRef]
56. Marzuki, A.; Hay, I.; James, J. Public participation shortcomings in tourism planning: The case of the Langkawi Islands, Malaysia. *J. Sustain. Tour.* **2012**, *20*, 585–602. [CrossRef]
57. Williams, A.; Shaw, G. Tourism and the environment: Sustainability and economic restructuring. *Tour. Environ. Sustain. Econ. Restruct.* **1998**, 49–59.

Article

Aging Society and the Selected Aspects of Environmental Threats: Evidence from Poland

Elżbieta Sobczak [1], Bartosz Bartniczak [2] and Andrzej Raszkowski [1,*]

[1] Department of Regional Economy, Wroclaw University of Economics and Business, Nowowiejska 3, 58-500 Jelenia Góra, Poland; elzbieta.sobczak@ue.wroc.pl

[2] Department of Quality and Environmental Management; Wroclaw University of Economics and Business, Nowowiejska 3, 58-500 Jelenia Góra, Poland; bartosz.bartniczak@ue.wroc.pl

* Correspondence: andrzej.raszkowski@ue.wroc.pl; Tel.: +48-606-262-335

Received: 29 April 2020; Accepted: 4 June 2020; Published: 6 June 2020

Abstract: The article addresses problems of population aging in Poland and the selected environmental hazards exerting a negative impact on seniors' health. The introduction presents the reasons underlying the above-mentioned aging process and provides the characteristics of the primary environmental threats. The next part covers the most important trends and indicators related to the demographic situation in the country. The core of the study is focused on presenting and interpreting the results of empirical research on the periodization of the population aging process in Poland, in the years 2004–2019, using the multidimensional statistical analysis method including, in particular, the data classification method. The key demographic factors differentiating the development phases of the population aging process include, in order of their significance: longer life expectancy of the population, narrowed gross reproduction rate, declining birth rate, and total migration balance. In addition, the article provides the analysis of the selected environmental threats' impact on population aging in Poland, among which the following were indicated: high temperatures, solar conditions, heavy rainfall, strong winds, droughts, and fires. Moreover, the process of longer life expectancy in fine health is essentially related to two issues: the level of medical care, with particular attention paid to check-ups and preventive measures, and the promotion of a healthy lifestyle. In summary, it should be highlighted that the elimination of all pollutants or the reasons of environmental hazards is not possible; however, the actions primarily focused on reducing the emission of harmful gases into the atmosphere and other forms of environmental pollution should definitely be taken.

Keywords: natural environment; sustainable development; society; aging; environmental hazards; periodization; multidimensional statistical analysis; Ward's method; Poland

---

1. Introduction

Population aging is a problem which affects the vast majority of European countries, including Poland. It can be viewed that this phenomenon represents a specific sign of our times—we live longer, whereas the fertility rate remains at a relatively low level. The activities focused on the intensification of pro-family policy and the allocation of parental benefits, which is also taking place in Poland, stand for natural remedial undertakings in this respect. Unfortunately, as the previous analyses have shown, such actions, even though expensive, are not very effective. The financial factor is significant; however, it is not the most important one in the context of family planning and the number of offspring. The approach to private life and professional careers of the members of Polish families seems to be more important nowadays. In addition to the typical social support and financial transfers, the promotion of a large family model, strengthened by the system of discounts and promotions offered by both public and private sector entities, can turn out effective.

The aforementioned aging is a process difficult to put a halt to, let alone reverse. Therefore, the care offered to seniors has to be enhanced. Special programs should be developed to support and activate the elderly as well as provide them with medical services at a satisfactory level. When investigating the positive aspects of the described aging situation, the process of increasing the average life expectancy in Poland can be recognized as one of them, which, indeed, results from the improvement in the level of provided health care over the years. Life expectancy continues to grow practically all the time since the beginning of the transformation period in the 90s. Effective activities carried out by the non-governmental organizations represent an important factor activating seniors [1]. These activities are perceived as an important socio-economic phenomenon manifested in creating development perspectives for the poorer part of the society [2] and participation in shaping the current socio-economic policy [3]. In other words, Non-Government Organizations (NGOs) in Poland, in addition to the business and local government sector, remain an inseparable component, acting as the creators of development processes. It should be noted that their activity is not yet as dynamic as those in the United States or Western European countries [4].

The purpose of the study is to present the phenomena and processes related to the problem of population aging in Poland from various perspectives. This subject matter is considered highly important and current in the subject literature. [5,6]. The added value of the presented research is addressing simultaneously the environmental threats affecting both life and health of the society and, in particular, of the elderly. The main focus of the article was placed on empirical research covering the issue of periodization regarding population aging in Poland, in the years 2004–2019, using the multidimensional statistical analysis method, mainly including the data classification method. The study has the following structure: an introduction, an overview of the selected publications, the aging society in numbers, an impact of selected environmental threats on the aging society, methodology, main results, a discussion, and conclusions. In summary, the conclusions and recommendations were presented based on the conducted studies and analyses.

2. Literature Review

Social stratification is a phenomenon correlated with population aging and poverty. It often affects older people who cannot cope with the rapidly changing market milieu in the country. It is not possible to talk about sustainable development [7] when a significant part of the society persists at risk of poverty. Social wealth and stability are not evidenced by a narrow group of the rich, but a broad and strong middle class capable of covering senior citizens with proper care and creating fair conditions for their existence. It should be emphasized that despite an undeniable socio-economic development of Poland [8], low unemployment rate at the level of 5.2% at the end of 2019, calculated according to the methodology of the Statistics Poland, the scale of stratification has not improved significantly. For example, a higher standard of living in large cities (in terms of Polish conditions) is not always followed by an improved functioning in smaller ones, inhabited by the majority of the country population. Social stratification itself is associated with a number of other adverse phenomena, e.g., poverty, social exclusion, addictions, crime, hereditary unemployment, low social activity, and entrepreneurship.

The already mentioned sustainable development [9,10] can be defined, for the purposes of this study, as the process of transformations which ensures meeting the needs of the present generation, taking into account intergenerational justice [11], as a result of performing integrated activities in the social, economic, environmental, spatial, institutional, and political dimensions [11–14]. The initiators of sustainable development include society, entrepreneurs, the world of science, and public authorities [15] with the support of good governance by NGOs [16]. Good governance stands for, in simple terms, undertaking and implementing correct decisions, involving the stakeholders, and paying attention to social interest and consensus. In this context, openness, participation, responsibility, efficiency, and coherence are of high importance. In other words, good governance represents effective strategic management in the public sector, taking into account participation and social consensus [17]. In terms of sustainable development, the natural environment remains its base, the economy stands for its tool, and well-being along with high quality of social life are the goals of taking respective actions [18]. The important components of the sustainable development concept, reducing the use of renewable resources to the level defined by the capacity of restoring them [19], include limiting the use of non-renewable resources within the scale, allowing their gradual replacement by the appropriate substitutes [20]. In addition, it is important to eliminate consistently the hazardous and toxic substances from economic processes and maintain emissions within the limits set by the carrying capacity in the environment. Biodiversity should be restored and holistically protected at the landscape, ecosystem [21], genic, and species levels.

Polish society is faced with aging in the times of environmental threats. Among the most important hazards the following can be listed: atmospheric air pollution, water pollution, waste pollution, soil degradation, still unsolved problem of asbestos in the natural environment, noise and vibration predominantly in urban areas [22], radiation hazard, increasing carbon dioxide level, and depletion of natural resources. In addition, the threats [23] either directly or indirectly related to climate change [24] include more frequent occurrence of high temperatures (global warming), sudden floods and inundations caused by heavy rainfall, increased incidence of intensive gales, and higher frequency of droughts and the resulting fires.

The aforementioned facts can be supplemented with the species threatened with extinction and the loss of biodiversity. The latter phenomenon has particularly negative consequences, because biodiversity is the basis of ecosystems' existence and functioning [25], which we all benefit from. Care for the natural environment and biodiversity is one of the basic factors for counteracting poverty in third world countries. Furthermore, biodiversity is the basis of sustainable development and economic growth in developed countries [26].

A separate problem is the phenomenon of deforestation [27], i.e., the process of shrinking share of forest areas in the given total area [28], usually as a result of anthropopressure, which is visible through an excessive economic use of forests or environmental pollution [29]. Deforestation can lead to floods, droughts, landslides, and mudslides. In Poland, due to the fact that the forest area is slowly expanding, this phenomenon seems to be under control.

Industry is one of the economy sectors focused on the exploitation and processing of natural resources on a massive scale into products intended to meet human needs. The above-mentioned industry is often indicated as the factor most responsible for environmental degradation. Its negative impact is actually wide and varied [30] and is listed as follows: emission and deposition of gaseous, dust, and aerosol pollutants; direct and indirect soil degradation, acidification, and alkalization; impact on the well-being and life of plants, animals, and humans; destruction of fixed assets and anthropogenic capital through accelerated corrosion and degradation of surface layers; heavy metals emission and deposition in the natural environment, primarily in food; pollution and degradation of surface and underground waters; degradation of sea and ocean waters; emission of electromagnetic and ionizing radiation; noise emissions; acoustic climate disturbances; degradation of non-economic values of the environment, landscape, recreation and leisure; solid waste emissions and surface destruction; land consumption of industrial production; industrial disasters posing high environmental risks [31]; excessive demand for raw materials, energy, and other natural resources; thermal pollution of the natural environment; degradation of basic natural capital, climate, ozone layer. Moreover, agriculture also has a negative impact on the level of pollution [32]. Using artificial fertilizers, plant protection substances, and products improving the structure of arable land affects the accumulation of toxic substances in the natural environment, e.g., pesticides, nitrates, arsenic, cyanides, and heavy metals such as mercury and lead.

As the above examples show, practically all the presented phenomena adversely affect both human life and health, also having a particularly negative impact on the elderly. The accumulation of the described phenomena requires taking up comprehensive remedial actions by public authorities [33] supported by the private sector and the already mentioned non-governmental organizations.

3. Aging Society in Numbers

Figure 1 presents basic information about the population dynamics and net migration in Poland in the years 2004–2018. The analysis of the natural increase rate shows that its value can be divided into three phases. In the years 2004–2005, it was negative. In the course of the next phase, covering the years 2006–2012, its values were positive. In the period 2013–2018, it presented negative values yet again. Taking into account the entire analyzed period, it can be observed that this indicator value was characterized by a downward trend.

The number of live births was systematically growing in 2004–2009. Next, in the period 2010–2013, it went down and then up again in the subsequent years. Taking into account the entire analyzed period, the discussed indicator was characterized by an upward trend. Its average annual value amounted to almost 386,000.

The death rate value was also characterized by an increasing tendency. Its average annual value was over 383,000. The value of net migration, in the years 2004–2014, showed a negative value. In the following years the indicator presented a positive value.

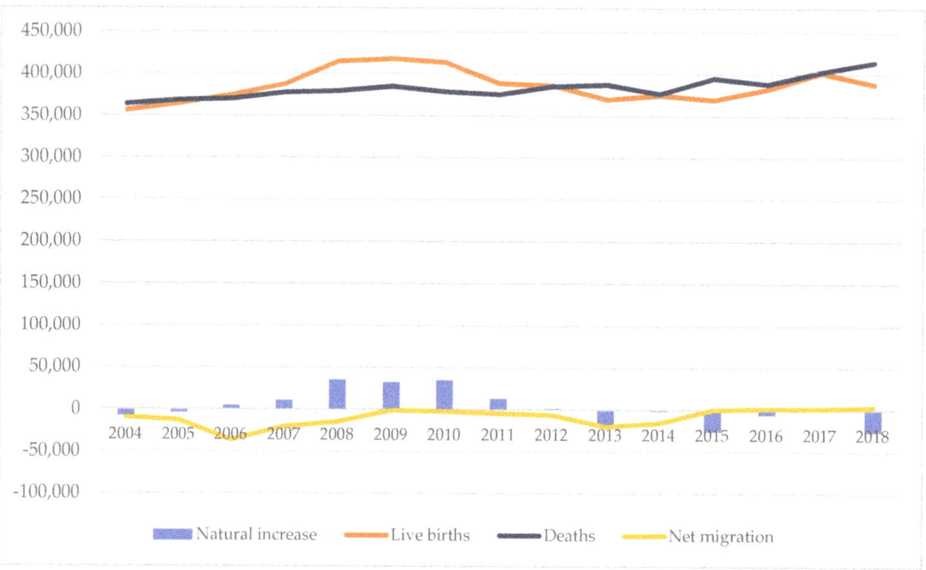

Figure 1. Natural increase rate and net migration in the years 2004–2018 [34].

An ongoing population aging in Poland can be observed by analyzing population structure based on economic age groups (Figure 2). The share of post-working age population was systematically increasing year by year. In the analyzed period, this increase amounted to as much as 6% points. In turn, the share decline was recorded in the other two groups. The share of the working age population dropped by 2.9% points, and the share of pre-working age population was lower by 3.1% points.

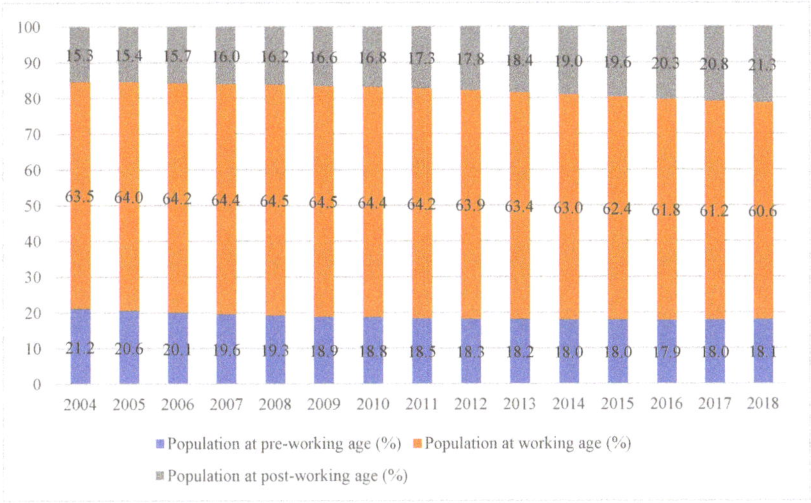

Figure 2. Population by economic age groups in the years 2004–2018 (%) [34].

Polish society aging was also confirmed by analysing population shares in individual age groups in 2004 and 2018 (Figure 3). In 2018 against 2004, the share of the population aged 55 and over definitely increased, whereas the share of the population aged 85 and over doubled.

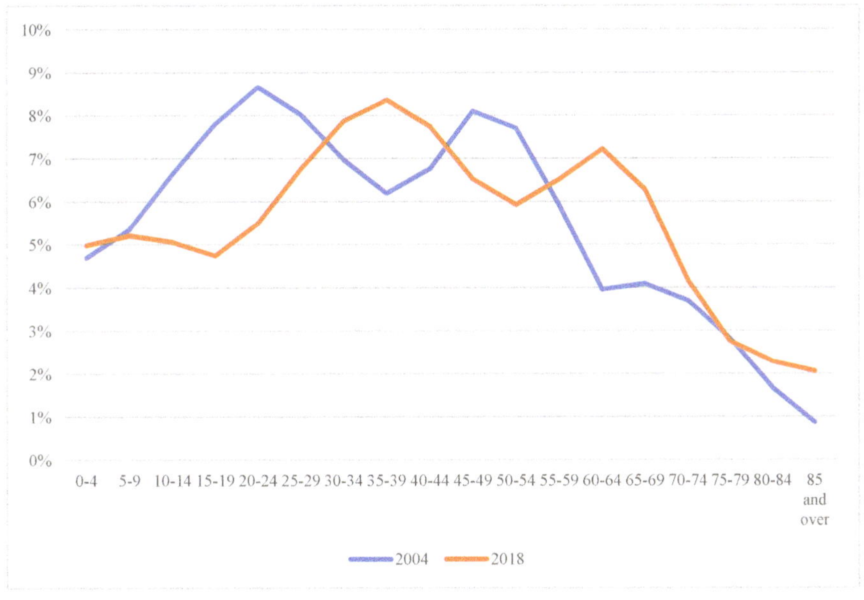

Figure 3. Share of the population by age groups in 2004 and 2018 [34].

Polish society aging can also be confirmed by analysing life expectancy at birth (Figure 4). This value went up for both women and men. In the case of women, life expectancy was 2.5 years and 3.1 years for men. It should be emphasized, however, that the maximum value for women was reached in 2016 and for men a year later. The values of both indicators show an upward trend. In the case of men, this increase was higher and amounted to almost 3 months on average per year, whereas for women the average annual increase was 2 months.

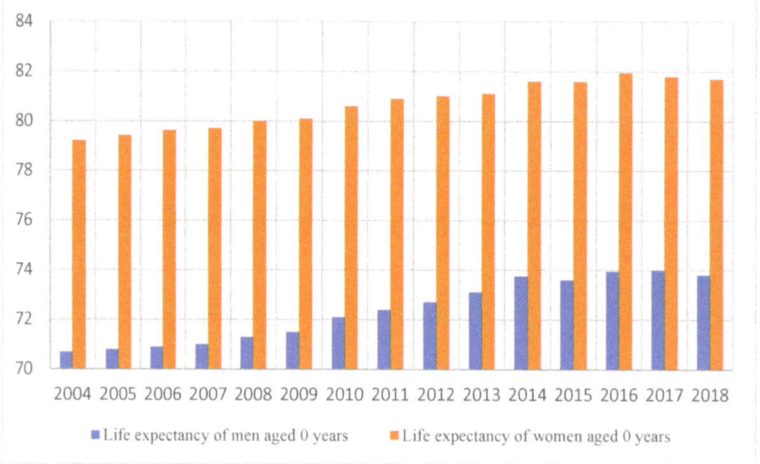

Figure 4. Life expectancy at birth in the years 2004–2018 [34].

Life expectancy of the population aged 65 and over was also increasing (Figure 5). In the analyzed period, in the case of women, this increase was 1.7 years, and 1.6 years for men. In both cases an upward trend was recorded.

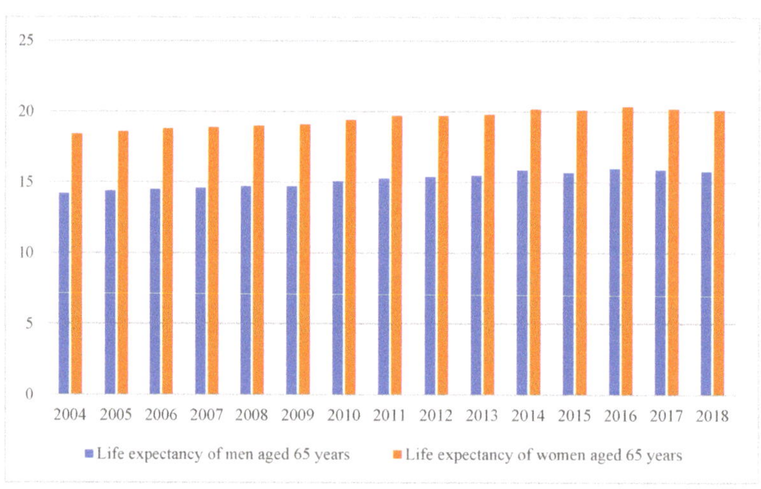

Figure 5. Life expectancy at the age of 65 in the years 2004–2018 [34].

The analysis of old-age dependency ratio provides very interesting information (Figure 6). This indicator is defined as the number of the elderly (aged 65 and over) per 100 persons aged 15–64. The value of this indicator was increasing every year. When comparing 2018 to 2004, an almost 40% increase, from 18.7 in 2004 to 26.1 in 2018, was recorded.

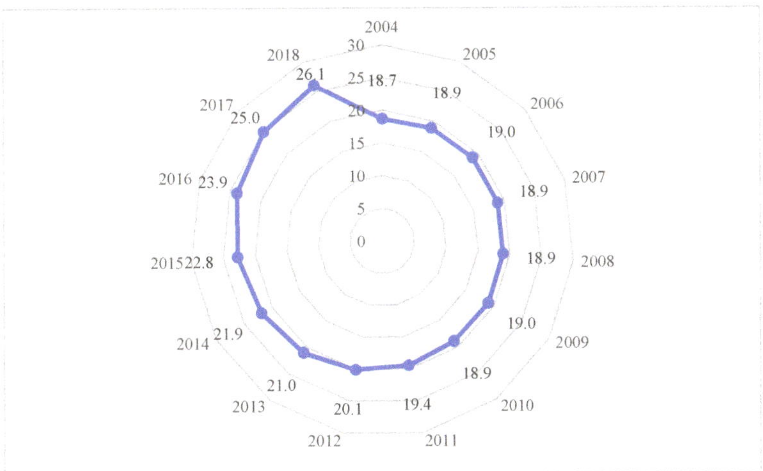

Figure 6. Old-age dependency ratio [34].

Similar conclusions regarding Polish population aging can be drawn by analysing the value of demographic aging rate, i.e., the share of people aged 65 and over in the total population (Figure 7). The demographic aging rate is a classical measure defining whether a given population is old. According to the UN recommendations presented at the end of the 1950s, a community where 6% of the population reached the age of 65 can be referred to as an old society. In the entire analyzed period, this indicator value for Poland was above 6%. The lowest value of 13.1% was achieved in 2004, and later its value was systematically growing. The increase reached 4.4% points.

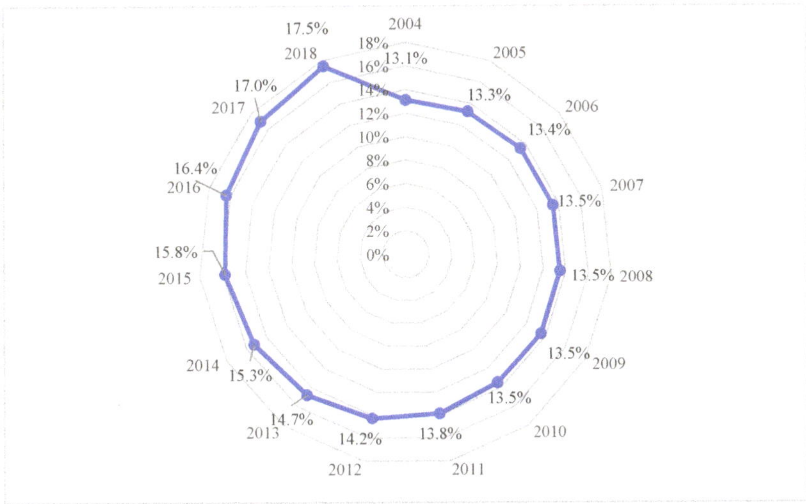

Figure 7. Demographic aging rate in the years 2004–2018 (%) [34].

Polish society aging is also shown based on the aging index analysis, i.e., the number of people aged 65 and over per 100 people aged 0–14 (Figure 8). The aging index is said to show the number of grandparents per 100 grandchildren. The increase in its value results from longer life expectancy as well as the shift—from a group of children to older groups—of people from the last baby boom which, in turn, is not compensated by the current births. The lower its value, the better protection for senior citizens. In other words, the number of children (aged 0–14) is a "counterbalance" in terms of the population aged 65 and over. In the years 2004–2013, fewer than 100 "grandparents" fell per 100 "grandchildren". The situation has been changing since 2014. In 2018, there were 115 "grandparents" were recorded per 100 "grandchildren."

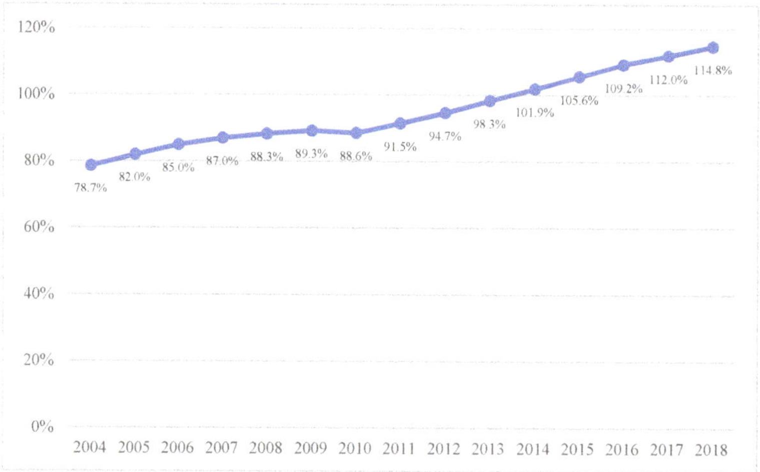

Figure 8. Aging index in the years 2004–2018 [34].

The degree of the aging process advancement regarding Polish society is demonstrated by the increasing share of post-working age population each consecutive year. The value of old-age dependency ratio is growing. The value of the demographic aging rate is also increasing, i.e., the share of people aged 65 and over in the total population. Polish society aging is also shown by the analysis of aging index, i.e., the number of people aged 65 and over per 100 people aged 0–14. Systematic human lifespan extension as well as low birth rate and the prevalence of deaths over births represent the essential demographic factors which determine population aging.

4. The Impact of Selected Environmental Threats on an Aging Society

Among the main phenomena occurring as a result of climate change the following are listed:

1. more frequent temperature extremes (increased number of heat days);
2. higher rainfall intensity which can cause floods at any time of the year (including primarily the so-called flash flooding of storm nature);
3. increased frequency and intensity of storms;
4. more frequent droughts and the related losses in agricultural production and higher risk of forest fires;
5. more frequent occurrence of temperatures oscillating around zero [35].

In Poland, temperature measurements are carried out at 32 measuring stations. The average annual air temperatures in 2017 at all measuring stations in Poland were higher than the average values for previous multi-year periods, starting from 1971.

The highest average annual air temperature was recorded in 2017 at the Wrocław weather station (10.4 °C), while the lowest at the station in Suwałki (7.5 °C).

Maximum air temperature for the multi-annual period 1971–2017 was recorded at the stations in Kalisz (38 °C) as well as Toruń, Wrocław, and Opole (37.9 °C). Minimum air temperature for this period was recorded at the stations in Bialystok (–35.4 °C), Terespol (–34.3 °C), and in Włodawa (–34.2 °C). The highest amplitudes of extreme temperatures for the multi-annual period 1971–2017 were registered at the stations in Białystok (the amplitude was 70.9 °C), Terespol (70.5 °C), Kielce (70.3 °C), and Włodawa (70.2 °C). The lowest extreme temperature amplitudes for the multi-annual period 1971–2017 were recorded at IMiGW (Polish Institute of Meteorology and Water Management) stations in Zakopane (where the amplitude was 59.9 °C), Zielona Góra (59 °C), and Śnieżka (56.7 °C). The information for the selected four measuring stations in 2004–2017 is presented in Table 1. Hel measuring station is situated in the north of Poland on the Baltic Sea, Śnieżka station is located in the south of Poland in the mountains, and the other two are placed in the central part of the country. Such choice of locations allows the presentation of selected data covering the entire country. The analysis of values for all four measuring stations shows that the average annual temperature is systematically rising in each station. It indicates the growing development tendency. The highest average annual increase of 0.08 °C was recorded in Warszawa, and the lowest at the average annual level of 0.04 °C in Lublin.

Table 1. Average air temperatures in the years 2004–2017.

Meteorological Stations	2004	2005	2006	2007	2008	2009	2010	2011	2012	2013	2014	2015	2016	2017
Hel	8.4	8.5	9.1	9.5	9.3	8.6	7.4	8.8	8.4	8.7	9.4	9.6	9.5	9.0
Warszawa	8.4	8.6	9	9.6	9.8	8.9	8	9.1	8.8	8.9	9.8	10.3	9.8	9.4
Lublin	7.8	7.8	8	8.8	8.9	8.2	7.5	8.3	8.1	8.1	9	9.4	8.7	8.4
Śnieżka	0.7	0.9	2.1	1.9	1.7	1.6	0.1	2	1.3	0.9	2.7	2.4	1.7	1.4

Source: authors' compilation based on [36–48].

Solar radiation studies have shown that the observed climate changes are reflected in the variability of solar conditions, especially in terms of extreme values. This requires considering their negative impact on the living conditions of the contemporary population, such as health and sometimes even life [36–48]. Solar radiation stimulates human body processes, intensely affecting the skin, internal organs and the nervous system [49]. The term insolation refers to the total time (hours) in a given period during which a specific place on the Earth's surface is directly radiated by the sunlight [36–48]. The information on insolation at four selected measuring stations in the years 2004–2017 is presented in Table 2.

Table 2. Insolation in the years 2004–2017.

Meteorological Stations	2004	2005	2006	2007	2008	2009	2010	2011	2012	2013	2014	2015	2016	2017
Hel	1790	1986	1993	1797	1923	1901	1856	2087	1951	1905	1778	-	2021	1903
Warszawa	2259	2460	2502	2305	2241	2258	2204	2418	2393	2234	2278	1931	1836	1351
Lublin	1738	1559	1714	1756	1772	1823	1689	1906	1786	1649	1828	2012	1872	1783
Śnieżka	1285	1273	1685	1550	1408	-	1406	1767	1625	1330	1411	1721	1374	1456

Source: authors' compilation based on [36–48].

Annual precipitation sums have been calculated taking into account the daily sums based on the selected IMiGW stations and checkpoints which reflect spatial diversity of the precipitation sums in the country. The data for the years 2004–2017 indicate that the amount of precipitation in three out of four measuring stations shows an increasing tendency (Table 3).

Table 3. Total annual precipitation in mm in the years 2004–2017.

Meteorological Stations	2004	2005	2006	2007	2008	2009	2010	2011	2012	2013	2014	2015	2016	2017
Hel	690	500	562	691	616	673	779	521	695	583	455	519	703	826
Warszawa	523	490	479	602	537	652	798	604	519	613	555	404	593	705
Lublin	590	559	521	662	649	681	751	502	503	650	790	532	698	612
Śnieżka	1036	1273	1072	1272	983	1213	1316	928	1008	1222	887	897	995	1258

Source: authors' compilation based on [36–48].

The information on the average wind velocity recorded at four measuring stations in the years 2004–2017 is presented in Table 4. The highest average wind velocity was recorded at Śnieżka measuring station. All of the phenomena described in the context of environmental risks can have an adverse effect and threaten both health and life of the community, with particular attention to seniors.

Table 4. Wind velocity in the years 2004–2017.

Meteorological Stations	2004	2005	2006	2007	2008	2009	2010	2011	2012	2013	2014	2015	2016	2017
Hel	4	3.8	3.6	4	3.9	3.6	3.6	3.9	3.7	3.5	3.3	4.2	4.2	4.5
Warszawa	3.9	3.8	3.6	4	3.9	3	3.2	3.1	3.1	3	3.2	3.5	3.4	3.5
Lublin	3.2	3.1	2.9	3.1	3.1	2.8	3	3.2	3.2	3.1	2.9	3	2.9	3
Śnieżka	14.5	12.9	14.4	14.4	15.3	10	10.6	11.3	11.2	10.3	10.5	11.4	10.5	11.9

Source: authors' compilation based on [36–48].

5. Materials and Methods

Demographic changes observed in Poland indicate a progressing process of the population aging. The purpose of this part of the study is to diagnose the respective changes recorded in Poland in the years 2004–2019 and also to provide answers to the following research questions:

1. Is it possible to identify the separate homogeneous development phases in the course of Polish population aging process?
2. What are the boundaries of the development phases resulting from the aging process in Poland?
3. Does the periodization of the population aging process allow the typology of the Polish population within the identified development phases?
4. What demographic factors determined population aging to the highest extent in each development phase?

The development phase is a sub-period of undetermined length, included in the analyzed period and characterized by the fact that all features included in periodization take on values more similar to each other in a given development phase than in relation to the adjacent sub-periods.

Multidimensional statistical analysis methods including, in particular, data classification methods were applied to carry out the adopted purpose. The analysis was based on the statistical information obtained from the Statistics Poland and Eurostat databases.

The study was conducted in accordance with the following research procedure:

1. Criteria selection for the periodization of the population aging process.
2. Normalization of the periodization criteria values (population aging indicators) to the range [0,1] using zero unitarization method based on the minimum value and range, allowing their

comparability by introducing additivity and unifying the orders of magnitude. The unitarization of variables was carried out according to the following formula:

$$z_{ti} = \frac{x_{ti} - minx_{ti}}{maxx_{ti} - minx_{ti}} \tag{1}$$

where: x_{ti}—value of the *t*-th period for *i*-th indicator (population aging indicators), z_{ti}—normalized value of the *t*-th period for *i*-th indicator. Those interested will find a lot of valuable information about the normalization of variables in: Kukuła [50], Walesiak [51], Jajuga, Walesiak [52], Zeliaś [53], and Milligan and Cooper [54].

3. The diversity of the analyzed periods, in terms of the population aging indicator values, was determined using the Euclidean squared distance.

4. The analyzed periods were subject to hierarchical classification by applying Ward's hierarchical clustering method. Its results were used to determine the number of development phases of the population aging process. The criterion of the first clear increase in integration distance was used to determine the number of classes by analysing the dendrogram of connections, integration distances, and classification stages. The description of classification methods is presented in the following studies: Ward [55], Johnson [56], Anderberg [57], Hartigan [58], Sneath, Sokal [59], Aldefender, Blashfield [60], Basiura, Sokołowski [61], and Everitt et al. [62].

5. The division of the analyzed period into development phases of the population aging was performed by applying k-means clustering.

6. The typology of development phases was carried out using mean values of the periodization criteria.

7. The initial set of diagnostic features—demographic factors differentiating development phases of the population aging process—were identified. The Hellwig's method for feature selection, modified by Sobczak [63] was used [64].

8. The key demographic factors were selected using correlation analysis.

The following indicators were used to measure the advancement phases of population aging:

-   MA—median age of population,
-   SA65+—demographic aging rate—share of population aged 65 and over as the percentage of total population (proportion of population aged 65 and over),
-   AI—aging index, the number of the population aged 65 and over per 100 people aged 0–14, i.e., the number of grandparents per 100 grandchildren (aging index),
-   TADR—total age dependency ratio, share of population aged 0–14 and aged 65 and more as the percentage of population aged 15–64 (population aged 0–14 and 65 and over to population aged 15–64),
-   OADR—old-age dependency ratio of the post-working age group, share of population aged 65 and over as the percentage of population aged 15–64, (old-age dependency ratio, population 65 and over to population aged 15–64),
-   YADR—young-age dependency ratio of the pre-working age group, share of population aged 0–14 as the percentage of population aged 15–64 (young-age dependency ratio, population aged 0–14 to population aged 15–64).

6. Results—Periodization of the Population Aging Process in Poland in the Years 2004–2019

Figure 9 illustrates the hierarchical classification results of the analyzed periods in terms of 6 applied indicators of population aging in Poland using spanning trees and integration distance diagrams with regard to classification stages. On their basis, a variant division of 16 years into four classes, representing relatively homogenous phases of population aging was developed.

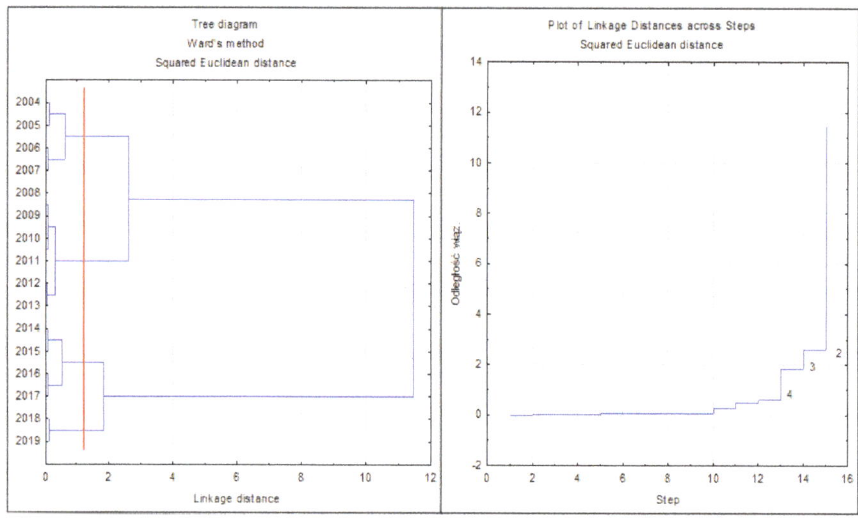

Figure 9. Dendrogram of connections, integration distances, and classification stages using Ward's method for the years 2004–2019 regarding measure values of population aging in Poland. Authors' compilation based on [65,66] applying STATISTICA 13.1 PL statistical package, StatSoft Polska Ltd., Cracow.

The analysis of the results obtained using Ward's method shows that, in the period 2004–2019, four development phases of the Polish population aging process should be distinguished. The next step of the research procedure consisted of dividing the analyzed period, using k-means clustering, into the relatively homogeneous sub-periods called development phases.

Figure 10 presents arithmetic means of the normalized values referring to the analyzed aging indicators, determined for the individual development phases. As Figure 10 shows, in the years 2004–2019, four successive relatively homogeneous sub-periods were selected: phase I—2004–2006, phase II—2007–2013, phase III—2014–2016, and phase IV. Phases I, II, and IV are characterized by the same length covering 3 years. Phase II turned out to be the longest—seven years.

The consecutive development phases were characterized by an increasing median age of population (MA), a larger share of older people aged 65 and over (SA65 +), a higher number of older people per child aged 0–14 (AI), as well as larger old-age dependency ratio of the post-working age group (OADR). This order was disturbed in the case of the total age-dependency ratio (TADR), because in the first of the sub-periods covering the years 2004–2006, it adopted a higher value than in the second phase falling in the period 2007–2013. In turn, the young-age dependency ratio of the pre-working age group (YADR) in phase I (2004–2006) was, by far, the highest among all the identified development phases in terms of population aging in Poland.

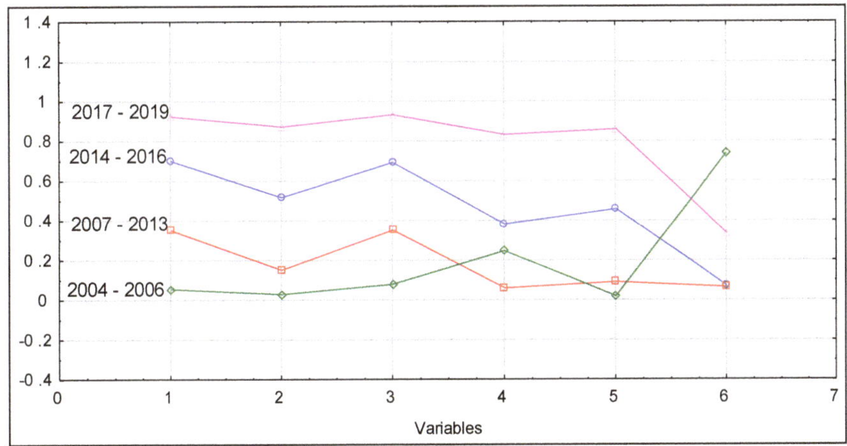

Figure 10. Mean values of the normalized population aging indicators in Poland in the identified development phases in the years 2004–2019. Authors' compilation based on [65,66] applying STATISTICA 13.1 PL statistical package, StatSoft Polska Ltd., Cracow.

Table 5 presents the typology of the Polish population in individual development phases using mean values of the indicators determining the advancement level and the specificity of the aging process.

Table 5. Development phases of the population aging process in Poland and the population types identified in the years 2004–2019.

Development Phases (Years)	Population Type	MA (Years)	SA65+ (%)	AI (%)	TADR (%)	OADR (%)	YADR (%)
		Mean Values of Indicators					
Phase I 2004–2006 (3)	Aging population with the potential of future labour resources	36.5	13.1	78.7	42.6	18.7	23.8
Phase II 2007–2013 (7)	Advanced in the aging process, with reduced but structurally unfavourable total dependency	37.9	13.7	89.4	40.9	19.3	21.6
Phase III 2014–2016 (3)	Highly advanced in the aging process, with increased and structurally unfavourable dependency	39.6	15.4	10.9	43.8	22.2	21.6
Phase IV 2017–2019 (3)	Old, presenting the highest dependency ratio of people in non-working age, with the increasing predominance of seniors	40.6	17.1	112.2	47.8	25.3	22.5

Source: authors' compilation based on [65,66].

Based on the data analysis presented in Table 5, it is noticeable that the individual development phases differ significantly in terms of mean values referring to Polish population aging indicators covered by the study. Different types of Polish population can be assigned to individual development phases. In the first phase of development, Polish population was defined as aging, but, presenting the potential of future labour resources. The average median age in the 2004–2006 sub-period was 36.5, the share of older people (aged 65 and over) exceeded 13%, and almost 79 grandparents fell per 100 grandchildren. These indicators adopted the lowest mean values in phase I. The mean value analysis of the total age dependency ratio (TADR), determining the economic dependency on the working population age group by the groups of non-working age population (pre- and post-working age jointly) and the structure of this indicator are of particular importance in determining the population type. The mean value of the total age dependency ratio for phase I is quite high and amounts to 42.6%

(the second development phase is characterized by its lower value); however, there is a favourable correlation between the components of TADR indicator. The average economic dependency on the post-working age group (OADR) is the lowest in the entire period under study and amounts to 18.7%. In turn, the dependency on the pre-working age group (YADR), approached as the potential of future labour resources and, therefore, positively assessed remains the largest and amounts to 23.8%.

The mean value of the total age dependency ratio for phase I is quite high and amounts to 42.6% (the development phase II is characterized by its lower value); however, there occurs a favourable correlation between the components of the TADR indicator. The average economic dependency on the post-working age group, i.e., old-age dependency ratio (OADR) is the lowest in the entire analyzed period and amounts to 18.7%. In turn, the economic dependency on the pre-working age group, i.e., young-age dependency ratio (YADR), approached as the potential of future labour resources and thus assessed positively, is the highest and amounts to 23.8%.

The development phase II is definitely the longest. It covers the period of seven years. It is a more advanced sub-period than the previous one regarding the population aging process in Poland. The sub-period 2007–2013 is characterized by the type of population advanced in the aging process, with a lower but structurally unfavourable total age dependency ratio. The average median age in this development phase is higher than in the case of phase I by almost 1.5 years. The demographic aging rate (SA65 +) went up by 0.6, and the aging index (AI) by as much as 10.7% points. The improvement in the mean value of the total age dependency ratio (a decline by 1.7% points) was accompanied by a deteriorating correlation between the partial indicators. However, this relationship continued to remain favourable in the development phase II, because the average dependency of the post-working age group was lower than the pre-working age group (the difference is over 2 % points).

In phase III, Polish population was defined as highly advanced in the aging process, characterized by higher and structurally unfavourable dependency. In the course of this three-year sub-period, the aging of the Polish population was significantly accelerated. The average median age increased by almost 2 years (1.7) compared to phase II and the old-age dependency ratio by 1.7% points. A significant rise in the mean value of the aging index (AI) was also recorded, i.e., almost 14 grandparents per 100 grandchildren. Phase III also presented an increase in the average total age dependency ratio by almost 3% points and, most importantly, for the first time in this phase, the economic dependency on the post-working age group was higher than with the pre-working age group. The assessment of this correlation between the partial indicators of the dependency ratio is negative.

In the final development phase IV, Polish population was defined as old and heavily dependent on the non-working age group with the majority of older people (TADR—47.8%). This phase was characterized by an increase in mean values of all population aging indicators. Moreover, in this sub-period, all mean indicator values turned out to be the highest. Median age reached the level of almost 41 years, the share of seniors in the population increased by 1.7% points, and the number of grandparents per 100 grandchildren grew by over 9. In the last aging phase, the average economic dependency on the working age population was very high as it amounted to almost 48% and increased by 4% points compared to the previous sub-period. The correlation between the partial indicators of economic dependency can also be assessed very negatively. The dependency of seniors on the working age group increased by more than 2% points and reached 25.3%. The pre-working age group dependency was lower (22.5%); however, it was also higher.

The next step of the research procedure was to identify the key demographic factors influencing the process of population aging. Substantive analysis was used to determine the initial set of demographic factors, having adapted that it should include diagnostic features describing the basic demographic phenomena affecting the population aging process. Based on the availability of statistical data, time range of the diagnostic features covers the years 2004–2018. A preliminary list of 10 demographic factors selected in this way is presented below:

(1)   DD—rate of demographic dynamics (number of births per 1 death),
(2)   NI—natural increase per 1000 population,

(3)   LB—live births per 1000 population,
(4)   TF—rate of total fertility (number of children per 1 woman),
(5)   GR—rate of gross reproduction (number of born female children per 1 woman of childbearing age),
(6)   LEF0—life expectancy at birth for females,
(7)   LEF65—life expectancy at 65 for females,
(8)   LEM0—life expectancy at birth for males,
(9)   LEM65—life expectancy at 65 for males,
(10) TNM—total net migration per 1000 population.

The purpose of this research stage was to identify the key factors of demographic development which differentiate the identified development phases of the population aging process in Poland. It was adopted that the features presenting the highest information value and not duplicating the same information should be selected. As a consequence, the key factors of demographic development should include the features strongly correlated with the population aging indicators and insignificantly correlated with each other. The reduction of the initial set of diagnostic features, characterizing demographic development and determining the population aging processes was of iterative nature. The feature most strongly correlated with all population aging indicators (maximum value of the modules' sum of the correlation coefficients referring to the given feature with all population aging indicators) was selected in the course of the first stage. In the second stage, the diagnostic features significantly correlated with the feature selected in the first stage were removed to eliminate the repeated information. The critical value of the correlation coefficient was set at r* = 0.641 (for $n = 15$ $\alpha = 0.01$). Steps one and two were repeated until the last diagnostic feature was eliminated.

The demographic factors determining the average life expectancy LEF0, LEF65, LEM0, and LEM65, showed a very strong positive correlation with all the indicators referring to population aging. Among them, male life expectancy rate (LEM0) turned out to be the demographic factor most closely correlated with the population aging indicators. The remaining indicators, determining the average life expectancy, were eliminated from further analysis as very strongly correlated with LEM0.

In the next iteration, gross reproduction was the demographic factor most closely correlated with population aging, whereas such features as live births per 100 population (LB) and total fertility (TF) were significantly correlated with it.

In subsequent iterations, natural increase (NI) and total net migration (TNM) were selected as the features most closely correlated with the aging indicators. The rate of demographic dynamics (DD) was rejected as strongly correlated with the natural increase.

Ultimately, after reducing the initial set of diagnostic features, the following characteristics were considered the key demographic factors, most differentiating the development phases of the population aging process in Poland in 2004–2019: natural increase (NI), gross reproduction (GR), average life expectancy at birth for males (LEM0), and total net migration (TNM).

Table 6 and Figure 11 present mean values of the key demographic development factors which differentiate the development phases of the population aging process in Poland.

The expected average life expectancy at birth for males, increasing in the subsequent development phases was by far the most determining demographic factor responsible for the aging process of the Polish population in 2004–2019.

It appeared that this variable was most strongly correlated with the population aging indicators. In addition, other demographic factors related to life expectancy or further life span were also strongly correlated with this demographic factor (correlation coefficients exceed 0,99) (LEF0, LEF65, LEM65).

Other demographic factors differentiating the individual development phases of the aging process, arranged by their decreasing significance are as follows: gross reproduction (GR), natural increase (NI) and total net migration (TNM).

Table 6. Mean values of demographic factors differentiating the development phases of the population aging process in Poland in the years 2004–2019.

Development Phases (Years)	Population Type	LEM0	GR	NI	TNM
		Mean Values of Demographic Factors			
Phase I 2004–2006 (3)	Aging population with the potential of future labour resources	70.80	0.60	−0.07	−0.47
Phase II 2007–2013 (7)	Advanced in the aging process, with reduced but structurally unfavourable total dependency	72.01	0.65	0.40	−0.26
Phase III 2014–2016 (3)	Highly advanced in the aging process, with increased and structurally unfavourable dependency	73.76	0.64	−0.30	−0.27
Phase IV 2017–2019 (3)	Old, presenting the highest dependency ratio of people in non-working age, with the increasing predominance of seniors	73.90	0.70	-0.35	0.05

Note: LEM0: average life expectancy at birth for males; GR: gross reproduction; NI: natural increase; TNM: total net migration. Source: authors' compilation based on [65,66].

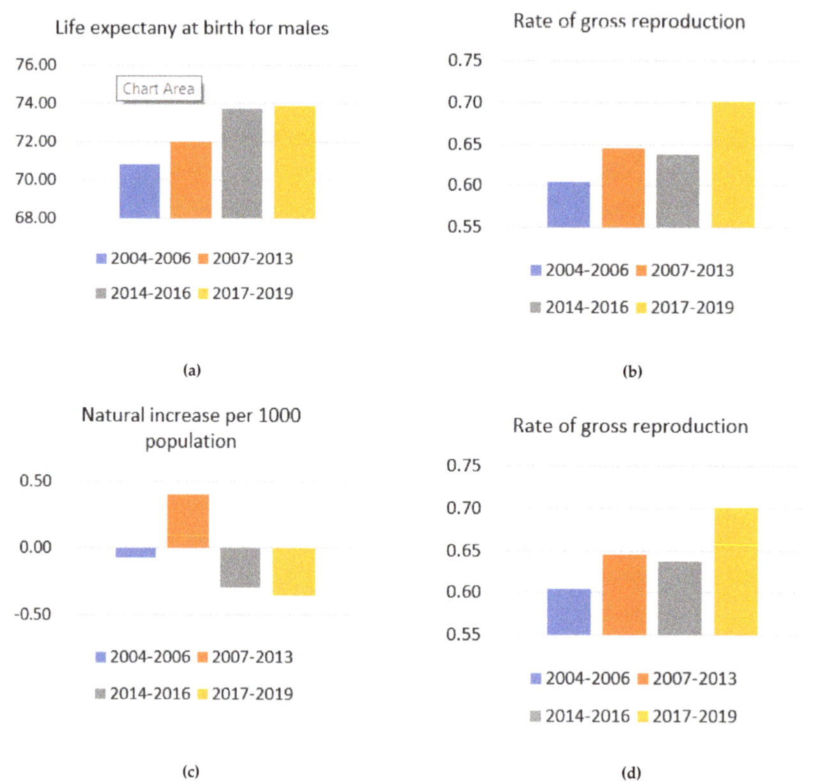

Figure 11. Mean values of demographic factors differentiating the development phases of the population aging process in Poland in the years 2004–2019 [65,66].

The rate of gross reproduction is determined by the population replacement rate ingrained in the society. The subsequent development phases of the aging process in Poland were characterized by the increasing mean values of the reproduction rate, which should be considered a favorable trend in demographic development. The exception is phase II, in the course of which this indicator value presented a slight decline. However, the values of GR rate in any of the distinguished development phases did not guarantee a simple population replacement, resulted in the continued narrowed reproduction and thus deteriorating age structure of the population in Poland. Live births (LB) and total fertility (TF) were strongly correlated with the simple reproduction rate.

Natural increase was another key factor determining the development phases of the population aging in Poland. Only during the longest, II development phase, covering the years 2007–2013, characterized by a slowdown in the population aging process, was the mean value of the natural increase positive and amounted to approx. 0.4%. The remaining phases showed a negative birth rate, the lowest one was recorded in the last phase.

The total net migration (TNM) is the last of the identified demographic factors influencing the development phases of the population aging process in Poland and also the one of the least importance. The first three development phases were characterized by a negative total migration balance (in phase II and phase III this ratio was very similar). In the fourth development phase, there was a slight immigration surplus over emigration. Population migrations have an impact on the demographic development, since usually it is young people who are leaving, hence the negative migration balance results in population aging.

7. Discussion

At this point, it is worth investigating the main causes of death in Poland [67]. Basically, they include cardiovascular diseases and malignant tumours. In 2016, they were responsible for approximately 70% of all deaths. Almost every second person in Poland dies from heart and vascular diseases, approximately 25% from cancer. Young people die more often as a result of accidents, suicides, and crimes. It can be adopted that men are at the highest risk of developing cardiovascular diseases and to a lesser extent, cancer. In the case of women, these proportions are slightly different. Malignant neoplasms are the most dangerous for those under the age of 75, and it is only at the oldest age that cardiovascular diseases become the number one threat. The health condition of Poles, which requires improvement, is primarily influenced by the growing epidemic of excess weight and obesity, smoking, lack of exercise, improper eating habits, and fairly high alcohol consumption among men. In general, in relation to the European Union countries, Poland is distinguished by a low level of physical activity of the population and a high prevalence of the smoking addiction among men. The systemic causes of the society's health situation, partly responsible for its aging process, were summarized highlighting, in particular, the insufficient funds, imprecise legal regulations, lack of care standards, and resource allocation mechanisms. In general, the system was identified as inappropriately adapted to the actual health needs of the society. The reasons for this situation are perceived in insufficient expenditure on health care compared to other European countries. Concerns are raised by the level of infant mortality in Poland, which is higher than the average one for the European Union. In Poland, each year in the years 2015–2017, before the end of the first year of life, 40 children died out of every 10,000 live births, whereas the average for the EU countries was 36 (in 2016). The population of Poland is still, on average, younger than the population of most European Union countries, but Eurostat forecasts indicate that this favourable difference will slowly disappear and in the middle of the current century both the median age and the percentage of people aged 65 and over will be higher in Poland than the average in the European Union Member States. In view of the aforementioned forecasts, there is now still time to take up the respective countermeasures.

Based on the conducted studies and analyses, a set of several recommendations aimed at supporting the functioning of seniors and counteracting environmental threats can be put forward:

1. Actions should be taken to reduce greenhouse gas emissions, which include thermomodernization of residential and public utility buildings, more extensive use of natural gas at the expense of coal including production plants, modernization of the heating networks, improving the energy efficiency of lighting, electronic equipment, the reduction of methane emissions, using public

transport, taking advantage of the renewable energy sources, more effective municipal waste and wastewater management, planting trees, recycling, improving infrastructure for cyclists and pedestrians, imposing stricter emission standards for internal combustion engines [68].

2. Investments are required to improve the condition of flood protection infrastructure [69,70] combined with an effective warning system against threats within the framework of crisis management in the local government units. Further investments in road infrastructure, the construction of ring roads resulting in the reduction of traffic noise and vibrations of building are needed. In terms of the groundwater exploitation, it is necessary to consistently take care of the hydrogeological balance, which is part of the environmental area of sustainable development.

3. In Poland, air pollution constitutes one of the essential environmental problems, which has a disastrous effect on senior citizens. This pollution type has a negative impact on health as a result of both long- and short-term exposure. It is caused by so-called low emission, which has its sources in domestic furnaces and local coal-fired boilers responsible for inefficient coal combustion and also in combustion transport (cars). It is easy to guess that the remedy to this problem is the transition from coal-fired buildings to more economical ones as well as the reduction of exhaust emissions.

4. Natural resources should be exploited in a rational and thoughtful way. Their stock is not unlimited. Great importance should also be attached to biodiversity which strengthens our national ecosystem. In addition, afforestation is a process that should be continued at least at the current intensity level.

5. Two issues are important in the process focused on extending lifespan of the society in full health: the level of medical care, with particular attention paid to check-ups and preventive measures and also the promotion of a healthy lifestyle [71]. The respective activities and financial resources should be allocated to prevention, education in the field of cardiovascular diseases, diabetes, obesity [72,73], and cancer. The country residents are living longer and longer, therefore it is necessary to activate older people but also to draw on their knowledge and life experience. NGOs can be helpful in this regard.

6. One of the most important ways to counteract the phenomenon of aging is to promote pro-family policy and the model of a large family. The elderly, in addition to avoiding the risk related to environmental hazards, should have access to high-quality medical care which should be allocated higher financial outlays. Ecological education [74] addressing the importance and impact of environmental changes on our life comfort is also helpful [75]. The education sector should be used in terms of teaching about the benefits of healthy eating habits, physical activity, good mental condition, or regeneration of older people.

8. Conclusions

The following conclusions can be drawn from the conducted empirical studies addressing periodization of the population aging process:

1. The development of the population aging process in Poland is not homogeneous. In the years 2004–2019, its four development phases were identified, which means that over the analyzed period, the recorded significant changes in the advancement of the population aging process occurred four times.

2. The turning points in the development of the population aging process in Poland coincide with the boundaries of the development phases defining various types of populations, differing in the degree and specificity of the aging process advancement related to Polish population. They fall at the turn of 2006/2007, 2013/2014, and 2016/2017.

3. The population aging process in Poland was systematically progressing in the course of the development phase I and II; however, it definitely accelerated during the development phase III and did not slow down in phase IV.

4. The population of Poland was characterized by a favourable structure of the economic dependency of working age population in the first two development phases. There occurred an excessive dependency on the pre-working age group over the post-working age group, amounting to over 5% points in the development phase I and slightly over 2% points in phase II. During phase III, the economic dependency structure of the working age population changed, for the first time, to an unfavourable one (the difference between the OADR and YADR indicators was -0.6% points. In phase IV, this negative phenomenon was strengthened (the difference between OADR and YADR went up to almost 3% points).

5. In the development phase IV, the population aging process was manifested by almost half of the population economically dependent on the productive age group, including more than a quarter of senior citizens. It definitely requires state intervention and undertaking actions aimed at changing both the level and structure of the demographic burden.

6. The key demographic factors differentiating the developmental phases of the population aging process in Poland in 2004–2019 include, in order of their significance, longer life expectancy of the population, narrowed gross reproduction rate, declining birth rate, and total migration balance.

7. The multidimensional statistical analysis methods used in the research proved to be useful for the purposes of identifying the development phases of the population aging process in Poland, for the population typology in non-homogenous sub-periods and for determining the key demographic factors differentiating the identified development phases of the population aging process in Poland.

The research scope adopted in this study has been limited to the macroeconomic scale covering the area of Poland. Three basic directions for further research can be formulated. The first of them is the analysis and assessment of the aging processes in Poland from a regional and local perspective—at the level of cities and villages, and an attempt to identify these factors which determine the occurring differences. Another important research direction is carrying out comparative analyses relating the situation of population aging in Poland, based on the national, regional, and micro-spatial perspective, to the analogical processes taking place in the European Union countries. The studies confronting the economic and social effects of Polish population aging with the assessment of the actions taken up by the state or regional and local authorities in terms of infrastructure adaptation, the development of health services, social welfare, as well as the allocation of financial outlays in the perspective of further increase in senior citizens' population may turn out valuable.

Climate change is an indisputable phenomenon. Despite that, however, it is very difficult to find statistical data that could clearly illustrate the level of this phenomenon. Polish public statistics lack data regarding, e.g., floods or droughts. The data on heavy rains or landslides are also missing. They are slowly being collected, but their vast majority raises reservations. The need to collect and analyze this type of data is necessary as the discussed phenomena affect people's health and lives. The elderly are particularly vulnerable in this respect.

Author Contributions: All authors designed the research, analyzed the data and wrote the paper. All authors have read and agreed to the published version of the manuscript.

Funding: The project is financed by the Ministry of Science and Higher Education in Poland under the program "Regional Initiative of Excellence" 2019–2022 project number 015/RID/2018/19 total funding amount 10 721 040,00 PLN.

Conflicts of Interest: The authors declare no conflicts of interest.

References

1. Doh, J.P.; Guay, T.R. Corporate social responsibility, public policy, and NGO activism in Europe and the United States: An institutional-stakeholder perspective. *J. Manag. Stud.* **2006**, *43*, 47–73. [CrossRef]

2. Cichowicz, E.; Rollnik-Sadowska, E. Inclusive Growth in CEE Countries as a Determinant of Sustainable Development. *Sustainability* **2018**, *10*, 3973. [CrossRef]

3. Hełdak, M.; Raszka, B. Evaluation of the Local Spatial Policy in Poland with Regard to Sustainable Development. *Pol. J. Environ. Stud.* **2013**, *22*, 395–402.

4. Smith, B. *Good Governance and Development*; Palgrave Macmillan: New York, NY, USA, 2007; ISBN 978-0230525665.

5. Leszko, M.; Zając-Lamparska, L.; Trempala, J. Aging in Poland. *Gerontologist* **2015**, *55*, 707–715. [CrossRef]

6. Robbins, J. Aging Societies, Civil Societies, and the Role of the Past: Active Aging beyond Demography in Contemporary Poland. *East Eur. Politics Soc.* **2020**. [CrossRef]

7. Stimson, R.; Stough, R.; Roberts, B. *Regional Economic Development. Analysis and Planning Strategy*; Springer: Heidelberg, Germany, 2006; ISBN 978-3540348290.

8. Raszkowski, A.; Bartniczak, B. Towards Sustainable Regional Development: Economy, Society, Environment, Good Governance Based on the Example of Polish Regions. *Transform Bus. Econ.* **2018**, *17*, 225–245.

9. Raszkowski, A.; Bartniczak, B. On the Road to Sustainability: Implementation of the 2030 Agenda Sustainable Development Goals (SDG) in Poland. *Sustainability* **2019**, *11*, 366. [CrossRef]

10. Raszkowski, A.; Bartniczak, B. Sustainable Development in the Central and Eastern European Countries (CEECs): Challenges and Opportunities. *Sustainability* **2019**, *11*, 1180. [CrossRef]

11. Starr, F. *Corporate Responsibility for Cultural Heritage: Conservation, Sustainable Development, and Corporate Reputation*; Routledge Studies of Heritage, Routledge: New York, NY, USA, 2013; ISBN 978-0415656191.

12. Johnston, P.; Everard, M.; Santillo, D.; Robert, K. Reclaiming the definition of sustainability. *Environ. Sci. Pollut. Res.* **2007**, *14*, 60–66.

13. Bartniczak, B.; Raszkowski, A. Sustainable Development in African Countries: An Indicator-Based Approach and Recommendations for the Future. *Sustainability* **2019**, *11*, 22. [CrossRef]

14. Bartniczak, B.; Raszkowski, A. Sustainable development in Asian countries–indicator-based approach. *Probl. Ekorozw. Probl. Sustain. Dev.* **2019**, *14*, 29–42.

15. Coulson, A.; Campbell, A. Local government in central and Eastern Europe: Introduction. *Local Gov. Stud.* **2006**, *32*, 539–541. [CrossRef]

16. Edwards, M.; Hulme, D. NGO Performance and Accountability. Introduction and Overview. In *Non-Governmental Organisations. Performance and Accountability. Beyond the Magic Bullet*; Edwards, M., Hulme, D., Eds.; Earthscan Publications: London, UK, 1995; ISBN 978-1853833106.

17. Agere, S. *Promoting Good Governance: Principles, Practices and Perspectives*; Commonwealth Secretariat: Management Service Training Division: London, UK, 2000; ISBN 978-0850926293.

18. Raszkowski, A.; Sobczak, E. Sustainability in the Baltic States: Towards the implementation of sustainable development goals (SDG). In *The 13th International Days of Statistics and Economics*; Löster, T., Pavelka, T., Eds.; Libuše Macáková, Melandrium: Prague, Czech Republic, 2019; pp. 1253–1262.

19. Bartniczak, B.; Raszkowski, A. Sustainable development in the Russian Federation–indicator-based approach. *Probl. Ekorozw. Probl. Sustain. Dev.* **2017**, *12*, 133–142.

20. Pawłowski, A. *Sustainable Development as a Civilizational Revolution: A Multidisciplinary Approach to the Challenges of the 21st Century*; CRC Press: Boca Raton, FL, USA; Taylor & Francis Group: London, UK, 2011; ISBN 978-0415578608.

21. Galic, N.; Schmolke, A.; Forbes, V.; Baveco, H.; Brink, P.J. The role of ecological models in linking ecological risk assessment to ecosystem services in agroecosystems. *Sci. Total Environ.* **2012**, *415*, 93–100. [CrossRef]

22. Heberle, L. Sustainable Urban Development: Local Strategies and Global Solutions. In *Local Sustainable Urban Development in a Globalized World. Urban Planning and Environment*; Heberle, L., Opp, S., Eds.; Ashgate Publishing: Farnham, UK, 2008; pp. 1–8. ISBN 978-0754649946.

23. Connelly, A.; Carter, J.; Handley, J.F.; Hincks, S. Enhancing the practical utility of risk assessments in climate change adaptation. *Sustainability* **2018**, *10*, 1399. [CrossRef]

24. Kelly, P.M.; Adger, W.N. Theory and practice in assessing vulnerability to climate change and facilitating adaptation. *Clim. Chang.* **2000**, *47*, 325–352. [CrossRef]

25. Michałowski, A. Ecosystem services in the light of a sustainable knowledge-based economy. *Probl. Ekorozw. Probl. Sustain. Dev.* **2012**, *7*, 97–106.

26. Salonen, A.; Siirilä, J.; Valtonen, M. Sustainable Living in Finland: Combating Climate Change in Everyday Life. *Sustainability* **2018**, *10*, 104. [CrossRef]

27. Bartniczak, B.; Raszkowski, A. Sustainable forest management in Poland. *Manag. Environ. Qual. Int. J.* **2018**, *29*, 666–677. [CrossRef]

28. Fischer, A.P.; Paveglio, T.; Brenkert-Smith, H.; Carroll, M.; Murphy, D. Assessing social vulnerability to climate change in human communities near public forests and grasslands: A framework for resource managers and planners. *J. For.* **2013**, *111*, 357–365. [CrossRef]

29. Schmithüsen, F.; Kaiser, B.; Schmidhauser, A.; Mellinghoff, S.; Perchthaler, K.; Kammerhofer, A. *Entrepreneurship and Management in Forestry and Wood Processing: Principles of Business Economics and Management Processes*; Routledge Explorations in Environmental Economics, Routledge: New York, NY, USA, 2014; ISBN 978-1138675230.

30. Czaja, S.; Bacla, A. *Ekologiczne Podstawy Procesów Gospodarowania*; Wydawnictwo Akademii Ekonomicznej: Wrocław, Polska, 2002; ISBN 8370118365.

31. Jones, R.N. An environmental risk assessment/management framework for climate change impact assessments. *Nat. Hazards* **2001**, *23*, 197–230. [CrossRef]

32. Staniszewski, J.; Czyżewski, A. Economic factors underpinning the structural genotypes of agriculture development in the European Union after 2004. *J. Agribus. Rural Dev.* **2018**, *50*, 445–454. [CrossRef]

33. Markowski, T. Territorial dimensions of integrated development policy—Expectations and challenges concerning planning and institutional systems. *Stud. Reg.* **2013**, *35*, 51–64.

34. Statistics Poland, Local Data Bank. Available online: https://bdl.stat.gov.pl/BDL/start (accessed on 20 April 2020).

35. *Polityka Ekologiczna Państwa 2030*; Ministerstwo Środowiska: Warszawa, Poland, 2019. Available online: https://www.gov.pl/web/srodowisko/polityka-ekologiczna-panstwa-polityka-ekologiczna-panstwa-2030 (accessed on 5 June 2020).

36. *Environment 2018*; Statistics Poland: Warsaw, Poland, 2018; p. 28. Available online: https://stat.gov.pl/en/topics/environment-energy/environment/environment-2018,1,10.html (accessed on 5 June 2020).

37. *Environment 2017*; Central Statistical Office: Warsaw, Poland, 2017; p. 104. Available online: https://stat.gov.pl/en/topics/environment-energy/environment/environment-2017,1,9.html (accessed on 5 June 2020).

38. *Environment 2016*; Central Statistical Office: Warsaw, Poland, 2016; p. 104. Available online: https://stat.gov.pl/en/topics/environment-energy/environment/environment-2016,1,8.html (accessed on 5 June 2020).

39. *Environment 2015*; Central Statistical Office: Warsaw, Poland, 2015; p. 104. Available online: https://stat.gov.pl/en/topics/environment-energy/environment/environment-2015,1,7.html (accessed on 5 June 2020).

40. *Environment 2014*; Central Statistical Office: Warsaw, Poland, 2014; p. 107. Available online: https://stat.gov.pl/en/topics/environment-energy/environment/environment-2014,1,5.html (accessed on 5 June 2020).

41. *Environment 2013*; Central Statistical Office: Warsaw, Poland, 2013; p. 91. Available online: https://stat.gov.pl/en/topics/environment-energy/environment/environment-2013,1,6.html (accessed on 5 June 2020).

42. *Environment 2012*; Central Statistical Office: Warsaw, Poland, 2012; p. 91. Available online: https://stat.gov.pl/en/topics/environment-energy/environment/environment-2012,1,4.html (accessed on 5 June 2020).

43. *Environment 2011*; Central Statistical Office: Warsaw, Poland, 2011; p. 91. Available online: https://stat.gov.pl/en/topics/environment-energy/environment/environment-2011,1,3.html (accessed on 5 June 2020).

44. *Environment 2010*; Central Statistical Office: Warsaw, Poland, 2010; p. 91. Available online: https://stat.gov.pl/en/topics/environment-energy/environment/environment-2010,1,2.html (accessed on 5 June 2020).

45. *Environment 2009*; Central Statistical Office: Warsaw, Poland, 2009; p. 91. Available online: https://stat.gov.pl/en/topics/environment-energy/environment/environment-protection-2009,1,1.html (accessed on 5 June 2020).

46. Statistics Poland. *Environment 2008*; Central Statistical Office: Warsaw, Poland, 2008; p. 99.

47. Statistics Poland. *Environment 2007*; Central Statistical Office: Warsaw, Poland, 2007; p. 97.

48. Statistics Poland. *Environment 2006*; Central Statistical Office: Warsaw, Poland, 2006; p. 95.

49. Błażejczyk, K.; Kunert, A. *Bioclimatic Principles of Recreation and Tourism in Poland*; Polish Academy of Sciences, Stanisław Leszczycki Institute of Geography and Spatial Organization: Warsaw, Poland, 2011; ISBN 978-83-61590-47-7.

50. Kukuła, K. *Metoda Unitaryzacji Zerowanej*; Wydawnictwo Naukowe PWN: Warszawa, Polska, 2000; ISBN 8301130970.

51. Walesiak, M. *Uogólniona Miara Odległości w Statystycznej Analizie Wielowymiarowej*; Wydawnictwo Akademii Ekonomicznej: Wrocław, Polska, 2006; ISBN 8370118186.

52. Jajuga, K.; Walesiak, M. Standardisation of data set under different measurement scales. In *Classification and Information Processing at the Turn of the Millennium*; Decker, R., Gaul, W., Eds.; Springer: Berlin/Heidelberg, Germany, 2000; pp. 105–112. ISBN 978-3540675891.

53. Zeliaś, A. Some Notes on the Selection of Normalisation of Diagnostic Variables. *Stat. Transit.* **2002**, *5*, 787–802.

54. Milligan, G.W.; Cooper, M.C. A Study of Standardization of Variables in Cluster Analysis. *J. Classif.* **1988**, *5*, 181–204. [CrossRef]

55. Ward, J.H. Hierarchical Grouping of Optimize an Objective Function. *J. Am. Stat. Assoc.* **1963**, *58*, 236–244. [CrossRef]

56. Johnson, S.C. Hierarchical clustering schemes. *Psychometrika* **1967**, *39*, 241–254. [CrossRef]

57. Anderberg, M.R. *Cluster Analysis for Application*; Academic Press: New York, NY, USA; San Francisco, CA, USA; London, UK, 1973; ISBN 978-1483191393.

58. Hartigan, J.A. *Clustering Algorithms*; John Wiley & Sons: New York, NY, USA, 1975; ISBN 978-0-471-35645-5.

59. Sneath, P.H.A.; Sokal, R.R. *Numerical Taxonomy. The Principles and Practice of Numerical Classification*; W.H. Freeman and Company: San Francisco, CA, USA, 1973; ISBN 978-0716706977.

60. Aldefender, M.S.; Blashfield, R.K. *Cluster Analysis*; Sage Press: Beverly Hills, CA, USA; Los Angeles, CA, USA, 1984; ISBN 978-0803923768.

61. Basiura, B.; Sokołowski, A. The clustering effectiveness in Ward's method. *Folia Oeconomica Crac.* **2007**, *46–47*, 53–64.

62. Everitt, B.S.; Landau, S.; Leese, M.; Stahl, D. *Cluster Analysis*; John Wiley & Sons: Chichester, UK, 2011; ISBN 978-9462094048.

63. Sobczak, E. *Segmentacja Rynków Zagranicznych*; Wydawnictwo Uniwersytetu Ekonomicznego we Wrocławiu: Wrocław, Polska, 2010; ISBN 978-8376950228.

64. Hellwig, Z. Wielowymiarowa analiza porównawcza i jej zastosowanie w badaniach wielocechowych obiektów gospodarczych. In *W: Metody i Modele Matematyczno-Ekonomiczne w Doskonaleniu Zarządzania Gospodarką Socjalistyczną*; Welfe, W., Ed.; Wydawnictwo Naukowe PWN: Warszawa, Polska, 1981; ISBN 978-8320800425.

65. Statistics Poland. Available online: https://stat.gov.pl/en/ (accessed on 14 April 2020).

66. Eurostat Database. Available online: https://ec.europa.eu/eurostat/data/database (accessed on 14 April 2020).

67. Wojtyniak, B.; Goryński, P. (Eds.) *Health Status of Polish Population and Its Determinants*; National Institute of Public Health, National Institute of Hygiene: Warsaw, Poland, 2018; ISBN 978-8365870162.

68. Füssel, H.-M.; Klein, R.J. Climate change vulnerability assessments: An evolution of conceptual thinking. *Clim. Chang.* **2006**, *75*, 301–329. [CrossRef]

69. Szewrański, S.; Kazak, J.; Szkaradkiewicz, M.; Sasik, J. Flood risk factors in suburban area in the context of climate change adaptation policies—Case study of Wroclaw, Poland. *J. Ecol. Eng.* **2015**, *16*, 13–18. [CrossRef]

70. Szewrański, S.; Świąder, M.; Kazak, J.; Tokarczyk-Dorociak, K.; Van Hoof, J. Socio-Environmental Vulnerability Mapping for Environmental and Flood Resilience Assessment: The Case of Ageing and Poverty in the City of Wroclaw, Poland. *Integr. Environ. Assesment Manag.* **2018**, *14*, 592–597. [CrossRef] [PubMed]

71. Kerr, J.; Rosenberg, D.; Frank, L. The role of the built environment in healthy aging: Community design, physical activity, and health among older adults. *J. Plan. Lit.* **2012**, *27*, 43–60. [CrossRef]

72. Marteau, T.M.; Hollands, G.J.; Fletcher, P.C. Changing human behavior to prevent disease: The importance of targeting automatic processes. *Science* **2012**, *337*, 1492–1495. [CrossRef] [PubMed]

73. Sallis, J.F.; Floyd, M.F.; Rodríguez, D.A.; Saelens, B.E. Role of built environments in physical activity, obesity and cardiovascular disease. *Circulation* **2012**, *125*, 729–737. [CrossRef]

74. Judson, G. *Engaging Imagination in Environmental Education: Practical Strategies for Teachers*; University of British Columbia Press: Vancouver, BC, Canada, 2017; ISBN 978-1926966755.

75. Ramanathan, V.; Han, H.; Matlock, T. Education Children to Bend the Curve: For a Stable Climate, Sustainable Nature and Sustainable Humanity. In *Children and Sustainable Development: Ecological Education in a Globalized World*; Battro, A., Léna, P., Sorondo, M., Braun, J., Eds.; Springer International Publishing: Cham, Germany, 2017; pp. 3–16. ISBN 978-3319471297.

 *sustainability*

Article

A Customized Decision Support System for Renewable Energy Application by Housing Association

Aleksandra Besser *, Jan K. Kazak, Małgorzata Świąder and Szymon Szewrański

Department of Spatial Economy, Faculty of Environmental Engineering and Geodesy, Wrocław University of Environmental and Life Sciences, Grunwaldzka 55, 50-357 Wrocław, Poland
* Correspondence: aleksandra.besser7680@gmail.com

Received: 5 July 2019; Accepted: 7 August 2019; Published: 13 August 2019

Abstract: One of the major problems in socio-environmental systems is the growing depletion of non-renewable resources and environmental degradation, resulting from inadequate environmental management and planning. Deepening environmental problems have forced countries to create management instruments that will help repair damage and support environmental protection efforts. The aim of this research is to develop a customized decision support system for the management of renewable energy based on the existing Geographic Information Systems (GIS). The proposed tool enables assessing the potential of solar energy production at the local scale, analyzing each rooftop. Due to the scale of the analyzed area and the details of the assessment, the tool is customized to the needs of housing associations. The system combines an existing GIS tool for calculating the solar radiation potential of rooftops (SOLIS) together with Tableau software that was used to aggregate and analyze data. In order to present the applicability of the developed tool, visualizations were prepared based on housing buildings managed by the "Biskupin" Housing Association in Wrocław (Poland) which is responsible for the management of 3415 residential premises. The created system based on spatial and environmental data will help to decide how to manage the available resources and the environment at the local scale while reducing the pressure on the environment. The tool allows for the aggregation, filtering and presentation of spatial data for the entire area of a housing association, as well as for a single building.

Keywords: decision support system; renewable energy; solar energy radiation; sustainable development; ArcGIS; Tableau

---

1. Introduction

As a result of the progress of civilization, structures of the natural system and various spheres of the Earth were transformed [1]. These transformations are the result of various activities and phenomena occurring in the social, economic and environmental sphere. The creation of new settlement units caused the increase in the population in a given area. Clusters of the population indirectly and directly use natural resources offered by the Earth to survive. As a consequence, the development of urban environments led to the depletion of non-renewable resources and even environmental degradation. Initially, it was exploited at a local scale until the 20th century, when the period of industrialization began. Increasing technological and scientific progress as well as civilization development resulted in increased work efficiency and increased environmental exploitation [2,3]. It was realized that the effects of human activities on the environment will not appear immediately, and their consequences will be felt by future generations [4].

From year to year, the demand for resources, i.e., energy [5], increases and this means that local authorities should look for alternative sources to ensure clean energy for their residents. In households, energy is most often used for heating rooms, water and cooking meals [6]. According to the energy consumption in households, the largest percentage of the population uses solid fuels, mainly hard coal and wood. The use of energy fuels is associated with the increased exploitation of non-renewable resources and, as a consequence, negative impacts on the air quality [7,8]. In order to reduce environmental pollution, many countries have decided to use renewable energy for energy purposes. One of the renewable energy sources with high potential is solar radiation, which is an inexhaustible resource. Solar energy systems are gaining popularity due to easy and universal access to the resource [9]. The use of solar technology such as photovoltaics using renewable sources in the urban environment will allow the use of clean energy, limiting the use of non-renewable sources such as coal for building heating [10].

Governments have introduced many programs [11], action plans [12,13] and documents at an international level [14,15] to prevent further, uncontrolled depletion of the natural environment [16]. However, one of the most urgent problems is to create a system that will allow the use of environmental resources (including energy) in a sustainable manner, ensuring the existence and development of humanity [17]. Such a system would require a reduction in socio-environmental vulnerability in order to guarantee the stability of processes driven by humans in the built environment reality [18]. The system should rationally supervise the management of environmental resources in order to provide future generations with the opportunity to live in a quality environment similar to the one in which it currently resides [19]. In order to improve the resources, management novel tools using the possibilities of data sources and computation abilities should be implemented at the local level [20]. That refers also to the domain of solar energy. Many solutions for solar radiation assessment are predesigned for regional studies [21,22] and, therefore, do not suit the needs of local stakeholders. Advanced IT tools, including decision support systems, are becoming increasingly popular and are often used in environmental management, including in the area of renewable sources [23], air quality [24], water ecosystem [25] or impurities from agriculture [26].

The use of decision support tools results in making more rational and effective decisions [27]. By calculating indicators describing a given problem and analyzing its relationships, we can graphically see and understand how the phenomenon would change overtime, what impact it would have on further development and, as a result, make the conscious decision [28]. The system itself does not respond to the problem. It processes the entered data and the user, based on his expert knowledge, checks the correctness of the data, prepares information, interprets the results and makes a subjective assessment of the variants [29]. In addition, the model uses various data sources, e.g., thematic maps, satellite images or spreadsheets, which are adapted to the needs of a given issue and require different methods and algorithms of recreation [30]. Adopting appropriate multi-criteria assumptions in decision support systems and adapting data to the spatial dimension in GIS systems would allow for creating a system with an integrated approach to planning and managing environmental resources [31,32].

Despite the fact that decision support systems are well known in the literature, the complex and diverse issues that could use these tools do not allow developing one universal solution. In order to convince users to apply such systems, there is a need to customize solutions according to the needs of users [32]. The technical support of local stakeholders in renewable energy implementation might also be helpful to combat an emerging problem of energy poverty [33] that decreases the quality of life of inhabitants in urban areas [34]. The issue of energy poverty is especially important in the context of ageing populations [28] despite the fact that long-term care expenses are constantly growing in Poland [35]. Urban renewal actions, also in the context of improving energy performance, will increase the value of properties [36–39], which would enable housing associations to achieve customer satisfaction by showing the costs together with benefits of these investments [40,41]. Therefore, there is a need to develop a flexible approach, matching the tool to the available data as well as the key drivers of the management process [31].

Therefore, the aim of this research is to develop a customized decision support system for renewable energy application based on the existing Geographic Information Systems (GIS). The system combines an existing GIS tool for calculating the solar radiation potential of rooftops (SOLIS) together with a business intelligence system (Tableau). The integration of these two computing environments enables enlarging the functionality of the developed solution [42]. The created system contains data, which can help managers to decide how to manage reducing energy consumption at the local level and the pressure on the environment. The decision support system allows aggregation, filtering and presentations along with the assessment of environmental data. As a customized solution has to be developed according to the needs of the user, the tool was designed based on the needs of one Polish housing association. The presentation of the applicability of the developed tool is shown in the case of housing buildings managed by the "Biskupin" Housing Association in Wrocław (Poland). The assumed impact of the developed tool application is to activate a housing community to reduce the consumption of electricity and heat through the use of modern solutions based on renewable energy sources, which will help to reduce the consumption of non-renewable resources. The model for the management of renewable energy is built in two parts—the first part, which contains analyses in a modified GIS model, and the analytical part, which includes the analysis of the obtained results of tests performed in the Tableau software (version 2018.3). In the first part, solar radiation was calculated in a particular month in 2018 in the ArcGIS program (version 10.5.1.) with the help of the Solar Area Radiation tool (SOLIS).

The direction of changes in the energy sector in Poland is not a new issue. Policies and action promoting renewables were undertaken and implemented by Poland before joining the European Union in 2004 [43]. However, the share of renewable energy is still relatively low. The target defined in the project entitled Energy Policy of Poland until 2040 aims to increase the share of renewable energy sources up to 21% in the whole energy sector in 2030. Achieving that goal requires doubling this source in one decade [44]. Therefore, a proposed model for energy resources management could also be relevant in this context.

2. Methodology of Environmental Management Decision Support Tool

The decision support system that has been created combines the functionality of the GIS program and Business Intelligence systems (BI), the combinability of which allows for better user functionality (Figure 1).

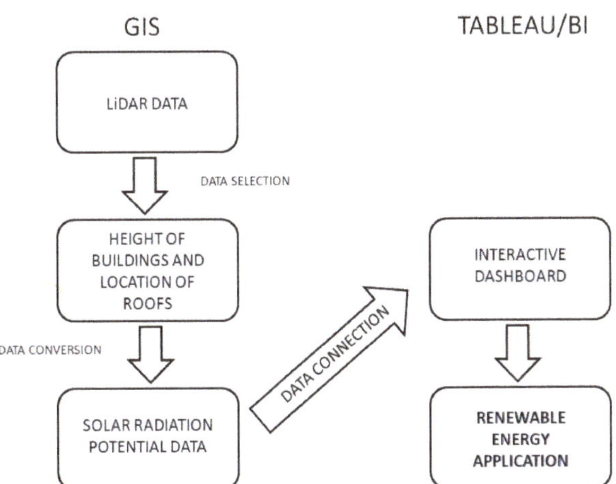

Figure 1. Framework of operation diagram. Source: Own elaboration.

The input data were represented by a Light Detection and Ranging (LiDAR) dataset, which was a base for the extraction of buildings together with height information. The input data were a basis for solar radiation calculations, and then connection with the BI tool. The use of spatial data in Tableau software allowed for interactive dashboard building. The result of this connection and dashboard building was a detailed model for renewable energy application.

2.1. Research Area

The test object in this work is the "Biskupin" Housing Associationin Wrocław (Poland), which has been operating since 1959. The properties included in the development area are located in the north-eastern part of Wrocław, in the Lower Silesia Voivodship.

The cooperative consists of 120 properties, including housing (58) located in the following districts: Ołbin (11), Biskupin-Sępolno-Dąbie-Bartoszowice (46), Grunwaldzki Square (2), and Nadodrze (3). The buildings located inOłbin and Grunwaldzki Square come from the pre-war period. Most real estate are located in Biskupin-Sępolno-Dąbie-Bartoszowice and were created in the period 1919–1935. They are mainly multi-family two-story terraced houses, as well as 4-or 6-storey detached houses. The buildings located in Gersona 5 and Olszewskiego 149–153 street are dated from the period 2010–2015 and they include multi-family buildings. The property on the Biskupin estate also includes buildings constructed of large, prefabricated concrete slabs, which were built in the 1960sand 1970s. The housing estate in Nadodrze dates back to the 19/20th century and has a quarterly development layout. The buildings were not so damaged during the war. In Nadodrze, the buildings are mainly high tenements and cobbled streets, whereas the housing development in Plac Grunwaldzki is from the pre-war period.

The height of the buildings depends on the location of the property and looks as follows:

- *Biskupin-Sępolno-Bartoszowice-Dąbie* district: medium-sized, 4–7-storey buildings predominate, with the height of buildings between 10 main A. Mickiewicza street, through 16meters in the area ofGerson and S.Sempołowska Streets, and 22 m in the vicinity of Canaletto Street;
- *Ołbin* district:average buildings dominate, 5 storeys with the height of buildings between 16m along the streets of E.Stein and B.Prusa, and 22 m along the Daszyńskiego and National Unity (org. Jedności Narodowej) Streets;
- *Grunwaldzki Square* (org. Plac Grunwaldzki) district: 8–10-storey buildings, 20–22 m height;
- Nadodrze district: 8–10-storey buildings with a height of 20–22 m.

In the case of service buildings and garages, the maximum area of individual real estate varies between 3 and 6m of a single-storey buildings. In addition, the co-operative also includes an accompanying infrastructure, such as waste containers, garages and service premises (62) (Figure 2). The analyzed area consists of 3415 residential premises, which are inhabited by approximately 6050 people. The area subject to the cooperative has in its resources in over 266,328 m^2 of land.

2.2. Model Framework

In order to create an environmental management model in relation to solar energy management, this work relied on the existing decision support tool to assess the solar potential—SOLIS, which is available online in the form of the ArcGIS tool [45]. The SOLIS was made in the ArcGIS program with the help of the visual model Builder geoprocessing editor. Available instruments that can be used to create a model can be found in the ArcToolbox catalog in Spatial Analyst. The basic tool in the model is Area Solar Radiation. The SOLIS model is based on a set of multipart patches in the form of 3D building data (so-called multipatch), which is characterized by a Z value, showing the height of each patch-the minimum height and the maximum height. In this work, this tool has been modified. Obligatory input data that should be included in the model are the height value and roof area. In the model for the "Biskupin" Housing Association, instead of multipatch data, numerical digital data were used in the form of a point cloud in the LAS format from aerial laser scanning (LAS stands for LiDAR data exchange file), from which information on the minimum and maximum height of individual buildings and roof surfaces with inclination were obtained. The proposed modification enlarges the group of potential users, as in many cases multipatch data as a postprocessing effect might be not available, while LiDAR as raw data might be possible to obtain. Besides the data format, the model has changed the order of attribute selection—the roof surface is suitable for the installation of photovoltaic panels. In the primary formula (Figure 3a), the selection of cells presenting the value of radiation, with an area larger than 2m^2, takes place before the calculation of the solar potential, while in the second model (Figure 3b), the selection of such areas with an area larger than or equal to 2m^2 is the last element of calculation.

Figure 2. Buildings included in the "Biskupin" Housing Association in Wrocław in particular settlements. Source: Own elaboration using ArcGIS.

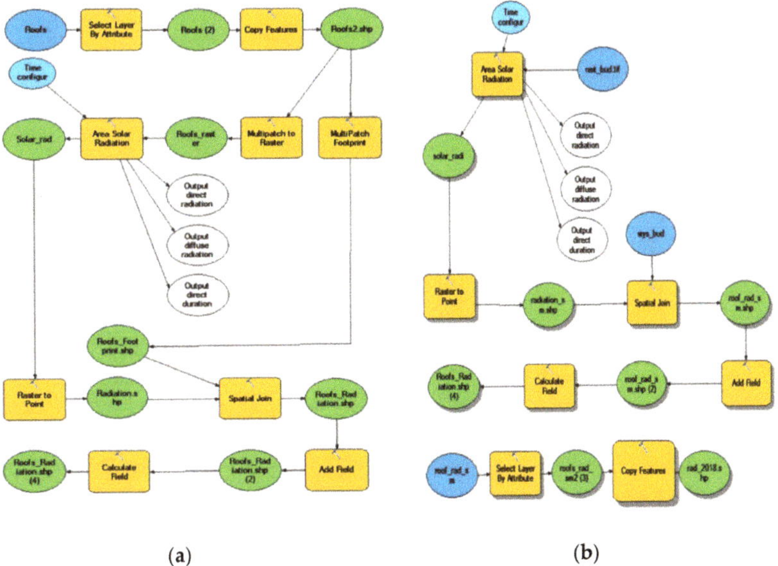

(a)   (b)

Figure 3. Workflow of model calculation: (**a**) presents the solar radiation potential of rooftops (SOLIS) calculation workflow; (**b**) presents the modified model calculation workflow. Own elaboration using ArcGIS.

2.3. Conversion of Data

The first stage of the work in the model was the development of data that will be the basis for calculating the amount of potential solar radiation with the division into individual months in 2018. Numerical altitude digital data were used for the analysis, which contains a point cloud from LiDAR (Figure 4). The following figure shows the digital spatial data in LAS format imported into ArcGIS. This data source was selected due to the fact that laser scanning technologies are being developed right now and already support a variety of domains with high-quality spatial data [46–48].

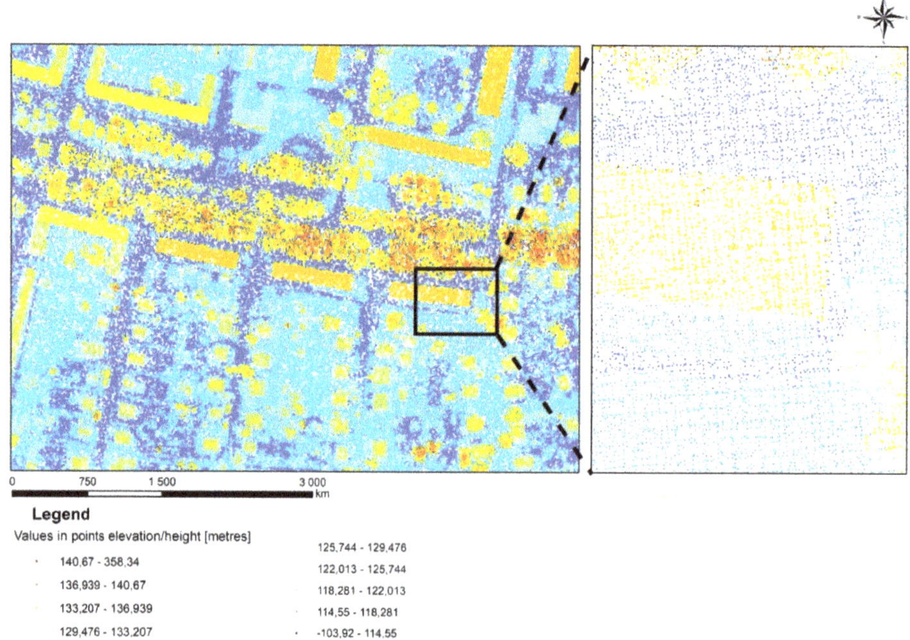

Figure 4. Input LiDAR data. *Source: Own elaboration using ArcGIS.*

In the analyzed area, urban development, the average density of points is 12 points/m^2. The data are available in LAS format and divided into the eight classes [49]. In order to obtain information about the minimum and maximum height of individual buildings from the LAS layer, it was necessary to select the data class named "6 BUILDING", which only shows buildings. The data conversion allows obtaining the altitude of the buildings, on the basis of which we can obtain data for the model.

3. Results

As a result of the analyses created in the model, vector layers were obtained with the value of average solar radiation per square meter of roof, as well as total radiation on the entire roof surface. Depending on the type, height, slope and exposure of individual roof fragments, the results of the analyses are different.

3.1. Solar Radiation

The obtained results did not take into account the actual weather conditions that appeared in 2018 and other objects that could obscure sunlight, like trees, and which affect the amount of actual solar radiation. Despite this, the results obtained constitute the basic information for the housing cooperative, with the possible value of the solar potential of the annual roofs of individual properties.

On this basis, it will be possible to find roofs with the highest potential, which will allow the decision to install photovoltaic panels that will achieve the highest possible energy efficiency. This will allow the housing cooperative to reduce the consumption of non-renewable resources such as coal or gas and to reduce the emission of pollutants from the boiler room, and instead allow the use of clean energy sources. The most numerous groups are residential buildings, where the average roof potential of which is estimated between 613,777 Wh/m^2 and 742,645 Wh/m^2—108 roof patches (group IV) (Figure 5).

Figure 5. Solar potential per 1m^2 of roof properties included in the "Biskupin" Housing Cooperative in Wroclaw. The values of solar radiation are shownusing a blue–red color palette. The dark blue represents buildings with the lowest value of solar potential, the dark red represents the highest values of solar potential. Source: Own elaboration using ArcGIS.

The total roof area in this category is 19,721.36 (23.33% of the total area of all roofs). The second most numerous group is roofs with annual solar potential in the range of 382,579 Wh/m^2–508,700 Wh/m^2–103 objects (group II)—the area of which is 18,808.41m^2 (22,25% of the surface of all roofs) Values of solar radiation potential at a similar level are also in the range 742,646 Wh/m^2–944,887 Wh/m^2 (group V)—the area of which is 19,171.90 m^2 At least for the large group, they constitute roof patches, and the amount of solar energy falling on their surface is 944,888 Wh/m^2–1,225,944 Wh/m^2 (group VI) and 41,196 Wh/m^2–382,578 Wh/m^2 (group I) during the year. The solar potential values can be different in neighboring buildings because, when calculating solar intensity, parameters such as the height of buildings, aspect, slope of roofs, size of roofs, number of roofs' patchesare taken into account. The values of individual parameters can give the total value of the solar potential other than in the neighboring building, which affects the final result of the intensity of solar radiation.

3.2. Model of Environmental Ecision Support Tool

Based on the obtained data, an interactive model of environmental management for the housing association was created in the Tableau program. The model includes a decision support tool for assessing the solar potential. This program is used to visualize the obtained data graphically and is easier to interpret. With the help of the tool, it will be possible to select the appropriate location of the solar installation by comparing the solar potential of individual buildings (Figure 6).

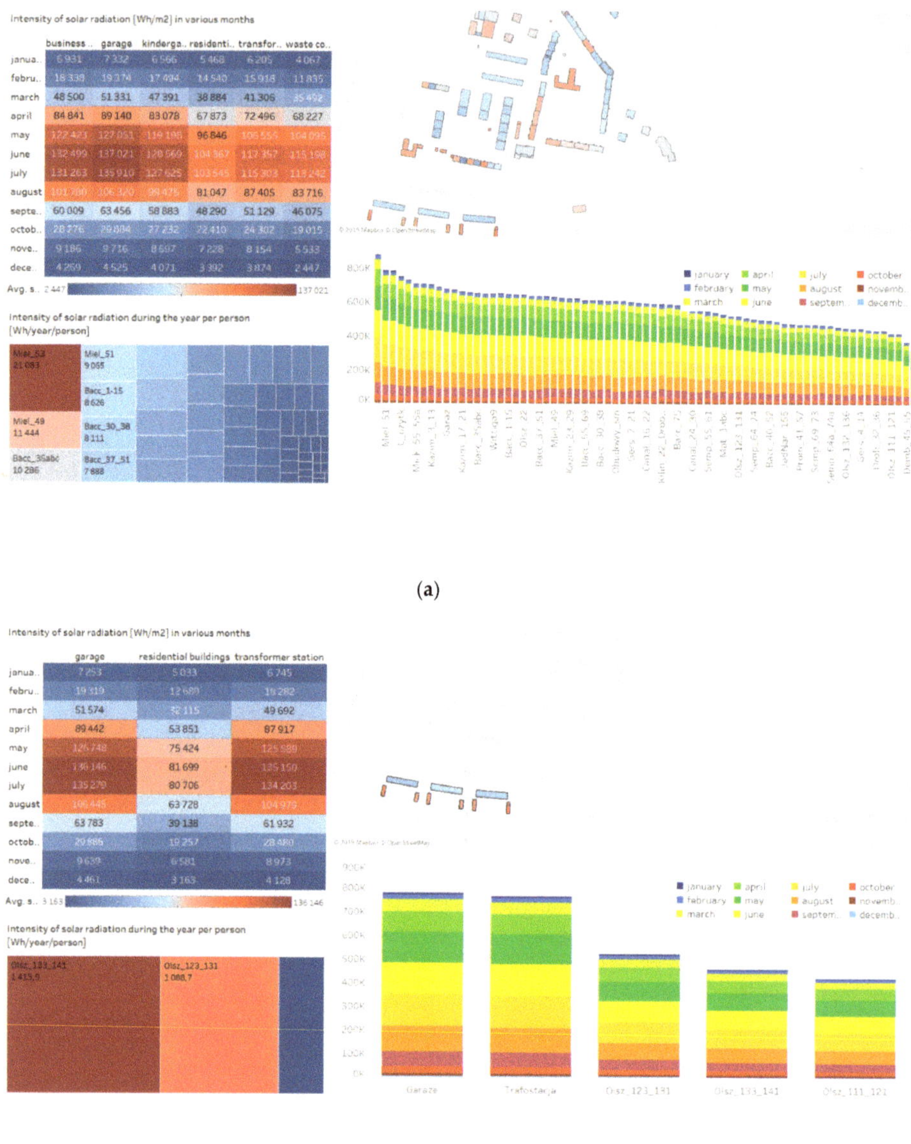

Figure 6. The customized interactive dashboard for renewable energy assessment: (**a**) General view; (**b**) selected buildings.Source: own elaboration using Tableau Software.

Using this model, the housing association is able to analyze the solar potential of individual properties during the year, and on their basis, decide on the introduction of a solar installation on roofs, the use of which would help reduce the consumption of electricity and heat. Demand for energy continues to grow year by year, which is why alternative sources are needed to manage the environment, which will help reduce excessive consumption of energy resources.

The model consists of three analyses and a map. The input data to the program were data in the form of a vector layer covering solar roof intensity values divided into months obtained in the ArcGIS program. The model is interactive, and, at the same time, it presents information on the sum of the total solar radiation intensity per year, per capita and solar potential depending on the type of real estate selected, group of real estate and a specific type of real estate. The model presents analyses regarding:

(1) The monthly solar potential for given types of buildings belonging to a housing association

The largest average solar potential in terms of the size of the building is garages and commercial premises. However, all properties have solar potential at a similar level. The total solar intensity in the month with the highest potential in June is for garages 137,021 Wh/m^2. In almost every type of property, the period of the highest potential of solar radiation occurs from April to September (Figure 7).

	business premises	garage	kindergarten	residential buildin..	transformer station	waste container
janua..	6 931	7 332	6 566	5 468	6 205	4 067
febru..	18 338	19 374	17 494	14 540	15 918	11 835
march	48 500	51 331	47 391	38 884	41 306	35 492
april	84 841	89 140	83 078	67 873	72 496	68 227
may	122 423	127 051	119 198	96 846	106 555	104 095
june	132 499	137 021	128 569	104 367	117 357	115 198
july	131 263	135 910	127 625	103 545	115 303	113 242
august	101 780	106 320	99 475	81 047	87 405	83 716
septe..	60 009	63 456	58 883	48 290	51 129	46 075
octob..	28 276	29 884	27 232	22 410	24 302	19 015
nove..	9 186	9 716	8 697	7 228	8 154	5 533
dece..	4 269	4 525	4 071	3 392	3 874	2 447

Avg. solar ra.. 2 447   137 021

Figure 7. Intensity of solar radiation [Wh/m^2] in various months. Source: own elaboration using Tableau Software.

The average intensity of solar radiation in residential buildings is smaller than in business premises. This is due, among other factors, to the fact that residential buildings have different roof slopes, and the business premises included in the cooperative have flat roofs on which the position of photovoltaic panels will allow for higher solar energy accumulation. In addition, the roofs of residential buildings are more divided into parts in relation to business premises. However, residential buildings are higher than business premises.Therefore, the differences in the solar potential are close to each other. In relation to other objects, residential buildings have the lowest average intensity of solar radiation in the most sunny period (May–August). The highest average solar potential is provided by garages, the value of which is higher by over 30,000 Wh/m^2 from residential buildings. Also, the kindergarten building as well the business premises havea higher solar potential per 1m^2 of roof. If photovoltaic panels are installed on the roofs of garages and waste containers, it will be possible to accumulate energy in the event of insufficient energy for nearby residential buildings.

(2) The annual solar radiation on all properties in the household association

The proposed model allows for verification of the solar potential for each building belonging to a housing association. The building located in Mielnickiego 13 street has the highest average potential of solar radiation—21,053kWh per year—which is three times the average energy consumption per person per year. Statistically, in Poland, a person consumes approximately 6111 kWh/year [50]. A total of 28 buildings have a lower potential per inhabitant than the statistical amount of energy consumption per capita and, in 29 cases, solar panels will be able to produce an amount of energy that satisfies the needs of residents (Figure 8).

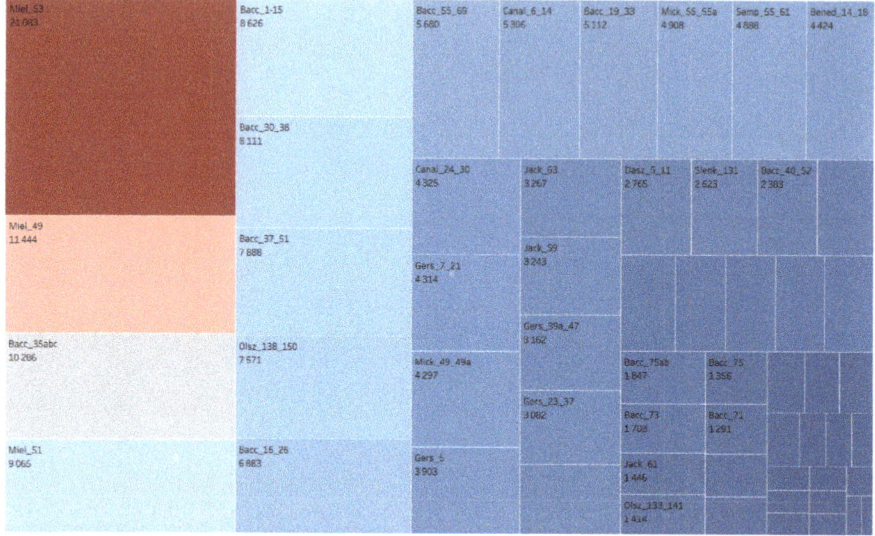

Figure 8. Intensity of solar radiation during the year per person [Wh/year/person]. Source: Own elaboration using Tableau software.

(3) The detailed annual intensity of solar radiation on all properties in the household association

The model allows for the detailed analysis of the annual intensity of solar radiation. Therefore, itis possible to verify the annual value of solar radiation per given property, as well as the verification of solar potential during each month. The graph below presents the values of solar intensity per $1m^2$ of real estate. The biggest potential in the year is the building located at Canallette 6-14 street, approx. 900,000 Wh/m^2, which is a value higher than the average potential of solar radiation intensity in the whole of Wrocław—828,924.8 Wh/m^2. The smallest potential was recorded on the roofs of real estates located in Gersona 4-14 street and Olszewskiego 132-136—approximately 400,000 Wh/m^2 (Figure 9). The highest annual solar potential is located in the building at Canalette 6-14. This is due to the fact that the building is high, has 5 storeys, with a total height of 12m. In addition, the roof of the property is flat, which allows for higher possibilities of accumulating solar energy. In the months with the highest intensity of solar intensity, the potential ranges from 80,000 to 100,000 Wh/m^2 of roof. The lowest solar potential occurs in January, November and December and its maximum value is up to 10,000 Wh/m^2. However, the building with the lowest solar intensity is the property located at Gerson 4-14. The maximum intensity in months with the highest solar radiation is approximately60,000–80,000 Wh/m^2. The roof of the property is more divided (due to existing chimneys), which affects the total value of solar intensity. The largest share of energy possible to accumulate is in the period from May to September. The difference in the amount of intensity in some months is even several times higher. An example of this is the value of solar radiation intensity in real estate with the highest potential. In January, the intensity value is 8313 Wh/m^2 and in June as much as 153,338 Wh/m^2.

3.3. Accumulated Energy vs. Energy Use

Taking into account the obtained results, the integration of SOLIS and Tableau would allow for more detailed analysis, i.e., verification of accumulated energy during year versus actual annual consumption of energy by inhabitants of given residential building (Table 1).

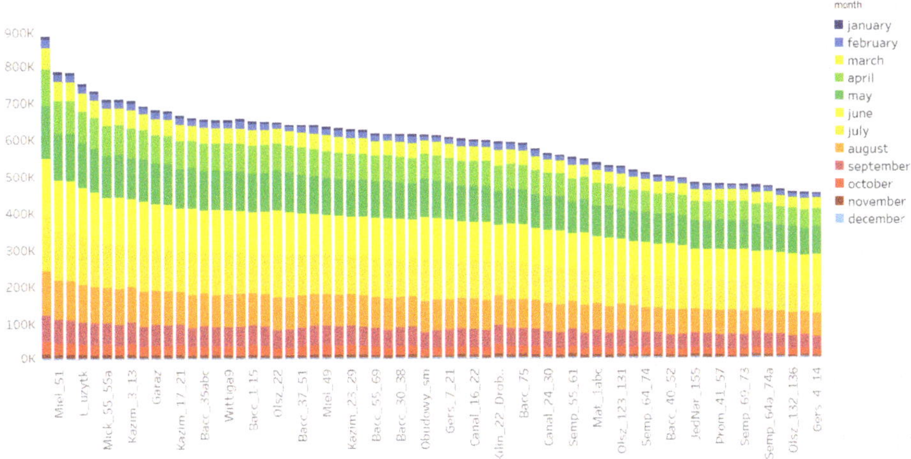

Figure 9. The intensity of solar radiation in particular months divided into the location of the property. Source: own elaboration using Tableau software.

Table 1. The comparison of accumulated energy to energy demand—examples from building inventory.

Real Estate Code	Annual Accumulated Energy in Wh	Annual Use of Energy in Wh	Balance in Wh
Olsz_111_121	271,141,195	51,332,400	+219,808,000
Olsz_123_131	688,801,124	48,888,000	+639,913,000
Olsz_132_136	705,721,916	71,498,700	+634,223,000
Olsz_133_141	949,865,835	49,499,100	+900,366,000
Olsz_138_150	715,400,977	76,998,600	+638,402,000
Prom_25_39	523,210,936	78,220,800	+444,990,000
Prom_41_57	803,637,760	98,998,200	+704,639,000
Semp_55_61	624,099,638	34,221,600	+589,878,000
Semp_63_67	501,570,235	69,054,300	+432,515,000
Semp_64_74	404,252,992	68,443,200	+335,809,000
Semp_64a_74a	391,366,621	62,332,200	+329,034,000
Semp_69_73	392,539,126	81,887,400	+310,651,000
Gers_23_37	952,738,626	99,609,300	+853,129,000
Gers_39a_47	970,299,234	95,942,700	+874,356,000
Gers_4_14	745,840,280	70,276,500	+675,563,000
Gers_5	844,439,276	16,1330,400	+683,108,000
Gers_7_21	773,564,185	42,777,000	+730,787,000
Jack_59	294,365,394	19,555,200	+274,810,000
Jack_61	151,276,653	21,388,500	+129,888,000
Jack_63	294,365,754	19,555,200	+274,810,000

When comparing the data obtained, it can be seen that each of the buildings has a higher solar potential than the current demand of residents for energy. By installing panels, the cooperative is able to produce almost 80% more energy than current demand.

4. Discussion

The solar radiation tool developed in this study supports the process of improving energy performance by selecting these locations which present the highest potential for energy production at the scale of individual buildings. By converting the values in the model, information was obtained on the potential of each type of building, which vary due to their technical parameters and the surrounding objects. Such a solution is not possible in every solar radiation tool.

The tool developed by Hofierka and Kanuk presents an example of how GIS-based approaches effectively calculate the solar radiation for analyzed neighborhood. However, it does not allow selecting a single object or group of objects to compare with deleting other objects from analysis [51]. Moreover, due to the development of remote sensing technologies [52,53], a wider spectrum of data should be possible to use in solar radiation assessments. Kaynak et al. recently developed their software which accepts only CityGML-based file formats as input data. However, they highlighted that the direction of future works is to enlarge the type of data format types that can be used [54]. The modification that we did in this research modify an existing tool (SOLIS) exactly in that aspect, by using LiDAR as an input data source. However, LiDAR data processing is not a simple task. Processing the point cloud, the user should obtain aneasy and ready-to-use database, but most often the data are considered in terms of the position and elevation of buildings as well as intensity when applied [55,56]. In addition, the currently available resolution of LiDAR data limits the definition of roofs, which means that roofs, chimneys, dormers, antennas are not taken into account in the analyses, the use of which would help to more accurately determine the area available on building facades [57,58]. The other crucial aspect is functionality from the point of view of the user. Wijeratne et al. in their work "Design and development of distributed solar PV systems: Do the current tools work?" analyzed 23 software and four smart phone/tablet applications according to their features. Only two solutions were categorized as user-friendly [59].

In the case of our tool, by incorporating business intelligence environment, it was possible to integrate visualization on the map together with charts. The panels that were used in the interactive dashboard were selected based on the suggestions by specialists from housing associations. The tools allow selecting any object (or group of objects) on any panel and highlight the assigned value on other panels on the dashboard. That allows quicker and user-friendly visual analyses which meet the needs of non-professionals in GIS, who are responsible for environmental management at the level of housing association. Finally, the success of the application is connected with the affordability of the tool. The review of over 200 solar design tools shows that only 12 solutions give the possibility of dynamic visualization (similar like the tool developed in our study) and only four of them are freeware [60]. That shows the limitation in the promotion of developed tools in practice. In the case of our dashboard, it has to be designed in a pay-ware version of Tableau software. However, the use of the dashboard can be supported for free by Tableau Public. The proposed solution shows the solar potential at the scale of individual buildings. It also allows for verification if the produced energy could support the energy needs of inhabitants in a given area. However, the crucial element in energy management is the ability to control the issue of energy peaks. Such elements can be incorporated into decision support systems [61]. Therefore, future works should also focus on that aspect. Moreover, the approach does not include other elements in the area, such as trees. Therefore, the assessment could be incomplete and a lack of this data could represent a further limitation and necessity for the development of the model. The renewable energy applications by housing associations could also be relevant in other research, as in the context of energy poverty, vulnerable groups to energy poverty (impoverished people or ageing society) [28,62], spatial management [36], and comprehensive environmental management system [63,64].

5. Conclusions

The developed customized decision support system for renewable energy application presented in the results proves that the aim of the study was fulfilled. Based on the obtained data, an environmental management model was created to assess the solar potential of roofs. The system combines an existing GIS tool for calculating the solar radiation potential of rooftops (SOLIS) together with a business intelligence system (Tableau). An interactive tool allows for the aggregation, filtering and presentation of data on the amount of solar intensity of each type of building in individual months, during the year and per capita. The environmental management model contains analyses in a modified model, and the analysis of the obtained results of tests. The tools used in this system result in making more rational and effective decisions. Thanks to the Tableau program, the property manager obtains information in an easy and transparent way. The created system of decision support contains data based on which it will be possible to reduce energy consumption and pressure on the environment at the local level using modern solutions in the energy sector. This tool was developed based on the cooperation between academics and practitioners from the housing association. Such an approach allowed including the information in interactive dashboards based on the needs reported by specialists in the environmental management domain.

The results obtained for the "Biskupin" Housing Association show that over 50% of the analyzed rooftops has sufficient potential for solar radiation intensity per capita, and a smaller part of the property would provide half of the energy demand. The use of photovoltaic installation will allow reducing the consumption of solid fuels in the boiler house and the consumption of natural resources used by association will decrease. In addition, by abandoning boilers fired with solid fuels, mainly coal, the emission of carbon dioxide into the air will be limited. The potential of the values obtained is sufficiently detailed for the cooperative's roof potential, because they are data on the value of the possible energy produced, which does not take into account the limited efficiency planned for the use of solar technology. Using an installation consisting of the placement of photovoltaic panels on roofs in areas of the highest solar radiation intensity allows for maximizing the energy benefits of real estate. It has both economic and environmental benefits for residents. A reduced consumption of natural resources means limiting the use of the environment. This has a positive effect both on the environment and the inhabitants of this neighborhood. In addition, a reduction in stoves fired with minerals will reduce the emission of carbon dioxide to the atmosphere and finally reduce the vulnerability of socio-environmental systems.

Author Contributions: Conceptualization, A.B. and S.S.; methodology, A.B. and S.S.; software, A.B., M.Ś. and S.S.; validation, A.B., M.Ś. and S.S.; formal analysis, A.B. and S.S.; investigation, A.B. and S.S.; resources, S.S.; data curation, A.B., M.Ś. and S.S.; writing—original draft preparation, A.B. and J.K.K.; writing—review and editing, J.K.K.; visualization, A.B.; supervision, J.K.K. and S.S.

Funding: This research was funded by the Department of Spatial Economy of the Wrocław University of Environmental and Life Sciences from statutory funds.

Acknowledgments: The publication has been prepared as a part of the Support Programme of the Partnership between Higher Education and Science and Business Activity Sector financed by City of Wrocław.

Conflicts of Interest: The authors declare no conflict of interest.

References

1. Antolak, M. *Wybrane Elementy Zarządzania Ochroną Środowiska w Polsce*; Mantis Olsztyn: Austin, Poland, 2013.
2. Rogall, H. *The Economics of Sustainable Development*; Springer-Verlag: New York, NY, USA, 2009; ISBN 9783895187650.
3. Szyszka, B. The Effectiveness of the Eco-Management and Audit Scheme EMAS in Organizations on Polish Territory. 2016. Poznań University of Economics and Business. Available online: http://www.wbc.poznan.pl/Content/393452/Szyszka_Beata-rozprawa_doktorska.pdf (accessed on 5 July 2019).
4. Caekelbergh, A.F. *Zintegrowane Zarządzanie Środowiskiem. Systemowe Zależności Między Polityką, Prawem, Zarządzaniem i Techniką*; Oficyna Wolters Kluwer Business: Alphen aan den Rijn, The Netherlands, 2013; ISBN 9788326438110.
5. Matthaios, S. *Minimizing Energy Consumption, Energy Poverty and Global and Local Climate Change in the Built Environment: Innovating to Zero*; Elsevier: Amsterdam, The Netherlands, 2018.

6. Central Statistical Office; Production Department, Ministry of Energy. *Energy Consumption in Households in 2015*; Statistics Public Establishment: Warsaw, Poland, 2017. Available online: https://stat.gov.pl/download/gfx/portalinformacyjny/en/defaultaktualnosci/3304/3/8/1/energy_from_renewable_sources_in_2015.pdf (accessed on 5 July 2019).

7. Żyromski, A.; Biniak-Pieróg, M.; Burszta-Adamiak, E.; Zamiar, Z. Evaluation of relationship between air pollutant concentration and meteorological elements in winter months. *J. Water L. Dev.* **2014**, *22*, 25–32. [CrossRef]

8. Kamińska, J.A. Probabilistic forecasting of nitrogen dioxide concentrations at an urban road intersection. *Sustainability* **2018**, *10*, 4213. [CrossRef]

9. Lobaccaro, G.; Croce, S.; Lindkvist, C.; Munari Probst, M.C.; Scognamiglio, A.; Dahlberg, J.; Lundgren, M.; Wall, M. A cross-country perspective on solar energy in urban planning: Lessons learned from international case studies. *Renew. Sustain. Energy Rev.* **2019**, *108*, 209–237. [CrossRef]

10. Gielen, D.; Boshell, F.; Saygin, D.; Bazilian, M.D.; Wagner, N.; Gorini, R. The role of renewable energy in the global energy transformation. *Energy Strateg. Rev.* **2019**, *24*, 38–50. [CrossRef]

11. European Commission. *EU Actions to Improve Environmental Compliance and Governance*; European Commission: Brussels, Belgium, 2018.

12. United Nations. Adoption of the Paris Agreement. In Proceedings of the Conference of the Parties on Its Twenty-First Session, Paris–Le Bourget, France, 30 November 2015.

13. Ministry of Economy. Energy Policy of Poland Until 2030. Energy Policy; 2009. Available online: https://www.iea.org/policiesandmeasures/pams/poland/name-24723-en.php (accessed on 5 July 2019).

14. European Commission. *Europe 2020 Strategy*; European Commission: Brussels, Belgium, 2010.

15. Transforming Our World: The 2030 Agenda for Sustainable Development. 2018. Available online: https://sustainabledevelopment.un.org/post2015/transformingourworld (accessed on 5 July 2019).

16. Bazan-Krzywoszańska, A.; Skiba, M.; Mrówczyńska, M.; Sztubecka, M.; Bazuń, D.; Kwiatkowski, M. Green energy in municipal planning documents. *E3S Web Conf.* **2018**, *45*, 00006. [CrossRef]

17. Poskrobko, B. Zarządzanieekologiczne w przedsiębiorstwiejakonarzędziewdrażaniaekorozwoju. In *MechanizmyiUwarunkowaniaEkorozwoju*; Publisher Białystok Politechnika Białostocka: Białostocka, Poland, 1996; ISBN 9788390556321.

18. Szewrański, S.; Świąder, M.; Kazak, J.K.; Tokarczyk-Dorociak, K.; van Hoof, J. Socio-Environmental Vulnerability Mapping for Environmental and Flood Resilience Assessment: The Case of Ageing and Poverty in the City of Wrocław, Poland. *Integr. Environ. Assess. Manag.* **2018**. [CrossRef] [PubMed]

19. Marcyniuk-Kluska, A. Environmental management in the scope of economic sustainable development. *Zeszyty Naukowe Uniwersytetu Przyrodniczego w Siedlcach* **2013**, *23*, 129–140.

20. Wang, T.; Kazak, J.; Han, Q.; de Vries, B. A framework for path-dependent industrial land transition analysis using vector data. *Eur. Plan. Stud.* **2019**, *27*, 1391–1412. [CrossRef]

21. Huld, T. PVMAPS: Software tools and data for the estimation of solar radiation and photovoltaic module performance over large geographical areas. *Sol. Energy* **2017**, *142*, 171–181. [CrossRef]

22. Sarmiento, N.; Belmonte, S.; Dellicompagni, P.; Franco, J.; Escalante, K.; Sarmiento, J. A solar irradiation GIS as decision support tool for the Province of Salta, Argentina. *Renew. Energy* **2019**, *132*, 68–80. [CrossRef]

23. Harper, M.; Anderson, B.; James, P.; Bahaj, A. Assessing socially acceptable locations for onshore wind energy using a GIS-MCDA approach. *Int. J. Low-Carbon Technol.* **2019**, *14*, 160–169. [CrossRef]

24. Dominik, K.; Skotak, K. The conception of decision support system for assessment and management of ambient air quality. *Inf. Syst. Manag.* **2012**, *1*, 305–317.

25. Giordano, R.; Pilli-Sihvola, K.; Pluchinotta, I.; Matarrese, R.; Perrels, A. Urban adaptation to climate change: Climate services for supporting collaborative planning. *Clim. Serv.* **2019**. [CrossRef]

26. Massei, G.; Rocchi, L.; Paolotti, L.; Greco, S.; Boggia, A. Decision Support Systems for environmental management: A case study on wastewater from agriculture. *J. Environ. Manag.* **2014**, *146*, 491–504. [CrossRef]

27. Dong, Y.; Miraglia, S.; Manzo, S.; Georgiadis, S.; Sørup, H.J.D.; Boriani, E.; Hald, T.; Thöns, S.; Hauschild, M.Z. Environmental sustainable decision making—The need and obstacles for integration of LCA into decision analysis. *Environ. Sci. Policy* **2018**, *87*, 33–44. [CrossRef]

28. Van Hoof, J.; Kazak, J.K. Urban ageing. *Indoor BuiltEnviron.* **2018**, *27*, 583–586. [CrossRef]

29. Kazak, J.K.; van Hoof, J. Decision support systems for a sustainable management of the indoor and built environment. *Indoor Built Environ.* **2018**, *27*, 1303–1306. [CrossRef]

30. Pérez-Pérez, R.; Benito, B.M.; Bonet, F.J.; Modele, R. An environmental model repository as knowledge base for experts. *Expert Syst. Appl.* **2012**, *39*, 8396–8411. [CrossRef]

31. Chang, N.B.; Parvathinathan, G.; Breeden, J.B. Combining GIS with fuzzy multicriteria decision-making for landfill siting in a fast-growing urban region. *J. Environ. Manag.* **2008**, *87*, 139–153. [CrossRef]

32. Rahman, M.A.; Rusteberg, B.; Gogu, R.C.; Lobo Ferreira, J.P.; Sauter, M. A new spatial multi-criteria decision support tool for site selection for implementation of managed aquifer recharge. *J. Environ. Manag.* **2012**, *99*, 57–61. [CrossRef]

33. Van Hoof, J.; Bennetts, H.; Hansen, A.; Kazak, J.K.; Soebarto, V. The living environment and thermal behaviours of older south australians: A multi-focus group study. *Int. J. Environ. Res. Public Health* **2019**, *16*, 935. [CrossRef]

34. Przybyła, K.; Kulczyk-Dynowska, A.; Kachniarz, M. Quality of Life in the Regional Capitals of Poland. *J. Econ. Issues* **2014**, *48*, 181–196. [CrossRef]

35. Przybyłowicz, A. The legal position of persons dependent on long-term care in the Republic of Poland. In *Long-Term Care in Europe*; Springer: Cham, Switzerland, 2018; ISBN 9783319700816. Available online: https://link.springer.com/chapter/10.1007/978-3-319-70081-6_10 (accessed on 5 July 2019).

36. Bieda, A. Urban renewal and the value of real properties. *Stud. Reg. I Lokal.* **2017**, *69*, 5–28.

37. Trembecka, A.; Kwartnik-Pruc, A. An analysis of the changes in the structure of allotment gardens in poland and of the process of regulating legal status. *Sustainability* **2018**, *10*, 3829. [CrossRef]

38. Trojanek, R.; Gluszak, M.; Tanas, J. The Effect of Urban Green Spaces on House Prices in Warsaw. *Int. J. Strateg. Prop. Manag.* **2018**, *22*, 358–371. [CrossRef]

39. Jacksohn, A.; Grösche, P.; Rehdanz, K.; Schröder, C. Drivers of renewable technology adoption in the household sector. *Energy Econ.* **2019**, *81*, 216–226. [CrossRef]

40. Palicki, S.; Rącka, I. Influence of Urban Renewal on the Assessment of Housing Market in the Context of Sustainable Socioeconomic City Development. In *International Summit, Smart City 360°*; Lecture Notes of the Institute for Computer Sciences, Social-Informatics and Telecommunications Engineering (LNICST); Springer: Cham, Switzerland, 2016; Available online: https://link.springer.com/chapter/10.1007/978-3-319-33681-7_76 (accessed on 5 July 2019).

41. Świtała, M.; Cichosz, M.; Trzęsiok, J. How to Achieve Customer Satisfaction? Perspective of Logistics Outsourcing Performance. *LogForum* **2019**, *15*, 39–51. [CrossRef]

42. Szewrański, S.; Kazak, J.; Sylla, M.; Świąder, M. Spatial data analysis with the use of ArcGIS and Tableau systems. In *Lecture Notes in Geoinformation and Cartography*; Springer International Publishing: Cham, Switzerland, 2017; pp. 337–349. Available online: https://link.springer.com/chapter/10.1007/978-3-319-45123-7_24 (accessed on 5 July 2019).

43. Wohlgemuth, N.; Wojtkowska-Łodej, G. Policies for the promotion of renewable energy in Poland. *Appl. Energ* **2003**, *76*, 111–121. [CrossRef]

44. Szewrański, S.; Bochenkiewicz, M.; Kachniarz, M.; Kazak, J.K.; Sylla, M. 2018 International Conference on New Energy and Future Energy System (NEFES 2018). Available online: https://iopscience.iop.org/article/10.1088/1755-1315/188/1/011001 (accessed on 5 July 2019).

45. Kazak, J.K.; Świąder, M. SOLIS—A novel decision support tool for the assessment of solar radiation in ArcGIS. *Energies* **2018**, *11*, 2015. [CrossRef]

46. Klapa, P.; Mitka, B. Application of terrestrial laser scanning to the development and updating of the base map. *Geod. Cartogr.* **2017**, *66*, 59–71. [CrossRef]

47. Dąbek, P.B.; Żmuda, R.; Szczepański, J.; Ćmielewski, B. Evaluation of water soil erosion processes in forest areas in the Western Sudetes using terrestrial laser scanning and GIS tools. *E3S Web Conf.* **2018**, *44*, 00026. [CrossRef]

48. Klapa, P.; Mitka, B.; Brożek, P. Inventory of Various Stages of Construction Using Tls Technology. In Proceedings of the 18th International Multidisciplinary Scientific GeoConference(SGEM2018), Informatics, Geoinformatics Remote Sensing, Albena, Bulgaria, 30 June–9 July 2018; Volume 18.

49. Centralny Ośrodek Dokumentacji Geodezyjneji Kartograficznej. Available online: http://www.codgik.gov.pl/index.php/zasob/numeryczne-dane-wysokosciowe.html (accessed on 5 July 2019).

50. Central Statistical Office. Available online: https://stat.gov.pl/download/gfx/portalinformacyjny/pl/defaultaktualnosci/5485/1/6/1/energia_2018.pdf (accessed on 5 July 2019).

51. Hofierka, J.; Kaňuk, J. Assessment of photovoltaic potential in urban areas using open-source solar radiation tools. *Renew. Energy* **2009**, *34*, 2206–2214. [CrossRef]

52. Klapa, P.; Mitka, B.; Zygmunt, M. Study into Point Cloud Geometric Rigidity and Accuracy of TLS-Based Identification of Geometric Bodies. *IOP Conf. Ser. Earth Environ. Sci.* **2017**, *95*, 032008. [CrossRef]

53. Mitka, B.; Klapa, P.; Piech, I. Comparative analysis of geospatial data received by TLS and UAV technologies for the quarry. In Proceedings of the 18th International Multidisciplinary Scientific GeoConference (SGEM2018), Informatics, Geoinformatics and Remote Sensing, Albena, Bulgaria, 30 June–9 July 2018.

54. Kaynak, S.; Kaynak, B.; Özmen, A. A software tool development study for solar energy potential analysis. *Energy Build.* **2018**, *162*, 134–143. [CrossRef]

55. Holmgren, J.; Persson, Å.; Söderman, U. Species identification of individual trees by combining high resolution LiDAR data with multi-spectral images. *Int. J. Remote Sens.* **2008**, *29*, 1537–1552. [CrossRef]

56. Agugiaro, G.; Remondino, F.; Stevanato, G.; De Filippi, R.; Furlanello, C. Estimation of solar radiation on building roofs in mountainous areas. *ISPRS-Int. Arch. Photogramm. Remote Sens. Spat. Inf. Sci.* **2013**, *38*, 155–160. [CrossRef]

57. Mainzer, K.; Fath, K.; Mckenna, R.; Stengel, J.; Fichtner, W.; Schultmann, F. A high-resolution determination of the technical potential for residential-roof-mounted photovoltaic systems in Germany. *Sol. Energy* **2014**, *105*, 715–731. [CrossRef]

58. Redweik, P.M.; Catita, C.; Brito, M.C. 3D local scale solar radiation model based on urban LiDAR data. *Int. Arch. Photogramm. Remote Sens. Spatial Inf. Sci.* **2011**, *XXXVIII-4/W19*, 265–269. Available online: https://doi.org/10.5194/isprsarchives-XXXVIII-4-W19-265-2011 (accessed on 5 July 2019).

59. Wijeratne, W.M.P.U.; Yang, R.J.; Too, E.; Wakefield, R. Design and development of distributed solar PV systems: Do the current tools work? *Sustain. Cities Soc.* **2019**, *45*, 553–578. [CrossRef]

60. Jakica, N. State-of-the-art review of solar design tools and methods for assessing daylighting and solar potential for building-integrated photovoltaics. *Renew. Sustain. Energy Rev.* **2018**, *81*, 1296–1328. [CrossRef]

61. Shen, J.; Cheng, C. A Generalized Decision Support System for Short-Term Scheduling of China's Big Hydropower Systems. In *World Environmental and Water Resources Congress 2015*; American Society of Civil Engineers: Reston, VA, USA; pp. 1872–1886. Available online: https://ascelibrary.org/doi/abs/10.1061/9780784479162.183?src=recsys (accessed on 5 July 2019).

62. Kurtyka-Marcak, I.; Hełdak, M.; Przybyła, K. The Actual Demand for the Elimination of Architectural Barriers among Senior Citizens in Poland. *Int. J. Environ. Res. Public Health* **2019**, *16*, 2601.

63. Dąbrowska, J.; Pawęska, K.; Dąbek, P.; Stodolak, R. The Implications of Economic Development, Climate Change and European Water Policy On Surface Water Quality Threats. *Acta Sci. Pol. Formatio Circumiectus* **2017**, *16*, 111–123.

64. Chintala, V.; Subramanian, K.A. A comprehensive review on utilization of hydrogen in a compression ignition engine under dual fuel mode. *Renew. Sustain. Energy Rev.* **2017**, *70*, 472–491. [CrossRef]

 *sustainability*

Article

The Use of Waste Biomass from the Wood Industry and Municipal Sources for Energy Production

Andrzej Greinert [1,*], **Maria Mrówczyńska** [2] and **Wojciech Szefner** [3]

[1] Institute of Environmental Engineering, University of Zielona Góra, 65-516 Zielona Góra, Poland
[2] Institute of Civil Engineering, University of Zielona Góra, 65-516 Zielona Góra, Poland;
    M.Mrowczynska@ib.uz.zgora.pl
[3] Lubuski Centre for Innovation and Agricultural Implementation Ltd., University of Zielona Gora, Co.,
    Kalsk 122, 66-100 Sulechów, Poland; w.szefner@loiiwa.com.pl
* Correspondence: A.Greinert@iis.uz.zgora.pl

Received: 9 May 2019; Accepted: 29 May 2019; Published: 31 May 2019

Abstract: Biomass can be used for the production of energy from renewable sources. Because of social resistance to burning crop plants, mixtures and pellets made from or including waste materials are a good alternative. The mixtures analyzed, prepared from wood and municipal waste, were characterized for their calorific values, 7.4–18.2 $MJ \cdot kg^{-1}$. A result, over 15 $MJ \cdot kg^{-1}$ was obtained for 47% of the quantities of mixtures being composed. It has been demonstrated that wood shavings and sewage sludge have a stabilizing effect on the durability of pellets. The emissions of acidic anhydrides into the atmosphere from the combustion of pellets from waste biomass were lower for NO, NO_2, NOx and H_2S than emissions from the combustion of willow pellets. Obtained emission results suggest the need to further optimize the combustion process parameters.

Keywords: energy from biomass; pellets; wood waste; municipal waste

1. Introduction

The production of energy by burning solid fuels as the main item in energy mixes is prevalent across many countries in the world, including the largest energy producers, such as the USA, China, India and most of the European Union (EU) countries. It seems that coal fuels will remain an important element of their economies for at least several dozen years [1]. The share of energy from renewable sources in final energy consumption in the EU-28 and in Poland in 2016 was 8.0% and 8.3%, respectively. In 2013–2016, there was an increase by 0.4% in the EU-28, but a decrease by 0.6% in Poland [2,3]. In 2016, the structure of primary energy production from renewable sources in the European Union included a share of 44.7% of solid biofuels and 4.7% of municipal waste in this stream but the results for Poland are different—70.7 and 0.9%, respectively. In 2016, in the EU, solid biomass had a 15.2% share in the structure of the consumption of household energy from individual energy carriers. A smaller index has been noted in Poland, 13.5% [3], and for the whole world, 10–14% [4]. The 27 EU member states have a high potential for waste biomass for energy applications, calculated at 8500 $PJ \cdot y^{-1}$ [5]. The use of biomass for energy production has a number of advantages, such as low costs and high availability, which lowers the costs of transport and reduces its environmental impact [4,6,7]. It is also important that biomass can be used both in the boilers of the power industry to generate heat and electricity as small individual heating installations [8]. The market for biomass pellets is growing systematically on a global scale. As a result, woody and herbaceous biomass will be more difficult to obtain [9]. Difficulties are also noted in the access to wood and agricultural residues in many places around the world, i.e., African countries [8].

Large amounts of waste with a significant potential for energy production are produced by the wood industry (logging and wood processing residues)—up to 27% of the wood mass. Up to 42% of this waste can be obtained for energy purposes [8]. Poland is a country with a relative high forest residues theoretical potential in the EU [5]. This waste contains 43–51% of carbon, and the heat generated by its combustion reaches 18.5–20 MJ·kg^{-1}, which makes it a valuable energy material [10,11]. An additional advantage is the low amount of ash produced during the combustion of the wood residues, ranging from 0.4 to 2.0%. The use of a mixture consisting of 80% of sewage sludge, 19% of wood dust and 1% of quicklime produced results that proved that a widespread use of this type of fuel was possible since the heat generated by its combustion was slightly over 13 MJ·kg^{-1}. Replacing wood dust with coal dust raised this value to 19 MJ·kg^{-1}. Attention is paid to the hygroscopy of this fuel and its susceptibility to crumbling under the influence of moisture [11]. Due to the high price of wood obtained from forests by the wood industry using typical methods, slash and waste material obtained by pruning trees and shrubs are used to produce pellets. Wood from short cultivation cycles of *Populus*, *Salix*, *Eucalyptus* and *Robinia* is also used [12]. This method of wood management is also used in agriculture in orchards [13,14]. Forest residues and straw are counted as the top two contributors of energy from the residual biomass in the EU—7000 PJ·y^{-1} [5]. Agriculture is the main source of fuel biomass in the world. In this context, apart from trees and shrubs, attention is paid to the cultivation of energy plants and the use of post-harvest waste for this purpose [6]. Among agricultural crops, the typical biomass used for energy production are cereals, miscanthus, mallow, rapeseed, sunflower, and Jerusalem artichoke, but also agricultural residues such as vine, shrub and fruit tree shoots, corn stalks, peanut and hickory shells [14–16]. According to Hamelin et al.'s elaboration, the straw theoretical potential in Poland is high compared with many other EU-27 countries [5]. Pellets from vine shoot biomass have a standard calorific value for a majority of biomass types—18 MJ·kg^{-1} [14]. Due to the large diversity of agricultural crops and residues, the possibility of their recovery for energy purposes varies from 19 to 75% of the total mass [8].

The third of the important biomass sources for modern energy production, after wood and agricultural products, is municipal waste. Solid municipal waste (residues from urban green areas, roadside vegetation, food, paper, textile) and sewage sludge are used for energy production, both as homogeneous fuels, as well as in mixtures with other biomass for co-combustion with coal. There is a lot of information about the difficulties in the management of municipal waste due to the significant differentiation of their properties [17]. Chen et al. [18] suggested using fuel in the form of granules from sewage sludge and wood dust in a proportion of 10:1, with a moisture content of 14.2–18.5%, in the form of granules with a diameter of 2 and 7 mm. The results (a calorific value of 21.8–23.4 MJ·kg^{-1}) were favorable and the fuel proved to meet the requirements of the Taiwanese company Taipower. Also, Jiang et al. [19] mentioned sewage sludge as an interesting material that could be a biomass binder in the production of pellets. According to these authors, the addition of sewage sludge reduces the energy needed to compress and extrude materials during the production of pellets, increases the density and hardness of pellets (reducing dust during transport and operation) and improves combustion parameters. The downside is an increase in the weight of combustion residues in comparison to pure wood and herbaceous biomass. Stabilized sewage sludge contains 40–70% of carbon in its dry matter, which means that these materials could be used for energy production [20,21]. Nevertheless, the process of preparing sewage sludge for combustion and co-combustion is so expensive that the rolling costs of this type of use are not very favorable—in Poland, 375–438 € for 1 Mg d.m., compared to 75–150 in agriculture and reclamation [20]. An additional problem is the amount of ash generated from this material, which is greater than in the case of most other fuels—an average of 36.4% for sewage sludge as against approx. 1% for wood, 6% for wheat straw and 19–22% for coal [22,23]. Therefore, the co-combustion of sewage sludge should be redefined to find an innovative method for the preparation of this material before its use for energy production. Apart from energy production, the new material has to give end users measurable financial benefits in the form of better furnace operating parameters. Yilmaz et al. [24] showed that the best results could be obtained by burning pellets with a size of 35 mm made from waste from plant oil production and sewage sludge. These authors noticed the high susceptibility of pellets to mechanical degradation under the influence of moisture and, for this reason, they recommended short-term storage under conditions that would counteract the moistening of the material. Jiang et al. [25] showed that better results could be obtained by burning pellets made from a mixture of sewage sludge and wood biomass than sewage sludge alone. As far as energy production is concerned, pellets are much more efficient than raw biomass. Increasing biomass density reduces transport costs and improves combustion parameters [26]. When pellets are formed, it is possible to control their composition, and when they are finally used, it is possible to automatically feed them to the furnace.

In the literature, there is little mention of the process of granulation/pelletization of energy materials with the use of municipal waste. Relatively few tests of this type have been conducted on sewage sludge for only a couple of years. Li et al. [27] presented optimum parameters for the production of pellets from biomass and sewage sludge (50 + 50%): pressure 55 MPa, temperature 90 °C and moisture in the material 10–15%. The energy needed to produce pellets using sewage sludge was 50% lower than the energy needed to produce pellets from pure biomass. Similar moisture parameters of material intended for pelletization were reported by Kliopova and Makariski [28]. Kijo-Kleczkowska et al. [23] obtained pellets with densities of (kg/m^3): 1089.2 (hard coal), 859.9 (sewage sludge), 802.6 (lignite), 363.1 (*Salix viminalis*), 898.1 (50% sewage sludge + 50% hard coal), 803.5 (50% sewage sludge + 50% brown coal), and 515.9 (50% sludge + 50% *Salix viminalis*). There were a number of changes in the process of combustion of particular solid biofuels after the addition of sewage sludge. Jiang et al. [19] analyzed the possibilities of pelleting energy materials using sewage sludge and noticed an increase in the density of pellets obtained while increasing pressure to 28, 41, 55, 69 and 83 MPa. A further increase in pressure to 110 MPa no longer caused any significant differences. The increase in density and hardness of pellets was also the resultant of the share of sewage sludge (from 20 to 80%). The temperature during pelletization should not exceed 110 °C, and the content of moisture in the input material should not exceed 15%. Higher temperatures and humidity reduce the density of pellets. If granulates/pellets based on limed sludge and other selected solid waste are to be used for co-combustion with biomass in heating furnaces, it is necessary to prepare a production technology that will make them easy to produce repeatedly and make them profitable to producers, easy to transport, easy to use precisely, effective during biomass combustion and safe. They should also generate ash that is easy to remove from furnaces and environmentally safe. García-Maraver et al. [29] noticed that the process of the preparation of heating pellets and their physical, physicochemical and chemical properties had a significant impact on the combustion process and emission parameters. Lehtikangas [30] stated that it was necessary to use physically stable materials in the combustion process due to the possibility of disturbing the operation of automatic fuel feeders and advanced systems of automatic furnace control by dust. Moreover, this author noted that the temperature in the furnace could increase significantly (due to the combustion of dusts), which would lead to the melting of combustion residues. According to Sarenbo and Claesson [31], the production of granulate has to be effective, the granulate binder must not adversely affect the properties of the whole aggregate, and the final chemical composition and stability of the aggregate has to be consistent with the recommendations for a particular type of use for the material. Thus, the problem arises of developing a technology that will make it possible to produce granulates/pellets with the desired characteristics and properties. During previous research projects dealing with the granulation/pelletization of materials, the problem of the durability of products and their homogeneity was noted.

Emissions from the combustion of carbonaceous materials may be an obstacle to a wider use of biomass for energy production [7]. In this respect, attention is paid to the emission of CO_2 to the atmosphere as one of the main gases affecting the greenhouse effect. Nevertheless, in comparison to the combustion of solid fossil fuels, the combustion of biomass leads to a reduction in CO_2 emissions to the atmosphere when co-combustion technology is used [6,16]. As far as the so called carbon footprint is concerned, biomass is neutral to carbon circulation in the environment [10]. Biomass combustion also results in the emission of gases other than CO_2 (including NOx, CO, SO_2, hydrocarbons) as well as dust polluting the atmosphere, especially when combustion does not take place in optimum conditions.

It is often mentioned in the literature that the emission of these gases to the atmosphere could be reduced by mixing fuel with lime and lime and dolomite dust, which in terms of chemical properties are mainly calcium compounds: $CaCO_3$, CaO, $Ca(OH)_2$ [7,31–33].

The study was carried out in a region with the largest forest cover in Poland—49.3% of the total area and 51.7% of the land area of the region, compared to 30.5% for the whole country. The wood industry plays one of the key roles in its economy—wood acquisition amounts to 3.6 $kdam^3$ per year [34]. Data on municipal waste management show that the amount of municipal waste that is combusted is still small, amounting to 19.4% of the total waste stream in the Lubusz region, compared to 24.4% in Poland [35]. Counting the current energy production from biomass in Poland in relation to the theoretical potential of this source (20–30% of the final energy consumption [5]), there is still a large reserve for activities intensifying this process. The use of waste from the wood industry as a fuel material allows to reduce the amount of this waste and reduce the energy demand from conventional sources. This approach also reduces the use of natural resources while meeting the energy needs of the population, so socio-environmental systems would be sustainable [36]. In order to further develop the thermal management of waste biomass, it is necessary to solve problems related to the emission of pollutants into the atmosphere. The aim of this study was to show the possibility of using waste generated in this region in large quantities as a renewable energy source. In many wastewater treatment plants, sewage sludge is still hygienized with lime. Materials prepared in this way are usually used as a soil improver. Therefore, the question was asked whether it could also be used as an improver in the biomass combustion process. For this purpose, it was considered whether it would be possible to use waste from the wood industry and limed municipal sewage sludge to produce solid fuel in the form of pellets with good thermal properties and at the same time ecologically safe. It was hypothesized that the addition of limed sewage sludge to fuel made from biomass would reduce the emission of such gases as acidic anhydrides.

2. Materials and Methods

A number of tests were carried out to find out whether it would be possible to produce durable pellets with good thermal properties from waste biomass. They included:

- Evaluation of input waste materials available in the region in large quantities;
- Preparation of mixtures of input materials and checking their thermal properties;
- Preparation of pellets from selected materials with the best properties in two groups—with lime and without lime;
- Testing pellet durability under high humidity conditions;
- Burning pellets under controlled conditions;
- The input materials for the tests were: straw, wood shavings, wood dust and sewage sludge hygienized with lime;
- Straw (S)—raw material obtained from a producer of straw pellets, homogeneous, crushed to a fraction of approx. 2 cm in length and packed in 10-kg bags; 100 kg of material was obtained for the needs of the experiments;
- Wood shavings (WS)—raw material obtained from a carpenter's workshop, material from debarked wood, non-homogeneous fraction from 1.5 to 3.5 cm; the material was not additionally sieved for the tests; 100 kg of material was obtained for the needs of the experiments;
- Wood dust (WD)—raw material obtained from a carpenter's workshop, material from debarked wood, homogeneous fraction, powder; the material was not additionally sieved for the tests; 100 kg of material was obtained for the needs of the experiments;
- Sewage sludge (SS)—from the Krosno Water Utility Company Ltd. (Krośnieńskie Przedsiębiorstwo Wodociągowo-Komunalne Sp. z o.o.); municipal sludge treated with lime in an innovative installation for simultaneous hygienization and granulation of sewage sludge, using lime (CaO) in a sediment to a lime ratio of 1:0.9 by weight; the material was pre-screened through a 3-mm

mesh sieve; the subscreen fraction was taken for testing; 100 kg of the material was obtained for the needs of the experiments.

For transport to the research place, the materials were packed in plastic bags, which were opened on the spot to avoid the phenomenon of organic material thermal degradation.

Each material was pre-homogenized to obtain a homogeneous mass with average properties. The homogenization process was carried out by hand, using a table for mixing substrates onto which a particular material was poured from the transport packaging. Each material was mixed for about 10 min. At the end of the homogenization process, 102 samples of materials (34 variants, 3 repetitions, 100 g each sample) were taken for laboratory analysis. In each case, the method of average pooled samples was used. The mixed samples consisted of 30 individual samples taken from the entire volume of the material.

The materials were analyzed in laboratory conditions in terms of:

- bulk density—by weight in 5 repetitions, from which the mean and standard deviations were calculated;
- the calorific value—using a calorimetric bomb Parr 6100, acc. to norm PN-C-04375-2:2013-07, in 3 repetitions, from which the mean and standard deviations were calculated;
- the total carbon content—using the Pregla–Dumas method, the samples were combusted in a pure oxygen environment and the resulting exhaust gases were automatically measured using a CHNS/O 2400 Series II PerkinElmer elemental analyzer. The measurements were conducted for weights of 1.5–2.5 mg in three repetitions, from which the mean and standard deviations were calculated;
- the content of heavy metals (Cd, Cr, Cu, Ni, Pb and Zn) was measured using the ICP-OES method and a Perkin Elmer ICP-OES Optima 8000 spectrometer, after wet mineralization in concentrated nitric acid in a Perkin Elmer Titan microwave mineralizer, in three repetitions from which the mean and standard deviations were calculated;
- the content of K, Na, Ca, Mg and Fe by ICP-OES was measured using a Perkin Elmer ICP-OES Optima 8000 spectrometer, after wet mineralization in concentrated nitric acid in a Perkin Elmer Titan microwave mineralizer, in three repetitions from which the mean and standard deviations were calculated;
- pH—potentiometrically in a mixture of solid material (air dry) and water in a proportion of 1:5 using a pH-meter InoLab, with a WTW SenTix 41 glass electrode.

In order to obtain a homogeneous material, the samples were homogenized by grinding.

In laboratory conditions, mixtures of input materials were prepared and further components were weighed as shown in Table 1. Each mixture was ground to obtain completely equal properties before further analysis.

Table 1. Composition of the mixtures of materials.

Mixture Number	Composition of Components	Proportion of Components
M1	WS + WD	1:1
M2	WS + S	1:1
M3	WS + SS	1:1
M4	WD + S	1:1
M5	WD + SS	1:1
M6	S + SS	1:1

Table 1. *Cont.*

Mixture Number	Composition of Components	Proportion of Components
M7	WS + WD + S	1:1:1
M8	WS + WD	4:1
M9	WS + S	4:1
M10	WS + SS	4:1
M11	WD + SS	4:1
M12	S + SS	4:1
M13	WS + SS	9:1
M14	WD + SS	9:1
M15	S + SS	9:1
M16	WS + WD + SS	7:2:1
M17	WS + S + SS	7:2:1
M18	WS + WD + SS	3:1:1
M19	WS + S + SS	3:1:1
M20	WS + WD + SS	6:1:3
M21	WS + S + SS	6:1:3
M22	WS + SS	7:3
M23	WD + SS	7:3
M24	S + SS	7:3
M25	WS + WD + S + SS	3:3:3:1
M26	WS + WD + S + SS	1:1:1:2
M27	WS + WD + S + SS	1:1:1:1
M28	WS + WD + S + SS	3:3:3:11
M29	WS + WD + S + SS	1:1:1:7
M30	WS + WD + SS	3:3:4

Note: S—straw, WS—wood shavings, WD—wood dust, SS—sewage sludge.

Each mixture was analyzed in laboratory conditions in terms of:

- the calorific value—calorimetrically in three repetitions, from which the mean and standard deviations were calculated;

- pH—potentiometrically in a mixture of solid material (air dry) and water in a proportion of 1:5 using a pH-meter InoLab, with a WTW SenTix 41 glass electrode;

- electrical conductivity (EC)—conductometrically in a mixture of air-dry solid material and water in a proportion of 1:5; using a Eutech Instruments Cyberscan PC300 and Elmetron CPC-411 devices with an EC-60 conductivity sensor;

- the subtotal content of selected components—after the hot dissolution of substrate samples (in a Perkin Elmer MPS microwave oven) in a mixture of hydrochloric and nitric acid (aqua regia), using the ICP-OES method (Perkin Elmer Optima 8000)—ISO 11466 (1995);

- the total carbon content—using the Pregla–Dumas method, the samples were combusted in a pure oxygen environment, and the resulting exhaust gases were automatically measured using a CHNS/O 2400 Series II PerkinElmer elemental analyzer. The measurements were conducted for weights of 1.5–2.5 mg in three repetitions, from which the mean and standard deviations were calculated.

Based on the results of these analyses, mixtures expected to bring the best results in terms of energy production were selected. A total of 10 kg of each selected mixture were prepared for pelleting. A pelletizer, ZLSP200, 7.5 kW made by Eko-Pal, with a capacity of approx. 80–100 kg h^{-1} was used for pelleting. The input materials were mixed in the right proportions and brought to a humidity of 12–13%. After wetting, the mixtures used for pelletization were rested for 24 h. A matrix with a diameter of 6 mm was used in the process of pelletization. The matrix was not lubricated with any substances. During this process, the rotation speed of the pelletizer was not regulated, and only the pressure of the roller on the matrix was corrected for better performance. Each of the mixtures intended for pelletization was passed through the device until pellets of the right quality had been obtained—in some cases, the process was repeated three times. After the pellets were obtained, the output material was rested for 24 h before packing. After each of the individual mixtures was pelletized, oat husks were passed through the pelletizer to clean the sieve before working with the next mixture. Each time, the first batch of pellets obtained from a new mixture was also discarded in order to get rid of the remnants of the oat husks in subsequent completed mixtures.

The pellets were combusted in fully controlled conditions to obtain information about the combustion process and exhaust emissions. The pellets were combusted in a prototype FORST boiler equipped with two burners: 24 and 48 kW (modified Forster Heizkessel, PE40 boiler), adapted for burning biomass, with an automatic pellet feeding and ash removal system. Combustion of individual pellets was carried out after preheating the furnace to its optimal temperature, using a standard willow tree pellet. Each of the test pellets was fed to the boiler for 4 h. The main combustion parameters: incineration temperature 700–800 °C, air flow 20 m^3 h^{-1}, mass fuel consumed 1,5 $kg \cdot h^{-1}$, exhaust gas temperature 141–162 °C, exhaust gas mass flow 30 $g \cdot s^{-1}$, O_2 in exhaust gas 11–14%, max. operating temperature 90 °C, max. operating pressure 2.5 bar, pellet calorific value 17.73 $MJ \cdot kg^{-1}$, pellet residual moisture 8–9%, and ash production 3.84% of the pellet mass. The biomass boiler was included into the heat production and distribution system at the Laboratory of Thermal Technologies in the Renewable Energy Center (REC). The laboratory research system is equipped with an advanced measuring system based on measuring devices made by E&H. The exhaust duct of the boiler is equipped with a measuring connection for chemical analysis of exhaust gases. Measurement data are registered by the building management system, Building Management System (BMS), using the Wonderware System Platform. Exhaust gas was measured continuously using a TESTO 350 Xl exhaust gas analyzer. Emission measurements were conducted after the combustion conditions were stabilized, i.e., in the so called high-temperature combustion phase [37].

The following analyses of the pellets were performed in laboratory conditions:

- static and dynamic tests of moisture absorption by the pellets;
- test of the mechanical strength of the pellets;
- pH and EC analyses of water extracts after 48 h of incubation.

The static test of moisture absorption was carried out in 250-ml glass flasks. An amount of 10 g of pellets were poured into a flask and 90 cm^3 of distilled water were added. After 48 h of incubation at room temperature, the samples were subjected to gravity dehydration on sterile gauze for 10 min. After that, a visual evaluation of the disintegration was carried out, the pellets were weighed and the leachate was collected to the volumetric flasks, then the mechanical strength of the pellets was examined.

The dynamic moisture absorption test was carried out in 250-cm^3 plastic bottles. An amount of 10 g of pellets were poured into the bottles and 90 cm^3 of distilled water was added. The samples were shaken on a mechanical stirrer at 60 rpm for 1 h and they were left for 24 h at room temperature. After 24 h, the samples were again subjected to mechanical shaking at 60 rpm for 1 h. The pellets were then subjected to gravity dehydration on sterile gauze for 10 min. Decomposition was assessed visually, the pellets were weighed, and the amount of leachate was measured using the method of quantitative gathering to the volumetric flasks. Then, the mechanical strength of the pellets was examined.

Due to the disintegration of some of the pellets during the static and dynamic tests, mechanical strength was tested only in the samples that did not disintegrate. The study consisted in dropping each pellet from a height of 1.5 m onto a concrete surface and a visual assessment of disintegration.

The pH and electrical conductivity test was carried out in water solutions obtained by gravitational drainage of the pellets, pH was measured potentiometrically (using a WTW Inolab ph level 1 pH meter, with a WTW SenTix 41 glass electrode) in the supernatant of a 1:2.5 dry solid material: water suspension and EC was measured conductometrically (using a Cyberscan PC300 Series conductor) in the water-saturated paste. In the samples that completely disintegrated, pH and electrical conductivity were measured in the suspension and in the leachate obtained by mechanical filtration of the suspension.

3. Results and Discussion

3.1. Initial Observations

Although the materials used in this study consisted of waste generated on a large scale by industrial and municipal plants, they were morphologically balanced (Figure 1). While mixing the input materials, it was observed that large amounts of dust appeared during the process of mixing wood dust and limed sewage sludge with other materials. This will certainly be a problem both during the transport of these materials to the sites where mixtures and pellets will be prepared, as well as during the preparation of the products themselves. Nevertheless, the use of materials with a diversified structure promised well for the formation of pellets. By using particles of varying equivalent diameter, fiber length and shape, a compact mass of the mixture can be obtained. As a result, the pellets should have a higher bulk density and mechanical durability.

Figure 1. Input materials used in the study—wood shavings, wood dust, straw and limed sewage sludge.

3.2. Properties of the Input Materials

For subsequent input materials, the following bulk density values were obtained (mean ± standard deviation): WS—110 ± 5.4 kg·m^{-3}, WD—172 ± 5.0 kg·m^{-3}, S—85 ± 2.2 kg·m^{-3}, SS—845 ± 4.7 kg·m^{-3}.

This is important information because of the need to transport the materials to the site where they will be utilized. Since the materials will eventually be combusted, it is essential that with the relatively low bulk density of straw and wood shavings, the limiting criterion for transport will be the volume of material that can safely be loaded onto a transport vehicle—without the risk of uncontrolled scattering.

Although wood dust has the highest bulk density among these materials, it will also be problematic in transport because it generates large amounts of dust.

All biomass materials used in this study had a similar carbon content—from 48 to 53% of dry mass. The content of dry matter in them was also similar—within 90–92%. Sewage sludge hygienized with lime was a significantly different material; its carbon content was approx. 14% with 76% of dry matter. The pH analysis showed that the pH of straw was slightly alkaline, the pH of sewage sludge hygienized with lime was strongly alkaline, and the pH of the other materials—wood shavings and wood dust—was acidic (Table 2). The potassium content was clearly higher in the straw than in the other analyzed components. Straw was also characterized by a higher content of calcium than wood shavings and wood dust. Of course, the sewage sludge hygienized with lime has outperformed the remaining materials in this respect many times. Sewage sludge involves considerably higher contents of Fe, Cr, Cu, Ni, and Zn than other materials (Table 2).

Table 2. Selected physical and physic-chemical properties of the input materials.

Material	C	Dry Mass	pH-H$_2$O	K	Ca	Mg	Na	Fe	Cd	Cr	Cu	Ni	Pb	Zn
	(%)			mg·kg^{-1}										
WS	49.1	92.5	5.1	287	364	101	13	nd	nd	nd	nd	nd	0.8	4.0
WD	52.8	92.0	5.5	234	298	82	11	nd	nd	nd	nd	nd	0.7	3.3
S	48.0	89.8	8.0	2180	1109	120	5	16	nd	nd	nd	nd	nd	3.0
SS	13.9	76.0	12.4	307	75,700	348	48	433	nd	1.5	14	0.7	0.9	44.7

Similar results were presented by Houshfar et al. [10]. They indicated that the average content of carbon was 51.4% for wood pellets, 53.4% for wood waste and 49.5% for straw. A significantly higher content of C in sewage sludge was reported in the literature—28.4% [37], 48.9% [10], 53–60% [21]. However, in contrast to this study, that sewage sludge was not limed. In these analyses, we also obtained a dry mass of wood material and straw similar to that presented in the analyses of the cited authors—90.4–93.5% and 88.3%, respectively [6]. However, the dry mass of sewage sludge was smaller than presented in the literature. Chen et al. [38] indicated a content of 91.5% of dry matter for sun-dried sediments, but this is only 16–19% of the mass of raw sewage sludge as determined by Pulka et al. [21].

For subsequent input materials, the following calorific values were obtained (mean ± standard deviation):

- WS—16883 ± 755 kJ·kg^{-1} (4029 ± 180 kcal·kg^{-1})
- WD—14550 ± 1027 kJ·kg^{-1} (3473 ± 245 kcal·kg^{-1})
- S—18549 ± 753 kJ·kg^{-1} (4427 ± 180 kcal·kg^{-1})
- SS—no data (the material did not burn)

The analysis of the calorific value of the input materials showed the highest value for straw (by 11.3% higher than the mean for the materials analyzed), lower for wood shavings (1.3% higher than the mean for the materials analyzed), whereas for wood dust it was lower by 12.7% than the mean for the materials analyzed. The calorific value of the materials indicated that the most promising ones at this stage appeared to be straw and wood shavings. Due to the high content of minerals in the limed sewage sludge, it turned out that this material could be problematic as a fuel additive. The results for wood biomass and straw are typical as compared to those described widely in the literature, where calorific values within 16.5–20.5 MJ kg^{-1} [33,39,40] are most frequently recorded. The calorific value of sewage sludge is lower, amounting to 11.37 MJ·kg^{-1} [38], but always for non-limed sewage sludge.

3.3. Properties of Fuel Mixtures

It was noted that the addition of sewage sludge hygienized with lime had a significant impact on the pH of materials, the content of dry matter and content of total carbon (TC). The content of dry matter and total carbon are important properties in terms of the combustion characteristics of solid fuel, and the low pH may additionally affect the resulting emissions. Since different input materials were used in different proportions for the preparation of subsequent mixtures, significant differences were noted in the content of carbon in the mixtures, what was expected (Table 3).

Table 3. Properties of the mixtures of materials.

Material	TC	Dry Mass	pH	K	Ca	Mg	Na	Fe	Cd	Cr	Cu	Ni	Pb	Zn
	%			mg·kg^{-1}										
M1	50.9	92.2	5.4	521	662	183	24	n/a	n/a	n/a	n/a	n/a	1.49	7.33
M2	48.1	91.3	5.7	2094	1490	170	10	8	n/a	0.21	n/a	n/a	0.56	6.50
M3	28.1	87.2	12.3	598	74,571	396	55	430	n/a	1.36	12.56	0.56	1.55	44.57
M4	51.4	91.9	9.7	2788	1390	252	23	24	n/a	n/a	n/a	n/a	0.22	7.09
M5	30.3	86.3	11.4	537	77,432	482	65	436	0.02	1.58	14.54	0.76	1.62	52.07
M6	28.4	86.5	12.3	3159	53,465	428	43	352	n/a	0.65	8.27	0.14	0.79	33.23
M7	51.1	92.6	6.4	1460	1068	216	13	8	0.00	0.17	n/a	n/a	0.82	6.35
M8	51.0	93.4	5.1	536	692	139	17	n/a	n/a	0.02	n/a	n/a	0.43	7.57
M9	49.0	92.2	5.8	1481	2075	162	9	14	n/a	n/a	n/a	n/a	n/a	5.40
M10	42.9	91.3	12.2	674	48,413	314	40	295	n/a	0.60	3.28	n/a	1.46	22.68
M11	45.1	91.0	10.6	616	33,886	367	49	255	n/a	n/a	1.14	n/a	0.29	20.71
M12	37.2	89.6	10.9	3423	30,235	375	27	219	n/a	0.32	n/a	n/a	0.37	19.09
M13	46.8	91.9	12.0	572	13,453	159	16	107	n/a	n/a	n/a	n/a	0.32	11.27
M14	48.3	91.9	10.6	487	14,831	288	36	141	0.02	1.11	n/a	0.02	6.31	13.64
M15	44.4	90.3	11.9	3243	13,033	230	15	97	n/a	1.61	n/a	0.28	n/a	11.23
M16	46.2	91.8	11.7	587	16,868	177	20	139	0.02	0.48	n/a	0.31	1.31	13.20
M17	45.6	91.4	11.8	888	10,835	142	13	61	n/a	n/a	n/a	n/a	0.98	9.12
M18	42.4	91.2	12.2	526	28,130	256	30	223	n/a	n/a	0.21	n/a	1.06	21.01
M19	41.7	92.8	12.2	1047	32,961	264	27	280	n/a	0.40	2.40	0.22	0.91	19.96
M20	37.6	90.7	12.3	551	43,583	287	34	333	0.01	0.44	1.48	n/a	1.34	25.55
M21	37.0	89.9	12.3	866	50,099	359	39	418	n/a	0.70	7.84	0.30	1.30	29.62
M22	38.2	90.7	12.3	551	19,318	157	19	86	n/a	n/a	n/a	n/a	0.21	10.58
M23	40.3	90.6	12.0	598	74,572	396	55	430	n/a	0.42	3.87	n/a	0.35	24.09
M24	37.5	88.0	11.7	3929	19,347	292	20	142	n/a	0.04	n/a	n/a	n/a	11.91
M25	46.2	91.4	12.0	1806	12,046	220	21	96	n/a	0.09	n/a	n/a	0.16	10.67
M26	35.5	88.7	12.3	1323	52,593	354	44	377	n/a	0.33	4.43	0.07	0.30	26.40
M27	40.5	90.6	12.3	1525	34,102	310	33	259	n/a	0.18	1.50	n/a	0.52	17.07
M28	30.2	87.4	12.3	835	79,491	464	58	634	0.00	1.88	12.91	0.45	1.14	42.96
M29	23.3	86.0	12.3	842	10,3626	491	65	720	n/a	2.21	24.51	0.88	1.34	50.38
M30	35.2	89.4	12.3	463	62,336	364	45	431	0.01	1.29	8.36	0.55	1.80	28.70

In order to use waste from the wood industry as well as agricultural and municipal waste for energy production, one of the basic research paths to take is to determine the chemical composition of these materials. In this way, it is possible to determine whether this waste contains an excessive number of components that could be dangerous to the environment. During combustion, some of them will be released from the boiler as gases, some as dust and the remaining part will be present in the ash. The presence of certain components, such as sodium, potassium, calcium, magnesium and iron, may be undesirable because they leave residue on the grate and other elements of the boiler, causing technical problems. The high content of Ca, Mg, Na and Fe resulted from the addition of sewage sludge hygienized with lime to wood as well as agricultural and organic materials (mixtures: M3, M5, M23, M28 and M29). The high content of potassium resulted from the composition of material solely based on straw or with the presence of straw in the mixtures. As far as the chemical composition of waste materials is concerned, much attention is usually paid to heavy metals. They are not transformed by combustion, so they remain unchanged in the dust and ash after the combustion process. At the same time, when organic matter burns, its input mass decreases since some of it is transformed into gaseous products and it is deprived of water, the content of heavy metals is concentrated in combustion residues and fly ashes. For this reason, the content of heavy metals in solid fuels should be low. The content of Cr was in the range of 1.11–2.21 mg·kg^{-1} in materials M3 (WS + SS; 1:1), M5 (WD + SS; 1:1), M14 (WD + SS; 9:1), M15 (S + SS; 9:1), M28 (WS + WD + S + SS; 3:3:3:11), M29 (WS + WD + S + SS; 1:1:1:7) and M30 (WS + WD + SS; 3:3:4). In all cases, the mixtures included sewage sludge. There was no significant Cu, Ni and Pb content in all mixtures—max. 24.51 mg Cu kg^{-1} in material M29, 0.88 mg Ni kg^{-1} in material M29 and 1.80 mg Pb kg^{-1} in material M30. As far as the content of Zn is concerned, a relatively higher content was noted in mixtures M3, M5, M28 and M29, where it was in the range of 42.96–52.07 mg kg^{-1} (Table 3).

The majority of wood and agricultural waste does not contain large amounts of heavy metals. However, there are problems with their content in municipal waste. The data in Table 2; Table 3 clearly indicate the sewage sludge as a potential source of heavy metals in biomass mixtures. The content of some heavy metals (Cr, Ni) may also result from the degradation of the hardened steel elements of the screw biomass feeder, the boiler itself or the degradation of the collecting probes [17].

Since the content of carbon in raw wood and wood waste was similar, as expected, mixing these materials with each other did not cause any significant differences. This property was at similar levels to those described by other authors [10,14,16,32].

The analysis of the calorific value of the mixtures showed significant differences between particular materials (Table 4). Ten highest calorific values were obtained for mixtures (in decreasing order): M1; M2; M9; M8; M4; M7; M17; M10; M19; and M13. The weakest of these mixtures reached a calorific value lower by 11.8% than the best mixture. In the case of some of the mixtures, relatively higher values of the standard deviation (5–9%) were obtained, which indicates that the materials did not burn completely despite their homogenization: M5 (WD + SS; 1:1), M7 (WS + WD + S; 1:1:1), M11 (WD + SS; 4:1), M15 (S + SS; 9:1), M16 (WS + WD + SS; 7:2:1), M24 (S + SS; 7:3) and M27 (WS + WD + S + SS; 1:1:1:1). Most of these mixtures contained sludge in their composition, which may mean that its presence interferes with the combustion process.

Table 4. Calorific value of the mixtures of materials (value ± SD).

Mixture No.	Calorific Value (kJ·kg^{-1})
M1	18,219 ± 244
M2	17,755 ± 658
M3	10,755 ± 322
M4	17,115 ± 153
M5	13,110 ± 1134
M6	12,004 ± 469
M7	16,843 ± 916
M8	17,615 ± 633
M9	17,688 ± 90
M10	16,117 ± 278
M11	13,815 ± 2763
M12	14,330 ± 369
M13	16,061 ± 407
M14	15,660 ± 154
M15	14,686 ± 971
M16	15,692 ± 864
M17	16,530 ± 280
M18	15,325 ± 129
M19	16,098 ± 279
M20	13,612 ± 631
M21	14,501 ± 407
M22	13,892 ± 468
M23	14,950 ± 588
M24	13,847 ± 1289
M25	15,810 ± 315
M26	11,295 ± 659
M27	12,914 ± 1116
M28	9527 ± 96
M29	7408 ± 160
M30	11,366 ± 314

Sewage sludge hygienized with lime reduces the calorific value of mixtures. This is expected both due to its lower calorific value in relation to other biomass [38] and a high proportion of mineral compounds. The results were in the range of 7–18 MJ kg^{-1}, i.e., within a typical range for different waste materials in developed countries (8.4–17 MJ kg^{-1}) and they were better than the results obtained in China (3–6.7 MJ kg^{-1})—Zhang et al. [39]. The correlation between the calorific value of the mixtures of materials and the TC content (Figure 2) and dry mass (Figure 3) was also analyzed. The calculation of correlation coefficients makes it possible to determine the linear relation between the data and at the same time to eliminate irrelevant information from the data sets [41]. The relations between the calorific value and the TC content (correlation r = 0.91) and the calorific value and dry mass (r = 0.86) were characterized by a high positive correlation. These relations, widely described in the literature [42], were expected. Ngangyo-Heya et al. [42] performed correlation analyzes for components of woody biomass without leaves, indicating the correlation of the calorific value with some of its properties. The highest positive correlation between the biomass calorific value, pH and lignin content have been noted.

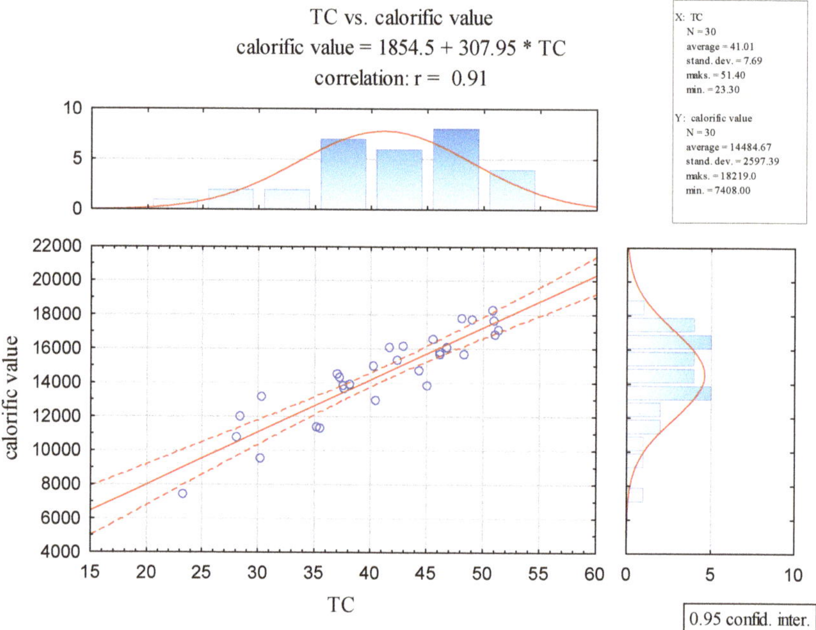

Figure 2. Graph of the correlation and dispersion of results; total carbon (TC) content vs. calorific value.

3.4. Pellet Durability Tests

After preliminary analyses of the input materials and mixtures, ten of them—the most promising ones, counting thermal properties (calorific value) and chemical composition (low heavy metal content)—were pelletized. In order to test the behaviour of the pellets under the influence of moisture, a static test was carried out – the reaction of the pellets in water and a dynamic test - the reaction of the pellets in water with mechanical mixing. Since it was expected that some of the components would be transferred into the solution, an analysis of the electrical conductivity (EC) and pH of the extracts was planned. The test results are presented in Table 5.

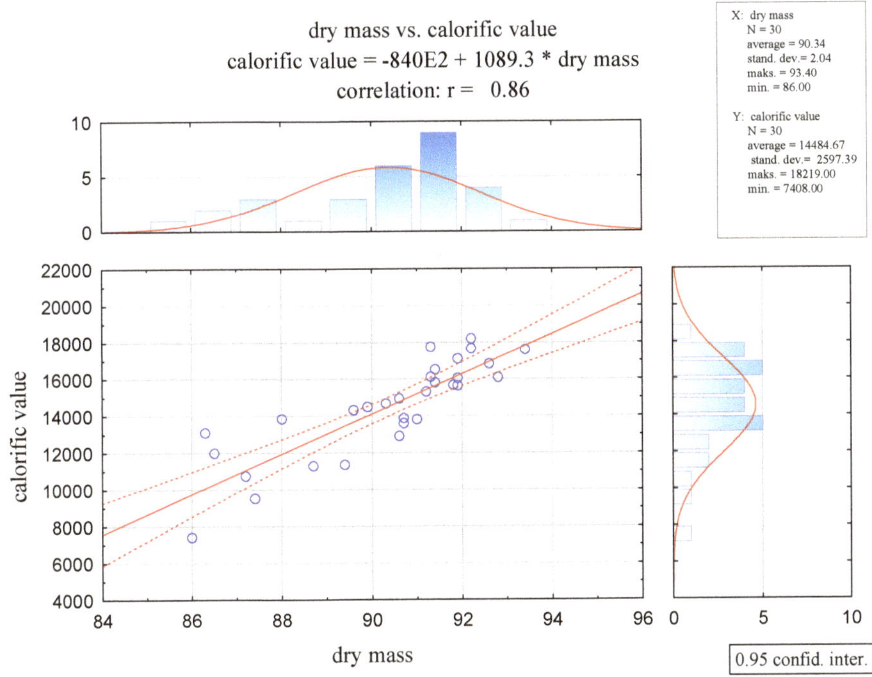

Figure 3. Graph of the correlation and dispersion of results; dry mass vs. calorific value.

Table 5. Static and dynamic test results for pellets from the selected mixtures.

Material Designation	Static Test: 2 d			Dynamic Test: 1 h Mixing 1, D Rest, 1 h Mixing		
	Disintegration	pH	EC	Disintegration	pH	EC
M3	13.552 g/76 ml	12.04	12.04	14.155 g/83 ml	12.26	7.04
M10	18.516 g/68 ml	11.96	2.77	19.563 g/75 ml	11.99	6.26
M13	22.040 g/68 ml	9.48	1.98	26.049 g/68 ml	10.05	3.87
M17	30.002 g/57 ml	9.07	1.49	15.442 g/74 ml	9.51	4.00
M18	16.521 g/72 ml	11.88	2.62	15.942 g/76 ml	11.54	6.25
M19	19.249 g/72 ml	8.73	2.16	21.016 g/77 ml	9.20	3.97
M20	16.069 g/72 ml	12.12	3.05	16.372 g/81 ml	12.15	6.92
M22	16.041 g/74 ml	12.10	2.93	15.114 g/79 ml	12.17	6.89
M30	20.636 g/68 ml	12.15	2.95	15.058 g/76 ml	12.21	6.81
MD	23.245 g/65 ml	5.25	0.50	11.700 g/67 ml	5.54	0.74

Most of the pellets disintegrated under the influence of water just a few hours after being soaked. After one day, mixtures M1, M2, M4, M7, M8, M9 and M16, as well as the pellets made from the input materials WS and S, disintegrated completely.

The durability of the pellets determined in the static and the dynamic tests was the same. The most durable were the pellets from mixtures M3 (WS + SS; 1:1), M10 (WS + SS; 4:1), M18 (WS + WD + SS; 3:1:1), M19 (WS + S + SS; 3:1:1), M20 (WS + WD + SS; 6:1:3) and M22 (WS + SS; 7:3). The pellets from mixtures M17 (WS + S + SS; 7:2:1) and M30 (WS + WD + SS; 3:3:4) and from pure WD, used as the input material, retained their structure to a lesser extent. The pellets from mixture M13 (WS + SS; 9:1) lost their structure in the dynamic test but retained it in the static test. Table 6 shows the structure of the pellets after the moisture absorption tests.

Table 6. Pellet structure after the moisture absorption tests.

Static Test	Dynamic Test	Static Test	Dynamic Test
M3		M10	
			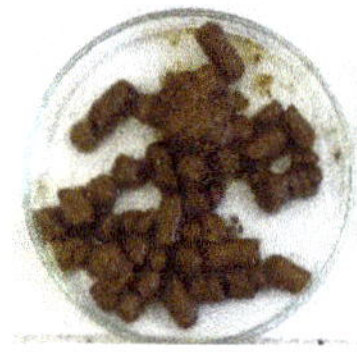
M13		M17	
	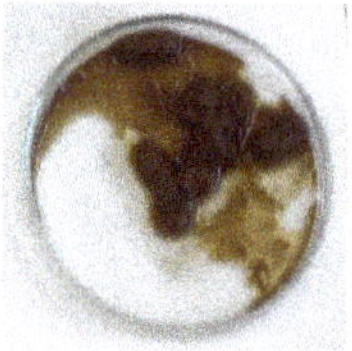		
M18		M19	

M20		M22	

M30		MD	
	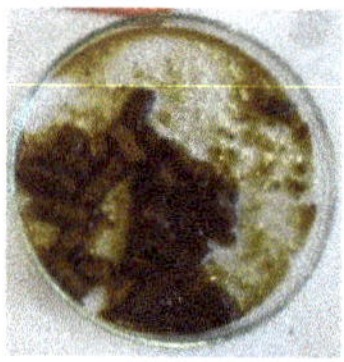	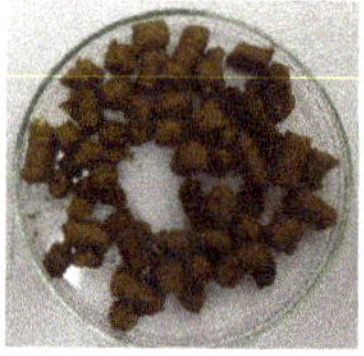	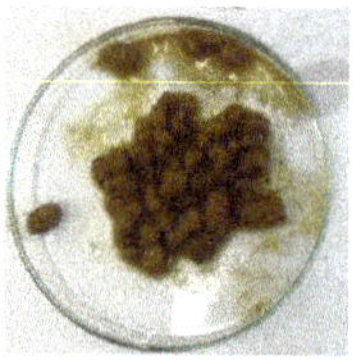

The tests showed that the combination of wood shavings (WS) with sewage sludge (SS) was good for the pellet structure. Sewage sludge hygienized with lime acted as a binder as long as it was present in the pellets in a proportion of at least 20%. When its share was lower, 10%, i.e., in mixture M13, the stabilization effect on the WS + SS pellets decreased significantly.

When the pellets were dropped three times from a height of $h = 1.5$ m, the results were as follows: the pellets from mixtures M18 (WS + WD + SS; 3:1:1), M19 (WS + S + SS; 3:1:1), M3 (WS + SS; 1:1), M22 (WS + SS; 7:3), M10 (WS + SS; 4:1) and M13 (WS + SS; 9:1) did not disintegrate. For this reason, it can be said that they should be good for transport in conditions of controlled air humidity and keep their durability. In this test, the WS + SS combinations also turned out to be the most durable.

3.5. Examination of the Combustion Process in Terms of Emissions Including the Possibility of Eliminating Acid Anhydrides

Pellets prepared from mixtures M2 (WS + S; 0.5:0.5) and M17 (WS + S + SS; 0.7:0.2:0.1)—Figure 4—and from energy willow for comparison (EW) were used for the combustion analysis in the test boiler.

Figure 4. Pellets prepared from mixtures M2 and M17.

The results showed that it was possible to reduce exhaust emissions by the addition of limed sewage sludge to biomass fuel (Table 7).

Table 7. Exhaust emissions during the combustion of the fuels analyzed.

Mixture	Ash Content	Exhaust Temp.	Pressure	Flow	External Temp.	CO	NO	NO$_2$	NOx	H$_2$S	H$_2$
	(%)	(°C)	(hPa)	(dm^3·min^{-1})	(°C)			(ppm)			
M2	2.1	100	1002	0.95	28.9	310	36	0.2	34	0.8	33
M17	5.5	156	1000	0.95	29.0	144	46	0.4	42	1.5	0
WE	1.5	120	1007	0.93	24.0	139	96	1.4	97	2	13

This information on exhaust emissions makes it possible to find differences between mixture M2 (without the addition of sewage sludge) and M17 (with the addition of 10% by weight of limed sewage sludge). The results are as follows:

- reduction in CO emissions by 115%;
- increase in NO emissions by 22%;
- increase in NO$_2$ emissions by 50%;
- increase in NO$_x$ emissions by 19%;
- increase in H$_2$S emissions by 47%;
- elimination of H$_2$ emissions.

The emissions of acidic anhydrides to the atmosphere from the combustion of pellets from both mixtures (M2 and M17) were lower than from the combustion of energy willow pellets:

- NO—by 63% for M2 and by 52% for M17;
- NO$_2$—by 63% for M2 and by 52% for M17;
- NOx—by 63% for M2 and by 52% for M17;

- H_2S—by 63% for M2 and by 52% for M17.

The decrease of CO content and the occurrence of the increased content of NO, NO_2, NO_x and H_2S in the exhaust gases was probably influenced by the burning conditions. Important elements of process disturbance could be the addition of a mineral fraction with sewage sludge hygienized with lime. The combustion of biomass pellets resulted in the accumulation of large amounts of bottom ash and slags in the combustion chambers of the boiler, which obstructed the air supply systems in the furnace over time. In literature, this phenomenon is described as a very important negative factor connected to the use of biomass for energy production [32].

In the phase of stabilized high temperature combustion, CO emissions from firewood are 250–500 ppm. Significantly higher emissions occur during the boiler warming up phase, up to 16,500 ppm, and boiler quenching, 3000 ppm. Chen et al. [38] compared the combustion of coal and sewage sludge and obtained three times less CO emissions and two times less CO_2 emissions when sewage sludge was used. This resulted in a reduction of CO and CO_2 emissions after the addition of sewage sludge to coal and a reduction of CO_2 emissions after the addition of sewage sludge to dried biomass of shiitake mushrooms. Monedero et al. [13] reported a more than 2-fold difference in CO emissions from the combustion of pellets from poplar wood (747 mg m^{-3}) and pine wood (331 mg m^{-3}). Polonini et al. [40] obtained CO emissions from the combustion of wood pellets that were similar to the emissions described in this study for M17 and EW pellets but by 40% lower than for M2. The concentration of oxygen in exhaust gases varies from 6 to 18% [37] depending on the phase. The composition of exhaust gases strongly depends on the conditions in which solid fuels are burnt, especially on combustion temperature and the primary excess air ratio. Under non-optimum conditions, e.g., when warming up a boiler, a gas with the following composition is formed: H_2, C_xH_y, H_2O, CO_2 and N_2, and NH_3, HCN, HOCN and NO in smaller amounts [10]. As a result of complete combustion, CO_2 and H_2O are formed and much more NOx. However, NOx emissions are relatively lower in comparison to wood pellets containing not much nitrogen and under optimum combustion conditions—a primary excess air ratio of 0.9–0.95 [10]. The concentration of NOx increases after the addition of sewage sludge to coal and to dried biomass of shiitake mushrooms. Monedero et al. [13] indicated that there were differences in NOx emissions depending on the type of wood that is burnt—poplar wood 229 mg m^{-3} and pine wood 97 mg m^{-3}. It is also possible to reduce CO and NOx emissions by adding calcium and magnesium compounds to poplar wood pellets—Ca,Mg-lignosulphonate. SO_2 and CH_4 emissions do not have such a clear tendency when additives from sewage sludge are used [38,43]. However, SO_2 increases after adding Ca,Mg-lignosulphonate to wood pellets [13].

4. Conclusions and Perspectives for Further Research

- The selected materials from waste biomass can be effectively used to produce energy. In perspective, pellets from wood materials mixture M1 (WS + WD; 1:1) can be a good variant of solid fuel. The addition of sewage sludge hygienized with lime (mostly mineral material) can disturb the process of biomass combustion, which resulted in incomplete fuel combustion. Wood, straw and sewage sludge mixture M17 (WS + S + SS; 7:2:1) turned out to be the best energy material prepared with the sewage sludge addition. It should be taken into account not to add more than 5% of lime to the whole fuel mixture, including the use of lime-treated sewage sludge by means of co-combustion.

- As an effect of sewage sludge addition, an increased pH and relatively high Ca, Mg, Na and Fe were found. The high potassium content was related to the construction of the material, which was solely based on straw, or the presence of straw in the mixtures. These elements will remain in the ash after the combustion of the pellets. As a result, the material rich in elements, being the desired fertilizer components, is created. Some problems can be connected with a high pH of ash (i.e., in the case of coniferous and heather cultivation) and high Ca, Mg and Na content relative to the other elements (ionic antagonism).

- Some materials containing lime-treated sewage sludge could have a relatively higher content of Cr. The evaluated situation shows the content of Cr lower than permitted by Polish law. However, the use of these materials for a long time as fertilizers may result in the accumulation of this undesirable metal in soil.
- A different problem is the durability of pellets prepared from various biomass mixtures. Wood shavings stabilized with sewage sludge hygienized with lime (mixtures M3, M13, M18, M19 and M22) are a good material for making durable pellets; from this point of view the proportion of sewage sludge in mixtures should be at least 20%.
- Due to the possible difficulties with burning a solid fuel material with a high content of mineral components, it is recommended to use mixtures M18 (WS + WD + SS; 3:1:1) and M19 (WS + S + SS; 3:1:1) for this purpose.
- The addition of sewage sludge to the biomass of M2, i.e., the production of material M17, resulted in a reduction of CO emissions and the elimination of H_2 emissions. At the same time, there was an increase in NO, NO_2, NOx, CO_2 and H_2S emissions. Further research is needed to optimize the combustion process parameters of individual biomass mixtures in terms of reducing the environmental impact of the occurring gaseous and residual products.

The combustion of waste biomass is difficult due to the highly varied composition of this fuel, both in terms of the combustion process and its environmental impact. The addition of a calcium-based mineral fraction to fuel, which is often described in the literature as effectively improving some parameters of boiler operation, is problematic, since it may cause some phenomena that are harmful to the environment. Studies should be continued to find an optimum composition of biomass mixtures to obtain a good calorific value with limited emissions of acidic anhydrides to the atmosphere. It would also be necessary to analyze how combustion technology (mainly the thermal characteristics of this process) affects the release of gases to the atmosphere in the case of fuels from pure biomass and biomass modified with lime. It should also be considered whether there is a significant difference in this respect between the addition of pure calcium carbonate and the addition of granulate from municipal sewage sludge hygienized with lime. It would also be necessary to check whether there would be any difference between the addition of municipal sewage sludge hygienized with lime in the form of granules and in the form of a non-granulated mixture.

Author Contributions: Conceptualization, A.G., M.M. and W.S.; Data curation, M.M. and W.S.; Formal analysis, A.G., M.M. and W.S.; Funding acquisition, A.G.; Investigation, A.G., M.M. and W.S.; Methodology, A.G., M.M. and W.S.; Project administration, A.G.; Supervision, A.G.; Validation, A.G. and M.M.; Visualization, M.M.; Writing—original draft, A.G., M.M. and W.S.; Writing—review & editing, A.G. and M.M.

Funding: This study was carried out as a project of the European Regional Development Fund. The Lubusz 2020 Operational Programme. Activity 1.2 Developing Enterprise—A Voucher for Innovation No. 015/2/2017 and 024/2/2017.

Conflicts of Interest: The authors declare no conflict of interest.

References

1. Tchapda, A.H.; Pisupati, S.V. A Review of Thermal Co-Conversion of Coal and Biomass/Waste. *Energies* **2014**, *7*, 1098–1148. [CrossRef]
2. Bazan-Krzywoszańska, A.; Mrówczyńska, M.; Skiba, M.; Sztubecka, M. Sustainable urban development on the example of the housing development of Zielona Góra (Poland), as a response to the climate policy of the European Union. In Proceedings of the 10th International Conference on Environmental Engineering, Vilnius, Lithuania, 27–28 April 2017. [CrossRef]
3. Energy 2018. *Statistical Analyses*; Statistics Poland: Warsaw, Poland, 2018; p. 28.
4. Islas, J.; Manzini, F.; Masera, O.; Vargas, V. Solid Biomass to Heat and Power. In *The Role of Bioenergy in the Emerging Bioeconomy: Resources, Technologies, Sustainability and Policy*; Academic Press: Cambridge, MA, USA, 2019; pp. 145–177.

5. Hamelin, L.; Borzęcka, M.; Kozak, M.; Pudełko, R. A spatial approach to bioeconomy: Quantifying the residual biomass potential in the EU-27. *Renew. Sustain. Energy Rev.* **2019**, *100*, 127–142. [CrossRef]

6. Antizar-Ladislao, B.; Turrion-Gomez, J.L. Decentralized Energy from Waste Systems. *Energies* **2010**, *3*, 194–205. [CrossRef]

7. Amaral, S.S.; de Carvalho, J.A., Jr.; Costa, M.A.M.; Pinheiro, C. Particulate Matter Emission Factors for Biomass Combustion. *Atmosphere* **2016**, *7*, 141. [CrossRef]

8. Mboumboue, E.; Njomo, D. Biomass resources assessment and bioenergy generation for a clean and sustainable development in Cameroon. *Biomass Bioenergy* **2018**, *118*, 16–23. [CrossRef]

9. Monteiro, E.; Mantha, V.; Rouboa, A. Portuguese pellets market: Analysis of the production and utilization constrains. *Energy Policy* **2012**, *42*, 129–135. [CrossRef]

10. Houshfar, E.; Løvås, T.; Skreiberg, Ø. Experimental Investigation on NOx Reduction by Primary Measures in Biomass Combustion: Straw, Peat, Sewage Sludge, Forest Residues and Wood Pellets. *Energies* **2012**, *5*, 270–290. [CrossRef]

11. Wzorek, M. Characterisation of the properties of alternative fuels containing sewage sludge. *Fuel Process. Technol.* **2012**, *104*, 80–89. [CrossRef]

12. Moya, R.; Tenorio, C.; Oporto, G. Short Rotation Wood Crops in Latin American: A Review on Status and Potential Uses as Biofuel. *Energies* **2019**, *12*, 705. [CrossRef]

13. Monedero, E.; Portero, H.; Lapuerta, M. Combustion of Poplar and Pine Pellet Blends in a 50 kW Domestic Boiler: Emissions and Combustion Efficiency. *Energies* **2018**, *11*, 1580. [CrossRef]

14. Toscano, G.; Alfano, V.; Scarfone, A.; Pari, L. Pelleting Vineyard Pruning at Low Cost with a Mobile Technology. *Energies* **2018**, *11*, 2477. [CrossRef]

15. Xu, J.; Chang, S.; Yuan, Z.; Jiang, Y.; Liu, S.; Li, W.; Ma, L. Regionalized Techno-Economic Assessment and Policy Analysis for Biomass Molded Fuel in China. *Energies* **2015**, *8*, 13846–13863. [CrossRef]

16. Cuellar, A.D.; Herzog, H. A Path Forward for Low Carbon Power from Biomass. *Energies* **2015**, *8*, 1701–1715. [CrossRef]

17. Lopes, E.J.; Queiroz, N.; Yamamoto, C.I.; da Costa Neto, P.R. Evaluating the emissions from the gasification processing of municipal solid waste followed by combustion. *Waste Manag.* **2018**, *73*, 504–510. [CrossRef] [PubMed]

18. Chen, W.-S.; Chang, F.-C.; Shen, Y.-H.; Tsai, M.-S. The characteristics of organic sludge/sawdust derived fuel. *Bioresour. Technol.* **2011**, *102*, 5406–5410. [CrossRef] [PubMed]

19. Jiang, L.; Yuan, X.; Xiao, Z.; Liang, J.; Li, H.; Cao, L.; Wanga, H.; Chen, X.; Zeng, G. A comparative study of biomass pellet and biomass-sludge mixed pellet: Energy input and pellet properties. *Energy Convers. Manag.* **2016**, *126*, 509–515. [CrossRef]

20. Kacprzak, M.; Neczaj, E.; Fijałkowski, K.; Grobelak, A.; Grosser, A.; Worwag, M.; Rorat, M.; Brattebo, H.; Almås, Å.; Singh, B.R. Sewage sludge disposal strategies for sustainable development. *Environ. Res.* **2017**, *156*, 39–46. [CrossRef]

21. Pulka, J.; Manczarski, P.; Koziel, J.A.; Białowiec, A. Torrefaction of Sewage Sludge: Kinetics and Fuel Properties of Biochars. *Energies* **2019**, *12*, 565. [CrossRef]

22. Kijo-Kleczkowska, A.; Środa, K.; Kosowska-Golachowska, M.; Musiał, T.; Wolski, K. Combustion of pelleted sewage sludge with reference to coal and biomass. *Fuel* **2016**, *170*, 141–160. [CrossRef]

23. Kijo-Kleczkowska, A.; Środa, K.; Kosowska-Golachowska, M.; Musiał, T.; Wolski, K. Experimental research of sewage sludge with coal and biomass co-combustion, in pellet form. *Waste Manag.* **2016**, *53*, 165–181. [CrossRef]

24. Yilmaz, E.; Wzorek, M.; Akçay, S. Co-pelletization of sewage sludge and agricultural wastes. *J. Environ. Manag.* **2018**, *216*, 169–175. [CrossRef]

25. Jiang, J.; Du, X.; Yang, S. Analysis of the combustion of sewage sludge-derived fuel by a thermogravimetric method in China. *Waste Manag.* **2010**, *30*, 1407–1413. [CrossRef]

26. Obernberger, I. Physical characteristics and chemical composition of solid biomass fuels. In *Script for the Lecture "Thermochemical Biomass Conversion"*; Eindhofen University of Technology, Department for Mechanical Engineering, Section Process Technology: Eindhoven, The Netherlands, 2003; (Chapter 3).

27. Li, H.; Jiang, L.-B.; Li, C.-Z.; Liang, J.; Yuan, X.-Z.; Xiao, Z.-H.; Xiao, Z.-H.; Wang, H. Co-pelletization of sewage sludge and biomass: The energy input and properties of pellets. *Fuel Process. Technol.* **2015**, *132*, 55–61. [CrossRef]

28. Kliopova, I.; Makarskiene, K. Improving material and energy recovery from the sewage sludge and biomass residues. *Waste Manag.* **2015**, *36*, 269–276. [CrossRef]

29. García-Maraver, A.; Popov, V.; Zamorano, M. A review of European standards for pellet quality. *Renew. Energy* **2011**, *36*, 3537–3540. [CrossRef]

30. Lehtikangas, P. Quality properties of pelletised sawdust, logging residues and bark. *Biomass Bioenergy* **2001**, *20*, 351–360. [CrossRef]

31. Sarenbo, S.L.; Claesson, T. Limestone and dolomite powder as binders for wood ash agglomeration. *Bull. Eng. Geol. Environ.* **2004**, *63*, 191–207. [CrossRef]

32. Shao, Y.; Wang, J.; Preto, F.; Zhu, J.; Xu, C. Ash Deposition in Biomass Combustion or Co-Firing for Power/Heat Generation. *Energies* **2012**, *5*, 5171–5189. [CrossRef]

33. Wang, Q.; Chen, J.; Han, K.; Wang, J.; Lu, C. Influence of $BaCO_3$ on chlorine fixation, combustion characteristics and KCl conversion during biomass combustion. *Fuel* **2017**, *208*, 82–90. [CrossRef]

34. *Statistical Yearbook—Lubuskie Voivodship 2017*; Statistical Office: Zielona Gora, Poland, 2017.

35. Environment 2018. *Statistical Analyses*; Statistics Poland: Warsaw, Poland, 2018.

36. Szewrański, S.; Świąder, M.; Kazak, J.K.; Tokarczyk-Dorociak, K.; van Hoof, J. Socio-Environmental Vulnerability Mapping for Environmental and Flood Resilience Assessment: The Case of Ageing and Poverty in the City of Wrocław, Poland. *Integr. Environ. Assess. Manag.* **2018**, *14*, 592–597. [CrossRef]

37. Bindig, R.; Butt, S.; Hartmann, I.; Matthes, M.; Thiel, C. Application of Heterogeneous Catalysis in Small-Scale Biomass Combustion Systems. *Catalysts* **2012**, *2*, 223–243. [CrossRef]

38. Chen, G.-B.; Chatelier, S.; Lin, H.-T.; Wu, F.-H.; Lin, T.-H. A Study of Sewage Sludge Co-Combustion with Australian Black Coal and Shiitake Substrate. *Energies* **2018**, *11*, 3436. [CrossRef]

39. Zhang, D.; Huang, G.; Xu, Y.; Gong, Q. Waste-to-Energy in China: Key Challenges and Opportunities. *Energies* **2015**, *8*, 14182–14196. [CrossRef]

40. Polonini, L.F.; Petrocelli, D.; Parmigiani, S.P.; Lezzi, A.M. Influence on CO and PM Emissions of an Innovative Burner Pot for Pellet Stoves: An Experimental Study. *Energies* **2019**, *12*, 590. [CrossRef]

41. To, W.M.; Lee, P.K.C.; Ng, C.T. Factors Contributing to Haze Pollution: Evidence from Macao, China. *Energies* **2017**, *10*, 1352. [CrossRef]

42. Ngangyo-Heya, M.; Foroughbahchk-Pournavab, R.; Carrillo-Parra, A.; Rutiaga-Quiñones, J.G.; Zelinski, V.; Pintor-Ibarra, L.F. Calorific Value and Chemical Composition of Five Semi-Arid Mexican Tree Species. *Forests* **2016**, *7*, 58. [CrossRef]

43. Falemara, B.C.; Joshua, V.I.; Aina, O.O.; Nuhu, R.D. Performance Evaluation of the Physical and Combustion Properties of Briquettes Produced from Agro-Wastes and Wood Residues. *Recycling* **2018**, *3*, 37. [CrossRef]

Article

Risk Analysis Related to Impact of Climate Change on Water Resources and Hydropower Production in the Lusatian Neisse River Basin

Mariusz Adynkiewicz-Piragas * and Bartłomiej Miszuk

Institute of Meteorology and Water Management—National Research Institute, ul. Podleśna 61, 01-673 Warszawa, Poland; bartlomiej.miszuk@imgw.pl
* Correspondence: mariusz.adynkiewicz@imgw.pl

Received: 22 April 2020; Accepted: 17 June 2020; Published: 21 June 2020

Abstract: Water resources are one of the most important issues affected by climate change. Climate scenarios show that in the upcoming decades, further climate change can occur. It concerns especially air temperature and sunshine duration, whose prognosis indicates a significant rising trend till the end of the century. The goal of the paper was the evaluation of water resources and hydropower production in the future, depending on climate scenarios with a consideration of risk analysis. The analysis was carried out on the basis of observation data for the Lusatian Neisse river basin (Poland) for 1971–2015 and climate projections till 2100 for the RCP2.6 and RCP8.5 (representative concentration pathways) scenarios. The results of the research showed that, especially in terms of RCP8.5, very high risk of decrease in water resources and hydropower production is expected in the future. Therefore, recommendations for mitigation of the possible effects are presented.

Keywords: water resources; climate change; environmental flow; hydropower production

---

1. Introduction

One of the most important problems related to future development of environment conditions are water resources. Such problems are noticeable especially in the regions of the mining industry, because of the high influence of mines on both ground and surface waters. The influence is noticeable especially in river valleys, which are one of the most vulnerable regions because of water ecosystems and water users, including hydropower stations. The basin of the Lusatian Neisse River, located in south-west Poland at the Polish–German–Czech border, is a good example of the regions where issues related to hydrological conditions are a crucial problem. Lusatian Neisse is a left tributary to Odra River (Figure 1). The region is characterized by significant variability in terms of altitude, relief and land use. The southern part of the basin is located in a mountainous area (Western Sudetes and their foreland), while the lowlands form the northern part. Total hypsometric differentiation of the region varies from 100 m a.s.l. (above sea level) in the north to over 1000 m a.s.l. in the south. Most of the region is comprised of agriculture and forest areas; however, an important role is played by industry, represented by opencast mines and hydropower stations. The Lusatian Neisse River forms a border between Poland and Germany, and the problem of water management is discussed within trans-border international commissions. Numerous opencast mines located in the Polish, German and Czech parts of the region significantly affect hydrological conditions. Furthermore, flooding of some of the former open mine pits, planned in the following decades with the use of water transfer from the Lusatian Neisse River, makes future water management extremely important.

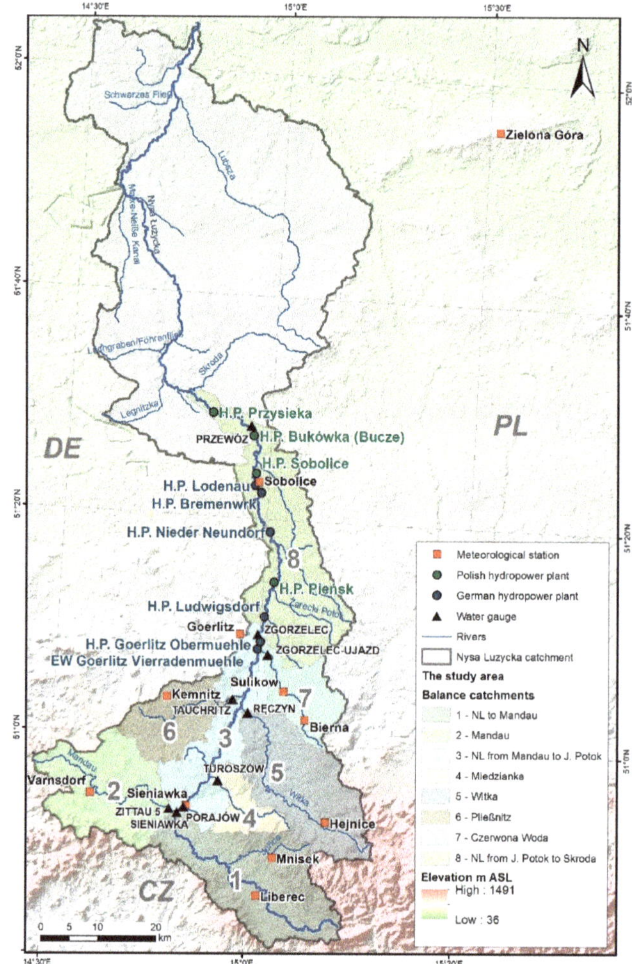

Figure 1. The Lusatian Neisse river basin with its eight sub-basins, the location of the meteorological stations of Zielona Góra and hydropower plants.

Besides mining, water resources in the discussed region are also affected by climate change. Increasing trends of air temperature and sunshine duration affect evaporation and consequently contribute to limitation of water resources. As a result, a deterioration of the ecological state and a decrease in water availability for various water users (including hydropower plants) can occur. Climate prognosis indicates further changes in the future, which can additionally affect water conditions. In the upcoming years, these changes can lead to frequent drought occurrence that have negative influences on forestry, agriculture and water availability [1,2]. In the case of the discussed region, analysis carried out within various EU projects (i.e., KLAPS—Climate change, air pollution and critical load of ecosystems in the Polish-Saxon border region; NEYMO—Lausitzer Neiße/Nysa Łużycka, Climatic and hydrological modeling, analysis and forecast; NEYMO-NW—Lausitzer Neiße/Nysa Łużycka—modeling of climate and hydrology, analysis and forecast of water resources in low-water conditions) showed that the availability of water resources had significantly decreased during last decades [3–6]. Additionally, climate prognoses indicate further negative trends that can result in serious difficulties at the end of the century [7,8]. Taking into consideration potential changes in water conditions, appropriate measures have to be undertaken in order to minimize the effects of climate change in the future. Numerous analyses showed that undertaking such measures related to reasonable management of water resources is a crucial issue in the case of many countries and in various climate zones [9–25].

Potential climate change and its impact on various sectors can be evaluated with the use of risk analysis. Such an analysis is usually based on a risk matrix that considers both the probability of changes of defined conditions and their consequences. The aspects of risk evaluation and risk matrixes were widely discussed in papers devoted to this problem in the fields of numerous sectors [26–40]. They may concern hydrological issues, like floods or droughts. Problems concerning water resources in Poland are one of the most crucial factors affecting society [2,41–44]. Consideration of environmental flow is also very important. It is a crucial factor in the context of ecological state. Climate changes and decreased water resources can cause serious problems in terms of this type of flow [45–48]. The changes can also negatively affect hydropower production. Analyses concerning hydropower stations show that energy production in the future is strongly dependent on climate scenarios [49–61].

Climate scenarios related to potential carbon dioxide concentration show that the climate may change in different ways, depending on socioeconomic development [62]. Such diversity can have a significant impact on water resources and hydropower production in the future. In the context of future climate change, one of the most important tasks consists of measures concerning downscaling and the use of regional climate models for climate projections and the evaluation of water resources and hydropower production. The measures are related to bias correction, which refers to the definition of a perturbation of the control time series in order to force some of their statistics closer to the historical ones [63]. It assumes that the bias between statistics of the model and data will not change in the future [63–65]. Delta change techniques assume that regional climate models are a good tool for the assessment of relative changes in the statistic between present and future, but that they do not assess absolute values [63]. In the case of precipitations projections, several bias correction techniques should be taken into consideration [66]. Various correction methods were used for the evaluation of changes in water conditions, i.e., in the river basin regions of Spain, Canada, Germany, Bulgaria and Greece [66–68]. It should also be emphasized that reduction of uncertainty related to climate projections could not have importance when the hydrological models have their own sources of uncertainty [67].

The main goal of the article was to evaluate the influence of potential climate change in the upcoming decades of the century on water resources and hydropower production with the use of risk assessment. The results of the analysis can be used in long-term planning activities focused on water use. They can also be useful for the evaluation of water resources in the context of currently operating hydropower stations located on the transborder Lusatian Neisse River.

2. Case Study and Data

In the case of climate conditions, meteorological data from the IMWM-NRI (Institute of Meteorology and Water Management—National Research Institute) stations of Zielona Góra (192 m a.s.l.), Sobolice (140 m a.s.l.), Sulików (215 m a.s.l.) and Bierna (270 m a.s.l.) were used. Zielona Góra is located about 60 km to the east of Lusatian Neisse River, while the other mentioned stations represent the lowlands (Sobolice) or the mountain foreland of the Isera Mountains (Sulików, Bierna). Meteorological data concerned the 1971–2015 period and considered precipitation (all of the stations) as well as air temperature and sunshine duration (Zielona Góra). The data were the basis for carrying out an analysis concerning trends for these variables. Besides the observation records, data from climate projections were also used in the context of the evaluation of possible climate change till 2100. Characteristics were carried out for two future periods: the near (2012–2050) and far future (2071–2100). Subsequently, the results were compared to the reference period (1971–2000). The climate projections were developed within the KLAPS and NEYMO projects by Climate and Environment Consulting Potsdam GmbH [7] on the basis of global models simulations: ECHAM5 MPI-OM (European Centre Hamburg Model) and MPI-ESM-LR (Max Planck Institute for Meteorology Earth System Model). In the case of downscaling, the regional climate model of WETTREG (weather-related regionalization method) was considered. It is a statistical model used for evaluations of climate projections for the region of Central Europe. The process of downscaling was connected with the evaluation of circulation conditions, a stochastic weather generator and the use of a statistical regression method. In terms of circulation conditions, WETTREG defines weather in different classes, depending on considered meteorological factors. The stochastic weather generator enables the development of various projections that are independent from each other and characterized by equal probability. Statistical regression is connected with calculations of parameters on the basis of the modeled simulations. 1971–2000 was considered as a base period for the projections' development. The period was used for model validation, comparing measurement and simulated data. Two types of climate scenarios were used: RCP2.6 and RCP8.5 (representative concentration pathways). They are new generation scenarios that are related to CO_2 concentration in the atmosphere. They represent different radiative forces in 2100, which are equal to 2.6 W/m^2 and 8.5 W/m^2, respectively. These scenarios are dependent on various social-economic aspects: world population, changes in GDP, use of fossil fuels and energy intensity. RCP2.6 is based on a principle that maximal emission will be noticed in 2010–2020 and the increase in air temperature will not be higher than 2 °C in comparison to the preindustrial era. On the other hand, for RCP8.5 a constant increase in emission is projected until 2100. In the case of RCP8.5, energy intensity and use of fossil fuels at the end of the century can be twice as higher as for RCP2.6 [69]. The scenarios of RCP2.6 and RCP8.5 were presented in one of the latest IPCC (Intergovernmental Panel on Climate Change) reports [62].

Climate analysis concerned changes in mean annual values of air temperature, sunshine duration and annual precipitations totals. The calculations and analysis for these sectors mainly concerned the periods of the near (2021–2050) and far (2071–2100) future. The results for these periods were compared with simulated values for the reference period (1971–2000).

In terms of hydrological conditions, discharges for RCP2.6 and RCP8.5 scenarios for the same periods were calculated. They were the basis for calculations of environmental flow according to Kostrzewa's formula [70], and were also used for the evaluation of unitary runoff. The hydrological modeling was carried out with the use of rainfall-runoff MIKE NAM software [71,72]. In the NEYMO project, the model was calibrated with the use of precipitation, evapotranspiration and discharges data. The model uses algorithms concerning the hydrological cycle within the basin. Such an approach enables simulations related to precipitations to outflow transformations. The analysis on hydrological conditions was based on sixteen rainfall-runoff models for balance sub-basins that were developed within MIKE NAM during the realization of the NEYMO project. The models concerned the whole region of the Lusatian Neisse basin and were developed and calibrated for observation data for the 1971–2000 period and subsequently validated for 2000–2010 data [4] (Table 1). They were examined for particular sub-basins. Finally, models developed for eight balance sub-basins were taken into consideration. Based on the results of the models, tendencies for discharges changes were analyzed for eight balance areas (Table 2). The results were summarized for Przewóz gauge, which is a closing gauge of the discussed balance region. The input for the models consisted of precipitation data and the meteorological variables necessary for the calculation of potential evapotranspiration mean, maximum and minimum air temperature, wind speed, sunshine duration and air humidity. Potential evapotranspiration was calculated with the use of ETO software, developed by the Land and Water Division of the FAO (Food and Agriculture Organization of the United Nations).

Table 1. List of measuring stations used to build and calibrate run-off models [4].

Sub-Basin Number	Name of Sub-Basin	Water Gauge	Precipitation	Evaporation
1	Nysa Łużycka from. Mandau river	Porajów	Liberec	Görlitz
2	Mandau	Zittau 5	Varnsdorf	Cottbus
3	Nysa Łużycka from Mandau river to Jędrzychowicki Potok river	Porajów, Sieniawka, Zgorzelec	Sieniawka	Görlitz
4	Miedzianka	Turoszów	Minsek	Görlitz
5	Witka	Ostróżno	Hejnice	Görlitz
6	Pließnitz	Tauchritz	Kemnitz	Görlitz
7	Czerwona Woda	Zgorzelec Ujazd	Sulików	Görlitz
8	Nysa Łużycka from Jędrzychowicki Potok to Skroda river	Zgorzelec, Przewóz	Sobolice	Görlitz

Table 2. Results of the assessment of MIKE NAM (Nedbor-Afstromnings-Model) models using Nash–Sutcliffe Efficiency (NSE) statistics and comparison of simulated and observed discharge in eight sub-basin areas.

Sub-Basin Number	Statistics NSE	Average Discharge [m^3/s]	
		Observed	Simulated
1	0.5	6.05	6.02
2	0.6	3.41	3.46
3	0.5	1.28	1.29
4	0.5	0.94	0.92
5	0.5	3.60	3.52
6	0.5	1.09	1.07
7	0.5	0.77	0.77
8	0.5	3.04	3.17

Results of calculations of discharges were also the basis for the evaluation of hydropower production in the future (Scheme 1). In the discussed region, it concerns Polish and German hydropower plants that are the main water users in the Lusatian Neisse river basin. The analysis presented in the paper is focused on changes in nominal energy production between the near (2021–2050) and far (2071–2100) future and the current period (2015–2020).

Both Polish and German hydropower plants were taken into consideration. Energy generated by hydropower plants depends on differences in water levels before and behind the turbines, the turbine's gullet, and the efficiency of the turbines, transmission and generator. There are eight German and three Polish hydropower stations located on the Lusatian Neisse River that provide energy for the region (Table 3). In the paper, analyses were carried out for nine of them, as in case of two German plants (Altes Sagewerk and Apelt Mühle) no data on the turbines' gullet were available.

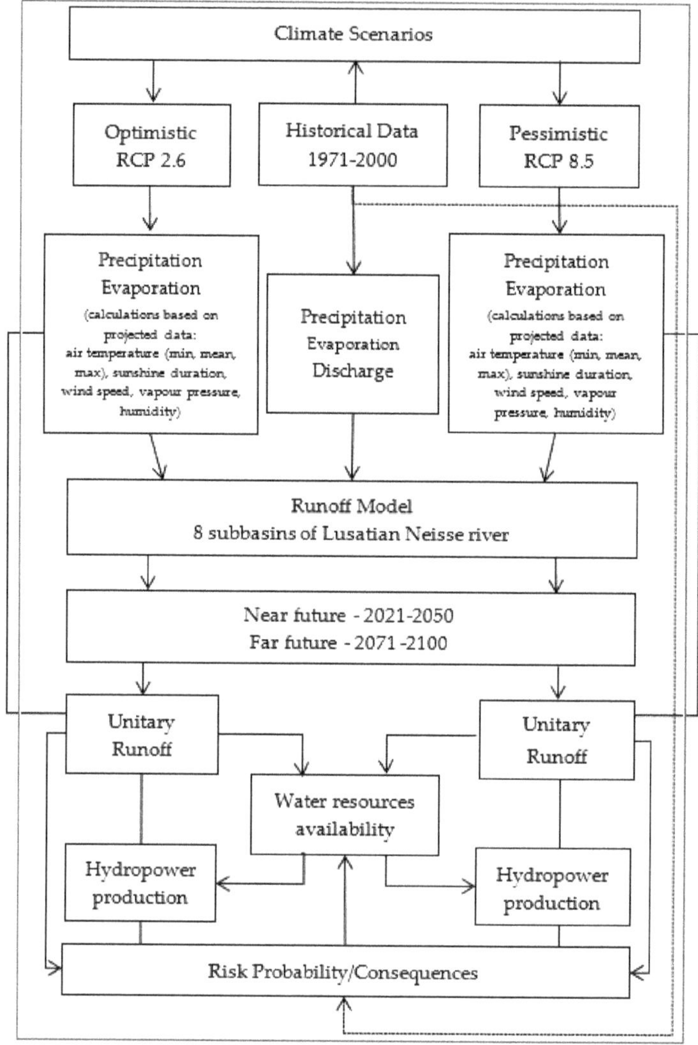

Scheme 1. Methodological approach related to the impact of climate change on water resources.

Table 3. Main features of the Polish and German hydropower station on the Lusatian Neisse River.

Hydropower Station	Turbine's Gullet [m³/s]	Differences in Water Level [m]	Efficiency Index	Nominal Hydropower Production [kWh/year]
Obermühle	7.3	1.80	0.84	948,522.78
Vierradmühle	10	2.70	0.86	1,995,424.63
Ludwigsdorf	22	2.46	0.86	3,999,717.82
Pieńsk	10	4.20	0.86	3,103,993.87
Nieder-Neundorf	24	3.64	0.86	6,456,307.25
Bremenwerk	22.4	3.00	0.86	4,966,390.20
Lodenau	24	4.70	0.86	8,336,440.68
Sobolice	22	3.60	0.80	5,444,879.62
Bukówka	20.6	5.50	0.83	8,081,297.89

Depending on possible changes in discharges in the future, calculations for potential power generation for RCP2.6 and RCP8.5 were carried out. They concerned the following formula:

$$N = 9.81 \cdot Q \cdot H \cdot \eta,$$

In the equation, N refers to generated power [kW], Q to the turbine's gullet [m^3/s], and H means differences in water levels before and behind the turbines [m]. Furthermore, η is an efficiency index and 9.81 refers to standard gravity [m/s^2]. The calculations were carried out on the basis of technical-exploitation data concerning the hydropower plants, which were presented in the paper devoted to the water balance of the Lusatian Neisse River [73]. The results of the analysis were shown for various multiannual future periods.

3. Methods

The main purpose of the analysis was to evaluate risk related to the impact of climate change on water resources and hydropower production. In this case, a risk matrix was used. It considers both the probability and consequences of climate change. Probability was evaluated on the basis of selected meteorological elements, while consequences were assessed for water resources and hydropower production.

Probability refers to changes in three meteorological variables—air temperature, sunshine duration and precipitations. For each of the variables, five classes of probability were selected (where 5 is the highest probability of change). In the case of probability evaluation, aspects of projected changes in the future were taken into consideration.

According to the settings of the projections, each scenario is characterized by the same probability of occurrence. In the case of RCP2.6, the tendency of climate change after 2050 can be different than in the first half of the century. It indicates that climate conditions can change in different ways depending on the period. Therefore, in order to evaluate the probability of changes in particular meteorological variables, the level of local warming (for air temperature) and local changes in sunshine duration and precipitation, totals in the future were taken into consideration. These particular levels of climate change were evaluated on the basis of projected data for two periods—2021–2050 and 2071–2100. The analysis was carried out for two considered scenarios that represent the same scenario family—RCP2.6 and RCP8.5. Thus, in terms of probability evaluation, the following factors were considered:

1. Projected air temperature/sunshine duration/precipitations for 2021–2050, according to RCP2.6
2. Projected air temperature/sunshine duration/precipitations for 2021–2050, according to RCP8.5
3. Projected air temperature/sunshine duration/precipitations for 2071–2100, according to RCP2.6
4. Projected air temperature/sunshine duration/precipitations for 2071–2100, according to RCP8.5

On the basis of the mentioned criteria, four levels of local warming and local changes in sunshine duration and precipitation totals (W1, W2, W3, W4) were calculated (Table 4). They are defined as differences between mean values for the near (2021–2050) and far (2071–2100) future and the reference period (1971–2000), considering both RCP2.6 and RCP8.5 scenarios. Each positive value of local climate change level (W) for air temperature or sunshine duration, in the comparison to the reference period of 1971–2000, increases the probability rate by 1 (Table 5). The more criteria are met, the higher is the probability of future changes in a given meteorological variable. Probability was assessed separately for air temperature and sunshine duration.

Table 4. Evaluation of levels of local warming (T), local changes in sunshine duration (SD) and precipitation totals (RR) on the basis of climate projections for RCP2.6 and RCP8.5 for 1971–2000, 2021–2050 and 2071–2100.

Level of Local Climate Change	Difference between Projected Data for Future Periods (2021–2050 and 2071–2100) and the Reference Period of 1971–2000
W1	Projected T, SD, RR for 2021–2050 (RCP2.6)—Projected T, SD, RR for 1971–2000
W2	Projected T, SD, RR for 2021–2050 (RCP8.5)—Projected T, SD, RR for 1971–2000
W3	Projected T, SD, RR for 2071–2100 (RCP2.6)—Projected T, SD, RR for 1971–2000
W4	Projected T, SD, RR for 2071–2100 (RCP8.5)—Projected T, SD, RR for 1971–2000

Table 5. Criteria for probability (P) evaluation related to changes in air temperature (T) and sunshine duration (SD).

P	Criteria Fulfilled	Criteria
1	None	Projected T, SD according to W1 > Projected T, SD for 1971–2000
2	One criterion	Projected T, SD according to W2 > Projected T, SD for 1971–2000
3	Two criteria	Projected T, SD according to W3 > Projected T, SD for 1971–2000
4	Three criteria	Projected T, SD according to W4 > Projected T, SD for 1971–2000
5	Four criteria	

In the case of precipitation, the criteria are reversed (Table 6). Because of the fact that a decrease in precipitations affects water availability, the probability of changes in precipitation increases when projected totals for the future periods are lower than the values simulated for 1971–2000. Four stations were considered with the purpose of evaluating precipitation, which gives 16 total results for potential change in precipitation totals in the future. The projected values were assessed according to the criteria presented in Table 6 in terms of how many results (in %) meet the criteria. In this case, a decrease rate of 5% was considered as a threshold. If projected totals are higher, comparable or slightly lower (<5%) than in 1971–2000, the criterion is not met.

Table 6. Criteria for probability (P) evaluation related to changes in precipitation totals (RR).

P	Criteria Fulfilled	Criteria
1	0%	Projected RR according to W1 < Projected RR for 1971–2000,
2	0–25%	Projected RR according to W2 < Projected RR for 1971–2000,
3	26–50%	Projected RR according to W3 < Projected RR for 1971–2000,
4	51–75%	Projected RR according to W4 < Projected RR for 1971–2000
5	76–100%	

Potential consequences of the impact of climate changes on selected sectors were also classified into five classes depending on the intensity of the effects. Consequences were evaluated separately for each of the sectors. Probability and consequences were the basis for risk evaluation. Probability multiplied by the ratio of consequences is equal to the risk value. A description of potential consequences is presented in Table 7, while dependence between probability and consequences is shown in a risk matrix.

The consequences rate for water resources and hydropower production shows how these sectors can be affected by projected air temperature, sunshine duration and precipitation changes. The range of consequences varies from minor (no impact of these meteorological variables) to very high (changes in climate conditions cause serious problems with water resources and hydropower production).

Table 7. Evaluation of consequences of the impact of meteorological variables on water resources and hydropower production.

		Evaluation of Consequences
1	Minor	No changes in water resources, no impact on hydropower production
2	Low	Low changes in water resources; low impact on hydropower production
3	Medium	Medium changes in water resources; medium impact on hydropower production
4	High	High changes in water resources; high impact on hydropower production
5	Very high	Very high changes in water resources; very high impact on hydropower production

In terms of evaluations of consequences for water resources, two factors were taken into consideration: future changes in mean maximal and minimal discharge, and environmental flow (Table 8). In the case of discharges, a potential decrease in the near and far future in comparison to the reference period was taken into consideration. Changes rate of 5%, 20%, 35% and 50% were considered as the boundary values. The threshold of the highest class (50%) was estimated on the basis of the results for the last decades of the century for other European rivers, according to the RCP8.5 scenario [74]. The impact of climate change on environmental flow was related to the projected number of mean discharges below environmental flow. If over 10% of the projected discharges were below environmental flow (changes over 10th percentile), consequences were classified as 4 or 5, depending on discharge changes.

Table 8. Criteria for the evaluation of consequences (C) for water resources.

C	Discharge Changes	Impact on Environmental Flow
1	Slight decrease (<5%)	No impact
2	Moderate decrease (5–20%)	No impact
3	Noticeable decrease (20–35%)	Minor impact (<10%)
4	High decrease (35–50%)	Noticeable impact (>10%)
5	Very high decrease (>50%)	Noticeable impact (>10%)

The evaluation of consequences for hydropower production was based on its possible changes in the future. In this case, differences (in %) between 2015–2020 and the near and far future were taken into consideration. Previous analysis showed that in the alpine region, where potential for hydropower production is very high, climate change can contribute to a decrease of over 25% in production in the following decades [50,52]. Therefore, for the purposes of this paper, the value of 25% was taken as a threshold for the evaluation of maximal level of consequences. If projected hydropower production in the future was more than 25% lower than nowadays, the maximum level of consequences was given (Table 9).

Table 9. Criteria for evaluation of consequences (C) for hydropower production.

C	Hydropower Production Changes
1	No changes or increase
2	Slight decrease (<5%)
3	Moderate decrease (5–15%)
4	High decrease (15–25%)
5	Very high decrease (>25%)

Based on the evaluated probability and consequences, a risk assessment was carried out. The evaluation of risk was carried out on the basis of the risk matrix (Table 10). The highest risk is noticed when both the probability and consequences are characterized by high values. It especially concerns situations with probability and consequences reaching 4 or 5, which eventually results in "very high" risk. Such a matrix enables risk evaluation for water resources and hydropower production under climate change conditions.

Table 10. Risk matrix presenting the probability of change in meteorological variables and consequences for water resources and hydropower production (red: very high risk; orange: high risk; yellow: medium risk; green: low risk) [27,39].

RISK		PROBABILITY				
		1	2	3	4	5
	5	5	10	15	20	25
	4	4	8	12	16	20
	3	3	6	9	12	15
	2	2	4	6	8	10
	1	1	2	3	4	5

4. Results

4.1. Changes in Climate Conditions

In the considered multiannual period (1971–2015), the mean annual air temperature in Zielona Góra was equal to 8.9 °C. The year 1996 was the coldest year (6.8 °C), while the highest values were observed in 2014 and 2015. In these years, mean annual air temperature reached 10.4 °C and 10.5 °C (Figure 2). The course of mean annual values for air temperature in the 1971–2015 period was characterized by an increasing, statistically important trend. The rate of changes in air temperature was equal to 0.34 °C/decade. The same rate was observed for mean annual values of maximum air temperature, while the increase in minimum temperature was equal to 0.33 °C/decade. In the considered period, the course for both maximum and minimum air temperature was characterized by statistical importance.

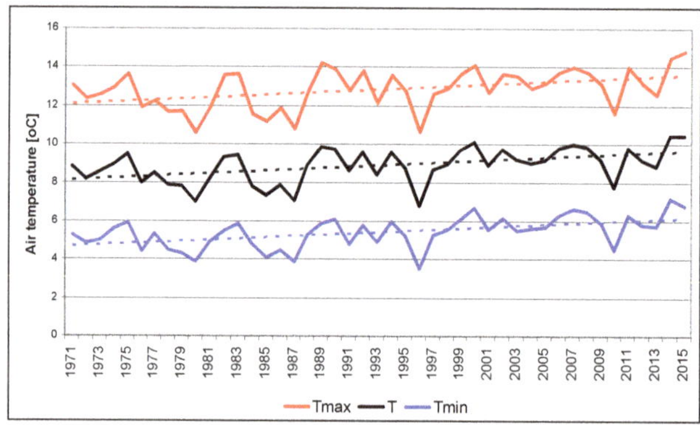

Figure 2. Mean annual values of maximum (Tmax), mean (T) and minimum (Tmin) air temperature with linear trends in 1971–2015 in Zielona Góra.

RCP scenarios project further changes in air temperature in the following decades (Figure 3). In terms of RCP2.6, prognosed data for Zielona Góra show an increase in air temperature for 2015–2100; however it is characterized by low intensity. Comparing the results for the near (2021–2050) and far future (2071–2100) with the reference period (1971–2000), air temperature in the near future in Zielona Góra can increase by about 1 °C. In the following decades, changes in air temperature are minor and are characterized by similar values to those of 2021–2050.

Scenario RCP8.5 is related to quite a different course of thermal conditions. In this case, a constant increase in air temperature is projected during 2015–2100. Air temperature values, calculated for the near and far future, are significantly higher when compared to the reference period. The differences are more noticeable than for RCP2.6. In 2021–2050, mean air temperature is higher than for the reference period by about 1.6 °C. The differences continue to increase for 2071–2100, when they reach 3.6 °C.

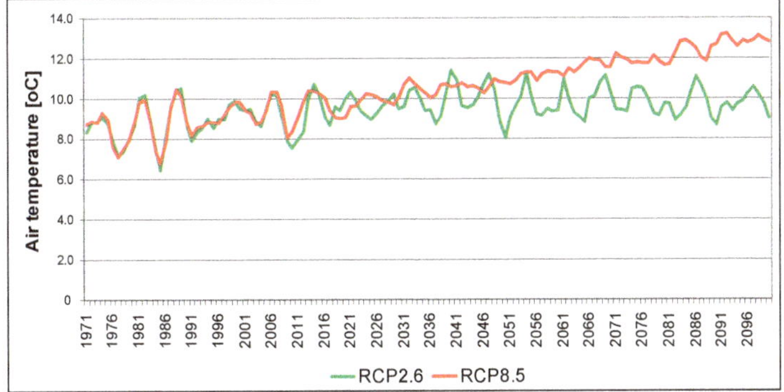

Figure 3. Projected air temperature values for 2015–2100 in Zielona Góra, according to RCP2.6 and RCP8.5 scenarios.

In terms of sunshine duration, the mean annual value in Zielona Góra is equal to 1552 h (hours). Observed changes in annual sunshine duration were comparable with air temperature. In 1971–2015, annual sunshine duration was characterized by an increasing trend with statistical importance. In the considered period, sunshine duration in Zielona Góra increased by about 80 h/decade. The highest sunshine duration was observed in 2015, when its annual value exceeded 2160 h. The lowest values were observed in 1977 (1166 h) (Figure 4).

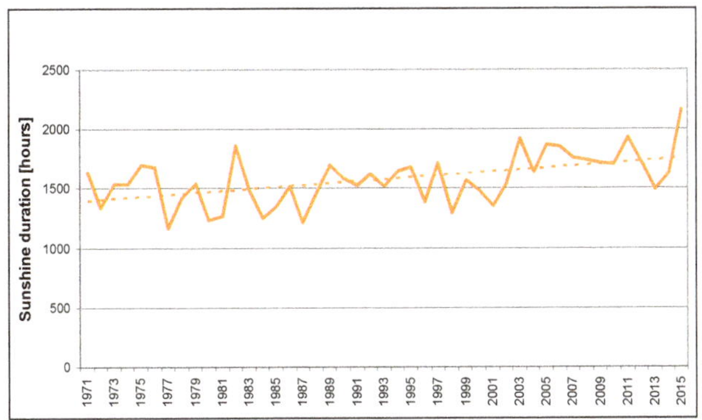

Figure 4. Annual sunshine duration with linear trend in Zielona Góra in 1971–2015.

A further increase in sunshine duration is also projected in terms of climate projections. Prognosed changes, similarly to air temperature, strongly depend on climate scenarios (Table 11). The most intensive changes are simulated for RCP8.5. In the case of the near future, predicted annual values of sunshine duration in Zielona Góra in 2021–2015 can be higher by about 180 h in comparison to the reference period. In the case of RCP2.6, the intensity of the changes is almost twice as low.

Simulations for RCP8.5 show that annual sunshine duration can additionally increase in the upcoming decades under RCP8.5 conditions. In 2071–2100, annual values of sunshine duration can be 340 h higher than in 1971–2000. In terms of RCP2.6, simulations show a slight decrease in sunshine duration in comparison to the near future. Therefore, such a tendency is similar to air temperature.

Comparing projected annual sunshine duration with the reference period, the far future is characterized by values exceeding those for 1971–2000 by 66 h.

Table 11. Differences in annual sunshine duration (hours) between the near (2021–2050) and far future (2071–2100) and the reference period (1971–2000) in Zielona Góra.

Scenario	2021–2050	2071–2100
RCP2.6	79.0	66.4
RCP8.5	179.6	339.8

Precipitation is characterized by high year-to-year variety. In the analyzed observation period (1971–2015), mean annual totals varied from 577 mm in Zielona Góra (lowlands) to 736 mm in Bierna (mountain foreland). The highest totals were noticed in 2010, when they exceeded mean values even by 50% (Bierna). On the other hand, 1982 and 2003 were the driest years, with precipitation totals reaching 50–70% of the mean multiannual totals. In the case of the course of annual precipitation totals, a slightly increasing trend was observed for all of the considered stations; however no statistical importance was noticed (Figure 5).

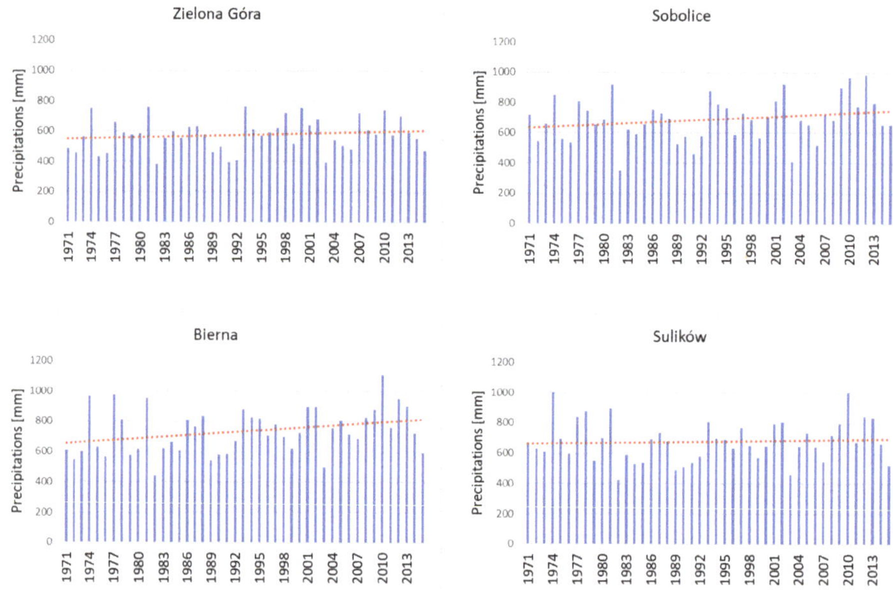

Figure 5. Annual precipitation totals and their trend for 1971–2015 in the Lusatian Neisse river basin.

Projections for precipitation, according to RCP2.6, do not show any significant changes in annual totals (Table 12). In the case of the considered stations, a slight increase (Zielona Góra and Sulików) or decrease (Bierna) were simulated for 2021–2050. However, the changes do not exceed 1%. The prognosis for the far future (2071–2100) does not indicate major changes either. Annual precipitation totals can be a little lower (Zielona Góra) or comparable (Sulików, Sobolice) to the values simulated for 1971–2000. The highest change was noticed for Bierna (for 2071–2100), where projected precipitations can be 2% lower than in the reference period.

More noticeable changes were modeled for RCP8.5. In this case, all the stations were characterized by lower precipitation totals. The most significant decrease was modeled for the station of Bierna, located in the southern part of the region, in the mountain foreland. The projections indicate that annual precipitation totals can be lower by about 5% in 2021–2015. An even higher decrease is simulated for the far future, when mean values in Bierna can be 13% lower than at the end of the 20th century. The remaining stations are also characterized by decreasing precipitation, especially for 2071–2100. In Sobolice and Sulików the totals can be lower by 5–8% in comparison to the reference period, while in Zielona Góra they are comparable with the near future.

Table 12. Projected changes in annual precipitation totals (%) between the near and far future (2021–2050; 2071–2100) and the reference period (1971–2000) in Zielona Góra (ZG), Sulików (SUL), Sobolice (SOB) and Bierna (BIE), according to the RCP2.6 and RCP8.5 scenarios.

Period	RCP2.6				RCP8.5			
	ZG	SUL	BIE	SOB	ZG	SUL	BIE	SOB
2021–2050	1	1	−1	0	−2	−1	−5	−1
2071–2100	−1	0	−2	0	−2	−8	−13	−5

4.2. Changes in Hydrological Conditions

One of the most important tasks concerning water resources is determining hydrological conditions. In this case, river discharges were projected for the 1971–2100 period. Similarly to the meteorological variables, simulations for hydrological conditions were carried out for three periods: 1971–2000 (reference period), 2021–2050 (near future) and 2071–2100 (far future).

The results showed that the direction and intensity of hydrological changes are dependent on changes in particular meteorological variables. In terms of the RCP2.6 scenario, negative changes in discharges are prognosed for the future periods. The changes concern a decrease in discharges, which consequently leads to a lower amount of water resources. In terms of the reference period, mean simulated discharges varied from 12.4 m^3/s to 13.4 m^3/s. In the near future, projected discharges reach 10.2–11.5 m^3/s, whereas for the far future, hydrological conditions seem to be a little more optimistic if compared to the near future—the mean discharge rate amounts to 10.4–12.3 m^3/s.

The results for RCP8.5 show that the discharge rate in the future can significantly fall, especially in the latest decades of the century. Basing on the models' principles, mean discharges for the reference period were classified into 11.2–13.5 m^3/s. In the case of the future periods, the simulations for 2021–2050 show that the values can be equal to 9.8–11.3 m^3/s. In the far future, a further significant decrease can occur. In this case, mean discharges can amount to 7.0–8.6 m^3/s. Therefore, minimum and maximum values of mean discharges, according to RCP8.5, can decrease by about 36–38% when compared to 1971–2000 (Table 13).

Table 13. Simulated maximum and minimum values for mean annual discharges [m^3/s] for the Przewóz gauge in the reference period (1971–2000), near future (2021–2050) and far future (2071–2100).

Period	RCP2.6		RCP8.5	
	Min	Max	Min	Max
1971–2000	12.4	13.4	11.2	13.5
2021–2050	10.2	11.5	9.8	11.3
2071–2100	10.4	12.3	7.0	8.6

The consequence of the changes in discharge rates in the future is also a modification of unitary runoff. The predicted tendency for unitary runoff is similar to the discharge changes presented above, and can vary depending on type of scenario. Simulated unitary runoffs for the most pessimistic realizations of RCP2.6 and RCP8.5 are presented in Figures 6 and 7.

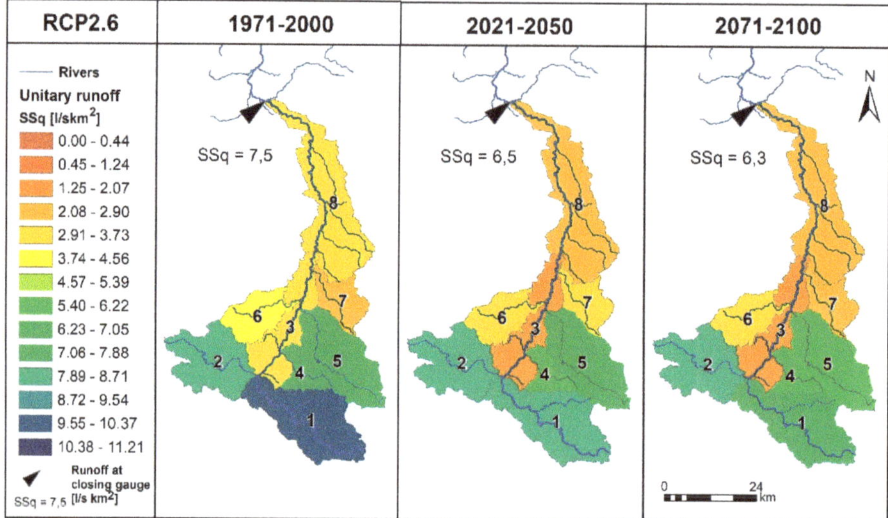

Figure 6. Simulated unitary runoff for the eight sub-basins of Lusatian Neisse river basin, according to a realization of R09 of the RCP2.6 scenario.

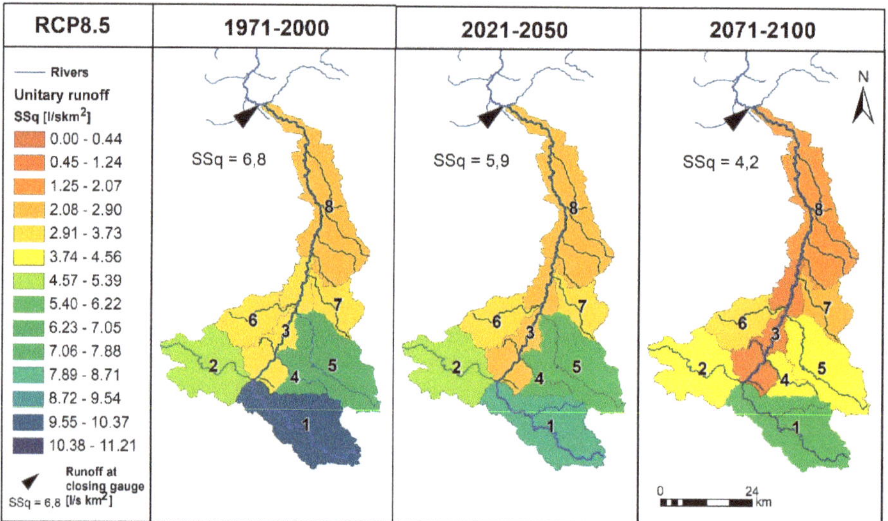

Figure 7. Simulated unitary runoff for the eight sub-basins of Lusatian Neisse river basin, according to a realization of R08 of the RCP8.5 scenario.

A decrease in discharges in the future can contribute to an increase in the number of situations when discharge rates can be lower than environmental flow. According to Kostrzewa's formula, the value of environmental flow for the Przewóz gauge is equal to 3.17 m³/s. Thus, considering projected discharge rates, calculations of frequency of mean monthly discharges below environmental flow for all of the considered scenarios and future periods were carried out. The results showed that a very high number of such situations can occur in the case of the RCP8.5 scenario, especially in the far future. The simulated mean frequency of mean monthly discharges that are below environmental flow for 2021–2050 is equal to 2.1% (RCP2.6) and 3.3% (RCP8.5). In the case of RCP8.5, for some of the runs of the scenario, it even amounts to over 6%.

For the period of the far future, the frequency of such discharges, in terms of RCP8.5, dramatically increases. The simulations project that over 10% of the mean monthly discharges can be below 3.17 m^3/s. Some of the runs for RCP8.5 show that the number of discharges below environmental flow can exceed 18% of all calculated discharges. In these terms, RCP2.6 is definitely a more optimistic scenario, as it projects the frequency of such critical discharges in the far future to be below 2% (depending on the runs, the increase in such a frequency can rise from 0.3% to 3.1%). That shows that in the context of water resources in the future, it is very important to keep climate change within the level considered in RCP2.6.

4.3. Changes in Hydropower Production

One of the most important issues related to water resources is hydropower production. Both German and Polish hydropower stations located on the Lusatian Neisse River play a crucial role in hydropower production. Calculations of potential annual hydropower production were carried out for current conditions and for the future. They were based on modeled discharge values for the periods of 2015–2020, 2021–2050 and 2071–2100 for two considered climate scenarios.

Taking into account changes in discharges evaluated on the basis of RCP2.6, hydropower production in the future can be equal or even higher than it is currently. In the near future, hydropower production at the power plants with large turbine gullets can increase by up to 5% in comparison to 2015–2020. In 2071–2100, it can additionally increase by 1–2% (Table 14). Quite different conditions are simulated for RCP8.5 (Table 15). In this case, a significant decrease in hydropower production is modeled, especially for the last decades of the century. In 2021–2050, the production can fall by 4–9% in comparison to 2015–2020, while for the far future (2071–2100) it can additionally decrease by even 25%. In the case of the last period, the hydropower stations with large turbine gullets can be characterized by energy production reaching only 34% of the nominal potential (Nieder-Neundorf, Lodenau). In Table 16 we present the projected values of hydropower production for particular hydropower stations and the ratio of its decrease between 2015–2020 and 2071–2100 for RCP8.5. According to these projections, almost all hydropower stations are characterized by a decrease of 31–34%. The hydropower station of Obermühle is the only one with a relatively low decrease, reaching 16%. This shows that in the case of the RCP8.5 scenario, potential energy supply by the hydropower stations can be seriously limited.

Table 14. Projected annual hydropower production in 2015–2020, 2021–2050 and 2071–2100 and its share [%] in nominal hydropower production at German and Polish hydropower plants on the Lusatian Neisse River, on the basis of mean discharges for the RCP2.6 scenario.

Hydropower Station	2015–2020		2021–2050		2071–2100	
	kWh/year	%	kWh/year	%	kWh/year	%
Obermühle	948,522.78	100	948,522.78	100	948,522.78	100
Vierradmühle	1,560,487.37	78	1,718,204.79	86	1,760,447.08	88
Ludwigsdorf	1,486,323.25	37	1,636,544.95	41	1,676,779.62	42
Pieńsk	2,694,622.68	87	2,966,966.40	96	3,039,909.65	98
Nieder-Neundorf	2,445,238.00	38	2,692,376.57	42	2,758,569.00	43
Bremenwerk	2,045,497.89	41	2,252,235.00	45	2,307,606.49	47
Lodenau	3,216,438.51	39	3,541,521.80	42	3,628,590.57	44
Sobolice	2,301,401.41	42	2,534,002.52	47	2,596,301.30	48
Bukówka	3,882,551.23	48	4,274,958.09	53	4,380,058.49	54

Table 15. Projected annual hydropower production in 2015–2020, 2021–2050 and 2071–2100 and its share [%] in nominal hydropower production at German and Polish hydropower plants on the Lusatian Neisse River, on the basis of mean discharges for the RCP8.5 scenario.

Hydropower Station	2015–2020		2021–2050		2071–2100	
	kWh/year	%	kWh/year	%	kWh/year	%
Obermühle	948,522.78	100	948,522.78	100	798,329.41	84
Vierradmühle	1,856,466.85	93	1,675,587.65	84	1,233,678.00	62
Ludwigsdorf	1,768,235.93	44	1,595,953.24	40	1,175,045.90	29
Pieńsk	3,103,993.87	100	2,893,375.85	93	2,130,293.90	69
Nieder-Neundorf	2,909,029.17	45	2,625,596.75	41	1,933,137.29	30
Bremenwerk	2,433,469.89	49	2,196,372.14	44	1,617,113.85	33
Lodenau	3,826,504.19	46	3,453,680.38	41	2,542,827.00	31
Sobolice	2,737,910.93	50	2,471,150.89	45	1,819,424.07	33
Bukówka	4,618,959.30	57	4,168,925.03	52	3,069,437.23	38

Table 16. Projected nominal hydropower production [%] at German and Polish hydropower plants on the Lusatian Neisse River and its decrease between the 2015–2020 and 2071–2100 periods, according to the RCP8.5 scenario

Hydropower Station	Nation	2015–2020 [%]	2071–2100 [%]	Decrease [%]
Obermühle	Germany	100	84	16
Vierradmühle	Germany	93	62	33
Ludwigsdorf	Germany	44	29	34
Pieńsk	Poland	100	69	31
Nieder-Neundorf	Germany	45	30	33
Bremenwerk	Germany	49	33	33
Lodenau	Germany	46	31	33
Sobolice	Poland	50	33	34
Bukówka	Poland	57	38	33

4.4. Risk Analysis

In terms of probability of changes in air temperature and sunshine duration in the future, climate scenarios usually indicate an increase in their values. Previously discussed differences in air temperature and sunshine duration between the future and current period showed that projected values for the future are in all cases higher than simulations for the reference period.

Mean projected annual air temperature for 1971–2015 for both RCP2.6 and RCP8.5 is equal to 8.8 °C. In case of future thermal conditions, projected mean annual air temperature for 2021–2050 and 2071–2100 amounts to 9.8 °C (for both periods) for RCP2.6 and 10.4 °C and 12.4 °C for RCP8.5. Thus, the level of local warming in all cases is characterized by positive values (Table 17). Considering the criteria presented in Table 4, it is equal to 1.0 °C (W1 and W3), 1.6 °C (W2) and 3.6 °C (W4). Consequently, the variable of air temperature, according to the criteria presented in Table 5, was given the highest (5) probability rank. It should be emphasized that an additional factor that confirms the high probability of changes in air temperature is its course for the observation period of 1971–2015. It is characterized by a noticeably increasing, statistically important trend. Furthermore, if such a trend continues, mean annual air temperature in the future will be similar to projected values in terms of RCP8.5. Thus, the current tendency of air temperature suggests a realization of the pessimistic scenario, related to a constant and high increase in air temperature by the end of the century.

Table 17. Projected levels of changes in air temperature in the future against 1971–2000, on the basis of the data for Zielona Góra.

Level of Changes in Air Temperature	[°C]
W1	1.0
W2	1.6
W3	1.0
W4	3.6

In case of sunshine duration, its mean annual values for 1971–2000, assessed on the basis of simulated data, are similar for both scenarios (1511–1519 h), whereas the projected annual sunshine duration for the near and far future is equal to 1590 and 1577 h for RCP2.6 and 1699 and 1859 h in the case of RCP8.5. Thus, similarly to air temperature, all of the levels of local changes in sunshine duration are characterized by positive values and vary from 66 h (W3) to 340 h (W4) (Table 18). Thus, the probability of changes in sunshine duration was also given the highest rank (5). Similarly to air temperature, sunshine duration, in terms of observation data for 1971–2015, was also characterized by an increasing, statistically important trend, which additionally proves that this variable can still increase in the future. If the current trend continues, an increase in this variable in the future can be even higher than simulated within the climate scenarios.

Table 18. Projected levels of changes in sunshine duration in the future in comparison to 1971–2000, on the basis of the data for Zielona Góra.

Level of Changes in Sunshine Duration	Hours
W1	79
W2	180
W3	66
W4	340

In the case of precipitation totals, predicted changes are not as significant as for air temperature and sunshine duration. Table 19 presents the changes for the stations representing the lowlands (Zielona Góra, Sobolice) and the mountain foreland (Sulików, Bierna) of the discussed region. In most cases, changes in precipitation are related to a decrease in relation to projected totals for 1971–2000. In the case of W4, changes reaching or exceeding 5% are expected for Sobolice, Sulików and Bierna. Additionally, the value of 5% is also reached for W2 in Bierna. Therefore, four of 16 results for the considered stations are characterized by a decrease in precipitation totals reaching at least 5%. As a result, the probability of negative changes in precipitation totals was given the rank of 2 (25%). Unlike air temperature and sunshine duration, the current increasing trend of precipitation totals (without statistical significance) for 1971–2015 does not correspond to the future projections. It confirms that future trends for precipitation are difficult to assess and are not as obvious as in the case of air temperature and sunshine duration.

Table 19. Projected levels of changes in precipitation totals in the future compared to 1971–2000, on the basis of data for Zielona Góra, Sobolice, Sulików and Bierna.

Level of Changes in Precipitation Totals	Zielona Góra [%]	Sobolice [%]	Sulików [%]	Bierna [%]
W1	1	0	1	−1
W2	−2	−1	−1	−5
W3	−1	0	0	−2
W4	−2	−5	−8	−13

Taking into consideration changes in air temperature, sunshine duration and precipitation, consequences for water resources availability and hydropower production were evaluated. Based on criteria presented in Table 8, mean maximal and minimal discharges projected within RCP8.5, especially for 2071–2100, can be more than 35% lower than in 1971–2000. Therefore, the total consequences rate for water resources was estimated at 4. In the case of future hydropower production, most of the hydropower plants in 2071–2100 can be characterized by a decrease in comparison to 2015–2020 exceeding 30% (Table 16). Thus, the rate of consequences for hydropower production, according to criteria presented in Table 9, is equal to 5. Considering the highest probability of changes in air temperature and sunshine duration in the future and potential consequences for water resources and hydropower (production, the risk value was estimated at 20 for water resources and at 25 for hydropower production (Table 20). Therefore, the highest risk rank was given to the considered sectors, which means that appropriate measures should be implemented in order to mitigate the effects of climate change for water resources and hydropower production. The potential impact of precipitation changes on both water resources and hydropower production is characterized by medium risk (8 for water resources and 10 for hydropower production) due to there being no significant changes simulated for the future periods.

Table 20. Risk matrix for consequences for water resources and hydropower production under the probability of changes in air temperature and sunshine duration.

Sector	Air Temperature			Sunshine Duration			Precipitation Totals		
	P	C	R	P	C	R	P	C	R
Water resources	5	4	20	5	4	20	2	4	8
Hydropower production	5	5	25	5	5	25	2	5	10

5. Discussion

Presented results on the water resources problems under climate change conditions show that significant changes are expected in the future in terms of water-related conditions. The analysis showed that a considerable increase in air temperature and sunshine duration in the remaining decades of the 21st century in the Lusatian Neisse river basin could be accompanied by a decrease in river discharges. In the near future (2021–2050), projected discharges for both scenarios amount to 13–17% and are about 2–5% higher in comparison to some projections for Swiss regions [53]. Projections for the far future (2071–2100) indicate a very high decrease in mean maximum and minimum discharges (36–38%), which would significantly contribute to limits in water availability. Projected discharges for the Lusatian Neisse River can be even lower than environmental flow, which may contribute to serious ecological problems. As much as 18% of discharges could be below environmental flow, especially according to RCP8.5. Therefore, the results for both water resources and ecology at the end of the century can be very noticeable. On the other hand, RCP2.6 shows a more optimistic scenario, with only 2% of discharges lower than environmental flow. Similar results were noticed for other regions, where effects related to environmental flow vary from negligible to disastrous [47].

A similar situation is observed for hydropower production by hydropower plants located on the Lusatian Neisse River. Under RCP2.6, hydropower production will still be kept at a normal level or even increase, while RCP8.5 indicates a significant decrease of as much as 34% of nominal production (in 2071–2100). The results for the near future (2021–2050) are comparable to some alpine regions for 2031–2050 [51]. In the Alps, hydropower production could increase by about 5% (the most optimistic scenario) or decrease by 7% (the most pessimistic scenario). In the region of the Lusatian Neisse river basin, huge energy problems can occur in the region at the end of 21st century if the pessimistic scenario of RCP8.5 is realized. Hydropower production between 2015–2020 and 2071–2100 can drop by over 30%, which is a higher rate than, e.g., in case of the Rhone River, where the maximal projected decrease between 2001–2010 and 2090–2100 reaches 25% [50].

Taking into consideration climate projections carried out for the RCP2.6 and RCP8.5 scenarios, it should be emphasized that results for these scenarios (especially for RCP8.5) are currently under discussion [75]. RCP8.5 was considered to be an extreme scenario, representing the 90th percentile of no policy baseline scenarios. It assumes an increase in coal use to become a major source of power generation; however the use of coal in the recent years has decreased [76]. Thus, the probability of RCP8.5 scenario realization seems to be smaller than was previously assumed. Nevertheless, current trends of some meteorological variables in the region of the Lusatian Neisse river basin (e.g., for air temperature) show similar tendencies to those modeled for RCP8.5. In the case of the discussed region, values simulated for future periods (on the basis of 1971–2000 data series) are comparable with the results of the trends carried out for measurement data for 1971–2015. Therefore, considering potential climate change in the future, the results of the extreme scenarios cannot be neglected. On the other hand, one of the latest reports of the IPCC [77] assumes a limitation of global warming to 1.5 °C in comparison to the preindustrial era, which is more optimistic than the limitation implemented for RCP2.6 (2 °C). Thus, taking into consideration the low probability of RCP8.5 realization and new opportunities related to a stronger limitation of global warming, the risks for water resources and hydropower production in the discussed region could be lower than presented in the paper.

6. Conclusions

The development of future conditions related to water resources and hydropower production is strongly dependent on climate scenarios. Current trends indicate that today's changes in discussed meteorological variables are more similar to the RCP8.5 scenario. Taking into consideration the presented results, increasing air temperatures and sunshine durations can negatively affect water resources in the future, even though precipitation projections do not show significant trends. The results of the analysis and current low water resources in Poland indicate that appropriate measures should be undertaken in order to use water resources in accordance with sustainable water management. Based on the presented results on meteorological and hydrological conditions as well as on climate projections, the following recommendations can be given in order to mitigate the impact of climate change on water resources in the region:

- Analysis of legal regulations and their adaptations. The most important adaptations measures refer to legal aspects, i.e., verification of legal water permits and their adaptation to a new Water Law. The permits should also consider pond economy and the aspect of hydropower stations.
- Water retention and sustainable use of water resources. Aspects of retention capacity of the basin and coordination of the flooding of former opencast mine pits should be taken into consideration. Furthermore, development of "green-blue" and "grey" water infrastructures would contribute to significant water savings.
- Changes in river valleys. In these terms, actions focused on the creation of buffer zones and reclamation of river valleys and riverbeds should be undertaken. In the reclamation process, the very important problem of invasive species should be considered.
- Changes in strategic documents of the local governments. The documents should consider the aspects of climate change in the context of their impact on the inhabitants and the economy. Such documents should be systematically updated. Strategies on sustainable water use should also be developed.
- Improvement in cooperation between bodies dealing with water management and water energy production. Such cooperation would enable the development of a common approach towards water management in terms of hydropower plants. Furthermore, an improvement in power plant technical measures would contribute to a higher efficiency of hydropower production.
- Increasing awareness in terms of current and future water resources. These types of measures concern promotional actions focused on the transfer of information to inhabitants, local authorities and persons dealing with water management in the region.

Author Contributions: Conceptualization, M.A.-P.; methodology, M.A.-P. and B.M.; software, M.A.-P.; validation M.A.-P.; formal analysis, M.A.-P. and B.M.; investigation, M.A.-P. and B.M.; resources, B.M.; data curation, M.A.-P. and B.M.; writing—original draft preparation, B.M.; writing—review and editing, M.A.-P.; visualization, M.A.-P. and B.M.; supervision, M.A.-P.; project administration, M.A.-P.; funding acquisition, M.A.-P. All authors have read and agreed to the published version of the manuscript.

Funding: The Ministry of Science and Higher Education (MNiSW) (Grant no. DS-W1/2017).

Conflicts of Interest: The authors declare no conflict of interest.

References

1. Gutry-Korycka, M.; Sadurski, A.; Kundzewicz, Z.W.; Pociask-Karteczka, B.; Skrzypczyk, L. Zasoby Wodne i Ich Wykorzystanie. *Nauka* **2014**, *1*, 77–98.

2. Kundzewicz, Z.W.; Piniewski, M.; Mezghani, A.; Okruszko, T.; Pińskwar, I.; Kardel, I.; Hov, Ø.; Szcześniak, M.; Szwed, M.; Benestad, R.E.; et al. Assessment of climate change and associated impact on selected sectors in Poland. *Acta Geophys.* **2018**, *66*, 1509–1523. [CrossRef]

3. Adynkiewicz-Piragas, M.; Zdralewicz, I.; Otop, I.; Miszuk, B.; Kryza, J.; Lejcuś, I.; Strońska, M.; Lunich, K.; Prasser, M.; Niemand, C. *Lausitzer Neise-Wasserressourcen in der Region. 2*; Broschüre im EU-Projekt NEYMO; Sachsisches Landesamt fur Umwelt, Landwirtschaft und Geologie: Dresden, Germany, 2014; Volume 139.

4. Adynkiewicz-Piragas, M.; Zdralewicz, I.; Otop, I.; Miszuk, B.; Kryza, J.; Lejcuś, I.; Strońska, M.; Lunich, K.; Prasser, M.; Niemand, C. *Nysa Łużycka–Klimat i Charakterystyka Region*; Sachsisches Landesamt fur Umwelt, Landwirtschaft und Geologie: Dresden, Germany, 2014; Volume 73.

5. Pluntke, T.; Schwarzak, S.; Kuhn, K.; Lünich, K.; Adynkiewicz-Piragas, M.; Otop, I.; Miszuk, B. Climate analysis as a basis for a sustainable water management at the Lusatian Neisse. *Meteorol. Hydrol. Water Manag.* **2016**, *4*, 3–11. [CrossRef]

6. Mehler, S.; Völlings, A.; Flügel, I.; Szymanowski, M.; Błaś, M.; Sobik, M.; Migała, K.; Werner, M.; Kryza, M.; Kryza, M.; et al. *Zmiany Klimatu w Regionie Granicznym Polski i Saksonii*; Sächsisches Landesamt für Umwelt, Landwirtschaft und Geologie: Dresden, Germany, 2014; Volume 85.

7. Kreienkamp, F.; Spekat, A.; Enke, W. *Klimaprojektionen im Projekt KLAPS. Publikationsreihe des EU-Projektes KLAPS–Klimawandel, Luftverschmutzung und Belastungsgrenzen von Ökosystemen im Polnisch-Sächsischen Grenzraum*; Sächsisches Landesamt für Umwelt, Landwirtschaft und Geologie: Dresden, Germany, 2013; Volume 3.

8. Schwarzak, S.; Völlings, A.; Surke, M.; Kryza, M.; Szymanowski, M.; Błaś, M.; Werner, M.; Sobik, M.; Migała, K.; Miszuk, B.; et al. *Projekcje klimatu, zanieczyszczenia powietrza i ładunki krytyczne w regionie granicznym Polski i Saksonii*; Sächsisches Landesamt für Umwelt, Landwirtschaft und Geologie: Dresden, Germany, 2014; Volume 90.

9. Hattermann, F.F.; Weiland, M.; Huang, S.; Krysanova, V.; Kundzewicz, Z.W. Model-Supported Impact Assessment for the Water Sector in Central Germany Under Climate Change—A Case Study. *Water Resour. Manag.* **2011**, *25*, 3113–3134. [CrossRef]

10. Torretta, V. Environmental and economic aspects of water kiosks: Case study of a medium-sized Italian town. *Waste Manag.* **2013**, *33*, 1057–1063. [CrossRef]

11. Kazak, J.K.; Szkaradkiewicz, M.; Sabura, K.; Oleszek, J. Changes in suburban development in the context of wastewater management. *Acta Sci. Pol. Form. Circumiectus* **2015**, *13*, 133–139. [CrossRef]

12. Torretta, V. The Sustainable Use of Water Resources: A Technical Support for Planning. A Case Study. *Sustainability* **2014**, *6*, 8128–8148. [CrossRef]

13. Stojanov, R.; Duží, B.; Daněk, T.; Němec, D.; Procházka, D. Adaptation to the Impacts of Climate Extremes in Central Europe: A Case Study in a Rural Area in the Czech Republic. *Sustainability* **2015**, *7*, 12758–12786. [CrossRef]

14. Szewrański, S.; Kazak, J.; Szkaradkiewicz, M.; Sasik, J. Flood risk factors in suburban area in the context of climate change adaptation policies–case study of Wrocław, Poland. *J. Ecol. Eng.* **2015**, *16*, 13–18. [CrossRef]

15. Röthlisberger, V.; Zischg, A.P.; Keiler, M. Identifying spatial clusters of flood exposure to support decision making in risk management. *Sci. Total. Environ.* **2017**, *598*, 593–603. [CrossRef]

16. Kazak, J.K.; Chruściński, J.; Szewrański, S. The Development of a Novel Decision Support System for the Location of Green Infrastructure for Stormwater Management. *Sustainability* **2018**, *10*, 4388. [CrossRef]

17. Malinowski, Ł.; Skoczko, I. Impacts of Climate Change on Hydrological Regime and Water Resources Management of the Narew River in Poland. *J. Ecol. Eng.* **2018**, *19*, 167–175. [CrossRef]

18. Szewrański, S.; Chruściński, J.; Kazak, J.K.; Świąder, M.; Tokarczyk-Dorociak, K.; Żmuda, R. Pluvial Flood Risk Assessment Tool (PFRA) for Rainwater Management and Adaptation to Climate Change in Newly Urbanised Areas. *Water* **2018**, *10*, 386. [CrossRef]

19. Szewrański, S.; Świąder, M.; Kazak, J.K.; Tokarczyk-Dorociak, K.; Van Hoof, J. Socio-Environmental Vulnerability Mapping for Environmental and Flood Resilience Assessment: The Case of Ageing and Poverty in the City of Wrocław, Poland. *Integr. Environ. Assess. Manag.* **2018**, *14*, 592–597. [CrossRef]

20. Chen, N.; Yao, S.; Wang, C.; Du, W. A Method for Urban Flood Risk Assessment and Zoning Considering Road Environments and Terrain. *Sustainability* **2019**, *11*, 2734. [CrossRef]

21. Kim, Y.; Newman, G. Climate Change Preparedness: Comparing Future Urban Growth and Flood Risk in Amsterdam and Houston. *Sustainability* **2019**, *11*, 1048. [CrossRef]

22. Piana, P.; Faccini, F.; Luino, F.; Paliaga, G.; Sacchini, A.; Watkins, C. Geomorphological Landscape Research and Flood Management in a Heavily Modified Tyrrhenian Catchment. *Sustainability* **2019**, *11*, 4594. [CrossRef]

23. Sojka, M.; Kozłowski, M.; Stasik, R.; Napierała, M.; Kęsicka, B.; Wróżyński, R.; Jaskuła, J.; Liberacki, D.; Bykowski, J. Sustainable Water Management in Agriculture—The Impact of Drainage Water Management on Groundwater Table Dynamics and Subsurface Outflow. *Sustainability* **2019**, *11*, 4201. [CrossRef]

24. Urbančíková, N.; Zgodavova, K. Sustainability, Resilience and Population Ageing along Schengen's Eastern Border. *Sustainability* **2019**, *11*, 2898. [CrossRef]

25. Zhou, Q.; Su, J.; Leng, G.; Peng, J. The Role of Hazard and Vulnerability in Modulating Economic Damages of Inland Floods in the United States Using a Survey-Based Dataset. *Sustainability* **2019**, *11*, 3754. [CrossRef]

26. Niczyporuk, Z.T. Safety Management in Coal Mines—Risk Assessment. *Int. J. Occup. Saf. Ergon.* **1996**, *2*, 243–250. [CrossRef] [PubMed]

27. Smolarkiewicz, M.; Smolarkiewicz, M.M.; Biedugnis, S. Matrix methods for risk management – Associated Matrices Theory. *Rocz. Ochr. Środowiska* **2011**, *13*, 241–252.

28. Ristic, D. A Tool for Risk Assessment. *Saf. Eng.* **2013**, *3*, 121–127. [CrossRef]

29. Wilkinson, M.E.; Quinn, P.F.; Hewett, C.J.M. The Floods and Agriculture Risk Matrix: A decision support tool for effectively communicating flood risk from farmed landscapes. *Int. J. River Basin Manag.* **2013**, *11*, 237–252. [CrossRef]

30. Iacob, V.S. Risk management and evaluation and qualitative method within the projects. *ECOFORUM* **2014**, *3*, 60–67.

31. Ale, B.; Burnap, P.; Slater, D. On the origin of PCDS—(Probability consequence diagrams). *Saf. Sci.* **2015**, *72*, 229–239. [CrossRef]

32. Ale, B.; Burnap, P.; Slater, D. Risk Matrix Basics. Available online: https://www.researchgate.net/publication/305889949_Risk_Matrix_Basics (accessed on 15 April 2020).

33. Dziadosz, A.; Rejment, M. Risk analysis in construction project–chosen methods. *Procedia Eng.* **2015**, *122*, 258–265. [CrossRef]

34. Roguska, E.; Lejk, J. Fuzzy risk matrix as a risk assessment method—A case study. 'SEE Tunnel: Promoting tunneling in SEE Region'. In Proceedings of the ITA WTC 2015 Congress 41st General Assembly, Lacroma Valamar Congress Center, Dubrovnik, Croatia, 22–28 May 2015.

35. Duijm, N.J. Recommendations on the use and design of risk matrices. *Saf. Sci.* **2015**, *76*, 21–31. [CrossRef]

36. Cobon, D.H.; Williams, A.A.J.; Power, B.; McRae, D.; Davis, P. Risk matrix approach useful in adapting agriculture to climate change. *Clim. Chang.* **2016**, *138*, 173–189. [CrossRef]

37. Peace, C. The risk matrix: Uncertain results? *Policy Pract. Health Saf.* **2017**, *15*, 131–144. [CrossRef]

38. Kowalski, J.; Polonski, M. Identification of risk investment using the riskmatrix on railway facilities. *Open Eng.* **2018**, *8*, 506–512. [CrossRef]

39. Napolitano, D.M.R.; Romero, M.; Sassi, R.J. Suport vector machines in projects risk classification. *Int. J. Dev. Res.* **2018**, *8*, 23915–23920.

40. Domínguez, C.R.; Martínez, I.V.; Peña, P.M.P.; Ochoa, A.R. Analysis and evaluation of risks in underground mining using the decision matrix risk-assessment (DMRA) technique, in Guanajuato, Mexico. *J. Sustain. Min.* **2019**, *18*, 52–59. [CrossRef]

41. Lubowiecka, T.; Wieczysty, A. Ryzyko w systemach zaopatrzenia w wodę. In *Ryzyko w Gospodarce Wodnej*; Maciejewski, M., Ed.; Monografia Komitetu Gospodarki Wodnej PAN: Warszawa, Poland, 2000; Volume 17, pp. 113–143.

42. Tchórzewska-Cieślak, B.; Pietrucha-Urbanik, K.; Szpak, D. Analysis of water consument safety arising from hazard in rural waterworks. *Inż. Ekolog.* **2016**, *48*, 208–213. [CrossRef]

43. Tchórzewska-Cieślak, B.; Pietrucha-Urbanik, K.; Zygmunt, A. Implementation of matrix methods in flood risk analysis and assessment. *Ekon. Środowisko* **2018**, *3*, 8–24.

44. Ozga-Zielinski, B.; Adamowski, J.; Ciupak, M. Applying the Theory of Reliability to the Assessment of Hazard, Risk and Safety in a Hydrologic System: A Case Study in the Upper Sola River Catchment, Poland. *Water* **2018**, *10*, 723. [CrossRef]

45. Dyer, F.; Eisawach, S.; Croke, B.; Griffiths, R.; Harrison, E.; Lucena-Moya, P.; Jakeman, A. The effects of climate change on ecologically-relevant flow regimeand water quality attributes. *Stoch. Environ. Res. Risk Assess.* **2014**, *28*, 67–82. [CrossRef]

46. Piniewski, M.; Laize, C.L.R.; Acreman, M.; Okruszko, T.; Schneider, C. Effect of Climate Change on Environmental Flow Indicatiors in the Narew Basin, Poland. *J. Environ. Qual.* **2014**, *43*, 155–167. [CrossRef]

47. Thompson, J.; Laize, C.; Green, A.; Acreman, M.; Kingston, D. Climate change uncertainty in environmental flows for the Mekong River. *Hydrol. Sci. J.* **2014**, *59*, 935–954. [CrossRef]

48. Salik, K.M.; Hashmi, M.Z.-U.-R.; Ishfaq, S.; Zahdi, W.-U.-Z. Environmental flow requirements and impacts of climate change-induced river flow changes on ecology of the Indus Delta, Pakistan. *Reg. Stud. Mar. Sci.* **2016**, *7*, 185–195. [CrossRef]

49. Hamududu, B.H.; Jjunju, W.; Killinkgtveit, A. Existing Studies of Hydropower and climate change: A review. In Proceedings of the Conference: Hydropower 2010 Tromsø, 6th International Conference on Hydropower. Hydropower supporting other Renewables, Tromsø, Norway, 1–3 February 2010.

50. Gaudard, L.; Gilli, M.; Romerio, F. Climate Change Impacts on Hydropower Management. *Water Resour. Manag.* **2013**, *27*, 5143–5156. [CrossRef]

51. Wagner, T.; Themeßl, M.; Schüppel, A.; Gobiet, A.; Stigler, H.; Birk, S. Impacts of climate change on stream flow and hydro power generation in the Alpine region. *Environ. Earth Sci.* **2016**, *76*, 4. [CrossRef]

52. Anghileri, D.; Botter, M.; Castelletti, A.; Weigt, H.; Burlando, P. A Comparative Assessment of the Impact of Climate Change and Energy Policies on Alpine Hydropower. *Water Resour. Res.* **2018**, *54*, 9144–9161. [CrossRef]

53. Ranzani, A.; Bonato, M.; Patro, E.R.; Gaudard, L.; De Michele, C. Hydropower Future: Between Climate Change, Renewable Deployment, Carbon and Fuel Prices. *Water* **2018**, *10*, 1197. [CrossRef]

54. Shu, J.; Qu, J.J.; Motha, R.; Xu, J.C.; Dong, D.F. Impacts of climate change on hydropower development and sustainability: A review. In Proceedings of the IOP Conference Series: Earth and Environmental Science, Bejing, China, 16–19 November 2017; Volume 163, p. 012126. [CrossRef]

55. Solaun, K.; Cerdá, E. Climate change impacts on renewable energy generation. A review of quantitative projections. *Renew. Sustain. Energy Rev.* **2019**, *116*, 109415. [CrossRef]

56. Teotónio, C.; Fortes, P.; Roebeling, P.; Rodriguez, M.; Robaina, M. Assessing the impacts of climate change on hydropower generation and the power sector in Portugal: A partial equilibrium approach. *Renew. Sustain. Energy Rev.* **2017**, *74*, 788–799. [CrossRef]

57. Bombelli, G.M.; Soncini, A.; Bianchi, A.; Bocchiola, D. Potentially modified hydropower production under climate change in the Italian Alps. *Hydrol. Process.* **2019**, *33*, 2355–2372. [CrossRef]

58. Molarius, R.; Keränen, J.; Schabel, J.; Wessberg, N. Creating a climate change risk assessment procedure: Hydropower plant case, Finland. *Hydrol. Res.* **2010**, *41*, 282–294. [CrossRef]

59. Alexandrovskii, A.; Tereshin, A.; Klimenko, V.; Fedotova, E.; Pugachev, R. Estimation of Hydropower Plants Energy Characteristics Change under the Influence of Climate Factors. *Adv. Eng. Res.* **2019**, *191*, 7–13.

60. Carless, D.; Whitehead, P.G. The potential impacts of climate change on hydropower generation in Mid Wales. *Hydrol. Res.* **2012**, *44*, 495–505. [CrossRef]

61. Savelsberg, J.; Schillinger, M.; Schlecht, I.; Weigt, H. The Impact of Climate Change on Swiss Hydropower. *Sustainability* **2018**, *10*, 2541. [CrossRef]

62. IPCC. Climate Change 2013: The Physical Science Basis. Contribution of Working Group I to the Fifth Assessment Report of the Intergovernmental Panel on Climate Change. Available online: https://www.ipcc.ch/report/ar5/wg1/ (accessed on 15 April 2020).

63. Pulido-Velazquez, D.; Collados-Lara, A.; García, F.J. Assessing impacts of future potential climate change scenarios on aquifer recharge in continental Spain. *J. Hydrol.* **2018**, *567*, 803–819. [CrossRef]

64. Watanabe, S.; Kanae, S.; Seto, S.; Yeh, P.J.-F.; Hirabayashi, Y.; Oki, T. Intercomparison of bias-correction methods for monthly temperature and precipitation simulated by multiple climate models. *J. Geophys. Res. Space Phys.* **2012**, *117*, 23114. [CrossRef]

65. Collados-Lara, A.; Pulido-Velazquez, D.; Pardo-Igúzquiza, E. An Integrated Statistical Method to Generate Potential Future Climate Scenarios to Analyse Droughts. *Water* **2018**, *10*, 1224. [CrossRef]

66. Räty, O.; Räisänen, J.; Ylhäisi, J.S. Evaluation of delta change and bias correction methods for future daily precipitation: Intermodel cross-validation using ENSEMBLES simulations. *Clim. Dyn.* **2014**, *42*, 2287–2303. [CrossRef]

67. Muerth, M.; St-Denis, B.G.; Ricard, S.; Velázquez, J.A.; Schmid, J.; Minville, M.; Caya, D.; Chaumont, D.; Ludwig, R.; Turcotte, R. On the need for bias correction in regional climate scenarios to assess climate change impacts on river runoff. *Hydrol. Earth Syst. Sci.* **2013**, *17*, 1189–1204. [CrossRef]

68. Lazoglou, G.; Anagnostopoulou, C.; Skoulikaris, C.; Tolika, K. Bias Correction of Climate Model's Precipitation Using the Copula Method and Its Application in River Basin Simulation. *Water* **2019**, *11*, 600. [CrossRef]

69. Van Vuuren, D.; Edmonds, J.; Kainuma, M.; Riahi, K.; Thomson, A.; Hibbard, K.; Hurtt, G.C.; Kram, T.; Krey, V.; Lamarque, J.-F.; et al. The representative concentration pathways: An overview. *Clim. Chang.* **2011**, *109*, 5–31. [CrossRef]

70. Kostrzewa, H. *Verification of Criteria and Inviolable Flow Volume for Polish Rivers*; Series: Water Management and Water Protection; Institute of Meteorology and Water Management; IMGW: Warsaw, Poland, 1977; Volume 208.

71. MIKE HYDRO Basin. User Guide, DHI. Available online: https://manuals.mikepoweredbydhi.help/2017/Water_Resources/MIKEHydro_Basin_UserGuide.pdf (accessed on 19 June 2020).

72. Shamsudin, S.; Hashim, N. Rainfall runoff simulation using MIKE11 NAM. *J. Civ. Eng.* **2002**, *15*, 1–13.

73. Dubicki, A. *Ilościowy i Jakościowy Bilans Wodnogospodarczy Nysy Łużyckiej*; IMGW O. Wrocław: Wrocław, Poland, 2001.

74. Lobanova, A.; Liersch, S.; Nunes, J.; Didovets, I.; Stagl, J.; Huang, S.; Koch, H.; López, M.D.R.R.; Maule, C.F.; Hattermann, F.; et al. Hydrological impacts of moderate and high-end climate change across European river basins. *J. Hydrol. Reg. Stud.* **2018**, *18*, 15–30. [CrossRef]

75. Hausfather, Z.; Peters, G.P. Emissions—The 'business as usual' is misleading. *Nature* **2020**, *577*, 618–620. [CrossRef] [PubMed]

76. Kummer, L. A Closer Look at Scenario RCP8.5, Climate Etc. 2015. Available online: https://judithcurry.com/2015/12/13/a-closer-look-at-scenario-rcp8-5/ (accessed on 14 May 2020).

77. IPCC, Global Warming of 1,5oC, An IPCC Special Report on the Impacts of Global Warming of 1.5 °C above Pre-Industrial Levels and Related Global Greenhouse Gas Emission Pathways, in the Context of Strengthening the Global Response to the Threat of Climate Change, Sustainable Development, and Efforts to Eradicate Poverty. 2018. Available online: https://www.ipcc.ch/sr15/ (accessed on 14 April 2020).

 *sustainability*

Article

Cycling as a Sustainable Transport Alternative in Polish Cittaslow Towns

Agnieszka Jaszczak [1,*], Agnieszka Morawiak [2] and Joanna Żukowska [2]

[1]   Department of Landscape Architecture, University of Warmia and Mazury, 10-719 Olsztyn, Poland
[2]   Faculty of Civil and Environmental Engineering, Gdansk University of Technology, 80-233 Gdansk, Poland;
      agnieszka.morawiak@pg.edu.pl (A.M.); joanna.zukowska@pg.edu.pl (J.Ż.)
*   Correspondence: agnieszka.jaszczak@uwm.edu.pl

Received: 24 May 2020; Accepted: 18 June 2020; Published: 20 June 2020

Abstract: It is well known that growing motor traffic in urban areas causes air pollution and noise which affects the environment and public health. It is hardly surprising then that cycling should be used as an alternative mode of transport, not just in major cities but also in smaller ones including those that are members of the Cittaslow network. Their approach is based on sustainable development, care for the environment and transport solutions which will support a healthy lifestyle, reduced energy consumption and fewer emissions. The objective of the article is to analyse how well cycling is used as a means of transport in Polish Cittaslow towns. For this purpose, an analysis was conducted to understand how towns use their transport space to ensure accessibility and road safety. Reference is made to revitalisation programmes of Cittaslow towns with focus on what has been done to improve and build cycle paths in each town and outside of it. The work uses the following research methods: analysis of the literature, analysis of documents, including analysis of road incidents and traffic count. It has been demonstrated that cycling infrastructure in the towns under analysis has been marginalised. As a result, recommendations and suggestions are given which may inform decisions on how to build and transform cycling infrastructure in Cittaslow towns and in similar towns in Poland and abroad.

Keywords: cycling; cycling routes; slow cities; traffic safety; sustainable mobility; urban planning

---

1. Introduction

Many high-growth regions worldwide have developed as a result of globalisation, urbanisation and sub-urbanisation processes, which also has an effect on transport and accessibility. People living in big conurbations have less and less time and make frequent and multiple trips during the day. The result is an increasing number of cars in cities. As a consequence, transport networks have to be built. This is a common problem, especially in city centres and dense developments [1–5]. Exhaust and CO_2 emissions rise, making cities increasingly more polluted [6]. In addition, motor transport increases noise levels which has a direct effect on people's health and well-being [7]. Given the strong need to save natural resources and care for the environment and people's health, it is important to operate urban transport policies that are based on alternative environmental solutions. These are largely related to cycling infrastructure which needs to be accessible and safe. While big cities understand this and have already initiated or completed relevant projects, smaller towns lack a clear plan for integrating transport services with networks of cycling roads and ensuring the availability of bicycles.

The purpose of the work was to analyse the structure of how bicycles are used and how Cittaslow towns develop their cycling infrastructure in the context of safety, a better quality of the environment and living standards, and reducing dependence on what are now limited sources of energy. The article describes the area of the research, research methods and the results. The discussion looks at the degree of bicycle usage, the availability of cycle paths and cycling infrastructure, an analysis of cyclist accidents and an analysis of the Revitalisation Programme of Cittaslow Towns as regards the planning of cycling infrastructure. Based on the results, guidelines and recommendations have been formulated, followed by a summary. The following were the assumptions:

- The research looks at the environmental needs (fewer sources of pollution, noise, a better energy mix), safety (reduction in cyclist accidents), pro-health aspects (better health and fitness of the population) and spatial and infrastructural needs (improved access to cycling infrastructure) of small towns.
- The research was carried out in small towns because infrastructure and safety analyses and data were not available for these sites, unlike Poland's major cities and conurbations. Cittaslow towns were selected on purpose due to their unique character and programmes which give priority to the environment, reduction in emissions, improved energy mix, better safety and health standards of the population.
- The research units were selected specifically for not having or having hardly any cycling infrastructure or cycle paths (per km).
- There is a close relation between a lack of infrastructure or poor access to infrastructure and cycling as well as road safety in towns and especially outside them (e.g., for commuters).
- The research takes account of the importance of daily cycling by residents of small towns and making sure that the infrastructure meets those needs rather than just those of tourism, a seasonal phenomenon (being part of cycling trails). For this reason, the research does not include major tourist trails and their tourist infrastructure; instead, the work focuses on sections in and around small towns.

2. Literature Review

2.1. Policy of Green Cycling Infrastructure in the Context of Increasing Urban Cycling

With a stronger focus on the strategy of green transport, the need to design transport space is of key importance [8–11]. We know from research that the biggest cities in Germany, Austria and Switzerland have significantly reduced the share of car trips in the last twenty-five years, despite high rates of motorization. The key to their success was a coordinated transport and spatial policy which has made car trips slower, less comfortable and more costly and, as a result, has discouraged people from taking them [12,13]. The policy has also contributed to better safety, comfort and accessibility of pedestrian and cyclist solutions, with German towns putting in a lot more work into promoting cycling [12]. While Poland has also seen a rise in cycling just as in other European countries, the scale of cycling and access to cycling infrastructure is disproportionately smaller than that in German-speaking countries, the Netherlands or Scandinavian countries. Despite that, the bicycle is becoming an alternative to the car or public transport [14], not just in major cities but more and more also in medium-sized and small towns. There are, however, no guidelines on how to plan for cycling space in small towns or how to link cycle paths into a network to ensure easy and safe cycling in the town and outside it. Cycling in towns has a lot to offer. One of the advantages includes time saved, especially where motorised traffic is very heavy [15]. Cycling helps to avoid congestion and allows riders to pick a route which is convenient for them. In addition, cycling as a means of transport helps to improve the environment [16–18]. Studies show that factors which are not related to transport are important. They include the geographic conditions or landscape along commuter routes and the role of the city-specific cycling culture [19]. As a consequence, the location of cycling routes matters a lot. Cycling routes should be ideally designed to run across green spaces in a city, such as parks, squares, leisure spots by the water and city forests [20]. Bicycles are cost-efficient [21], do not emit dangerous substances, reduce the use of dwindling resources of energy and can help to fight lifestyle diseases. Increasing interest in cycling as a means of transport could help reduce traffic congestion and carbon emissions, to which the use of motor vehicles makes a large and inequitably distributed contribution [22]. Another reason why it makes sense to have an urban "pro-cycling policy" and the related economic, social and environmental benefits is the city bike system [23,24]. As shown in research by Scotini and colleagues [25], a comprehensive urban cycling policy and the introduction of a cycling transport system can also help to create more jobs.

2.2. The Importance of Cycling for Health

Cycling as a means of transport is good for people's health and changes city dwellers' health and well-being [6,26–29]. It is clear from research that cycling has a positive effect on cardio-respiratory performance, especially in youth. Even a daily commute helps to reduce the risk of cancer, respiratory and heart diseases, overweight and obesity [22,30,31]. Cycling on a daily basis, e.g., commuting for half an hour and more, may replace physical activity without spending the extra time on physical exercise [32–34]. What is more, if used as a daily physical activity, such mobility is cheap, simple and works for people in different states of physical fitness and age. More cycling would bring additional public health benefits thanks to a reduced use of the car which, by the same token, would reduce air pollution, noise and general road traffic risks [34,35].

2.3. Road Safety in the Context of the Risk of Cyclist Accidents

How likely people are to cycle depends largely on how unsafe cycling is due to a lack of cycling infrastructure (cycle paths). Research conducted by Gutierrez and colleagues [36] in South America shows that the risk of an accident or assault may outweigh the economic savings and health benefits of cycling [37]. The United States represents a similar case where cycling is a safety problem because the infrastructure and access to cycle paths is not ensured [34,38]. Using examples from Japan, the US and Canada, Reynolds and colleagues [39] show that the type of road infrastructure has an effect on cyclist safety and risk of injury. They emphasise that cyclists feel the least safe at multi-lane junctions such as roundabouts, on pavements or combined routes. Cyclists do feel safe, however, on dedicated cycling routes or cycle paths, marked as such. A study in Dublin showed that cyclists perceive cycling around Dublin as less safe compared to using other modes of transport and their sense of safety drops when they realise they could be involved in a bus accident [40].

Sometimes safety and using cycle paths may be related to people's cultural and social backgrounds, as demonstrated in a study of four culturally different urban regions in the United Kingdom [41] and in the United States [38]. There may also be geographic and climatic conditions [42,43]. Polish studies conducted in 2015 by the Ministry of Sport and Tourism show that nearly half of people who regularly cycle to work do not feel safe cycling on a street (definitely or rather not safe) and more than one third of respondents believe the cycling infrastructure is not good enough to ensure a safe trip [44]. Comparative studies carried out by Pucher [34] in Germany and the Netherlands show that if applied, a long-term cycling strategy and cycling infrastructure policy, combined with educating people on alternative forms of transport, can in fact achieve good results such as fewer cyclist accidents and decreased accident fatality rates. This shows that better safety should be a key objective for promoting cycling to minimize the negative consequences of cyclist accidents and eliminate the barriers of a lack of safe infrastructure [33–45].

2.4. The Importance of Building Cycling Infrastructure in Cittaslow Towns

While urban cycling looks optimistic within major cities, it has very little bearing on small towns. Although small towns and rural areas are now better able to manage their transport and cycling space, the cycling infrastructure is still underdeveloped. This may be because cycling is not a popular mode of transport and funding is limited. Cittaslow network towns offer some ideas on ways to organise free movement, "slow" living and "liveability". Cittaslow as a movement goes back to 1998. The association has its seat in Orvieto, in Italy's Umbria region. Italian towns became a model for other European and then worldwide towns. Eventually, an international movement was formed to improve the quality of life of inhabitants, celebrate a "free" and healthy lifestyle and conserve the natural and cultural environment of small towns [46,47]. Towns applying for membership declare that they will work to meet the qualification requirements such as creating people-friendly places. Cittaslow network towns agree to plan public space, develop transport infrastructure which will offer convenience and quality of life to residents and ensure that it is accessible for disabled and older people. Slow means a sustainable urban development which plans for living and transport space that will improve living conditions and ensure better access to public functions for residents [48–50]. While these ideas are sadly not implemented often, Cittaslow towns understand the need for changing the way they think about transport and swapping the car for the bicycle. Proposals to improve transport space or build new transport infrastructure should take account of accessibility, better access and road safety which includes cycle safety.

2.5. Traffic Counts in Poland in Relation to the Legal Background

In accordance with Article 20.15 of the Public Roads Acts of 21 March 1985 (Journal of Laws 2020, item 470), road authorities are required to measure road traffic periodically [51]. Traffic counts are Poland's primary source of road traffic information. The data support how road infrastructure is managed, planned and developed, including decisions to build and upgrade infrastructure, withdraw projects and eliminate some elements of roads. Regular traffic studies in Poland began in the interwar period in the 20th century, concentrating on cities and major routes when they were particularly busy. Rural studies began in 1926 and were conducted three times every four years. In 1965, the traffic count system was harmonised into measurements run regularly every five years, as set out in the guidelines of the United Nations Economic Commission for Europe. In 1985, for economic reasons the study was carried out on national roads only. Since 2000, following Poland's new administrative division, traffic counts have been conducted every five years separately for national and regional roads [52]. Traffic counts cover the following vehicles: motorcycles, passenger cars and minibuses, light trucks, trucks with and without trailers, buses, tractors and bicycles. Bicycles using regional roads in the region of Warmia and Mazury have only been included since 2010.

3. Materials and Methods

3.1. The Study Area

Associated in the Cittaslow network, the towns under analysis are found in Warmia and Mazury, a region commonly known as the Green Lungs of Poland. The region is mostly agricultural and has a strong focus on environmental protection. With very little industry and services, the region is popular with people from Poland's big cities as a place to relax or have second homes (mostly summer homes). Twenty-two towns from Warmia and Mazury have joined the association, representing different environmental, spatial, economic and demographic conditions (as of 2020). The international network includes towns whose population does not exceed 50,000. In Warmia and Mazury, most of the towns are small with a population of up to 10,000 and slightly more; only two have a population of more than 20,000. The area they occupy ranges from 2.16 km (Bisztynek) to 16.84 km (Lubawa), see Figure 1.

Figure 1. The Warmia and Mazury region and towns associated in Cittaslow Poland. Source: Own elaboration based on www.invest.warmia.mazury.pl, www.Cittaslowpolska.pl and Google Maps.

Considering the availability of traffic data, Cittaslow towns were divided into two groups. The first group includes towns in which cycle traffic was measured on selected sections of streets in built-up areas. They are Braniewo, Biskupiec, Bartoszyce, Dobre Miasto, Działdowo, Gołdap, Lidzbark Warmiński, Lidzbark, Lubawa, Nidzica, Nowe Miasto Lubawskie, Orneta and Reszel. Group two includes towns in which traffic was measured on selected sections of roads outside of the Cittaslow towns which pass through them. They are Bisztynek, Barczewo, Górowo Iławeckie, Jeziorany, Olsztynek, Pasym, Ryn and Wydminy. Because ADT (average daily traffic) in Sępopol is not available, the town is not covered by this analysis (Table 1).

Table 1. Division of the towns by sections of urban and rural roads.

Sections in Built-Up Areas		Sections Beyond the	
Town	**Street**	**Town**	**Road Section**
Braniewo	Elbląska	Bisztynek	Bartoszyce-Bisztynek
Biskupiec	Kościuszki	Barczewo	Olsztyn-Barczewo
Bartoszyce	Kętrzyńska	Górowo Iławeckie	Górowo Iławeckie-Bartoszyce
Dobre Miasto	Łużycka	Jeziorany	Dobre Miasto-Jeziorany
Działdowo	Olsztyńska	Olsztynek	Olsztynek-Zgniłocha
Gołdap	Paderewskiego	Pasym	Olsztyn-Pasym
Lidzbark Warmiński	Olsztyńska	Ryn	Sterławki Wielkie-Ryn
Lidzbark	Piaski	Wydminy	Kąp-Wydminy
Lubawa	19 Stycznia		
Nidzica	1-go Maja		
Nowe Miasto Lubawskie	Wojska Polskiego		
Orneta	1-go Maja		
Reszel	Słowiańska		

Source: Own elaboration of authors.

To analyse the Supra-Local Revitalisation Programmes of Cittaslow Towns for their cycling infrastructure projects, fourteen and then nineteen towns were selected which had their programme developed in 2015, 2016–2017 and 2019. The towns are: Barczewo, Biskupiec, Bisztynek, Dobre Miasto, Działdowo, Gołdap, Górowo Iławeckie, Lubawa, Nidzica, Nowe Miasto Lubawskie, Olsztynek, Pasym, Reszel, Ryn; and from 2019: Bartoszyce, Działdowo, Lidzbark and Orneta.

3.2. Methods of Cycle Counts in 2010 and 2015

Average daily traffic was measured and counted by ZDW Olsztyn (Provincial Road Administration) [53] and GDDKiA (General Directorate for National Roads and Motorways) in 2010 and 2015 according to the "Guidelines of general traffic counts on regional roads" [54,55] and the "Guidelines for the organisation and completion of general traffic counts on national roads" [56,57] for 2010 and 2015 respectively, published by the Department of Roads and Motorways of the Ministry of Infrastructure and Development and the GDDKiA. Traffic counts on road sections in the towns were conducted manually, semi-automatically and automatically (on national roads). Cycle traffic was measured for entire road cross-sections (road, pavements and cycle paths) (Figure 2).

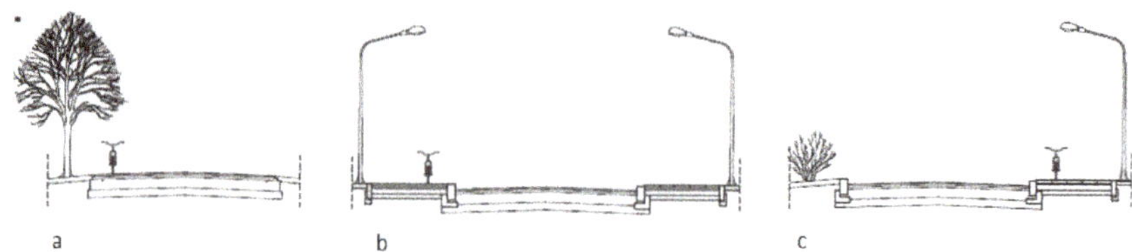

Figure 2. Cross-section of a road with shoulders (**a**), street cross-section with a pavement and cycle path (**b**), street cross-section with a cycle path and shoulder (**c**). Source: Own elaboration of authors.

Three categories of measurement points were used. The first category is type P and defines measurement points where counts are conducted during all hours. Data from these sections help to determine coefficients rates for enlarging samples for those sections which are not studied during all hours (W type). Type M includes sections which run across the towns which also have counts conducted during all hours. Type W are other sections where counts are conducted during limited hours. The annual count includes five so-called daily periods (X_1–X_4, X_6) and one night period (X_5) which is only included in type P and M points. In the case of type P and M, a count during all hours means 16 hours from 6:00 am to 10:00 pm and limited hours for W category means eight hours from 8:00 am to 4:00 pm. Night counts were conducted from 10:00 p.m. to 6:00 a.m. The ADT for particular road sections was calculated for specific measurement points. The formula below was applied both to motor vehicles and bicycles (in the case of bicycles the total number of motor vehicles was replaced with the total number of bicycles) and was applied to type P and M points [54–57].

$$ADT = R_N + \frac{M_R \times N_1 + 0.85\,M_R \times N_2 + M_N \times N_3}{N}\,(\text{veh.}/\text{day}) \tag{1}$$

where:

ADT—total average daily traffic of motor vehicles;
M_R—average daily traffic on workdays (from Monday to Friday, 6:00 a.m.–10:00 p.m.);
$0.85M_R$—average daily traffic on Saturdays and days before public holidays (6:00 a.m.–10:00 p.m.);
M_N—average daily traffic on Sundays and public holidays (6:00 a.m.–10:00 p.m.);
R_N—average night traffic (10:00 p.m.–6:00 a.m.);
N_1—number of workdays in 2010 and 2015;
N_2—number of Saturdays and days before public holidays in 2010 and 2015;
N_3—number of Sundays and days before public holidays in the year;
N—number of all days in the year.

The values of M_R, M_N and R_N were calculated as follows:

$M_R = \frac{1}{3} \times (X_1 + X_2 + X_4)$ (1, 2, 4 measurements on workdays)
$M_N = \frac{1}{2} \times (X_3 + X_6)$ (3, 6 measurements on Sundays and public holidays)
$R_N = X_5$

To calculate ADT for type W points, eight-hour traffic was converted into sixteen-hour traffic using the sample extension coefficient from type P points assigned to type W points (for each type W point a type P point should be assigned located on the road that has the same number).

$$r_{ij} = \frac{X_{ij}}{Y_{ij}}$$

where:

measurement i at point j;

X_{ij}—number of total motor vehicles between 6:00 am–10:00 pm (measurement i at point j);

Y_{ij}—number of total motor vehicles between 8:00 am–04:00 pm (measurement i at point j).

In the next stage of calculating ADT for point W, night traffic was calculated based on the assigned type P points. The final ADT for type W points was calculated using the formula in Equation (1) just like for points type P and M.

The 2010 and 2015 General Traffic Counts, which include ADT for cycling, helped to identify differences between Cittaslow towns' cycle traffic over a period of five years by road sections and outside Cittaslow towns. In addition, a point was made about the relation between cycle traffic and availability of cycle paths and cyclist safety based on cyclist accidents. The results show how the Cittaslow towns were able to deliver on cycling infrastructure in a period of five years and what needs to be done to develop the cycling infrastructure and promote the bicycle as a means of transport in the towns under analysis.

3.3. Analysis of Cyclist Accidents

Cyclist accidents were analysed using accident data from the National Police in Warsaw System of Accident and Collision Data. The statistics is delivered in annual national "Road Accident Reports", while detailed data including cyclist data is passed on to road authorities via the Regional Police. The analysis used cyclist accident and collision data from the database of the Regional Roads Authority in Olsztyn [52].

3.4. Analysis of the Supra-Local Revitalisation Programme of Cittaslow Towns for Their Cycling Infrastructure

The operating goals of the revitalisation programme are designed to improve the quality of the environment. They address issues such as better quality of the environment, increased environmental awareness and pro-environmental attitude of the communities and promoting environmentally friendly means of transport [53]. In addition, the Supra-Local Revitalisation Programme of Cittaslow towns is consistent with the following investment priorities set out in priority axes of the Warmia and Mazury Regional Operational Programme for the Years 2014–2020: 7b "Increase regional mobility by linking secondary and tertiary hubs with TEN-T infrastructure, including multimodal hubs"—Axis 7 Transport Infrastructure [58]. The study analysed all projects which were submitted to the revitalisation programme in 2015 and selected only those projects which tackled the revitalisation or construction of new infrastructure in the town. Because the towns developed their individual revitalisation programmes in the subsequent years (2016–2017) that are partly based on the 2015 programme, an analysis was carried out of the planning of cycling infrastructure projects and how they have been delivered. The first programme [59] and the individual programmes (2016–2017) were developed for fourteen Cittaslow towns. The last stage was to analyse the supplemented Supra-Local Revitalisation Programme of Cittaslow Towns 2019 (second programme), [60]. Four more towns joined the programme and were included in the study conducted at that time. Below are project topics, measures and locations as regards the planning of cycling infrastructure for fourteen towns (according to the first programme), Table 2.

Table 2. Planned projects involving cycling infrastructure in Cittaslow towns (as of 2015).

Town	Title of Comprehensive Project (PK), Sub-Measure (P) and Location (L)
Barczewo	PK: Develop the Old Town's public space P1: Prepare cycling and walking routes P2: Plan leisure infrastructure in the centre of the Old Town, prepare a cycling and walking route L: Old Town
Biskupiec	PK: Regeneration of deprived spaces P1: Build cycle roads L1: Link the existing cycle path at Warszawska street with the second existing path around Kraks Mały lake. L2: Floriańska, Topiel, Syreny, Warmińska, Poznańska, Polna, Bohaterów streets. L3: Along Hubalczyków street across the post-military area. L4: Link the existing cycle path in the part at Wiosenna street with the route around the adjacent retention pond
Bisztynek	PK: Improve population mobility by adding functional and aesthetic features into public space P1: Build a cycling and educational trail P2: Make avenues and streets people-friendly by building cycle roads—improve the mobility of the Bisztynek population, using an environmentally friendly form of transport L: the streets of Kąpielowa, Słoneczna, Obwodowa, Struga, Kajki, Wiktora, Moniuszki, Konopnickiej, Grunwaldzkiej, Szkolna, Polna, Pl. Wolności, Słoneczna, Nowego Osiedla, Orzeszkowej and Sportowa
Dobre Miasto	PK: Regenerate deprived public spaces P: Modernise, upgrade and build pedestrian and car shared zones and pedestrian and cyclist shared zones to ensure safety L: Sites along the river Łyna
Gołdap	PK: No main project in the revitalisation programme to address cycling infrastructure, additional projects to build a well-lit cycle path and implementation of the city bike project L: From the city centre to the health resort district
Górowo Iławeckie	PK: Improve the site by creating a leisure and sensory park P1: Build a path for pedestrians and cyclists L: Park at Staw Garncarski pond PK: Improve the area in the valley of the river Młynówka—project on reserve list P: Build a walking and cycling path and link the site to Eastern Poland Cycling Routes and other sites in the town L: City Lagoon, valley of the river Młynówka
Lidzbark Warmiński	PK: No main project in the revitalisation programme to address cycling infrastructure
Lubawa:	PK: Revitalisation of selected sites to meet the need for exercise therapy for socially excluded people P: Build a walking and cycling path Zalew—Lipy L: Kopernika street across parts of Lipowa street and Pielgrzyma street to Kupnera street
Nidzica	PK: No main project in the revitalisation programme to address cycling infrastructure
Nowe Miasto Lubawskie	PK: Integration and engagement of socially excluded people in the revitalised area through recreation and physical activity P: Build a cycle road on the railway embankment in place of the disused Brodnica—Iława train line and implement the city bike project L: City Park, site in the post-railway area
Olsztynek	PK: Build a health path "Healthy body, healthy spirit" P: Build a health path and implement the city bike project L: Urban and sub-urban areas
Pasym	PK: No main project in the revitalisation programme to address cycling infrastructure
Reszel	PK: Restore the natural, architectural and functional features of City Park. P: Build a health path L: City Park
Ryn	PK: No main project in the revitalisation programme to address cycling infrastructure

Source: Prepared by the author based on Cittaslow Supra-Local Revitalisation Programme of Cittaslow Towns of 2015.

When analysing the Supra-Local Revitalisation Programme of Cittaslow Towns of 2015 [59] and subsequent documents from 2016–2017 and 2019 [60], the following issues were considered:

- the degree to which planned cycling infrastructure projects were changed in subsequent years;
- the scale of planned cycling infrastructure projects, including their location and length of cycle paths;
- priority given to cycling infrastructure projects over other projects;
- the degree of connectivity with existing infrastructure and its size in the town and outside it;
- availability of cycling infrastructure, especially in suburban areas;
- the goals of cycling infrastructure projects including social, environmental, health and transport goals.

4. Results and Discussion

All of the Cittaslow towns in Warmia and Mazury analysed in this article have a strong and sadly unused potential for planning and building cycling infrastructure. This is particularly important in the case of the small towns of Warmia and Mazury because cycling there should reflect the local needs while drawing on models from countries which use the bicycle as a basic means of daily transport, whether in towns or outside of them. The bicycle is becoming an inseparable tool for sports or spending free time but it should also become one of the forms of commuting. Many people living in suburbs or in the countryside work in a Cittaslow town. As we know from Biostat's "Transport behaviour of Polish people" study, 20.9% of respondents are not happy with the number of public transport services between where they live and neighboring towns. In addition, approximately 8.4% of respondents said they did not have a car [61]. As a consequence, for some people cycling may become the only alternative for moving between towns. Therefore, an important stage in the study was to analyse cycle traffic on urban and rural sections.

4.1. Results of Studies into the Number of Bicycles Used

Cycling is a great alternative in urban spaces and for travelling between towns. It is very difficult, however, to identify the number of bicycles in Poland. Bicycles do not have to be registered under Polish law. To get a better understanding of the data and numbers, a 2011 report of the Central Statistical Office (GUS) will be presented which is based on the National Census (NSP) run every ten years, and on regular surveys. The survey looks at a household's possession of durable items (the bicycle) and overall number of households (an NSP survey). According to the National Census, there were about 13,567,999 households in 2011 in Poland, with about 515,857 in the region of Warmia and Mazury [62]. In addition, regular surveys showed that in 2011 an average of 62.9% households in Poland had a bicycle with 60.5% in the region of Warmia and Mazury [63]. It is safe to say that in all of Poland in 2011 there were about 8,534,270 bicycles, with about 312,090 bicycles in Warmia and Mazury in 2011. As we can see from an analysis of cycling used as a means of transport in the Cittaslow towns, cycle traffic increased over the five years between 2010 and 2015. In one case only—in the town of Biskupiec, traffic in 2015 fell compared to 2010 by about 31 bicycles per day. The highest increase in cycle traffic was recorded in the town of Działdowo where in 2015 there were 279 bicycles per day, which is more than double the 2010 figures which were 104 bicycles per day (Figure 3).

Figure 4 shows results that differ from the ones above. The distribution of increasing and decreasing traffic between 2010 and 2015 is equal. On the road sections Sterławki Wielkie–Ryn, Dobre Miasto–Jeziorany, Górowo Iławeckie–Bartoszyce and Olsztyn–Barczewo bicycles per day increased, while on the road sections Kąp–Wydminy, Olsztyn–Pasym, Olsztynek–Zgniłocha and Bartoszyce–Bisztynek a decrease was recorded. The biggest difference in cycle traffic was recorded on the section Kąt–Wydminy where in 2010 there were 71 bicycles per day and only 39 in 2015. The biggest increase in volumes was recorded on the section Dobre Miasto–Jeziorany, however, it was only an extra 16 bicycles.

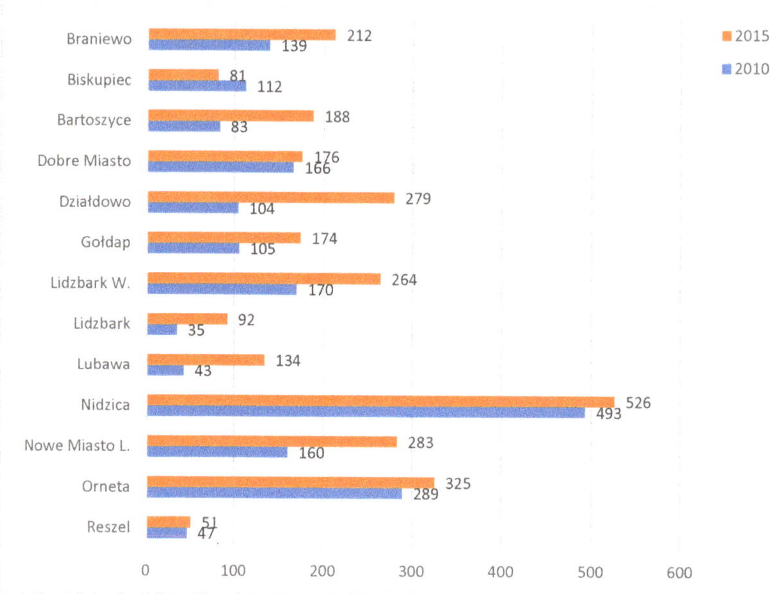

Figure 3. Comparison of ADT 2010 and 2015, built-up area in a Cittaslow town. Source: Own elaboration of authors based on ZDW Olsztyn [53].

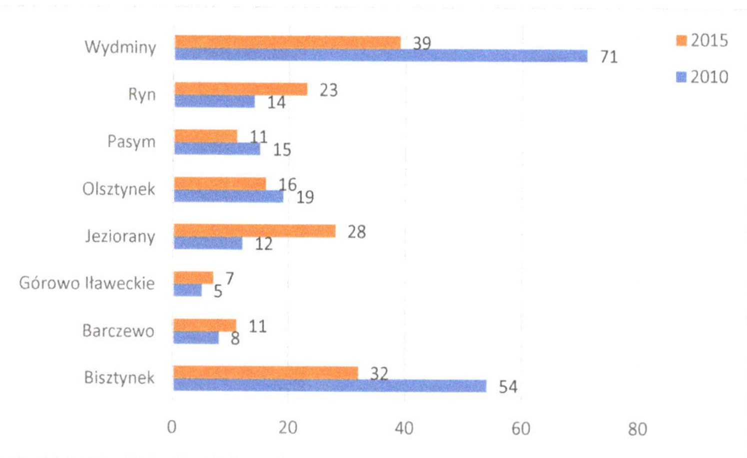

Figure 4. Comparison of ADT 2010 and 2015, road section outside Cittaslow towns. Source: Own elaboration of authors based on ZDW Olsztyn [53].

To recap, it should be said that while cycling increases, it does so only on road sections in built-up areas of Cittaslow towns. This may be mainly because the bicycle is gaining popularity as a means of urban transport and is related to availability of cycling infrastructure. Outside of built-up areas cycling has decreased which suggests that cyclists do not feel safe when they use the same road with the ever-increasing volume of motor vehicles and trucks (Figures 5 and 6).

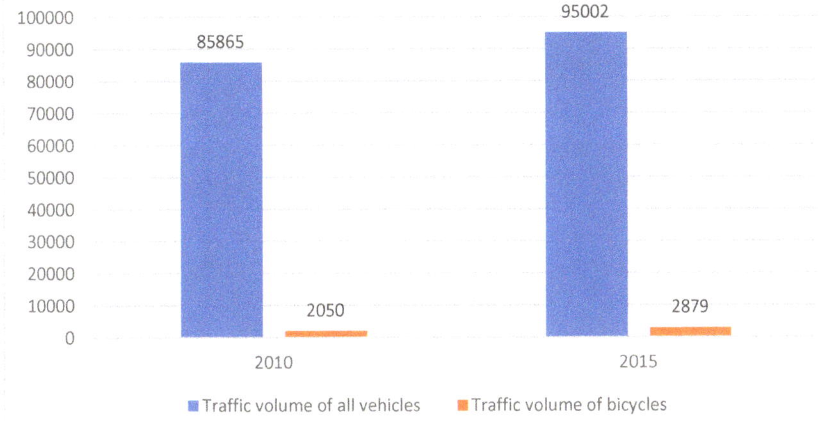

Figure 5. Comparison between traffic volumes for all vehicles and cycle traffic on regional roads. Source: Own elaboration of authors based on ZDW Olsztyn [53].

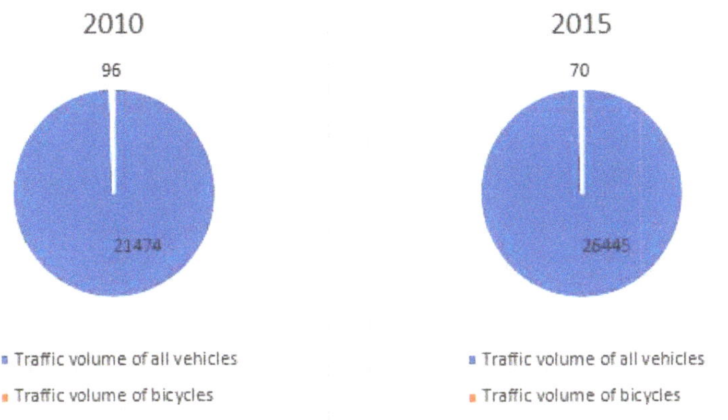

Figure 6. Comparison between traffic volumes for all vehicles and cycle traffic on national roads. Source: Prepared by the author based on GDDKiA Olsztyn data [64].

The charts show that cycling continues to be a fraction of overall traffic. In 2010, the share of bicycles in overall traffic of all vehicles using regional roads in Cittaslow towns accounted for a mere 2.39%, and 3.03% in 2015. This is even more striking in the case of national roads in Cittaslow towns, where in 2010 the relation between bicycles and all vehicles was 0.44% and in 2015 only 0.26%.

4.2. Results of Studies into Cycle Path and Cycling Infrastructure Coverage

According to the Local Data Bank in Poland, in 2011 there were 5782.8 km of cycle roads and in 2015 there were 10,797.2 km, an increase of about 100% (5014.4 km) [60] (Figure 7). The data show cycle roads which are managed by municipalities, counties and regional governments. This means they are roads within a road, roads that are separate from the main road and the pavement and are a walking and cycling zone. The data looks at the length of one-way roads. In the case of roads which are on both sides of the roadway, the length is calculated separately. The list does not include roads which are tourist routes [62]. The local data bank has data for the years 2011–2019 but there is no data for 2010. The years 2011–2015 show that the number of cycle paths rose in Poland each year which suggests that in 2010 cycle road length was less than 5000 km.

Analysis of road infrastructure data shows that cycle traffic has grown the most on pedestrian and cycle paths or cycle paths built from 2010 to 2015 (Bartoszyce, Działdowo, Lidzbark Warmiński, Lubawa, Nidzica, Orneta) (see Table 3). In addition, cycling also intensified in three towns which do not have cycling infrastructure. They are Braniewo, Gołdap and Nowe Miasto Lubawskie. They are, however, important tourist destinations with Green Velo, Poland's longest signed cycle route, passing through Braniewo and Gołdap.

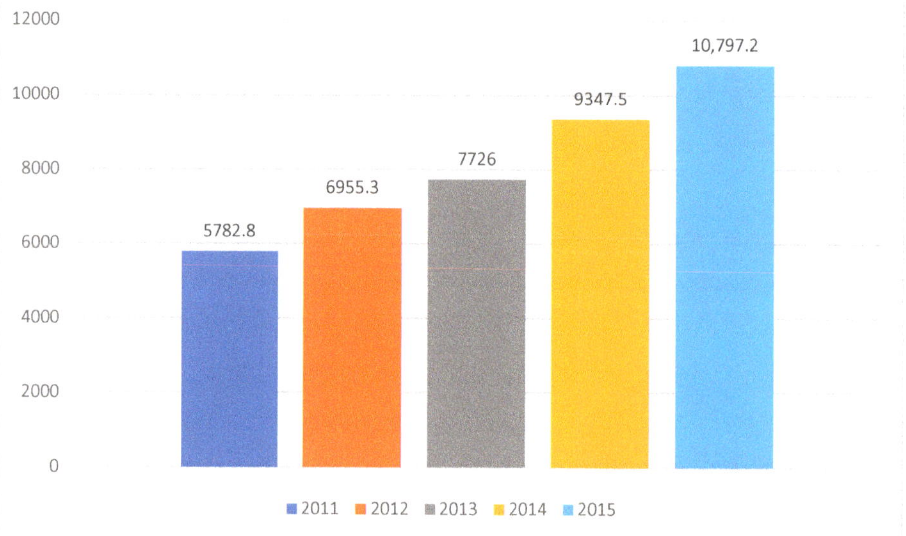

Figure 7. Comparison of the length of cycle roads in Poland 2011–2015. Source: Own elaboration of authors based on Bank Danych Lokalnych bdl.stat.gov.pl [62].

Table 3. Cycle traffic and existing cycling infrastructure. Source: Own elaboration of authors based on ZDW Olsztyn, GDDKiA Olsztyn.

No	Town	Road Category *	Study Area **	ADT 2010	ADT 2015	Cycling Infrastructure
1	Braniewo	P	T	139	212	No dedicated infrastructure, cyclists use the pavement or road
2	Biskupiec	P	T	112	81	No dedicated infrastructure, cyclists use the pavement or road
3	Bartoszyce	P	T	83	188	In 2012 a walking and cycling zone was built, about 2 km long
4	Dobre Miasto	P	T	166	176	No dedicated infrastructure, cyclists use the pavement or road
5	Działdowo	P	T	104	279	In 2014 about 2.5 km of cycle paths were built
6	Gołdap	P	T	105	174	No dedicated infrastructure, cyclists use the pavement or road
7	Lidzbark Warmiński	P	T	170	264	In 2014 about 0.6 km of a walking and cycling zone were built
8	Lidzbark	P	T	35	92	About 1.2 km of cycle paths were built
9	Lubawa	P	T	43	134	In 2014 about 1.0 km of a walking and cycling zone was built
10	Nidzica	P	T	493	526	In 2014 about 6 km of a walking and cycling zone were built
11	Nowe Miasto Lubawskie	P	T	160	283	No dedicated infrastructure, cyclists use the pavement or road
12.	Orneta	P	T	289	325	In 2014 about 2.0 km of a walking and cycling zone and 0.15 of a cycle path were built
13	Reszel	P	T	47	51	No dedicated infrastructure, cyclists use the road
14	Bisztynek	N	OT	54	32	No dedicated infrastructure, cyclists use the pavement or road
15	Barczewo	N	OT	8	11	No dedicated infrastructure, cyclists use the road
16	Górowo Iławeckie	P	OT	5	7	No dedicated infrastructure, cyclists use the road
17	Jeziorany	P	OT	12	28	No dedicated infrastructure, cyclists use the road
18	Olsztynek	N	OT	19	16	No dedicated infrastructure, cyclists use the road
19	Pasym	N	OT	15	11	No dedicated infrastructure, cyclists use the road
20	Ryn	P	OT	16	20	No dedicated infrastructure, cyclists use the road
21	Wydminy	P	OT	71	39	No dedicated infrastructure, cyclists use the road

* Provincial road (P), National road (N), ** in town (T), outside town (OT).

4.3. Analysis of Cyclist Accidents

According to National Police road accident statistics in 2010 [65], cyclists were involved in 3918 road accidents out of a total of 38,832 of all accidents (about 10%). There were 290 cyclist fatalities and 3806 cyclist injuries. Cyclists were responsible for 1588 accidents, of which as many as 1328 occurred in a built-up area and only 260 in a non-built-up area. There was a higher share of fatal accidents outside towns compared to fatal accidents in urban areas, with every fourth accident ending in death which represents 25% (65 people in 260 accidents). In built-up areas, every 17th accident was fatal (78 people in 1328 accidents—6%) [65].

The statistics in 2015 [66] are as follows: Cyclists were involved in 4634 road accidents out of a total of 32,967 accidents (about 14%), with as many as 300 cyclists killed and 4111 injured. Most accidents occurred in a built-up area—4011, and there were 623 accidents in non-built-up areas. Just as in 2010, the share of fatalities in non-built-up areas was higher than in built-up areas and represented about 20% (126 fatalities in 623 accidents). In 4011 accidents in built-up areas 174 people were killed, which accounted for only 4% (Figure 8).

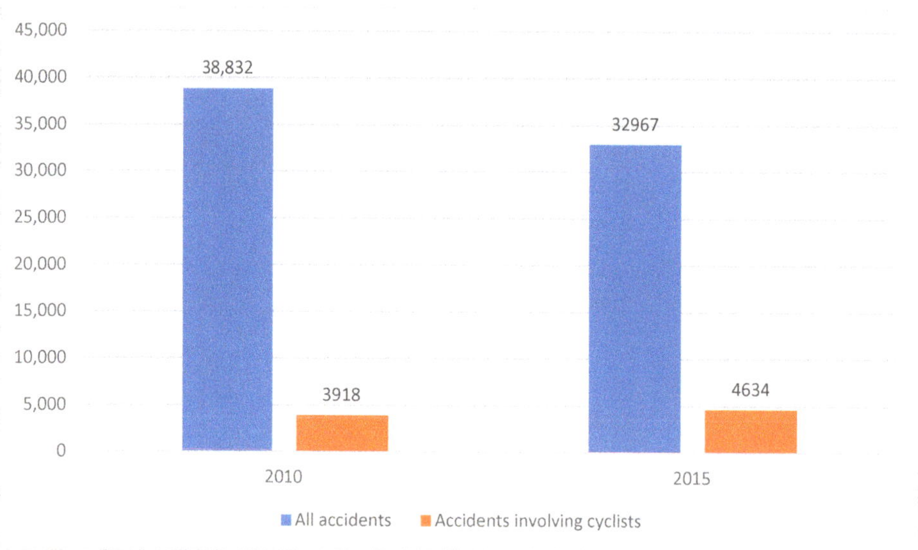

Figure 8. Comparison between cyclist accidents and overall accidents in 2010 and 2015—general data—Poland. Source: Own elaboration of authors based on data of the National Police [65,66].

In the case of roads in Cittaslow towns, cyclist accidents are as follows. In 2010, on regional roads passing through the Cittaslow towns there were 15 cyclist incidents, of which 8 were accidents (including fatality accidents) and 7 were collisions. In relation to overall incidents on sections of those roads, there were 691, of which 77 were accidents and 614 were collisions, cyclist collisions represent about 1.1% of all collisions but cyclist accidents account for as much as 10% of all accidents. In 2015, in Cittaslow towns (on regional roads) there were 15 incidents, of which only 4 were accidents (injury or death). The total number of that year's incidents on the analysed roads amounted to 513, of which 63 were accidents; cyclist collisions represented only about 2.4% (11 cyclist collisions out of 450 of all collisions) and cyclist accidents represented 6.3% (4 accidents out of a total of 63) [64], Figure 9.

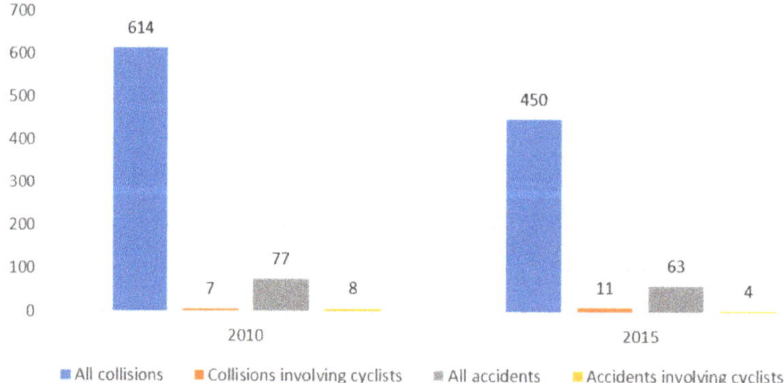

Figure 9. Comparison between cyclist accidents and overall accidents—regional roads of Cittaslow in 2010 and 2015. Own elaboration of authors based on ZDW in Olsztyn [53].

A comparison between cyclist accidents in Poland and cyclist accidents on regional roads in Cittaslow Warmia and Mazury towns shows that cyclist accidents in 2010 and 2015 accounted for a very small percentage of all accidents. Analysis of cyclist incidents in relation to all incidents in Cittaslow towns on regional roads also shows a small share, which is a puzzling result. This can be explained by analysing the share of other vehicles compared to bicycles (bicycles represented only about 2.39% in 2010 and 3.03% in 2015 of all traffic on regional roads), the underdeveloped cycling infrastructure (most of the cycling infrastructure can be found in and around towns, and outside towns cyclists mainly use the road) and a sense of vulnerability (the bicycle compared to other motor vehicles, including trucks). As a consequence, cycling is still an unpopular means of transport.

4.4. Results of Analysis of the Supra-Local Revitalisation Programme of Cittaslow Towns for Its Cycling Infrastructure Planning (First Programme from 2015, Individual Programmes 2016–2017, Second Programme from 2019)

It was found that under the First Revitalisation Programme of Cittaslow Towns of 2015, cycling infrastructure projects (both new and improved infrastructure) represented an important part of all projects (including social and economic projects). Project were designed to increase physical activity by encouraging the residents of the towns to cycle (in Barczewo, as an example). Social integration and a better environment were important factors in all projects. What is more, some of the infrastructure work was to be delivered by local entrepreneurs and job seekers or trainees (in Biskupiec, as an example). Project objectives included protection of the environment, in particular the conservation of wildlife (in Nowe Miasto Lubawskie, as an example) and CO_2 reduction by giving up using cars and changing the energy mix. In some cases, cycling infrastructure was to run through conservation areas with plans to improve or reproduce historic roadways (in Bisztynek, as an example). Despite a strong emphasis on cycling infrastructure in the programme objectives, the majority of cycle path projects were designed as an addition to other activities such as public utility projects. While there is nothing wrong with combining projects, this led to a shortening of cycling routes and increased spending on non-cycling elements such as leisure infrastructure, parks and other uses. On the other hand, the additional functions along the routes may attract more cyclists, including families who will take rides to specific leisure sites.

The next conclusion from the analysis of cycling infrastructure projects is that there are no cycle paths (in Barczewo, as an example) or they are fragmented. The majority of the towns only have fragments of signed and prepared paths (with cycle lanes), but cycling, if any is done, happens on pavements or on the road. This is associated with the risk of a collision and crashing into a pedestrian or a car. Only some of the projects plan connections into existing routes or provision of good transport links in the town. In addition, the links into suburbs or villages are poor or non-existent, which may be a problem for the daily commutes by bicycle (a matter of safety), making the choice of bicycle as a means of transport less likely (in Biskupiec, as an example). With no plans to build links into areas outside the town, the original railway embankments remain an untapped potential for cycling (Nowe Miasto Lubawskie was the only town to suggest that), see Table 4.

Only a small number of the projects are designed to offer the city bike and build bike stations and cycle parking. Only three towns (Gołdap, Nowe Miasto Lubawskie, Olsztynek) have offered to do this. This is quite important because big cities in Poland have bike systems and renting bikes is very popular with both residents and tourists.

A key finding in the analysis of all the documents from the three periods (2015–2017, 2019) is that there are no measures to prepare cycling infrastructure. While the Supra-Regional Revitalisation Programme of Cittaslow Towns of 2015 included such projects, their number fell gradually in the subsequent years. The programme of 2019 does not include such projects or it gives them little importance. Although initially given priority, the projects are mere additions to other projects (e.g., in Bisztynek, Lubawa, Nowe Miasto Lubawskie, Olsztynek). It is safe to say that cycling infrastructure has been marginalised and previous plans will not be delivered at this stage. In the

Second Programme (2019) only one of the cycle path projects was given priority (Górowo Iławeckie), a few are supplementary and the majority are not even included in the Programme. Results of the analysis are shown in Table 5.

Table 4. Connecting planned paths with other parts of the town and existing cycling paths (as planned in the 2015 programme).

	Town	Better Links to Other Parts of the Town	Links into Existing Cycle Paths, Including Supra-Local Paths
1	Barczewo	Y	N
2	Biskupiec	Y	N
3	Bisztynek	Y	-
4	Dobre Miasto	Y	N
5	Gołdap	Y	N
6	Górowo Iławeckie	Y	N
7	Lidzbark Warmiński	-	-
8	Lubawa	Y	N
9	Nidzica	-	-
10	Nowe Miasto Lubawskie	Y	N
11	Olsztynek	Y	N
12	Pasym	-	-

Y: yes, N: no, -: not applicable, no project/activity.

Table 5. Analysis of projects related to cycling infrastructure in Cittaslow cities (planned, for implementation, not implemented).

	Town	Location	Objective Involving Cycling Infrastructure *	MP/SP/NP RPC **	MP/SP/NP LRP ***	MP/SP/NP RPC ****
1	Barczewo	2	Improve the pedestrian and cycle path in the Old Town	MP	SP	NP
			Design of a pedestrian and cycle path with infrastructure along the river Pisa (2.5 km)	MP	SP	SP
2	Bartoszyce	-	-	No data	No data	N
3	Biskupiec	4	Planning of new cycle path in 4 locations in the town	MP	No data	SP (1 km of path)
4	Bisztynek	1	Design of a cycling and educational route in an area of revitalisation, 4.2 km	MP	NP	NP
5	Dobre Miasto	1	Design of pedestrian and cycle paths along the river Łyna, 4.5 km (including pedestrian walks)	No data	No data	NP
6	Działdowo	-	-	No data	No data	NP
7	Gołdap	1	Build a well-lit cycle path and implement city bike project	SP	NP	NP
8	Górowo Iław.	2	Build a pedestrian and cycle path in the park	SP	NP	NP
			In the river Młynówka valley	SP	MP	MP (0.7 km of path)
9	Jeziorany	-	-	No data	No data	NP
10	Lidzbark	-	Bike rental	No data	No data	SP
11	Lidzbark Wam.	2	None, in 2015 a section of a cycling route was built, part of the North East Poland cycling route (measure not part of the Programme)	NP	NP	NP
12	Lubawa	1	Build a pedestrian and cycle path Zalew–Lipy, cycle path 3.05 km pedestrian and cycle path 0.4 km	MP	NP	NP

Table 5. *Cont.*

	Town	Location	Objective Involving Cycling Infrastructure *	MP/SP/NP RPC **	MP/SP/NP LRP ***	MP/SP/NP RPC ****
13	Nidzica	-	No planned cycle paths. The city centre is to be closed for motor traffic	NP	NP	NP
14	Nowe Miasto Lub.	2	Build a cycle path on post-railway site	M	S	N
			Build a bike station (bike rental)	M	S	N
15	Olsztynek	1	Build a cycle path and buy bikes	M	S	N
16	Orneta	-	-	No data	No data	N
17	Pasym	-	No projects to build path infrastructure, pedestrian paths only	N	N	N
18	Reszel	1	Build pedestrian and cycle paths in the city park, 1.5 km long	M	M	S
			Improve the quality of cycle paths in the city centre	S	N	N
19	Ryn	-	No projects to build cycle paths, improving lakeside pedestrian paths only	N	N	N

* including selected elements such as road infrastructure, bridges, car parks, stops, sport facilities, outdoor gyms, playgrounds, city furniture, lighting, CCTV, information, education, therapeutic elements, roadside vegetation, security; MP: Main project, SP: Supplementary Project, NP: no project; ** RPCT—Regional Revitalisation Programme of Cittaslow Town (2015); *** LPCT—Local Revitalisation Programme of Cittaslow Town (2016–2017); **** RPCT—Regional Revitalisation Programme of Cittaslow Towns (2019).

5. Guidelines and Recommendations for the Development of Cycling Infrastructure and for Road Safety

5.1. Specific Guidelines Based on the Research

5.1.1. Guidelines for Cycling Policy

- Cooperation between administrative bodies and road authorities and other bodies responsible for road infrastructure to develop guidelines for safe cycling infrastructure and cyclist safety on the road. Formulate local guidelines for road authorities, for example by consulting city authorities. Roads in Poland and in Warmia and Mazury are typically managed by a number of bodies which creates problems with planning. Having local guidelines for safe infrastructure and cyclist safety for each town would improve and accelerate the planning and construction of cycle paths as part of all roads.

- Develop cycling policy strategies for Cittaslow towns to take account of their unique character and demand for cycling infrastructure of people who commute between towns. The strategies should be based on a diagnosis of the individual towns and analysis of local conditions. As an example, the strategies of Braniewo, Bartoszyce and Gołdap and their suburban villages should be related to the towns' close proximity to the state border with the Kaliningrad Region (Russia). This is because a cycle link must be ensured between these towns and the border crossing, the workplace for many commuters. On the other hand, this opens up new opportunities for tourism. The towns of Barczewo (and the areas outside it), Dobre Miasto and Olsztynek should focus their strategies on providing links to jobs and services in the regions' biggest towns (with the distance as an important criterion).

- Develop and correlate other cycling infrastructure projects locally, regionally and nationally, making sure that there is a general cycle path strategy rather than isolated solutions (this is a common problem in Warmia and Mazury, especially in small towns, but the rest of Poland is no different); a comprehensive approach is needed to include existing sections and add new ones.

- Reinstate the original projects from the Supra-Local Revitalisation Programme of Cittaslow Towns of 2015 and generate new cycling infrastructure projects by including them in the new programme. Some of the priority actions would be to implement the city bike project in Gołdap, Lidzbark and Nowe Miasto Lubawskie and develop similar projects in the other towns. Furthermore, the pavement along the regional road in Dobre Miasto to be constructed in 2020 should be correlated with a new cycle path. It is important to plan pavements and cycle paths at the same time as opposed to planning pavements only.

5.1.2. Guidelines for Planning Infrastructure and Environmental Solutions

- Connect existing paths in the towns into a uniform transport system to include sections of national, regional, county, municipal and local roads. This requires the cooperation of the Regional Roads Authority and the General Directorate for National Roads and Motorways with town mayors to ensure a joint effort and cost-sharing when building new or upgrading existing cycling infrastructure. Decisions to link cycle paths should be based on local guidelines previously agreed to between road authorities and city authorities. Joint implementation and cost-sharing may be part of the guidelines.
- Plan and build new paths as dedicated roadside/street lanes (in and outside the town) and separate them from pavements or crossings as a separate lane (especially in towns). It is clear that densely developed parts of the towns under analysis cannot accommodate dedicated cycle paths. As a result, the proposal is to convert one of the pavements into a cycle path like this: convert the pavement on one side of the Łużycka street in Dobre Miasto and the Wojska Polskiego street in Nowe Miasto Lubawskie into a cycle path.
- Connect cycle paths in the town and extend them into rural areas. This should be a priority objective for all the towns under analysis.
- Plan for accompanying cycling infrastructure and path development through landscaping, especially where public spaces are concerned such as parks, squares, plazas e.g., in Biskupiec, Dobre Miasto, Lidzbark Warmiński, Gołdap, Bartoszyce.

5.1.3. Guidelines for Improving Safety and Accident Statistics

- Conduct detailed cycle traffic counts as separate from the General Traffic Count. Cycle traffic counts should be conducted at least twice a year—at the beginning and end of the cycling season; traffic should be studied off-season to understand whether cycling is a seasonal or year-round activity.
- Monitor, analyse and improve road infrastructure in high risk sites for cycling. Analyse cyclist accident databases, put up CCTV in accident sites and analyse the data collected to identify the cause and develop solutions.
- Separate cycle paths clearly from other road users (especially from cars and pedestrians), primarily in towns and outside of them. Apply physical barriers between cycle traffic and motor traffic such as greenery or road barriers, separate pedestrians from cyclists by raising pavements in relation to cycle paths or by separating the two surfaces.

5.2. General Recommendations

- Conduct an analysis of good practice from Cittaslow towns worldwide which have successfully implemented and are running cycling infrastructure and safety projects. This solution can be easily taken up in the form of joint initiatives between Polish Cittaslow towns and transport and planning (which includes cycle path development) specialists from Cittaslow towns in, for example, the Netherlands, Germany, the Nordic Cittaslow Network.

- Correlate educational programmes on safety with the educational activities of schools, police, municipal guards, public administration and all entities involved in transport, road traffic and health.
- Joint activities to promote cycling as a means of transport for all Cittaslow towns using thematic meetings and events for the communities. Promote active leisure for families with the bicycle as a means of transport.
- Create a system to enable the free transport of bicycles by other means of transport over longer distances. This includes rail transport to carry bicycles free of charge from Braniewo, Nidzica, Olsztynek, Pasym, Barczewo to Olsztyn, the region's biggest town. It also includes a proposal to allow people to cycle from the town or village to a train station and then continue the trip by train (a solution commonly used in the countries of Western Europe, not so much in Poland).
- Prepare specimens of materials for paths and path surfaces to follow the local character and environmental requirements (e.g., paved or bituminous paths). Keep all types and colours of surfaces the same in Cittaslow towns, ensuring a more intuitive cycling.
- Identify ways to use the potential of the environment in ensuring that cycling routes pass through attractive countryside.
- Plan for green areas which are best run along paths with trees, shrubs and low growth. Using low growth in dedicated sections in Bartoszyce and Lidzbark Warmiński to physically separate bicycles from cars would replace road barriers which are not as aesthetic. A similar solution has already been applied in Działdowo.
- Use distinct horizontal marking for cyclist crossings on main roads (preferably red marking), primarily in non-built-up areas. Horizontal marking of the same colour as the cycle path (preferably red) every time the path crosses the road.
- Use additional vertical marking in places where cyclists cross the road (including fluorescent marking) to warn other road users.
- Where cyclist crossings cross minor roads, keep cyclist crossings away from the main road and run them perpendicularly to the minor road, "deflecting" the cycling route before the cycle crossing.
- Use elements of traffic calming, especially in the town centre and areas of heavy motor traffic. To get motorists to slow, use raised cycle crossings and narrower cross-sections by designating cycle paths that are made distinct from the road by separators and colour (red).
- Use contraflow lanes physically separated from the road as separate cycle lanes to allow cyclists to ride "against the flow" in towns in specific areas.

6. Conclusions

The article addresses the topic of cycling infrastructure to be introduced in Cittaslow towns in the region of Warmia and Mazury, in line with Cittaslow statutes and strategies which focus on the environment and community issues. It also aims to analyse the share of cycling as a means of transport in towns which are members of this organisation and in other small towns across Poland. There are no previous publications related to this topic. The authors have analysed transport space and its use in the context of availability and road safety and in relation to the projects included in revitalisation programmes.

It was established in the research that cycle paths in the Cittaslow towns only have a limited presence on national and regional roads. The existing cycling infrastructure is not adequately developed. Existing cycle paths are not connected with one another, the routes are incomplete and cycling on the road is difficult and dangerous.

The results of the analyses in this work have led us to the following conclusions. As regards the number of bicycles, it is important to study cycle traffic in more detail. As well as conducting general traffic counts every five years, more frequent counts should be carried out to measure cycle traffic specifically and how it relates to other road users. This will help to analyse the needs of cycle traffic and cycling infrastructure and to understand the safety of cyclists and other road users. As we can see from analysis of cycling infrastructure, cycling increases where new cycle paths and roads have been built, i.e., usually in urban areas. The results of infrastructure analysis in this work show that while cycle paths across Poland are growing in number (although not so much in Cittaslow towns), rural and out-of-town infrastructure is scarce which translates into very little cycle traffic. In the case of the analysed Cittaslow suburban areas, potential cyclists probably do not feel safe there. Please note that the problem of feeling that cycling on the road is unsafe is not confined to Poland only. The work gives examples from other countries which suggest a similar problem. Given the characteristics of Cittaslow towns defined in the organisation's programme, i.e., safety and better living standards, it is clear that unless the cycling infrastructure is improved in the towns and outside them, these assumptions will not be met fully.

The results of cyclist accident analysis on the roads we have studied show that there are not many accidents involving cyclists. While this may be viewed as a positive result, it may not in fact be the case when we consider the low share of cycling in overall traffic, the underdeveloped cycling infrastructure and the resulting sense of feeling unsafe when cycling outside of dedicated cycling infrastructure.

To make the picture complete, there needs to be an adequate plan for cycling infrastructure development and a determined effort to deliver that plan. It is clear from the analysis of the Revitalisation Programme of Cittaslow Towns regarding cycling infrastructure, that in 2019 cycling projects were made secondary despite the priority they were given in the 2015 programmes which aimed to promote cycling as a way to increase levels of physical activity, improve the environment, protect nature and reduce CO_2 emissions. The article has demonstrated that there is a close relation between abandoning plans to improve and build more cycle paths and cyclist safety, especially on rural roads.

It is claimed in the article that bicycles should be used as a basic and everyday means of transport, especially by residents of suburban areas and not just for tourism.

Author Contributions: Conceptualization, A.J.; methodology, A.J. and A.M.; analysis, A.J., J.Ż. and A.M.; investigation, A.J., J.Ż. and A.M.; writing preparation and visualization, A.J. and A.M. All authors have read and agreed to the published version of the manuscript.

Funding: This research received no external funding.

Conflicts of Interest: The authors declare no conflict of interest.

References

1. Kristianova, K. Public space connectivity and walkability in suburban neighborhoods—Case studies from Nitra and Košice, Slovakia. In Proceedings of the International Multidisciplinary Scientific GeoConference: SGEM: Surveying Geology & Mining Ecology Management, Albena, Bulgaria, 28 June–6 July 2016; pp. 675–682. [CrossRef]
2. Kwiatkowski, A. Urban Cycling as an Indicator of Socio-Economic Innovation and Sustainable Transport. *Quaest. Geogr.* **2018**, *37*, 23–32. [CrossRef]
3. McLeod, S.; Scheurer, J.; Curtis, C. Urban Public Transport: Planning Principles and Emerging Practice. *J. Plan. Lit.* **2017**, *32*, 223–239. [CrossRef]
4. Kazak, J.; Chalfen, M.; Kamińska, J.; Szewrański, S.; Świąder, M. Geo-Dynamic Decision Support System for Urban Traffic Management. In *Dynamics in GIscience, Proceedings of GIS Ostrava, Ostrava, Czech Republik, 21–23 March 2018*; Lecture Notes in Geoinformation and Cartography; Ivan, I., Horák, J., Inspektor, T., Eds.; Springer: Cham, Switzerland, 2018.
5. Kazak, J.; Garcia Castro, D.; Swiader, M.; Szewranski, S.Z. Decision Support System in Public Transport Planning for Promoting Urban Adaptation to Climate Change. *IOP Conf. Ser. Mater. Sci. Eng.* **2019**, *471*. [CrossRef]
6. Fraser, S.D.S.; Lock, K. Cycling for transport and public health: A systematic review of the effect of the environment on cycling. *Eur. J. Public Health* **2011**, *21*, 738–743. [CrossRef] [PubMed]

7. Aletta, F.; Van Renterghem, T.; Botteldooren, D. Influence of Personal Factors on Sound Perception and Overall Experience in Urban Green Areas. A Case Study of a Cycling Path Highly Exposed to Road Traffic Noise. *Int. J. Environ. Res. Public Health* **2018**, *15*, 1118. [CrossRef] [PubMed]

8. Hull, A.; O'Holleran, C. Bicycle infrastructure: Can good design encourage cycling? *Urban Plan. Transp. Res.* **2014**, *2*, 369–406. [CrossRef]

9. Sarjala, S. Built environment determinants of pedestrians' and bicyclists' route choices on commute trips: Applying a new grid-based method for measuring the built environment along the route. *J. Transp. Geogr.* **2019**, *78*, 56–69. [CrossRef]

10. Fishman, E. Cycling as transport. *Transp. Rev.* **2016**, *36*, 18. [CrossRef]

11. Li, H. Study on Green Transportation System of International Metropolises. *Procedia Eng.* **2016**, *137*, 762–771. [CrossRef]

12. Buehler, R.; Pucher, J.; Gerike, R.; Götschi, T. Reducing car dependence in the heart of Europe: Lessons from Germany, Austria, and Switzerland. *Transp. Rev.* **2017**, *37*, 4–28. [CrossRef]

13. Frondela, M.; Vance, C. Cycling on the extensive and intensive margin: The role of paths and prices. *Transp. Res. Part A Policy Pract.* **2017**, *104*, 21–31. [CrossRef]

14. Silva, C.; Teixeira, J.; Proenca, A.; Bicalho, T.; Cunha, I.; Aguiar, A. Revealing the cycling potential of starter cycling cities: Usefulness for planning practice. *Transp. Policy* **2019**, *81*, 38–147. [CrossRef]

15. OECD/ITF. *Transition to Shared Mobility—How Large Cities Can Deliver Inclusive Transport Services*; Raport of the International Transport Forum: Paris, France, 2017; pp. 1–54. Available online: www.itf-oecd.org (accessed on 5 May 2020).

16. Kristianova, K.; Marcinkova, D. Cestné komunikácie ako verejný priestor a cestná zeleň. In *Perspektivy Území III: Veřejné Prostory a Prostranství*; ČVUT: Prague, Czech Republic, 2015; pp. 50–55. Available online: http://adminu.lhosting4.cz/Include/Data/getfile.php?id=1756&db=uzemieu (accessed on 10 May 2020).

17. Cervero, R.; Duncan, M. Walking, Bicycling, and Urban Landscapes: Evidence from the San Francisco Bay Area. *Am. J. Public Health* **2003**, *93*, 1478–1483. [CrossRef] [PubMed]

18. Sturiale, L.; Scuderi, A. The Evaluation of Green Investments in Urban Areas: A Proposal of an eco-social-green Model of the City. *Sustainability* **2018**, *10*, 4541. [CrossRef]

19. Cervero, R.; Denman, S.; Jin, Y. Network design, built and natural environments and bicycle commuting: Evidence from British cities and towns. *Transp. Policy* **2019**, *74*, 153–164. [CrossRef]

20. Campos-Sánchez, F.S.; Valenzuela-Montes, L.M.; Abarca-Álvarez, F.J. Evidence of Green Areas, Cycle Infrastructure and Attractive Destinations Working together in Development on Urban Cycling. *Sustainability* **2019**, *11*, 4730. [CrossRef]

21. Fernández-Heredia, A.; Monzón, A.; Jara-Díaz, S. Understanding cyclists' perceptions, keys for a successful bicycle promotion. *Transp. Res. Part A Policy Pract.* **2014**, *63*, 1–11. [CrossRef]

22. Ogilvie, D.; Egan, M.; Hamilton, V.; Petticrew, M. Promoting walking and cycling as an alternative to using cars: A systematic review. *BMJ* **2004**, *329*, 763–776. [CrossRef]

23. Cohen, B.; Kietzmann, J. Ride On! Mobility Business Models for the Sharing Economy. *Organ. Environ.* **2014**, *27*, 279–296. [CrossRef]

24. Drużyńska, P.; Knysak, J.; Świąder, M.; Kazak, J. Assessment of the Effectiveness of the Wrocław City Bike. *Acta Sci. Pol. Adm. Locorum* **2016**, *15*, 33–45.

25. Scotini, R.; Skinner, I.; Racioppi, F.; Fusé, V.; Bertucci, J.D.O.; Tsutsumi, R. Supporting Active Mobility and Green Jobs through the Promotion of Cycling. *Int. J. Environ. Res. Public Health* **2017**, *14*, 1603. [CrossRef] [PubMed]

26. Meenar, M.; Flamm, B.; Keenan, K. Mapping the Emotional Experience of Travel to Understand Cycle-Transit User Behavior. *Sustainability* **2019**, *11*, 4743. [CrossRef]

27. Dill, J. Bicycling for Transportation and Health: The Role of Infrastructure. *J. Public Health Policy* **2009**, *30*, 95–110. [CrossRef] [PubMed]

28. Maas, J.; Verheij, R.A.; Spreeuwenberg, P. Physical activity as a possible mechanism behind the relationship between green space and health: A multilevel analysis. *BMC Public Health* **2008**, *8*, 206. [CrossRef]

29. Li, A.; Huanga, Y.; Axhausen, K.W. An approach to imputing destination activities for inclusion in measures of bicycle accessibility. *J. Transp. Geogr.* **2020**, *82*, 102566. [CrossRef]

30. Oja, P.; Titze, S.; Bauman, A.; De Geus, B.; Krenn, P.; Reger-Nash, B.; Kohlberger, T. Health benefits of cycling: A systematic review. *Scand. J. Med. Sci. Sports* **2011**, *21*, 496–509. [CrossRef] [PubMed]

31. Lee, I.-M.; Shiroma, E.J.; Lobelo, F.; Puska, P.; Blair, S.N.; Katzmarzyk, P.T. Effect of physical inactivity on major non-communicable diseases worldwide: An analysis of burden of diseases and life expectancy. *Lancet* **2012**, *380*, 219–229. [CrossRef]

32. Bauman, A.E.; Reis, R.S.; Sallis, J.F.; Wells, J.C.; Loos, R.J.; Martin, B.W. Correlates of physical activity: Why are some people physically active and others not? *Lancet* **2012**, *380*, 258–271. [CrossRef]

33. Götschi, T.; Garrard, J.; Giles-Corti, B. Cycling as a Part of Daily Life: A Review of Health Perspectives. *Transp. Rev.* **2016**, *36*, 45–71. [CrossRef]

34. Pucher, J.; Dijkstra, L. Promoting Safe Walking and Cycling to Improve Public Health: Lessons from the Netherlands and Germany. *Am. J. Public Health* **2003**, *93*, 1509–1516. [CrossRef]

35. Lakerveld, J.; Woods, C.; Hebestreit, A.; Brenner, H.; Flechtner-Mors, M.; Harrington, J.M.; Kamphuis, C.B.M.; Laxyi, M.; Luszczynska, A.; Mazzocchi, M.; et al. Advancing the evidence base for public policies impacting on dietary behaviour, physical activity and sedentary behaviour in Europe: The Policy Evaluation Network promoting a multidisciplinary approach. *Food Policy* **2020**. [CrossRef]

36. Gutierrez, M.; Cantillo, V.; Arellana, J.; Ortúzar, J. Estimating bicycle demand in an aggressive environment. *Int. J. Sustain. Transp.* **2020**, 1–14. [CrossRef]

37. Ortúzar, J. Sustainable Urban Mobility: What Can Be Done to Achieve It? *J. Indian Inst. Sci.* **2019**, *99*, 683–693. [CrossRef]

38. Fallon Mayers, R.; Glover, T.D. Whose Lane Is It Anyway? The Experience of Cycling in a Mid-Sized City. *Leis. Sci.* **2019**. [CrossRef]

39. Reynolds, C.C.; Harris, M.A.; Teschke, K.; Cripton, P.A.; Winters, M. The impact of transportation infrastructure on bicycling injuries and crashes: A review of the literature. *Environ. Health* **2009**, *8*, 47. [CrossRef]

40. Lawson, A.R.; Ghosh, B.; Pakrashi, V. Quantifying the perceived safety of cyclists in Dublin. *Proc. Inst. Civ. Eng. Transp.* **2015**, *168*, 290–299. [CrossRef]

41. Aldred, R.; Jungnickel, K. Why culture matters for transport policy: The case of cycling in the UK. *J. Transp. Geogr.* **2014**, *34*, 78–87. [CrossRef]

42. Helbich, M.; Böcker, L.; Dijst, M. Geographic heterogeneity in cycling under various weather conditions: Evidence from Greater Rotterdam. *J. Transp. Geogr.* **2014**, *38*, 38–47. [CrossRef]

43. Nosal, T.; Miranda-Moreno, L.F. The effect of weather on the use of North American bicycle facilities: A multi-city analysis using automatic counts. *Transp. Res. Part A Policy Pract.* **2014**, *66*, 213–225. [CrossRef]

44. Kachaniak, D.; Reducha, M.; Skrzyńska, J. *Raport z Badania na Temat Uwarunkowań do Podejmowania Transportowej Aktywności Fizycznej Polaków*; TNS POLSKA dla Ministerstwa Sportu i Turystyki Rzeczpospolitej Polskiej: Warszawa, Poland, 2015; p. 36. Available online: https://www.msit.gov.pl/pl/sport/badania-i-analizy/aktywnosc-fizyczna-spol/575 (accessed on 10 May 2020).

45. Bíla, M.; Bílová, M.; Müller, I. Critical factors in fatal collisions of adult cyclists with automobiles. *Accid. Anal. Prev.* **2010**, *42*, 1632–1636. [CrossRef]

46. Zawadzka, A.K. Making Small Towns Visible in Europe: The Case of Cittaslow Network—The Strategy Based on Sustainable Development. *Transylv. Rev. Adm. Sci.* **2017**, 90–106. [CrossRef]

47. Jaszczak, A. The Future of Cittaslow Towns. *Archit. Kraj.* **2015**, *1*, 70–81.

48. Wierzbicka, W.; Farelnik, E.; Stanowicka, A. The Development of the Polish National Cittaslow Network. *Olszt. Econ. J.* **2019**, *14*, 113–125. [CrossRef]

49. Farelnik, E.; Stanowicka, A. Smart city, slow city and smart slow city as development models of modern cities. *Olszt. Econ. J.* **2016**, 11. [CrossRef]

50. Jaszczak, A.; Kristianova, K. Social and Cultural Role of Greenery in Development of Cittaslow Towns. *IOP Conf. Ser. Mater. Sci. Eng.* **2019**, *603*, 032028. [CrossRef]

51. Ustawa z Dnia 21 Marca 1985 r. o Drogach Publicznych (Dz. U. 2020 poz. 470). Available online: http://prawo.sejm.gov.pl/isap.nsf/DocDetails.xsp?id=wdu19850140060 (accessed on 10 April 2020).

52. Available online: www.gddkia.gov.pl (accessed on 15 April 2020).

53. ZDW Olsztyn. Available online: http://www.zdw.olsztyn.pl/strona-glowna/aktualnosci.html?start=64 (accessed on 15 April 2020).

54. Opoczyński, K. *Wytyczne Pomiaru Ruchu na Drogach Wojewódzkich w 2010 Roku*; Generalna Dyrekcja Dróg i Autostrad: Warszawa, Poland, 2009. Available online: https://www.gddkia.gov.pl/userfiles/articles/g/GENERALNY_POMIAR_RUCHU_2010/0.1.1.2_Wytyczne_DW_GPR_2010.pdf (accessed on 15 March 2020).

55. Kowalski, K.; Kaplar, I.; Maśkiewicz, J.; Dobrzyński, Ł.; Wojdyński, R. *Wytyczne Generalnego Pomiaru Ruchu w 2015 na Drogach Wojewódzkich*; Generalna Dyrekcja Dróg Krajowych i Autostrad: Warszawa, Poland, 2014. Available online: https://www.gddkia.gov.pl/userfiles/articles/g/generalny-pomiar-ruchu-w-2015_15598/Wytyczne%20GPR%20na%20dr%20woj%20w%202015%20r_zakceptowane_ok.pdf (accessed on 15 March 2020).

56. Opoczyński, K. *Wytyczne Organizacji i Przeprowadzenia Generalnego Pomiaru Ruchu w 2010 na Drogach Krajowych*; Załącznik do Zarządzenia nr 59 z Dnia 12.10.2009; Generalna Dyrekcja Dróg Krajowych i Autostrad: Warszawa, Poland, 2009. Available online: https://www.gddkia.gov.pl/userfiles/articles/g/GENERALNY_POMIAR_RUCHU_2010/0.1.1.1_Wytyczne_GPR_2010.pdf (accessed on 15 March 2020).

57. Kowalski, K.; Kaplar, I.; Maśkiewicz, J.; Dobrzyński, Ł.; Wojdyński, R. *Metoda Przeprowadzenia Generalnego Pomiaru Ruchu w Roku 2015*; Generalna Dyrekecja Dróg i Autostrad: Warszawa, Poland, 2014. Available online: https://www.gddkia.gov.pl/userfiles/articles/z/zarzadzenia-generalnego-dyrektor_13901/zarzadzenie%2038%20Zalacznik_b_Metoda%20GPR%202015.pdf (accessed on 15 March 2020).

58. Regional Operational Programme of the Voivodeship of Warmia and Mazury 2014-2020. Available online: https://ec.europa.eu/growth/tools-databases/regional-innovation-monitor/policy-document/regional-operational-programme-voivodeship-warmia-and-mazury-2014-2020 (accessed on 15 April 2020).

59. Ponadlokalny Programu Rewitalizacji Miast Cittaslow. 2015. Available online: https://www.wmarr.olsztyn.pl/s/images/stories/Pliki/2015_06_08_Ponadlokalny_program_rewitalizacji_sieci_miast_Cittaslow.pdf (accessed on 10 February 2020).

60. Ponadlokalny Program Rewitalizacji Miast Cittaslow. 2019. Available online: https://cittaslowpolska.pl/images/PDF/PPR_08_2019.pdf (accessed on 10 February 2020).

61. Raport "Transportowe Zwyczaje Polaków" dla Busradar.pl, January 2019. Available online: www.transport-publiczny.pl (accessed on 10 March 2020).

62. Bank Danych Lokalnych Głównego Urzędu Statystycznego. Available online: www.bdl.stat.gov.pl (accessed on 18 May 2020).

63. Łysoń, P.; Barlik, M.; Lewandowska, B.; Siwiak, K. *Zeszyt Metodologiczny*; Badanie Budżetów Gospodarstw Domowych; Główny Urząd Statystyczny, Departament Badań Społecznych: Warszawa, Poland, 2018. Available online: https://stat.gov.pl/download/gfx/portalinformacyjny/pl/defaultaktualnosci/5486/10/2/1/zeszyt_metodologiczny_badanie_budzetow_gospodarstw_domowych.pdf (accessed on 15 April 2020).

64. GDDKiA Olsztyn. Available online: www.gddkia.gov.pl (accessed on 15 April 2020).

65. Symon, E. *Wypadki Drogowe w Polsce w 2010r*; Zespół Profilaktyki i Analiz Biura Ruchu Drogowego Komendy Głównej Policji: Warszawa, Poland, 2011. Available online: http://statystyka.policja.pl/st/ruch-drogowy/76562,wypadki-drogowe-raporty-roczne.html (accessed on 10 March 2020).

66. Symon, E. *Wypadki Drogowe w Polsce w 2015r*; Wydział Opiniodawczo-Analityczny Biura Ruchu Drogowego Komendy Głównej Policji: Warszawa, Poland, 2016. Available online: http://statystyka.policja.pl/st/ruch-drogowy/76562,wypadki-drogowe-raporty-roczne.html (accessed on 10 March 2020).

 *sustainability*

Article

Selected Elements of Technical Infrastructure in Municipalities Territorially Connected with National Parks

Alina Kulczyk-Dynowska * and Agnieszka Stacherzak

Department of Spatial Economy, Wrocław University of Environmental and Life Sciences, Grunwaldzka 53, 50-357 Wrocław, Poland; agnieszka.stacherzak@upwr.edu.pl
* Correspondence: alina.kulczyk-dynowska@upwr.edu.pl; Tel.: +48-71-320-5409

Received: 30 March 2020; Accepted: 11 May 2020; Published: 14 May 2020

Abstract: The article addresses the problem of selected technical infrastructure elements (e.g., water supply, sewage, gas networks) in municipalities territorially connected with Polish national parks. Therefore, the research refers to the specific areas: both naturally valuable and attractive in terms of tourism. The time range of the research covers the years 2003–2018. The studied networks were characterized based on the statistical analysis using linear ordering methods; synthetic measures of development were applied. It allowed the ranking construction of the examined municipalities in terms of the development level of water supply, sewage, and gas networks. The results show that the period 2003–2018 was characterized by a development of the analyzed networks in the vast majority of municipalities. Thus, the level of anthropopressure caused by the presence of local community and tourists in municipalities showed a decline. It is worth emphasizing that the infrastructure investments are carried out comprehensively. Favoring investments in the development of any of the abovementioned networks was not observed.

Keywords: local development; municipalities; nature protection; national park; technical infrastructure

---

1. Introduction

National parks represent a widely recognized spatial forms of nature conservation, which cover valuable natural areas [1]. There are 23 of national parks in Poland and their total area is only 315 thousand ha (1% of the country area). However, the values they offer make them interesting not only for the nature specific reasons but also in terms of the development of economic, local, and regional systems.

The pursuit of preserving natural heritage, combined with the physical functioning of human beings in space—and bearing in mind that protected areas are not the closed ones—requires appropriate technical infrastructure, which also minimizes the effects of anthropopressure in both the protected and the adjacent urbanized areas. It is all the more important that, apart from the local residents, tourists take advantage of the described spaces. Anthropopressure is the environmental effects of using natural resources to meet people's needs [2] and results from the impact exerted by both the local community and its visitors. It should be emphasized that the function of tourists in individual municipalities territorially connected with national parks has a different intensity [3] and in the case of parks it has to result from the nature conservation function [4]. It is important to keep in mind that the value streams between the protected area and its surroundings are subject to an ongoing exchange process [5], hence the quality of the broadly approached municipal technical infrastructure has a huge impact on national parks. The fact that nature does not respect administrative boundaries is reflected in Polish legislation. In order to improve the conditions for the protection of fauna, national parks are surrounded by buffer zones within which wild game protection zones are created. Although buffer zones are created by way of regulation issued by the minister competent for the environment, the protection of wild game remains within the responsibility of the director of a given national park [4].

Technical infrastructure consists of many elements (see Section 2). Its basic components have direct impact on reducing environmental pressure and ensure the sanitary safety of its users include the water supply and sewage network. The gas network allows apartment heating to be free from pollution, as the effect of solid fuel combustion is supplementary. Therefore, the long-term characteristics of technical infrastructure aimed at environmental protection (e.g., sewage, water supply networks) and gas network were adopted as the research purpose.

The following research hypothesis was adopted: "In the municipalities territorially connected with national parks, the pursuit towards reducing anthropopressure through the development of water supply, sewage, and gas networks is observed".

The applied statistical methods are described in detail in the methodological notes. It should, however, be noted that the authors decided to apply synthetic development measures (SDMs). These measures allow for the construction of rankings of the analyzed objects (municipalities) and also the performance of subsequent comparative analyzes. The choice of SDMs resulted from a long tradition of their application and high usability level [6–12].

The research was carried out in the period of 2003–2018 and was based on the data provided by Statistics Poland: Local Date Bank (for details, see methodological notes). The research period derives from the statistical data availability.

2. Infrastructure: The Context of Attractive Protected Areas in Terms of Tourism

The concept of technical infrastructure is quite extensive. Traditionally, it covers transport, communication, power engineering, irrigation, drainage, water supply, sewage, telecommunication, and other devices [13,14]. Following the new approach, it also covers the so-called green infrastructure relevant for the adaptation to climate change [15,16]. For the purposes of environment protection, water supply, sewage, and gas networks are still perceived as luxuries in developing countries. Broadly defined technical infrastructure not only provides comfort and safety to people, but it also aims at minimizing anthropopressure, especially in naturally valuable areas, and its design should remain as neutral in terms of the landscape as possible [17,18].

In Polish conditions, infrastructure investments are within the public sector domain; their significant part is implemented directly by the municipalities [19,20]. The majority of Polish national parks are territorially connected with rural municipalities. Hence, it is worth emphasizing that in rural areas the characteristic feature of the discussed investments is the requirement of cooperation involving local authorities and socio-occupational organizations of farmers, individual farmers, advisory services, and local government administration [21].

An important nuance of the research on technical infrastructure is the fact that some national parks and also the municipalities connected with them represent popular tourist destinations, and are thus affected by the phenomenon of mass tourism [22–24] and the related problems characteristic for Poland's most popular places [25]. A growing interest in both education and tourism in national parks has been observed for several years. It is extremely important that in the context of the World Tourism Organization (UNWTO) reports, it is indicated that there is an ongoing increase in tourism-oriented activity and, thus, tourism has become the foundation of local development [26]. A growing number of people that visit a given space automatically translates into a higher burden on infrastructure and demand for resources. The problem of water supply and sewage network efficiency and the available drinking water resources is of the utmost importance. At this point it is worth highlighting that Poland is a very poor country in terms of water [27,28].

3. Methodological Remarks

The research was initiated with a library query. Due to the fact that the term "technical infrastructure" is complex and covers many elements of the conducted research, it was limited to the selected aspects as the most important for environment protection, i.e., water supply, sewage, and gas networks.

During our research, statistical tools were used that allowed us to obtain results useful for presenting conclusions and recommendations. It was decided to apply analytical methods, including the linear ordering method—i.e, synthetic development measures (SDMs). It was adopted that the analyzed municipalities form one set of 117 objects. SDM was developed in three examined areas: the water supply network (SDMwater), the sewage network (SDMsewage), and the gas network (SDMgas).

It should be emphasized that not all municipalities reported the existence of the analyzed infrastructure elements. The following three municipalities reported no sewage network: Giby, Krasnopol, and Kobylin-Borzymy. As many as 17 municipalities reported no gas network: Sosnowica, Stary Brus, Urszulin, Kamienica, Szczawnica, Zawoja, Lutowiska, Bargłów Kościelny, Lipsk, Grajewo, Radziłów, Jedwabne, Wizna, Giby, Krasnopol, Nowy Dwór, and Szczawnica. To maintain SDM comparability in the abovementioned cases, the data were supplemented with zero values. These municipalities were finally assigned the last, equivalent position.

The identification of municipalities territorially connected with national parks was the first step of the conducted research procedure.

The construction of SDM started with determining variables for all three measures. Next, the variables were unified for the entire period simultaneously, i.e., for the years 2003–2018. Using the standardized sum method, SDM was developed with a weight system (a common development pattern was adopted for the entire analyzed period). It allowed us to determine in each analyzed year the ranking positions of municipalities developed for individual SDMs and to compare the changes in terms of the positions taken by the municipalities in these rankings.

The following indicators were defined for the purposes of determining SDMwater:

- accessibility index of social water supply network (DSwater)

$$DSwater = \frac{length\ of\ water\ supply\ distribution\ network\ in\ km}{number\ of\ municipality\ residents} \tag{1}$$

- adjusted accessibility index of social water supply network (SDSwater)

$$SDSwater = \frac{length\ of\ water\ sypply\ distribution\ network\ in\ km}{number\ of\ municipality\ residents + (number\ of\ tourists\ using\ accommodation : 365)} \tag{2}$$

- accessibility index of spatial water supply network (DPwater)

$$DPwater = \frac{length\ of\ water\ supply\ distribution\ network\ in\ km}{municipality\ area\ in\ km2} \tag{3}$$

- index of population using water supply network (Lwater)

$$Lwater = \frac{number\ of\ people\ using\ water\ supply\ network}{number\ of\ municipality\ residents} * 100 \tag{4}$$

To determine SDMsewage the following indicators were defined:

- accessibility index of social sewage network (DSsewage)

$$DSsewage = \frac{length\ of\ sewage\ network\ in\ km}{number\ of\ municipality\ residents} \tag{5}$$

- adjusted accessibility index of social sewage network (SDSsewage)

$$SDSsewage = \frac{length\ of\ sewage\ network\ in\ km}{number\ of\ municipality\ residents + (number\ of\ tourists\ using\ accommodation : 365)} \tag{6}$$

- accessibility index of spatial sewage network (DPsewage)

$$DPsewage = \frac{length\ of\ sewage\ network\ in\ km}{municipality\ area\ in\ km2} \tag{7}$$

- index of population using sewage network (Lsewage)

$$Lsewage = \frac{number\ of\ people\ using\ sewage\ network}{number\ of\ municipality\ residents} * 100 \tag{8}$$

It should be clarified that the calculation of the adjusted SDSwater and SDSsewage indexes was intended to capture both water supply and sewage networks' usage by tourists. The authors are aware of the imperfections inherent in the proposed measures, as they do not cover seasonal fluctuations or hikers (i.e., people not using accommodation) [29], but the availability of statistical data and the simultaneous striving to maintain comparability of the research results for all 117 municipalities did not allow for a different construction. Statistics Poland provides the total number of tourists for the entire year. Dividing this value by 365 days allowed us to obtain the average number of tourists each day of the year. The number of residents, according to the Statistics Poland data, is constant for all days of the year. Therefore, adding these values allows showing the adjusted number of people using the networks under study and, hence, presenting the synthetic measure of social accessibility of the analyzed networks.

To determine SDMgas the following indicators were defined:

- index of population using gas network (Lgas)

$$Lgas = \frac{number\ of\ people\ using\ gas\ network}{number\ of\ municipality\ residents} * 100 \tag{9}$$

- index of population heating the apartment with gas (Ogas)

$$Ogas = \frac{number\ of\ people\ heating\ the\ apartment\ with\ gas}{number\ of\ municipality\ residents} * 100 \tag{10}$$

Due to the fact that Statistics Poland only provides data regarding the length of the functioning gas network for the years 2003–2006, the calculation of analogical indicators, as in the case of the previous two SDMs, was not possible. The data on the population heating their apartments with gas also needed to be supplemented. Statistics Poland presents data only for the period 2005–2018, so the data covering the years 2003–2004 were adopted at the level of the data for 2005.

For all three SDMs, the Statistics Poland [30] data were used to calculate the listed indicators. All of them were considered stimulants without veto threshold, which means that the municipalities that presented high values of the aforementioned indicators were assessed as the units with the highest-ranking positions.

Unitarization procedure was performed after determining indicators for each SDM [31]. The unitarization covering values of the features adopted for the study was carried out according to the following formula:

$$Z_{jit} = \frac{X_{jit} - minX_{jit}}{maxX_{jit} - minX_{jit}} \tag{11}$$

notes:

x: feature value

j: j variable, where j = (1, … , p)

i: object (municipality), where I = (1, … , N),

N for each SDM = 117

T: time (year), where t = (2003, 2004, … , 2018)

Unitarization resulted in obtaining values in the range [0,1]. There was no need to harmonize variables' character (preference function) as they were all stimulants in each SDM. SDM was calculated using the standardized sum method [32]. The value of SDM for the analyzed municipalities was calculated using Equation (12):

$$\text{SDM}_{it} = \frac{1}{p} \sum_{j=1}^{p} z_{ijt} \, (i = 1, \dots, N)(t = 2003, \dots, 2018) \tag{12}$$

notes:

SDM: value of the non-model synthetic measure in an object (municipality) and

p: number of features.

For all four SDMs presented above, the highest value is equivalent to the most favorable situation. In the final phase, ranking positions were assigned to the analyzed municipalities and comparisons were made regarding the position determined by the analyzed SDMs.

It should be emphasized that for the purposes of the presented data readability, the tables present data for the selected years, i.e., 2003 and 2018 (the first and the last year of the study).

4. Level and Transformations of the Selected Technical Infrastructure Elements in the Municipalities Territorially Connected with National Parks

The analysis of the studied areas' location allows stating that national parks include coastal, lake, lowland, upland, and mountain areas (see Figure 1).

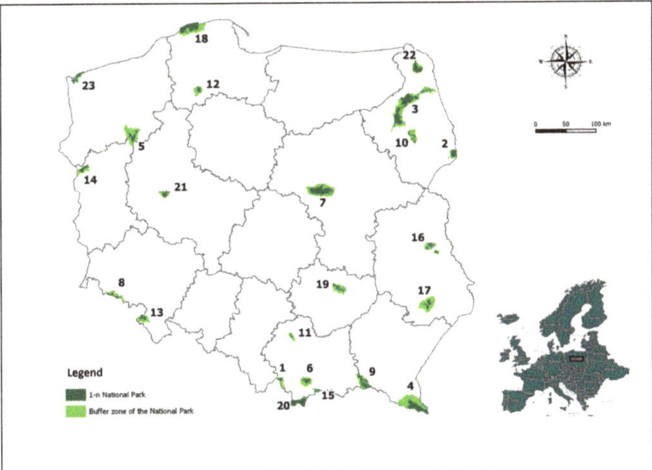

Figure 1. The location of Polish national parks and their buffer zones against the background of the administrative division of the country (voivodship level). Legend: Buffer zone: the area in which protection zones for wild game are created. 1. Babia Góra National Park, 2. Białowieża National Park, 3. Biebrza National Park, 4. Bieszczady National Park, 5. Drawno National Park, 6. Gorce National Park, 7. Kampinos National Park, 8. Karkonosze National Park, 9. Magura National Park, 10. Narew National Park, 11. Ojców National Park, 12. Bory Tucholskie National Park, 13. Stolowe Mountains National Park, 14. Warta Mouth National Park, 15. Pieniny National Park, 16. Polesie National Park, 17. Roztocze National Park, 18. Słowiński National Park, 19. Świętokrzyski National Park, 20. Tatra National Park, 21. Wielkopolska National Park, 22. Wigry National Park, and 23. Wolin National Park.

The administrative division in Poland identifies municipalities, districts, and voivodships. Voivodships correspond to the second level of geocoding in the European Union, the so-called

Nomenclature of Territorial Units for Statistics (NUTS). Due to their importance for the correct identification of the detailed location of individual national parks, the boundaries of NUTS 2 are presented in Figure 1.

The identification of municipalities territorially connected with Polish national parks is presented in the table below (Table 1). It should be emphasized that national parks are territorially connected with as many as 117 municipalities (11 of them have the status of an urban municipality, 31 the status of an urban-rural municipality, and 75 the status of a rural municipality).

Table 1. Municipalities territorially connected with national parks in Poland.

No.	National Park	Municipalities Territorially Connected Municipality	No. of Municipalities
1	Babia Góra	Jabłonka (2), Lipnica Wielka (2), Zawoja (2)	3
2	Białowieża	Białowieża (2), Narewka (2)	2
3	Biebrza	Bargłów Kościelny (2), Dąbrowa Białostocka (3), Goniądz (3), Grajewo (2), Jaświły (2), Jedwabne (3), Lipsk (3), Nowy Dwór (2), Radziłów (2), Rajgród (3), Suchowola (3), Sztabin (2), Trzcianne (2), Wizna (2)	14
4	Bieszczady	Cisna (2), Czarna (2), Lutowiska (2), Ustrzyki Dolne (3)	4
5	Tuchola Forest	Brusy (3), Chojnice (2)	2
6	Drawno	Bierzwnik (2), Człopa (3), Dobiegniew (3), Drawno (3), Krzyż Wielkopolski (3), Tuczno (3)	6
7	Gorce	Kamienica (2), Mszana Dolna (2), Niedźwiedź (2), Nowy Targ (2), Ochotnica Dolna (2)	5
8	Stołowe Mountains	Kudowa-Zdrój (1), Lewin Kłodzki (2), Radków (3), Szczytna (3)	4
9	Kampinos	Brochów (2), Czosnów (2), Izabelin (2), Kampinos (2), Leoncin (2), Leszno (2), Łomianki (3), Stare Babice (2)	8
10	Karkonosze	Jelenia Góra (1), Karpacz (1), Kowary (1), Piechowice (1), Podgórzyn (2), Szklarska Poręba (1)	6
11	Magura	Dębowiec (2), Dukla (3), Krempna (2), Lipinki (2), Nowy Żmigród (2), Osiek Jasielski (2), Sękowa (2)	7
12	Narew	Choroszcz (3), Kobylin-Borzymy (2), Łapy (3), Sokoły (2), Suraż (3), Turośń Kościelna (2), Tykocin (3)	7
13	Ojców	Jerzmanowice-Przeginia (2), Skała (3), Sułoszowa (2), Wielka Wieś (2)	4
14	Pieniny	Czorsztyn (2), Krościenko nad Dunajcem (2), Łapsze Niżne (2), Szczawnica (3)	4
15	Polesie	Hańsk (2), Ludwin (2), Sosnowica (2), Stary Brus (2), Urszulin (2), Wierzbica (2)	6
16	Roztocze	Adamów (2), Józefów (3), Zamość (2), Zwierzyniec (3)	4
17	Słowiński	Główczyce (2), Łeba (1), Smołdzino (2), Ustka (2), Wicko (2)	5
18	Świętokrzyski	Bieliny (2), Bodzentyn (3), Górno (2), Łączna (2), Masłów (2), Nowa Słupia (2)	6
19	Tatra	Bukowina Tatrzańska (2), Kościelisko (2), Poronin (2), Zakopane (1)	4
20	Warta Mouth	Górzyca (2), Kostrzyn nad Odrą (1), Słońsk (2), Witnica (3)	4
21	Wielkopolska	Dopiewo (2), Komorniki (2), Mosina (3), Puszczykowo (1), Stęszew (3)	5
22	Wigry	Giby (2), Krasnopol (2), Nowinka (2), Suwałki (2)	4
23	Wolin	Międzyzdroje (3), Świnoujście (1), Wolin (3)	3
	Sum		**117**

Legend: (1) urban municipality, (2), rural municipality, and (3) urban-rural municipality.

As it has been indicated in methodological notes, SDMs were used to describe the analyzed elements of technical infrastructure. The research results for the first and the last year of the study are presented in the table below (Table 2).

Table 2. Synthetic development measures (SDMs) of water supply, sewage, and gas network for the municipalities territorially connected with national parks. Data covers the years 2003 and 2018.

Municipality Name	Water Supply Network				Sewage Network				Gas Network			
	2003		2018		2003		2018		2003		2018	
	SDM	L	SDM	L	SDM	L	SDM	L	SDM	L	SDM	L
Adamów (2)	0.1533	101	0.2779	90	0.0019	110	0.0069	114	0.0497	45	0.1518	42
Bargłów Kościelny (2)	0.7158	1	0.7719	1	0.0511	91	0.0552	111	0.0000	62	0.0000	73
Białowieża (2)	0.3421	53	0.3891	64	0.2672	10	0.4331	12	0.0000	62	0.0140	61
Bieliny (2)	0.2542	76	0.5088	26	0.0817	71	0.3012	44	0.0000	62	0.0198	58
Bierzwnik (2)	0.3653	42	0.3806	69	0.0136	103	0.2781	51	0.0000	62	0.0000	73
Bodzentyn (3)	0.3586	47	0.4461	43	0.0547	89	0.2633	58	0.0000	62	0.0000	73
Brochów (2)	0.3383	57	0.4266	51	0.0878	64	0.2042	75	0.0008	58	0.0204	57
Brusy (3)	0.3506	50	0.4145	56	0.1667	34	0.3389	29	0.0003	61	0.0003	72
Bukowina Tatrzańska (2)	0.1469	102	0.1419	109	0.0833	68	0.2162	72	0.0000	62	0.0004	70
Chojnice (2)	0.5153	7	0.5011	27	0.2461	13	0.3087	39	0.0000	62	0.0775	49
Choroszcz (3)	0.3849	35	0.3991	61	0.1204	52	0.1874	79	0.0175	49	0.0761	50
Cisna (2)	0.0975	112	0.2115	105	0.1278	46	0.2688	56	0.0000	62	0.0000	73
Czarna (2)	0.1618	98	0.2450	97	0.0472	92	0.0822	108	0.0285	48	0.0265	55
Człopa (3)	0.3067	67	0.3766	71	0.1657	35	0.2346	63	0.0000	62	0.0000	73
Czorsztyn (2)	0.2332	81	0.2570	94	0.3132	3	0.3891	18	0.0000	62	0.0000	73
Czosnów (2)	0.1816	93	0.4747	33	0.0444	94	0.3638	24	0.4539	20	0.5986	11
Dąbrowa Białostocka (3)	0.4265	25	0.5338	20	0.1286	45	0.1457	92	0.0000	62	0.0000	73
Dębowiec (2)	0.1750	95	0.2490	95	0.0243	101	0.2089	74	0.5666	9	0.5481	16
Dobiegniew (3)	0.3147	65	0.4100	59	0.2320	17	0.3650	23	0.0000	62	0.0000	73
Dopiewo (2)	0.4325	22	0.4275	50	0.1272	47	0.4025	14	0.2802	30	0.8027	6
Drawno (3)	0.2603	75	0.3695	74	0.1842	25	0.2481	61	0.0876	42	0.1819	41
Dukla (3)	0.2151	86	0.2676	92	0.0792	73	0.2258	68	0.5286	11	0.5271	18
Giby (2)	0.4419	20	0.4871	31	0.0000	112	0.0000	115	0.0000	62	0.0000	73
Główczyce (2)	0.2927	71	0.2997	89	0.1238	49	0.2138	73	0.0000	62	0.0000	73
Goniądz (3)	0.3385	56	0.3705	72	0.0713	78	0.1255	98	0.0000	62	0.0000	73
Górno (2)	0.3789	37	0.4424	44	0.0107	105	0.2654	57	0.0000	62	0.0016	68
Górzyca (2)	0.3748	41	0.3914	63	0.1932	23	0.3666	22	0.0312	47	0.1375	43
Grajewo (2)	0.3776	38	0.6573	5	0.0078	106	0.0073	113	0.0000	62	0.0000	73
Hańsk (2)	0.2707	74	0.4482	42	0.1755	30	0.1649	85	0.8488	3	0.8487	5
Izabelin (2)	0.1893	91	0.3816	68	0.0025	109	0.3375	30	0.8488	3	0.8487	5
Jabłonka (2)	0.3598	45	0.2315	103	0.2114	19	0.5479	5	0.0000	62	0.0000	73
Jaświły (2)	0.4459	19	0.5930	12	0.0691	79	0.1788	83	0.0000	62	0.0000	73
Jedwabne (3)	0.1624	97	0.3310	83	0.0576	86	0.0788	109	0.0000	62	0.0000	73
Jelenia Góra (1)	0.3936	32	0.4846	32	0.3097	4	0.4002	17	0.5540	10	0.5291	17
Jerzmanowice-Przeginia (2)	0.4577	17	0.4903	30	0.0000	112	0.1449	93	0.4597	18	0.5797	12
Józefów (3)	0.3432	52	0.3880	65	0.0227	102	0.1415	95	0.2092	33	0.3169	33
Kamienica (2)	0.1584	99	0.2351	100	0.0898	62	0.3249	32	0.0000	62	0.0000	73
Kampinos (2)	0.5493	3	0.6108	10	0.0827	69	0.2225	70	0.0005	60	0.0191	59
Karpacz (1)	0.3513	49	0.3777	70	0.2203	18	0.5094	7	0.5924	6	0.6000	10
Kobylin-Borzymy (2)	0.5315	5	0.5678	15	0.0000	112	0.0000	115	0.0000	62	0.0000	73
Komorniki (2)	0.4306	23	0.4408	46	0.1841	26	0.3736	20	0.4629	17	0.9347	2
Kostrzyn nad Odrą (1)	0.3216	64	0.3608	78	0.2807	8	0.3030	43	0.5828	7	0.6777	8
Kościelisko (2)	0.1555	100	0.3235	84	0.1806	27	0.2773	52	0.0000	62	0.0004	71
Kowary (1)	0.3234	62	0.3506	80	0.2405	15	0.3111	38	0.5197	12	0.4977	21
Krasnopol (2)	0.2005	89	0.4190	54	0.0000	112	0.0000	115	0.0000	62	0.0000	73
Krempna (2)	0.2295	82	0.2329	101	0.0114	104	0.1885	78	0.0010	57	0.0000	73
Krościenko nad Dunajcem (2)	0.2029	88	0.2416	98	0.1635	37	0.2835	46	0.0000	62	0.0000	73
Krzyż Wielkopolski (3)	0.3487	51	0.4379	47	0.1945	21	0.2433	62	0.0000	62	0.0000	73
Kudowa-Zdrój (1)	0.3059	68	0.3499	81	0.2875	7	0.3197	35	0.4704	16	0.4437	26
Leoncin (2)	0.1199	108	0.3126	87	0.0617	83	0.1178	103	0.0007	59	0.0000	73
Leszno (2)	0.3768	39	0.4205	53	0.0967	60	0.1281	97	0.2855	29	0.3793	31
Lewin Kłodzki (2)	0.2188	84	0.2453	96	0.1736	31	0.3432	27	0.3587	24	0.3167	34
Lipinki (2)	0.0000	117	0.0097	116	0.1485	41	0.4353	11	0.4592	19	0.5662	14
Lipnica Wielka (2)	0.0234	115	0.0000	117	0.1759	29	0.5095	6	0.0000	62	0.0013	69
Lipsk (3)	0.1728	96	0.5606	16	0.1074	55	0.1224	101	0.0000	62	0.0000	73
Ludwin (2)	0.4960	10	0.5926	13	0.0886	63	0.1811	81	0.0526	44	0.1262	45
Lutowiska (2)	0.2061	87	0.1972	106	0.1939	22	0.3167	36	0.0000	62	0.0000	73

Table 2. *Cont.*

Municipality Name	Water Supply Network 2003 SDM	L	2018 SDM	L	Sewage Network 2003 SDM	L	2018 SDM	L	Gas Network 2003 SDM	L	2018 SDM	L
Łapsze Niżne (2)	0.1009	111	0.1244	110	0.3000	5	0.3112	37	0.0000	62	0.0000	73
Łapy (3)	0.3404	55	0.3624	77	0.2627	11	0.3353	31	0.0110	51	0.0589	53
Łączna (2)	0.3857	34	0.4422	45	0.0062	107	0.1893	77	0.0081	53	0.0215	56
Łeba (1)	0.4872	11	0.5336	21	0.3757	2	0.4014	16	0.0000	62	0.0848	47
Łomianki (3)	0.1271	107	0.4738	34	0.1200	53	0.3629	25	0.9253	2	0.9368	1
Masłów (2)	0.2253	83	0.3834	67	0.0389	96	0.3040	42	0.0046	54	0.0632	52
Międzyzdroje (3)	0.2942	70	0.3082	88	0.2588	12	0.3243	33	0.3557	25	0.6014	9
Mosina (3)	0.3338	58	0.3634	75	0.0844	67	0.3080	40	0.1841	36	0.5079	20
Mszana Dolna (2)	0.0805	113	0.0764	114	0.0572	87	0.2700	55	0.3626	23	0.3882	28
Narewka (2)	0.4584	16	0.6556	6	0.1656	36	0.3800	19	0.0000	62	0.0051	64
Niedźwiedź (2)	0.1149	109	0.2129	104	0.0000	112	0.1810	82	0.3293	26	0.4262	27
Nowa Słupia (2)	0.2453	79	0.4715	35	0.0629	82	0.1997	76	0.0000	62	0.0056	62
Nowinka (2)	0.3410	54	0.6930	3	0.0290	100	0.6646	1	0.0000	62	0.0000	73
Nowy Dwór (2)	0.5759	2	0.7458	2	0.1214	51	0.1237	99	0.0000	62	0.0000	73
Nowy Targ (2)	0.1810	94	0.1240	111	0.0870	65	0.2318	65	0.0460	46	0.0806	48
Nowy Żmigród (2)	0.1295	105	0.0879	113	0.0032	108	0.2825	47	0.4476	21	0.4858	24
Ochotnica Dolna (2)	0.0102	116	0.0122	115	0.0538	90	0.4214	13	0.0000	62	0.0000	73
Osiek Jasielski (2)	0.1315	104	0.1572	108	0.0014	111	0.2242	69	0.5168	13	0.5643	15
Piechowice (1)	0.3963	30	0.3632	76	0.1769	28	0.2293	67	0.4945	15	0.4747	25
Podgórzyn (2)	0.3589	46	0.4253	52	0.1033	58	0.4627	10	0.2563	32	0.2969	36
Poronin (2)	0.2425	80	0.2679	91	0.2443	14	0.2848	45	0.0142	50	0.0142	60
Puszczykowo (1)	0.4567	18	0.5361	18	0.1234	50	0.5900	2	0.6116	5	0.7240	7
Radków (3)	0.3308	60	0.3699	73	0.0852	66	0.2804	50	0.1281	38	0.1218	46
Radziłów (2)	0.3948	31	0.4922	29	0.0726	77	0.1553	89	0.0000	62	0.0000	73
Rajgród (3)	0.1281	106	0.4939	28	0.0734	75	0.1090	105	0.0000	62	0.0000	73
Sękowa (2)	0.1416	103	0.2316	102	0.1672	33	0.2815	49	0.2950	28	0.3694	32
Skała (3)	0.4269	24	0.4599	38	0.1271	48	0.3614	26	0.3098	27	0.4947	22
Słońsk (2)	0.4048	28	0.4489	40	0.2321	16	0.2494	60	0.0086	52	0.0532	54
Smołdzino (2)	0.3650	43	0.4681	37	0.0462	93	0.1579	87	0.0000	62	0.0000	73
Sokoły (2)	0.4704	14	0.5220	23	0.1048	56	0.1560	88	0.0027	56	0.0029	67
Sosnowica (2)	0.3296	61	0.4485	41	0.1479	42	0.1587	86	0.0000	62	0.0000	73
Stare Babice (2)	0.3901	33	0.5397	17	0.1604	39	0.4865	8	0.9378	1	0.9033	3
Stary Brus (2)	0.5380	4	0.6060	11	0.0783	74	0.0823	107	0.0000	62	0.0000	73
Stęszew (3)	0.3749	40	0.4130	57	0.1034	57	0.3396	28	0.3719	22	0.5795	13
Suchowola (3)	0.5263	6	0.6893	4	0.0920	61	0.1437	94	0.0000	62	0.0000	73
Sułoszowa (2)	0.5002	9	0.5284	22	0.0000	112	0.4681	9	0.1622	37	0.2711	37
Suraż (3)	0.4802	13	0.5090	25	0.1710	32	0.2221	71	0.0000	62	0.0000	73
Suwałki (2)	0.4384	21	0.6501	7	0.0378	97	0.2772	53	0.0000	62	0.0000	73
Szczawnica (3)	0.1929	90	0.2607	93	0.1539	40	0.3216	34	0.0000	62	0.0000	73
Szczytna (3)	0.3329	59	0.3192	85	0.1336	44	0.1220	102	0.2062	34	0.2085	39
Szklarska Poręba (1)	0.3806	36	0.4121	58	0.1895	24	0.3672	21	0.5036	14	0.5105	19
Sztabin (2)	0.4809	12	0.6429	8	0.0555	88	0.0763	110	0.0000	62	0.0000	73
Świnoujście (1)	0.2911	72	0.3162	86	0.2730	9	0.3078	41	0.5811	8	0.4867	23
Trzcianne (2)	0.3109	66	0.4315	49	0.0596	85	0.1547	90	0.0000	62	0.0000	73
Tuczno (3)	0.3543	48	0.3851	66	0.2111	20	0.2749	54	0.0000	62	0.0000	73
Turośń Kościelna (2)	0.5130	8	0.5119	24	0.1114	54	0.2304	66	0.1153	39	0.2604	38
Tykocin (3)	0.3643	44	0.4365	48	0.0682	80	0.0828	106	0.0000	62	0.0000	73
Urszulin (2)	0.4039	29	0.6118	9	0.0442	95	0.2824	48	0.0000	62	0.0000	73
Ustka (2)	0.4648	15	0.5361	19	0.4554	1	0.5753	4	0.0000	62	0.0682	51
Ustrzyki Dolne (3)	0.2171	85	0.2399	99	0.1357	43	0.1492	91	0.0030	55	0.0039	66
Wicko (2)	0.2752	73	0.3979	62	0.1020	59	0.1346	96	0.0000	62	0.0050	65
Wielka Wieś (2)	0.4222	27	0.4711	36	0.0652	81	0.5776	3	0.7242	4	0.8895	4
Wierzbica (2)	0.1850	92	0.5700	14	0.0822	70	0.1673	84	0.0000	62	0.0000	73
Witnica (3)	0.2537	77	0.3339	82	0.0816	72	0.2320	64	0.2735	31	0.3872	30
Wizna (2)	0.2462	78	0.4164	55	0.0734	76	0.1229	100	0.0000	62	0.0000	73
Wolin (3)	0.2963	69	0.4079	60	0.1625	38	0.1844	80	0.1123	40	0.1831	40
Zakopane (1)	0.3233	63	0.3508	79	0.2885	6	0.4023	15	0.0622	43	0.1347	44
Zamość (2)	0.1088	110	0.1902	107	0.0336	98	0.1150	104	0.1873	35	0.3877	29
Zawoja (2)	0.0751	114	0.1021	112	0.0309	99	0.0496	112	0.0000	62	0.0000	73
Zwierzyniec (3)	0.4252	26	0.4567	39	0.0605	84	0.2598	59	0.0898	41	0.3074	35

Notes: (1) urban municipality, (2) rural municipality, and (3) urban-rural municipality; SDM—synthetic development measure value; L—position based on SDM positions from 1 to 10 are marked in gray, indicating the highest level of development regarding the studied phenomenon among the analyzed municipalities.

The municipalities ranked among the top 10 in the analyzed period, based on the rankings taking into account the values of three SDMs, are presented in the table below (Table 3).

Table 3. Ranking leaders SDMwater SDMsewage and SDMgas in the years 2003–2018.

SDM$_{water}$	SDM$_{sewage}$	SDM$_{gas}$
Bargłów Kościelny (**entire period**) Chojnice (2003–2004) Grajewo (2004; 2014–2018) Jaświły (2005–2010) Kampinos (2003–2012; 2018) Kobylin-Borzymy (2003–2010) Ludwin (2003–2013) **Narewka (2013–2018)** **Nowinka (2011–2018)** Nowy Dwór (**entire period**) Stary Brus (2003–2017) Suchowola (**entire period**) **Sułoszowa 2003** Suwałki (2005–2018) Sztabin 2005–2018 Turośń Kościelna (2003–2004) Urszulin (2011–2018)	Białowieża (2003; 2011–2014) Czorsztyn (2003–2012) Dobiegniew (2006–2008) Jabłonka (2010–2018) **Jelenia Góra (2003–2010; 2013)** **Karpacz (2014–2018)** **Kostrzyn nad Odrą (2003–2006)** **Kowary (2009–2011)** Kudowa Zdrój (2003–2004) **Lipinki (2015)** Lipnica Wielka (2013–2018) Łapsze Niżne (2003–2005) Łapy (2005–2008) Łeba 2003–2013) **Międzyzdroje (2004–2009)** **Narewka (2009)** **Nowinka (2014–2018)** Ochotnica Dolna (2013–2014) Podgórzyn (2009–2012; 2014–2018) **Puszczykowo (2004–2018)** Skała (2012) Stare Babice (2015–2018) **Sułoszowa (2016–2018)** **Świnoujście (2003; 2007–2008)** Ustka (**entire period**) **Wielka Wieś (2010–2018)** Zakopane (2003–2013)	Czosnów (2014–2017) Dębowiec (2003–2006) Dopiewo (2007–2018) Izabelin (**entire period**) **Jelenia Góra (2003–2006; 2008; 2012–2013)** **Karpacz (2003–2013; 2016; 2018)** Komorniki (2007–2018) **Kostrzyn nad Odrą (2003–2013; 2017–2018)** **Kowary (2012–2013)** **Lipinki (2014–2015)** Łomianki (**entire period**) **Międzyzdroje (2015–2018)** Osiek Jasielski (2014) **Puszczykowo (2003-2007, 2009–2011; 2014–2018)** Stare Babice (**entire period**) **Świnoujście (2003–2011)** **Wielka Wieś (entire period)**

Legend: the years in brackets show periods in which municipalities were ranked among the top ten. Municipalities in bold are listed in at least two columns.

It is characteristic that some of the municipalities are listed among the leaders of the rankings developed based on various SDMs. This suggests that infrastructural investments are implemented comprehensively. If a municipality has the respective financial resources, it invests simultaneously in the construction of the three analyzed networks. No regularity can be identified regarding the location of the municipalities-leaders. These municipalities are characterized by a different status (urban, rural, and urban-rural) and are connected with different national parks.

The analysis indicates a relative stability of SDMwater leaders. In the entire research period, this group included 17 municipalities. The municipalities connected with Biebrza National Park (NP), i.e., Bargłów Kościelny, and Nowy Dwór, throughout the entire studied period, were ranked as the first and the second, respectively. The comparison of positions at the beginning and at the end of the analyzed period shows that 66 examined municipalities recorded a lower, 46 a higher, and 5 maintained their position. The majority of municipalities showed changes in their ranking position. Only 40% of the municipalities changed their position by a one-digit value. In terms of growth, Leipzig (connected with Biebrza NP) was the dominating one (increased by 80 positions in the ranking), whereas the largest drop (by 58 positions) was recorded by Jabłonka municipality (Babia Góra NP).

The absolute growth in SDMwater value, calculated as the difference in SDMwater value in 2018 (analyzed) and 2003 (baseline), indicates that the measure value dropped only in 12 municipalities: Jabłonka, Nowy Targ, Nowy Żmigród, Piechowice, Lipnica Wielka, Chojnice, Szczytna Lutowiska, Bukowina Tatrzańska, Dopiewo, Mszana Dolna, and Turośń Kościelna. It means that the development of water supply network was recorded in the vast majority (90%) of municipalities, levelling off the increase in the number of users.

Positive changes in the value of SDMwater were primarily the consequence of a longer distribution network; as many as 106 municipalities extended their water supply network. Łomianki municipality was the leader in this respect (the length of the functioning distribution network increased by 144 km). It should be noted, however, that this municipality is territorially connected with Kampinos NP and located near the country capital. Łomianki—in a sense—was affected by the suburbanization phenomenon, resulting from the residential housing pressure of Warsaw community.

The observations during the period 2003–2018 allow stating that the infrastructure that provides access to water is developing in the vast majority of the analyzed municipalities. This phenomenon should definitely be considered a positive one.

This analysis indicates that in the case of SDMsewage the leadership changes were much greater than in relation to SDMwater. In the analyzed period, 27 municipalities were listed among the top ones. In the Ustka municipality, which is connected with Słowiński, NP was the leader and for most part the studied period was ranked first. The second position was undisputedly taken by the Puszczykowo municipality located near Poznań metropolis (Wielkopolska NP).

The comparison of positions occupied at the beginning and at end of the research period shows that 68 analyzed municipalities recorded lower and 48 higher, while one maintained its ranking position. The majority of municipalities were characterized by significant changes in their ranking positions. Only 23% of them changed their position by a one-digit value. In terms of growth, Sułoszowa municipality connected with Ojców NP dominated (increase by 103 places in the ranking) and the largest drop (by 58 places) was noted for Szczytna municipality (Stołowe Mountains National Park).

The absolute growth in SDMsewage value, calculated as the difference in SDMsewage value in 2018 (analyzed) and 2003 (baseline), indicates that the measure value dropped only in three municipalities: Szczytna, Hańsk, and Grajewo. Due to the fact that the sewage network does not exist in all the municipalities of Giby, Krasnopol, and Kobylin-Bokuje, (they occupied ex aequo in the last position in the ranking), it can be stated that in approximately 95% of the analyzed municipalities in the sewage network development leveled off the increase in the number of users.

Positive changes in the value of SDMsewage resulted mostly from the increase in the length of sewage network. As many as 106 municipalities recorded this network extension. The Jabłonka municipality, which is connected with Babia Góra NP, was the leader in this respect (the length of the network increased by 106 km).

The observations made for the period 2003–2018 allow stating that the sewage infrastructure is under development in the vast majority of the analyzed municipalities. This phenomenon should definitely be considered a positive one.

The analysis shows that in the case of SDMgas, the group of leaders included 17 municipalities, which was identical for SDMwater. Keepin in mind that as many as 17 units did not have a gas network during the study period (thus ranked ex aequo at the last position), it can be adopted that the variability in this respect was slightly higher than in the case of SDMwater. The unquestionable ranking leaders were: Łomianki municipality (first or second ranking position throughout the entire analyzed period) and Stare Babice municipality (first or second position, and incidentally, in 2008, fourth in the ranking) connected with Kampinos NP, and located in the vicinity of Warsaw.

The comparison of positions from the beginning and the end of the analyzed period shows that 87 examined municipalities recorded a decrease and 25 an increase, while five maintained their position. In total, 60 municipalities changed their position by two-digit values, which constituted a slight majority. In terms of growth, Dopiewo municipality connected with Wielkopolska NP was the dominating one (increase by 24 ranking positions) and the largest drop (73 places) was recorded by the Krempna municipality (Magura NP).

The absolute growth in SDMgas value, calculated as the difference in SDMgas value in 2018 (analysed) and 2003 (baseline), indicates that the measure value dropped in 15 municipalities: Świnoujście, Lewin Kłodzki, Stare Babice, Kudowa-Zdrój, Jelenia Góra, Kowary, Piechowice, Dębowiec, Radków, Czarna, Dukla, Krempna, Leoncin, and Izabelin oraz Brusy.

Bearing in mind that the gas network is nonexistent in the municipalities of Sosnowica, Stary Brus, Urszulin, Kamienica, Szczawnica, Zawoja, Lutowiska, Bargłów Kościelny, Lipsk, Grajewo, Radziłów,

Jedwabne, Wizna, Giby, Krasnopol, Nowy Dwór, and Szczawnica (they occupied ex aequo the last ranking position), it can be stated that the development of a gas network was recorded in almost 73% of the analyzed municipalities, which leveled off the increase in the number of users.

Positive changes in SDMgas value derived mainly from a larger number of populations using gas networks. Due to the absence of data on the length of a functioning network, it can be presumed that not only the number of connections but the length of the network increased. The development of the gas network in the analyzed municipalities should be assessed positively.

5. Discussion

It is difficult to indicate the research comparable to the one presented in the article. The authors are aware of this situation and the reasons for no comparability of the studies on Polish national parks with the national parks in other countries have already been described in detail [3]. Although national parks are known worldwide, this term is associated with different security regimes in various countries, as well as organizational and legal differences resulting from the functioning forms of such parks and also the rights and entitlements of local authorities. These differences often result from just the size of the park. However, the above does not change the fact that Polish national parks represent an important link in the protection of European nature and also an important destination for domestic tourist traffic.

The specificity of protected areas means that from both natural and economic perspectives it is important to properly understand the message presented on the Federation of Nature and National Parks of Europe (EUROPARC Federation) website: "nature knows no boundaries" [33]. Nature protection requirements are not synonymous with the need to eliminate a human being from the protected space. The research results indicate that the function of nature conservation, as well as the economic functions (including tourism), are not mutually exclusive [34]. However, the communities residing in the municipalities territorially connected with national parks, the investors operating within the discussed area and also tourists have to comply with stricter environmental standards.

Users make space evolve, as it changes physically and functionally. This refers to both urban space [35,36], rural areas [37,38], and valuable natural areas the least changed by a man. Therefore, in the context of the presented article, this phenomenon applies to the area of national parks and the areas of municipalities connected with them.

The EUROPARC Federation clearly emphasizes that sustainable tourism is desirable for both parks and people (in the sense of local communities and tourists). At the same time, it should be emphasized that the concept of sustainable tourism is still open and widely discussed. Even though there is a consensus regarding the principle that an ongoing and sustainable development of tourism is such a method for doing business and organizing social life, which ensures both the development and preservation of the environmental values along with improving the quality of people's lives, there are still many detailed interpretations of the discussed concept [39,40].

Balancing the tourist function not only takes time but it remains a complicated process [33,40]. The significance of the aforementioned issue is also strengthened by the fact that 2017 was announced the International Year of Sustainable Tourism for Development.

To sum up, it can be stated that technical infrastructure is indispensable not only for the development of economic initiatives (including the who influences the multifunctional and sustainable development of municipalities. It is natural, then, that technical infrastructure has to be supplemented by social infrastructure (these problems—although very interesting—are not the subject matter of this article).

6. Conclusions

The conducted empirical research allowed us to achieve our defined research goals. We were forced to adjust the adopted indicators to the available statistical data. Despite that, we managed to develop measures that allowed for a comprehensive and measurable approach to the analyzed problems.

The collected results allowed us to conclude that, between 2003–2018, the development of the analyzed technical infrastructure elements were recorded in the vast majority of municipalities. Therefore, the level of anthropopressure declined, which was caused by the local community and tourists in municipalities within the most valuable natural areas.

The largest percentage of municipalities that were characterized by an increase in synthetic development measures were referred to sewerage network research. As many as 95% of the analyzed municipalities recorded an increase in the absolute value of SDMsewage in the period 2003–2018. Slightly lower values were true for the water supply network, in the case of which development was observed in 90% of municipalities. The poorest—although not to be considered bad—refer to the gas network. In total, 73% of the studied municipalities recorded development in this area.

In view of the above, the research hypothesis put forward at the beginning of the article should be adopted and it should be recognized that the development of water supply, sewage, and gas networks is observed in the municipalities territorially connected with national parks.

The interpretation of the collected results (SDMwater, SDMsewage, and SDMgas) highlight an important nuance: in the set of 117 units there are both urban municipalities which, in the past, played the role of voivodship capitals (Jelenia Góra), and also rural municipalities inhabited by less than 2000 people (Lewin Kłodzki, Cisna). Hence, the settlement and population system as well as the wealth of local governments in the analyzed municipalities are very different. It is highly positive that despite the abovementioned differences, the municipalities remain connected not only by their tourist attractiveness and unique nature, but also by striving to protect it. Its measurable expression is the identified development of the analyzed technical infrastructure elements, which is highly important in the context of aiming at sustainable tourism development in the naturally valuable areas.

Summing up the presented discussion it should, yet again, be emphasized that the processes occurring in the environment or the exchange of value streams between specific spaces do not respect the administrative boundaries of the protected area. Therefore, the functioning of the analyzed technical infrastructure is extremely important for the nature protection of national parks.

Author Contributions: Conceptualization: A.K.-D.; Data curation: A.K.-D. and A.S.; Formal analysis: A.K.-D. and A.S.; Investigation: A.K.-D.; Methodology: A.K.-D.; Resources: A.K.-D.; Supervision: A.K.-D.; Visualization: A.S.; Writing—review & editing: A.K.-D. and A.S. All authors have read and agreed to the published version of the manuscript.

Funding: This research was funded by University of Environmental and Life Sciences (no external funding).

Conflicts of Interest: The authors declare no conflict of interest.

References

1. UNEP-WCMC and IUCN. *Protected Planet Report 2016*; UNEP-WCMC and IUCN: Cambridge, UK; Gland, Switzerland, 2016.

2. Kennish, M.J. Anthropogenic Impacts. In *Encyclopedia of Estuaries*; Encyclopedia of Earth Sciences Series; Kennish, M.J., Ed.; Springer: Dordrecht, The Netherlands, 2016; pp. 29–35.

3. Kulczyk-Dynowska, A.; Bal-Domańska, B. The National Parks in the Context of Tourist Function Development in Territorially Linked Municipalities in Poland. *Sustainability* **2019**, *11*, 1996. [CrossRef]

4. The Act of 16 April 2004 on Nature Conservation Journal of Laws of 2004, No. 92, Item 880. Available online: http://prawo.sejm.gov.pl/isap.nsf/download.xsp/WDU20040920880/U/D20040880Lj.pdf (accessed on 4 February 2020).

5. Koda, E.; Sieczka, A.; Osinski, P. Ammonium concentration and migration in groundwater in the vicinity of waste management site located in the neighborhood of protected areas of Warsaw, Poland. *Sustainability* **2016**, *8*, 1253. [CrossRef]

6. Hellwig, Z. Zastosowanie metody taksonomicznej do typologicznego podziału krajów ze względu na poziom ich rozwoju oraz zasoby i strukturę wykwalifikowanych kadr/The application of taxonomic method in the typological division of countries regarding their development level and resources and structure of qualified personnel. *Stat. Rev.* **1968**, *4*, 307–327.

7. Strahl, D. Propozycja konstrukcji miary syntetycznej/The proposal of statistical measure construction. *Stat Rev.* **1978**, *25*, 205–215.

8. Walesiak, M. *Uogólniona Miara Odległości w Statystycznej Analizie Wielowymiarowej/Generalized Distance Measure in a Statistical Multidimensional Analysis*; University of Economics in Wrocław Press: Wrocław, Poland, 2006.

9. Marti, R.; Reinelt, G. *The Linear Ordering Problem, Exact and Heuristic Methods in Combinatorial Optimization*; Springer: Berlin/Heidelberg, Germany, 2011.

10. Kukuła, K. Propozycja budowy rankingu obiektów z wykorzystaniem cech ilościowych oraz jakościowych. *Metod. Ilościowe w Badaniach Ekonomicznych* **2012**, *13*, 5–16.

11. Kukuła, K.; Luty, L. Jeszcze o procedurze wyboru metody porządkowania liniowego. *Stat. Rev.* **2017**, *64*, 163–176.

12. Manly, B.F.J.; Navarro Alberto, J.A. *Multivariate Statistical Methods*; CRC Press/Taylor& Francis Group: Boca Raton, FL, USA, 2017.

13. Górz, B.; Kurek, W. Variations in technical infrastructure and private economic activity in the rural areas of Southern Poland. *GeoJournal* **1998**, *46*, 231–242. [CrossRef]

14. Feldman, M.P.; Florida, R. The geographic sources of innovation: Technological infrastructure and product innovation in the United States. *Ann. Assoc. Am. Geogr.* **1994**, *84*, 210–229. [CrossRef]

15. Kazak, J.; Chruściński, J.; Szewrański, S. The Development of a Novel Decision Support System for the Location of Green Infrastructure for Stormwater Management. *Sustainability* **2018**, *10*, 4388. [CrossRef]

16. Kiełkowska, J.; Tokarczyk-Dorociak, K.; Kazak, J.; Szewrański, S.; Van Hoof, J. Urban Adaptation to Climate Change Plans and Policies–the Conceptual Framework of a Methodological Approach. *J. Ecol. Eng.* **2018**, *19*, 50–62. [CrossRef]

17. Krajewski, P. Monitoring of Landscape Transformations within Landscape Parks in Poland in the 21st Century. *Sustainability* **2019**, *11*, 2410. [CrossRef]

18. Heilig, G.K. Multifunctionality of landscapes and ecosystem services with respect to rural development. In *Sustainable Development of Multifunctional Landscapes*; Helming, K., Wiggering, H., Eds.; Springer: Berlin/Heidelberg, Germany, 2003. [CrossRef]

19. Rogowska, M. Technical infrastructure investments in municipality—A theoretical outline. *Bibl. Reg.* **2010**, *10*, 185–198.

20. Furmankiewicz, M.; Macken-Walsh, A.; Stefańska, J. Territorial governance, networks and power: Cross-sectoral partnerships in rural Poland. *Geogr. Ann. Ser. B Hum. Geogr.* **2014**, *96*, 345–361. [CrossRef]

21. Mickiewicz, A.; Wawrzyniak, B. The Importance of Technical Infrastructure for the Development of Rural Areas. *Res. Pap. Wrocław Univ. Econ.* **2011**, *166*, 483–493.

22. Przybyła, K.; Kulczyk-Dynowska, A. Transformations of Tourist Functions in Urban Areas of the Karkonosze Mountains. In *IOP Conference Series: Materials Science and Engineering*; IOP Publishing: Bristol, UK, 2001; Volume 245, p. 072001. [CrossRef]

23. Walas, B.; Fedyk, W.; Pasierbek, T.; Nemethy, S. Diagnosis of functioning of National Parks in Poland in their socioeconomic environment. *Ekonomiczne Problemy Turystyki* **2018**, *43*, 69–79. [CrossRef]

24. Zawilińska, B.; Mika, M. National parks and local development in Poland: A municipal perspective. *Hum. Geogr. J. Stud. Res. Hum. Geogr.* **2013**, *7*. [CrossRef]

25. Szromek, A.; Kruczek, Z.; Walas, B. The Attitude of Tourist Destination Residents towards the Effects of Overtourism—Kraków Case Study. *Sustainability* **2020**, *12*, 228. [CrossRef]

26. World Tourism Organization. *UNWTO Annual Report 2017*; UNWTO: Madrid, Spain, 2018; p. 16. [CrossRef]

27. Mioduszewski, W. Czy Polska jest krajem ubogim w wodę? *Gospodarka Wodna* **2008**, *5*, 186–193.

28. Orlińska-Woźniak, P.; Wik, P.; Gębala, J. Water availability in reference to water needs in Poland. *Meteorol. Hydrol. Water Manag. Res. Oper. Appl.* **2013**, *1*. [CrossRef]

29. Warszyńska, J.; Jackowski, A. *Podstawy Geografii Turyzmu/Basics in Tourism Geography*; PWN Publishers: Warsaw, Poland, 1978.

30. Statistics Poland, Local Data Bank. Available online: https://bdl.stat.gov.pl/BDL/start (accessed on 4 February 2020).

31. Kukuła, K. Zero unitarisation method as a tool in ranking research. *Econ. Sci. Rural Dev.* **2014**, *36*, 95–100.

32. Kowalewski, G. *Metody Porządkowania Liniowego*; Dziechciarz, J., Ed.; Ekonometria, Wyd. Akademii Ekonomicznej: Wrocław, Poland, 2002; pp. 287–304.

33. Available online: www.europarc.org (accessed on 3 February 2020).

34. Kulczyk-Dynowska, A. *Parki Narodowe a Funkcje Turystyczne i Gospodarcze Gmin Terytorialnie Powiązanych*; Wyd. Uniwersytetu Przyrodniczego we Wrocławiu: Wrocław, Poland, 2018.
35. Słodczyk, J. *Przestrzeń Miasta i Jej Przeobrażenia*; Studia i Monografie; Uniwersytet Opolski: Opole, Poland, 2003; Volume 298.
36. March, L.; Martin, L. (Eds.) *Urban Space and Structures*; Cambridge University Press: Cambridge, UK, 1972.
37. Woch, F.; Woch, R. Zmiany użytkowania przestrzeni wiejskiej w Polsce. *Infrastruktura i Ekologia Terenów Wiejskich* **2014**, *I/1*. [CrossRef]
38. Stacherzak, A.; Hełdak, M.; Hajek, L.; Przybyła, K. State Interventionism in Agricultural Land Turnover in Poland. *Sustainability* **2019**, *11*, 1534. [CrossRef]
39. Petrevska, B.; Terzi'c, A.; Andreeski, C. More or Less Sustainable? Assessment from a Policy Perspective. *Sustainability* **2020**, *12*, 3491. [CrossRef]
40. Wight, P.A. Supporting the principles of sustainable development in tourism and ecotourism: Government's potential role. *Curr. Issues Tour.* **2002**, *5*, 222–244. [CrossRef]

 *sustainability*

Article

Sustainable Green Roof Ecosystems: 100 Years of Functioning on Fortifications—A Case Study

Łukasz Pardela [1,*], Tomasz Kowalczyk [2], Adam Bogacz [3] and Dorota Kasowska [4]

[1] Institute of Landscape Architecture, Wrocław University of Environmental and Life Sciences, 50-357 Wrocław, Poland

[2] Institute of Environmental Protection and Development, Wrocław University of Environmental and Life Sciences, 50-363 Wrocław, Poland; tomasz.kowalczyk@upwr.edu.pl

[3] Institute of Soil Science and Environmental Protection, Wrocław University of Environmental and Life Sciences, 50-357 Wrocław, Poland; adam.bogacz@upwr.edu.pl

[4] Department of Botany and Plant Ecology, Wrocław University of Environmental and Life Sciences, 50-363 Wrocław, Poland; dorota.kasowska@upwr.edu.pl

* Correspondence: lukasz.pardela@upwr.edu.pl

Received: 2 April 2020; Accepted: 5 June 2020; Published: 9 June 2020

Abstract: Green roofs have received much attention in recent years due to their ability to retain rainwater, increase urban diversity, and mitigate climate change in cities. This interdisciplinary study was carried out on three historical green roofs covering bunkers in Wrocław, located in southwestern Poland. It presents the results of a three-year investigation of the water storage of these roofs. The study also presents soil conditions and spontaneous vegetation after their functioning for over 100 years. The soils covering the bunkers are made of sandy, sandy-loam, and loamy-sand deposits. This historical construction ensures good drainage and runoff of rainwater, and is able to absorb torrential rainfall ranging from 100 to 150 mm. It provides suitable conditions for vegetation growth, and forest communities with layers formed there. In their synanthropic flora, species of European deciduous forests dominate, which are characteristic of fresh or moist and eutrophic soils with a neutral reaction. Some invasive species, such as *Robinia pseudoacacia*, *Padus serotina*, and *Impatiens parviflora*, also occur with high abundance. Nowadays, historical green roofs on fortifications, although they have lost their primary military role, are of historical and natural value. These roofs can promote the nonmilitary functions of historical fortifications in order to strengthen the ties between nature and heritage. Protecting and monitoring historical green roofs should be included in the elements of the process of sustainable development and the conservation of these structures in order to mitigate climate change in the outskirts of the city. For this, it is necessary to ensure proper conservational protection, which, in addition to maintaining the original structure, profiles, and layout of the building, should include protection of their natural value.

Keywords: green infrastructure; technogenic soil; soil water retention; synanthropic flora; urban vegetation; heritage protection; fortified landscape

---

1. Introduction

The development of urban areas often negatively affects the natural environment in terms of soil erosion or the reduction of available water resources for plants [1]. These changes may also occur in historical areas [2], an example of which may be historical fortifications (hereinafter referred to as HF). Protecting their cultural, historical, and natural values is a new challenge in the face of climate change [3], since HF no longer exist in the environmental conditions for which they were designed. This applies, inter alia, to earthen fortifications with soil used in their construction (e.g., embankments, traverses, etc.), which are an integral element of HF. An example of such earthen coverings are green roofs, most often made up of sandy deposits that tend to lose water that is collected within due to low retention. The more frequent occurrence of prolonged dry periods may cause the health of plants to deteriorate and may even lead to the loss of certain sensitive species grown on such green roofs. Therefore, learning about their construction and how they function is important for maintaining integrity within historical urban landscapes [4]. Likewise, despite supporting sustainable urban development [5], these heritage sites can be used as "living laboratories to implement mitigation strategies and highlight better mitigation and adaptation practices" [3]. Furthermore, according to Fluck and Wiggins [6], the preservation of heritage and the historical character of a landscape has a positive effect on communities, while the ways in which heritage is managed can lead to a better understanding of the effects of climate change in other areas.

Currently, many historical fortifications, together with their vegetation, constitute "green islands" among the new housing estates of many cities. They may be threatened by urban sprawl, not only in their immediate vicinity, but also within the borders of the plots of land belonging to former fortress. In addition to the growing interest among developers looking to build residential buildings on the grounds of historical fortifications, there is also a growing group of city dwellers seeking opportunities to actively protect monuments, often with their own hands. Practice shows, however, that providing deteriorating monuments with proper care and use (i.e., adapting them to new functions) can be a challenge for new hosts looking for a proper way to maintain genius loci.

On the other hand, professional practice reveals that in Poland, in the process of renovating and adapting the fortifications built at the turn of the 19th and 20th century, investors saw an alternative to historic green roofs in the form of creepers or new roofing made of sheet metal. This came to light as a result of work conducted, including the partial or complete replacement of damp coursing or the existing ventilation system. With full replacement or the introduction of insulation, preservation of the original green roofs is not practiced. The result is a new, turf roof, also using roll-out grass.

From a conservational perspective, the valuable elements of historical green roofs are their original material and profile (layout) of the buildings covered in this way, including their integration with the surrounding landscape. Their historical value stems from the preservation of as many original elements of the fortifications as possible. Currently, according to the conservational recommendations (the procedure of conservational conventions) in Poland, when a building requires a thorough replacement of damp coursing, the emphasis is on restoring green roofs using couch grass (*Elymus repens*), for instance. However, every such case is individually considered due to various forms of conservation protection, fortress schools, types of buildings and their construction, the building materials used, and the historical layers found in such fortifications.

Environmental aspects related to climate change often fade to the background, opening the way to dangers in the form of "idealized scenery," resulting from the premise of restoring historic monuments to their former military glory of yesteryear. After some time, counterproductive practices come to light involving the degradation of historical sites, often resulting from a desire to tidy them up. Degradation may affect the whole area or part of it (e.g., earthworks and green roofs on buildings) regardless of the form of their conservational protection. This can last for years or can take the form of one-off incidents. As a result, there are changes in the form of soil erosion, water retention, and disturbance in plant coverage (Figure 1).

Figure 1. Examples of work that cause soil and plant ecosystem degradation of historical green roofs: (**a**,**c**) Infantry shelter before and after restoration; (**b**,**d**) infantry shelter before and during restoration works with a new green roof, Wrocław, Poland. 2012 and 2016, sourced from the author Łukasz Pardela's archive.

Harmful practices resulting from increased anthropopressure include the elimination of earthen forms permanently connected with fortifications (e.g., earth coverings often considered to be unnecessary), replacing historical soil cover with present-day extensive green roofs, the hardening of road surfaces, soil exchange (e.g., foundations for new roads), the cutting of all vegetation causing excessive insolation of structures (exposure of earthen forms), and improper maintenance involving the raking and removal of fallen leaves instead of composting or leaving them in place as litter. Such actions are not conducive to mitigating the effects of burgeoning climate change. Despite the preservation of heritage values such as authenticity and integrity, they often pose a greater threat than just leaving the building or other military structure alone while systematically monitoring its condition.

This research studies green roofs built on historical infantry shelters, commonly referred to as "bunkers." Their role and significance for urban areas in terms of habitat conditions and retention potential are worth learning more about. HF are an example of historic buildings that were designed in conjunction with the surrounding landscape. This meant taking into account unfavorable humidity conditions in the form of both excess precipitation and water shortages, which is why these areas have demonstrated resistance to fluctuations in soil humidity. The fortification buildings would not have stood out too much in the surroundings, and so were "greener" than the surrounding landscape, e.g., during a drought. This also applied to the vegetation cover, whose species composition was similar to that occurring in adjacent areas. Green roofs on fortifications were often elements of the fortifications elevated above the surrounding area, making them visible from the ground and from the air. In addition, the long period that elapsed since the construction of the fortifications and, above all, the loss of their original functions and the related lack of maintenance, causes the accumulation of organic matter (litter) on an unprecedented scale, e.g., in urban parks. Similarly, abandoning the cultivation of greenery, along with progressive succession, has caused an increase in woody vegetation, which partly limits the direct influence of sunlight on the vegetation of green roofs. Green roofs on fortifications are also an example of the use of natural substrates, on which vegetation has been developed throughout the decades via natural succession (including what is not currently recommended, e.g., in Belgium and the Netherlands for roofs (e.g., *Acer platanoides*), which is difficult to observe in the case of green roofs on urban buildings.

On the other hand, the research results can help to highlight values that for years have lain beyond the main trends of research on the functioning of green roofs and technogenic soils in the city, as well as the transition from a rural to a suburban landscape. The inspiration to undertake this topic was the ambition and impulse to promote the nonmilitary functions of HF in urban areas in order to strengthen the ties between nature and heritage. Nowadays, HF, as well as historical parks and gardens, can influence climate change in a condensed form over a limited area [2]. In the case of this research, green roofs have sheltered HF for over a century. Since their construction, these roofs have blended into the agrarian landscape [7,8]. After time, some of the green-roofed shelters lay on the city peripheries, and today they mimic natural ecosystems in these areas. Therefore, even understanding the short-term performance of these historical green roofs in different seasons of the year and weather conditions is of great importance within the process of adapting diverse areas in cities to climate change. This is also connected with developing biodiversity strategies for cities, especially highly polluted ones such as Wrocław, located in southwestern Poland [9].

Consequently, there is an important role for cultural heritage sites to play in terms of adapting to future climatic impacts [6]. This interdisciplinary research into heritage, interrelated with historical and natural environments, supports sustainable development, the management of cultural heritage, and the mitigation of climate change in cities.

Therefore, through interdisciplinary research, this paper presents the following:

- The characteristics of technogenic soils formed on the roofs of historical fortifications based on field studies;
- The present vegetation growing on shelter green roofs with some of the species' ecological characteristics;
- Defined directions for the development, restoration, and protection of historic green roofs in the city peripheral areas in the context of the challenges posed by a changing climate.

1.1. Historical Green Roofs

Green roofs sheltering historical fortifications, as a result of a long military engineering tradition, have been used in fortifications for centuries [10]. They are similar to present-day "earthen sheltered" buildings with green roofs, which are beneficial to wildlife and the environment [11]. In terms of fortifications, green roofs are "living roofs" on top of bunkers and can be defined broadly as engineered constructions that include environments suitable for well-adapted plant species [12]. Therefore, their construction is similar to intensive green roofs [12,13]. As artificial habitats, like any constructed ecosystem, they evolve over time [13].

The transformation of green roofs has been the subject of numerous studies over the years [14–17]. A good proportion of the research concerned the historical origins and evolution of intensive green roofs in many countries [18–20]. Simultaneously, many authors indicate the need to focus on researching "real-life" examples of green roofs in order to determine the benefits arising from deeper substrates in different weather conditions [21] and vegetation dynamics on green roof performance and design [13], as well as biodiversity conservation [22–24].

Nonetheless, the most commonly examined facilities are green roofs related to civil architecture, including historical [18–27]. This also includes engineering facilities—for example, the water filtration plant near Zurich (Wollishofen) in Switzerland built in 1914 [28,29]. Research related to green roofs on HF, by contrast, is rather rare. However, the technogenic soils constructed on 19th century heritage fortifications have been the subject of some research. In a study by Jankowski et al. (2013), three soil formations located on three different fortified heritage sites (fort, infantry shelter, and ammunition storage) were selected for detailed soil formation analysis [30], while a study was carried out by Pardela and Kowalczyk (2018) to estimate the variation of soil water retention in six soil formations of a single infantry shelter heritage site [31].

1.2. How Green Roofs Function in an Urban Environment

The ecosystem services (ESS) and disservices (EDS) of green roofs have been widely discussed in numerous works [14,32,33]. ESSs include the improvement of storm water quality and quantity, a reduction in the rate and volume of runoff [21], mitigation of the urban island heat effect [17], facilitation of energy efficiency by reducing maintenance costs, fire retardation, or enhanced noise reduction [34]. Green roofs may also improve air quality and carbon sequestration [35], support wildlife and habitat biodiversity, enhance rooftop agriculture and roof life, increase educational opportunities, and boost health [19]. EDSs include, for example, stormwater quality and quantity with the leaching of nitrates (nitrogen sink) and soluble carbon, which decreases the runoff water quality [36]. For some buildings, thermal mass increased by thick substrate layers on the building and the downward movement of heat leads to higher cooling costs [37]. If green roofs are located too high above the ground, they can offer ecological traps for many species [38] or attract nuisance wildlife [39].

In addition to the ongoing "present-day" functionality of green roofs, the lost historical functions related to the green roofs on HF are worth mentioning and include: hiding flat roofs against aerial observation, draining rain water from the building structure, protecting flat roofs (e.g., the top layer features tar waterproofing), retarding fire (which may be caused by enemy shelling), and limiting the damage inflicted by artillery grenades or shrapnel.

1.3. Green Roofs on Shelters in Historical Guidelines

Unfortunately, little is known about the details of soil layers of many of the green roofs on historical shelters. From general guidelines for laying earthen covers in the German Empire shortly before 1900, it is known that the thickness of green roofs historically ranged from 30 to 50 cm [40,41]. After 1900, the earthen covering of concrete shelters consisted of a 20–30 cm layer. This provided the roof insulation with sufficient protection against insolation and shielded flat roofs from air reconnaissance (from balloons and, later, planes). A guideline from 1905 indicated that on a (green roof) covering, it was possible to sow grass species on a bed of clay mixed with sand [42]. This provision also indicated that flat roofs should be maintained with herbaceous vegetation without introducing any shrubs. Similarly, the trampling and destruction of forest litter on earthworks was forbidden.

2. Materials and Methods

2.1. Study Sites

The study was conducted in the Wroclaw city area (Figure 2). The rainfall in Wrocław is highly variable: the annual total rainfall ranges between 318 and 892 mm, while the average annual precipitation in the 20th century was 583 mm. The average annual temperature is 9 °C, and the annual temperature amplitude is 19.2 °C. Winters are short (65 days) and mild, with frequent periods of warming in February of up to 10–15 °C.

Three historical infantry shelters (more accurately referred to as 'Infanterie Raum') (Table 1, Figures 3 and 4), whose construction is described in literature regarding fortifications, were selected for the research [43,44].

The studied shelters (referred to as S-1, S-2, and S-3) are a representative group of military fortifications with well-preserved green roofs. Shelters S-1 and S-2 are the central points of the infantry works (Infanterie Stützpunkt), which consisted of smaller shelters and a permanent shooting position. Shelter S-3 is a single facility partially surrounded by an embankment. Their origins date back to the quarter-century preceding the First World War, when a ring of shelters was built around the city as part of the fortification of the former German city of Breslau (now Wrocław). After the Second World War, Wrocław found itself in Polish territory [8]. The historical surroundings of the researched facilities are presented in Figure 4. The largest changes in the landscape and land cover in the surroundings of the studied facilities took place after 1947. As a result, some of the areas around the fortifications were allocated as allotment gardens for workers (now family allotment gardens) from divided agricultural land, and after the political transformations of 1989, new housing developments appeared, along with expanded road and flood protection infrastructure.

Around the described shelters, we can now find various soil units, including Fluvisols, Luvisols, Czernozems, and Podzols [45]. Adjacent vegetation consists mainly of garden flora, abandoned fields, roadsides, levees and retention ponds, and other ponds of ruderal areas. Further away, some fragments of oak-hornbeam and alluvial forests are preserved. In the vicinity of S-1, there is a landfill with the remains of bonfires.

Shelters S-1 and S-2 with roofs are currently located on the outskirts of the city of Wrocław on low hills bordering the floodplain of the Widawa River, protected against flood waters thanks to embankments serving as shooting positions. These shelters rise above the surrounding area. A pair of ponds are located at the S-2 facility, where rainwater from the entire facility was drained. In turn, shelter S-3, with a green roof is recessed in the ground below the surrounding area and is located near railway lines, whose access it was supposed to defend. In the 1960s, allotment gardens were created nearby, and still exist there today. All the researched facilities were abandoned by the army and civil defence in the early 1990s. Facilities S-1 and S-3 are somewhat secluded, while S-2 is fenced and guarded by a local association. An illegal landfill was created in the immediate vicinity of S-3, which is regularly cleared out. In recent years, there has also been growing anthropopressure in the form of the organization of illegal obstacle courses (S-1) and the organization of nighttime events (S-3). The flat roofs of these shelters constituting the structure that holds the substrate of green roofs were made in two ways. The earlier type of the flat roof was 3 m thick in a three-layer (brick-sand-concrete) structure (which is the case for S-1 and S-2). In this case, the outer part of the roof was concrete with granite aggregate with waterproofing made from hydraulic cement. The later type of flat roof featured a monolithic concrete flat roof that was 2.20 m thick, with tar waterproofing (S-3). Both types of roofs sloped slightly towards the outside of the building. Currently, the shelters under study are under conservation protection and are the property of the municipality of Wrocław or the Treasury. The buildings are open to the public. One (S-2) is currently under conservation and restoration works to serve the local community.

Table 1. Summary of the characteristics of each green roof covering the shelters.

Study Site	Location	Historical Name	Year Installed	Approx. Size (m^2)	Age (when Surveyed)
Shelter S-1	51°9′1.934″ N 17°5′13.829″ E	Infanterie Stützpunkt 4	1891	488	128
Shelter S-2	51°9′53.118″ N 17°2′37.137″ E	Infanterie Stützpunkt 6	1891	488	128
Shelter S-3	51°4′26.654″ N 17°4′23.994″ E	Infanterie Raum 20	1900	420	119

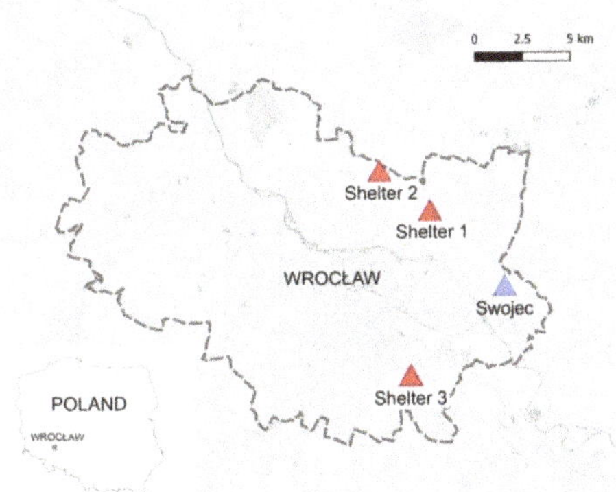

Figure 2. Location of the shelters and the Swojec station within the borders of City of Wrocław, Poland.

Figure 3. The green roof shelters in winter, 2016, and spring, 2019. From left: Shelter S-1 (**1a–c**), S-2 (**2a–c**), and S-3 (**3a–c**). Sourced from the author Łukasz Pardela's archive.

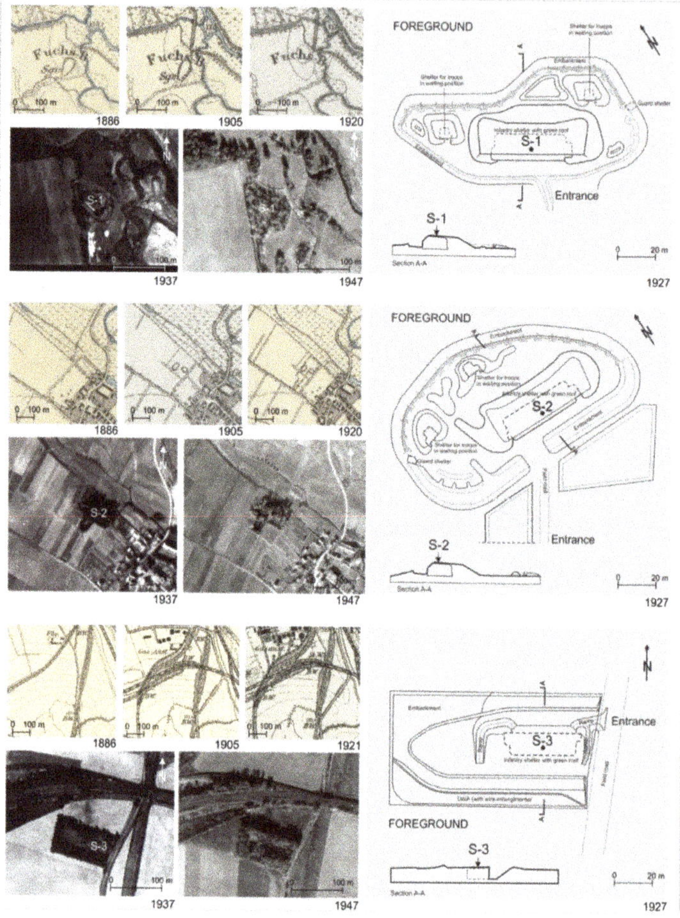

Figure 4. Historical landform and land use around the studied facilities in the years 1886–1947 [8]. The fragments of the topographic maps Meßtischblätter 2828 (4868) and 2892 (4968) come from the collection of Staatsbibliothek zu Berlin—Preussischer Kulturbesitz ©. Aerial photographs (1937 and 1947) were taken from the authors' archives and the state archives in Wrocław © (reference number 328/II; 244, 239, 281). Drawings were made based on documentation of the League of Nations Archives in Geneva (1927) and field measurements.

2.2. Analytical Methods

2.2.1. Historical Records

In order to identify the historical appearance, construction and maintenance of the study sites, documents and references from historical archives and libraries in Poland, Germany, and Switzerland were collected. In particular, technical instructions and documents from the 19th to the 20th centuries were selected and interpreted. Historical topographic maps from 1860 to 1938, as well as aerial and satellite photos from 1926 to 2018 were also analyzed for any changes in landform and land cover, both within the plots of the former fortress and in their immediate vicinity. Due to the military significance of the facilities, they were usually not marked on topographic maps (Figure 4), nor have historical drawings or logbooks detailing the construction and cultivation of the fortifications survived.

2.2.2. Soil Studies

Within the scope of the fieldwork conducted, the profile of the soil on the surface of the shelters was described. The morphological features were characterized, with particular emphasis on the: color, structure, and presence of redoximorphic features and artefacts. Samples were taken from the soil levels for laboratory analysis. In order to determine the retention capacity, samples were taken in Kopecki metal rings with a volume of 100 cm^3. The retention properties of the soil samples were analyzed in the pF value (soil moisture tension, pF = log(h) cm H$_2$O) 0–2 and pF 2–2.7 range using Eijkelkamp sand and kaolin-sand [46]. Soil retention properties in the 3.2–4.2 pF range were measured in Richard's high-pressure chambers [47].

The textures of the soils were determined using an aerometric sieve method in accordance with the PN-R-04032 standard [48]. The names of the soil granulometric groups were determined on the basis of the Polish Soil Science Society's classifications [49]. The soil reactions were potentiometrically measured in H$_2$O (water-soil ratio of 1:2.5) [50]. The soil types and subtypes were determined using the Polish soil classification [51]. International soil units were determined using the classification of the World Reference Base for Soil Resources (FAO-WRB) [45].

2.2.3. Soil Water Storage

Analysis of soil moisture distribution on the shelters was performed on the basis of measurements of three field profiles located in central points of the green roofs. Soil sampling and measurements were made with an Eijkelkamp auger. In the areas where moisture was to be checked, investigative borings were drilled down to the flat roof. Soil moisture was measured using the direct method, with a handheld meter for measuring soil volumetric moisture, based on the time domain reflectometry (TDR) FOM/mts reflectometric technique with an FP/mts probe. The probe was placed in holes made with a hand drill as required. Systematic moisture measurements were taken once a month on average from April to October in 2017–2019. The meteorological conditions during the period of moisture measuring were based on data from the Swojec station (51°4′25.9″ N, 17°4′22.8″ E, Bartnicza Street) (Figure 2, Tables 2 and 3) located, like the facilities under the research objects, on the outskirts of the city. At the same time, visual observation was carried out and the necessary photographic documentation was completed.

Table 2. Periodic and annual atmospheric precipitation, P (mm) in the hydrological years of 2017–2019 in the context of a longer period according to the Wrocław-Swojec station.

Years	P (mm)			
	XI-X	XI-IV	V-X	IV-IX
2017	668	219	449	442
2018	415	134	281	260
2019	535	196	339	347
2001–2010	587	216	371	366

Table 3. Average periodic and annual air temperature T (°C) in the hydrological years of 2017–2019 in the context of a longer period according to the Wrocław-Swojec station.

Years	T (°C)			
	XI-X	XI-IV	V-X	IV-IX
2017	9.7	3.0	16.3	15.8
2018	10.8	4.0	17.6	18.1
2019	10.5	4.4	16.7	16.5
2001–2010	9.5	3.3	15.8	15.9

2.2.4. Floristic Studies

Floristic observations were carried out in August during 2019. The occurrence of vascular plant species and their abundance were investigated throughout the entire roof area of each shelter (S-1-S-3). The names of the species were provided by Mirek et al. [52] and their abundances were expressed using a four-point scale. Habitat conditions (i.e., light, soil moisture, trophy, and soil acidity) were rated for the recorded species according to Zarzycki et al. [53].

3. Results

3.1. Soil Studies

The shelter soils consist of sandy, sandy-loam, and loamy-sand deposits spread over a concrete base. The level of the concrete was sometimes covered with wood tar. The thickness of the topsoil (humus horizon) ranged from 14 to 18 cm was mainly caused by diverse microrelief, uneven accumulation of organic matter, and anthropogenic influences (Figure 5). The results of the soil texture analysis indicate an increase in the content of clay and silt particles in soils on the green roofs of S-2 and S-3 shelters in relations to the soil on the green roof of S-1. The soil texture composition offers good drainage and easy runoff for rainwater. Sometimes deeper, more compact soil horizons manifested hydromorphic features. The topsoil is characterized by the presence of anthropogenic admixtures in the form of slag, bricks, mortar fragments, ceramics, and glass, as well as metal debris and nails. The addition of artefacts to the soil was observed from the time of constructing green roofs on the bunkers to the present day. The plant litter is classified as moder-mull type of forest humus and mainly consists of the remains of maple, black locust, and oak leaves. The slightly acidic reaction of the litter often indicates the presence of admixtures made up of building debris. The described soils fall within the broad category of Anthropogenic soils, of the Technogenic soil type and the Konstruktosol subtype with a solid technogenic layer [51]. The soils were examined and classified as Isolatic Dystric Arenic Technosols and Isolatic Eutric Arenic Calcic Technosols [45].

In the studied mineral soil levels, the pH ranged from 4.5 (S-1) to 7.2 (S-3), which indicates a strongly acidic to neutral pH. The pH of the plant litter ranged from 5.0 to 6.6, meaning strongly or slightly acidic.

In the case of the green roofs covering shelters S-1 and S-2, crushed brick and granite aggregate constituted the basic building material used for erecting the shelters. The soil levels in the shelters also contain glass, the remains of malacofauna and *Robinia pseudoacacia* seeds. In turn, the flat roof on top of shelter S-3 is made entirely of concrete and its waterproofing layer is made of wood tar. The green roof over this shelter features an admixture of fine granite fragments, originating from the aggregate used for concrete, and also charcoal.

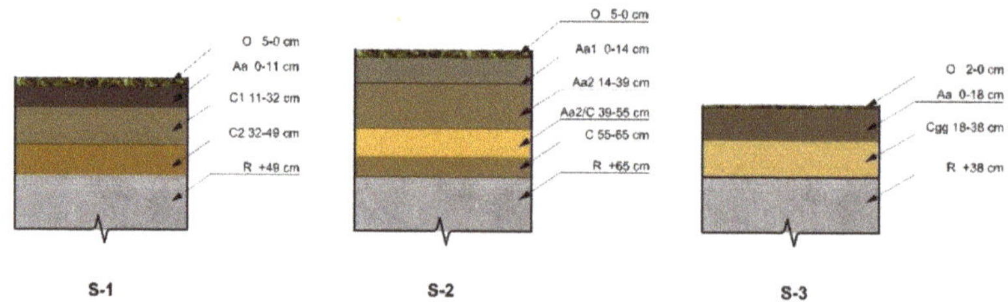

Figure 5. The depth of the soil horizons on the green roofs: O, organic horizon; A, humus horizon; C, parent rock horizon; AC, mix of the horizons; R, lithic rock (ecranic); a, anthropogenic material; g, gleyic properties. The colors of the horizons are shown (gray, brown, yellow, etc.) (description of horizon symbols [51]); S-1, S-2, and S-3 are symbols of the shelters.

3.2. Soil Water Storage

The soil granulometric composition was determined to be diverse. This feature clearly affects the retention capacity of the soil levels and the availability of water for plants. The profile on shelter S-1 represents sandy soil, which translates into the least water retention capacity. However, the water resources available for vegetation on this edifice did not differ greatly and, furthermore, did not fall below the drought period capacity (DPC) over the long term, just like the other green roofs. On shelters S-2 and S-3, the green roof soils were made up of clay in sands, with the S-3 profile containing the most silt. As a result, these roofs have a proportionally greater retention capacity. However, the profile on shelter S-2 revealed the deepest depletion of soil moisture resources of all the edifices studied. Below is a summary of the soil moisture reserves in the context of meteorological conditions between 2017 and 2019 (Figure 6).

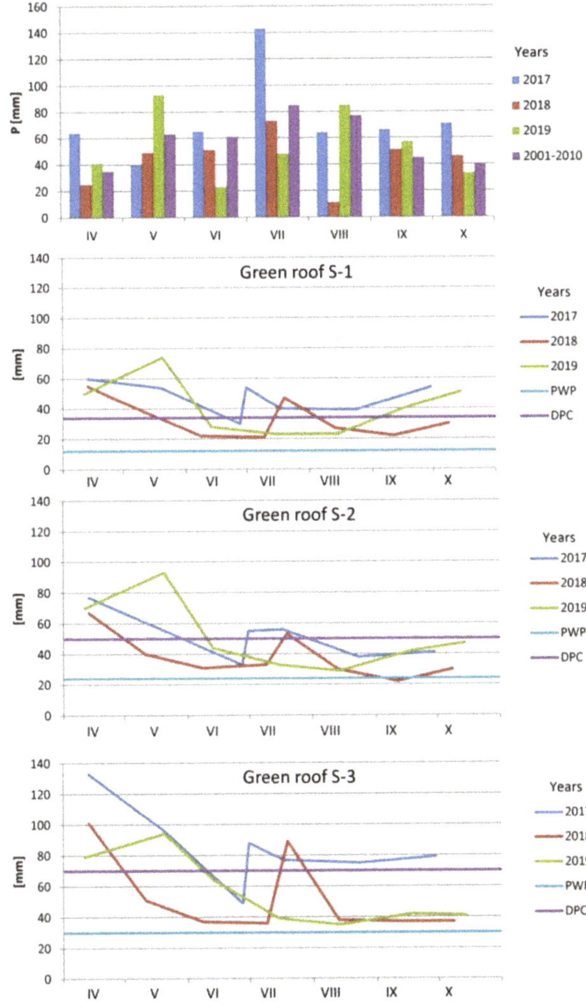

Figure 6. Changes in moisture in the soils on green roofs in the context of the storage capacity of these soils.

In 2017, the rainfall was 14% (+76 mm) above the 2001–2010 mean, while the temperature remained at the average level. April was accompanied by high soil moisture (close to field water capacity (FWC), pF = 2.0). By June, the moisture reserves had decreased, dipping below DPC for a short time (drought period capacity, pF = 2.85), but then increased after high rainfall in July and remained stable until the end of the growing season.

In 2018, the annual rainfall amounted to 71% (−106 mm) of the decade average. The annual temperature was 1.5 °C higher, rising to 2.2 °C during the growing season (indicating intensive evapotranspiration). This is slightly below the FWC (moisture reserves were only noted at the beginning of the growing season). By June, humidity had decreased almost to PWP (permanent wilting point at pF = 4.2). In July, after torrential rainfall, there was a short-term, pronounced increase in reserves (above DPC), and from August (precipitation 14% of the average, temperature of +2.6 ° C) until the end of the measurement period, the inventory fell again to a level close to the PWP.

In 2019, rainfall amounted to 95% (−19 mm) of the 2001–2010 average, while the annual temperature was 0.6 °C higher, which even rose to 3.7 °C in June (very high evapotranspiration). The highest (near FWC) moisture reserves were recorded in May after heavy rainfall. From June until the end of July, the reserves fell again to just above the PWP. After heavy rainfall in August, there was a noticeable increase in the reserves (close to the FWC), which continued on a slight upward trend until the end of the measurement period.

Analyzing the impact of the immediate environment on the results obtained, it can be concluded that the facilities as a whole are in of themselves roof buffers. At the same time, the roofs, at least abiotically (in terms of soil, and humidity), are independent of the rest of the building. Effective atmospheric precipitation (reaching the soil) and evapotranspiration predominately influence changes in soil retention.

3.3. Floristic Characterization of the Green Roofs with Reference to Habitat Conditions

The green roofs of military facilities are artificial habitats of anthropogenic origin, so they host synanthropic flora. After 100 years of functioning, forest communities with layers developed on each shelter (S 1-3). The list of recorded vascular plant species is presented in Table 4, along with their population abundances and ecological indicator values. In all phytocenoses, the dominant element of the tree layer was *Robinia pseudoacacia*. The regular pattern of individual specimens of this species implies that their juveniles were planted there, shortly after the shelters were constructed. *R. pseudoacacia* is a North American species with invasive traits. Subdominants include *Quercus robur* and *Acer platanoides*, which are native elements of deciduous forest communities. The shrub layers mainly consist of native species from deciduous forest and shrub communities that most likely settled there in a process of spontaneous succession. However, invasive *Padus serotina* is largely prevalent, and specimens of ornamental plants can also be found. In these layers, the juvenile stages of tree layer species are present. The herb layer vegetation differs in composition, as it developed under the tree canopy or in the canopy gaps. Under the tree canopy, tree seedlings dominate in the phytocenoses on all shelters. In the case of S-1, this stratum consists of the smallest number of the herbaceous plants, particularly in the canopy gaps, where only *Stellaria media* occurs with very small abundance. The herbaceous vegetation of S-2 and S-3 is richer and consists of hornbeam-oak forest species under the canopy, while species of ruderal and meadow communities occur in the gaps. Among the herbs, invasive *Impatiens parviflora* is present on each shelter, and this species is subdominant for S-1 and S-2.

Analysis of ecological indicator values showed that the vast majority of species from all the green roofs prefer fresh and moist soils, although the dominant *R. pseudoacacia* has the lowest moisture requirements and can even tolerate dry soils. Of all the shelters, the phytocenosis on S-2 has the highest share of species (i.e., shrubs and herbs from the canopy gaps) that can grow in dry soil conditions. This fact correlates with the deepest depletion of soil moisture resources disclosed in the soil on this shelter. In all communities, species of rich (eutrophic) soils dominate. Among the trees and shrubs, there are only a few species that can grow in moderately poor (mesotrophic) soils. Likewise, the indicator species of soils with a high humus and nitrogen content, for instance, *Chelidonium majus* and *Alliaria petiolata* also occur in all communities. These indicate the relatively high fertility of all the habitats, which results both from the depth of the plant litter and the humus horizons as well as the moder-mull type of forest humus, which is slightly acidic. The majority of species growing on all the roofs prefer neutral soils (pH 6.0–7.0), although some can grow in alkaline or moderately acidic soils. This corresponds with the results obtained for the plant litter pH. In all communities, species that prefer half shade and moderate light dominate, but plants that prefer full light conditions are well represented among the herbs growing in the canopy gaps. The light conditions (shade or light) strongly determine the differences in the occurrence of forest species relative to segetal/ruderal and meadow species in the herb layers.

Table 4. The list of species found on the three shelter roofs with their population abundances and ecological indicator values. Abundance is indicated according to the following scale: +—very few individuals, 1—a few individuals or clumps, 2—frequent but not dominant, 3—dominant or co-dominant. The ecological indicators are as follows: W—soil moisture, 2—dry, 3—fresh, 4—moist, 5—wet; Tr—trophy, 3—mesotrophic, 4—eutrophic, 5—very rich soil; R—acidity 2—acidic, 3—moderately acidic, 4—neutral, 5—alkaline; L—light, 2—moderate shade, 3—half shade, 4—moderate light, 5—full light; *—no data.

Species	Shelter			Ecological Indicator			
	1	**2**	**3**	**W**	**Tr**	**R**	**L**
Tree layer							
Populus alba L. dead	+			3-4	4	5	4
Quercus robur L.	+	+	+	3-4	3-4	3-4	4
Robinia pseudoacacia L.	1	3	2	2-3	3	3-5	4
Acer platanoides L.	3	3	2	3	3-4	4	4
Acer pseudoplatanus L.			2	3/4	4	3-5	3
Tilia cordata Miller			1	3	4-3	4-3	3
Tilia platyphyllos Scop.	+			3	4	5-4	3
Shrub layer							
Populus tremula L. juveniles	+	+		3	3	3	3
Quercus robur L. juveniles			1	3-4	3-4	3-4	4
Ulmus minor Miller juveniles	+	1		2-4	4	4	3
Crategus laevigata (Poir.) DC.		1	2	3-4	3-5	4-5	4-5
Crategus monogyna Jacq.		+		3-4	3-5	3-5	3-5
Padus avium Mill.		2		4	4	4-5	3
Padus serotina (Ehrh.) J.Agardh juveniles	1	2	1	3	3	3-4	3-4
Rosa canina L. dead		+		3-4	3-5	3-4	4-5
Rubus plicatus W. et N.		2	2	3-4	3	2-4	4-5
Robinia pseudoacacia L. juveniles			1	2-3	3	3-5	4
Acer platanoides L. juveniles	+	+	+	3	3-4	4	4
Acer pseudoplatanus L. juveniles			1	3/4	4	3-5	3
Cornus mas L.		1		*	*	*	*
Syringa vulgaris L.			+	*	*	*	*

Table 4. *Cont.*

Species	Shelter			Ecological Indicator			
	1	**2**	**3**	**W**	**Tr**	**R**	**L**
Sambucus nigra L.	2	1		3-4	4-5	4	(5)4-3
Sympharicarpos albus (L.) Blake		+		*	*	*	*
Liana							
Humulus lupulus L.			+	4-5	4-5	4-5	3
Parthenocissus quinquaefolia (L.) Planchon	+	2		*	*	*	*
Herb layer under the tree canopy							
Quercus robur L. seedings		3	2	3-4	3-4	3-4	4
Ulmus minor Miller seedlings		1		2-4	4	4	3
Chelidonium majus L.	2	1	1	3	4-5	4-5	3-4
Alliaria petiolata (Bieb.) Cav. et Grande	1	1	1	3/4	5	4	3
Crataegus monogyna Jacq seedlings	1			3-4	3-5	3-5	3-5
Geum urbanum l.			2	3-4	3-4	4-5	2-3
Padus serotina (Ehrh.) J. Agardh seedlings	1	2		3	3	3-4	3-4
Robinia pseudoacacia L. seedlings	2	2		2-3	3	3-5	4
Acer platanoides L. seedlings	3	2	1	3	3-4	4	4
Acer pseudoplatanus L. seedlings			1	3/4	4	3-5	3
Impatiens noli-tangere L.			1	4	4	4-5	2-3
Impatiens parviflora DC.	2	2	1	3	4	4	4-2
Tilia cordata Miller seedlings			1	3	4-3	4-3	3
Viola reichenbachiana Jord. ex Boreau		1	1	3	4-3	4-3	2-3
Lamium album L.			2	3	4	4	5-4
Sambucus nigra L. seedlings	2			3-4	4-5	4	(5)4-3
Dactylis polygama Horv.		2	2	3	3-4	4	3
Deschampsia cespitosa (L.) P.B.		2	1	4	3-4	3-4	3-5
Poa nemoralis l.		1		2-3	3	4-5	3
Herb layer in canopy gaps							
Urtica dioica L.		+	+	3-4	4-5	4	2-5
Fallopia dumetorum (L.) Holub		1	1	3	4	3-4	3
Rumex obtusifolius L.			+	3-4	4-5	3-5	3-5
Chenopodium album L.			+	3	4-5	4	5
Stellaria media (L.) Vill.	+		+	3-4	4-5	4	5
Geranium robertianum L.			1	3	3-4	4	2-3
Galium aparine L.			2	4-3	4-5	4	5-4
Ballota nigra L.		1		3	4-5	4	4
Galeopsis pubescens Besser			1	3-4	4-5	3-4	4
Conyza canadensis (L.) Cronq		1		2-3	3	3-4	5
Erigeron acris L.		1		2	3	4-5	5
Solidago gigantea Aiton		1	1	3-4	4		4-5

Table 4. *Cont.*

Species	Shelter			Ecological Indicator			
	1	2	3	W	Tr	R	L
Arrhenatherum elatius (L.) P.B. ex J. et C.Presl			3	3	4	4-5	4
Bromus inermis L.		2		2-3	3	4-5	5
Bromus sterylis L.			2	2	3	4	5
Dactylis glomerata L.		1		3	4-5	4-5	4
Festuca arundinacea Schreber		2		3-4	4	4	4
Poa pratensis L.		2	3	3	4	4	4
Poa trivialis L.			2	4	4	4	4
Total number of species	19	32	31				

4. Discussion and Data Limitations

The research on green roofs was conducted on the basis of regular observations on the shelters. Laboratory tests formed an essential complementary element. This is particularly important when protecting historic fortifications in situ where, as a result of many years of negligence and natural succession processes, the green roofs have changed over time. The exploratory nature of the research presented here has an advantage over studies carried out under controlled conditions on green roof models on a reduced scale. Field studies also provide an opportunity to confront the collected fragmented and not numerous archived documentation (e.g., drawings and aerial pictures) with the current facts. Due to the long period of operation (over 100 years), this documentation is often outdated. In addition, the adopted observation period, as well as the number of roofs tested, enabled observation of the periods of change occurring after different rain events in existing soil conditions with diverse vegetation scenarios. Furthermore, technical limitations precluded monitoring the microclimate directly at the facilities, but the research relied on data from a meteorological station located in the suburbs of Wroclaw, representative of the roofs studied. We are aware that such data would be extremely useful. Despite these limitations, some valuable results were obtained because very rarely have HF-related facilities been the subject of interdisciplinary environmental research.

It should be noted that the thickness of the levels on the green roofs depends mainly on how the builders covered them. Over time, this thickness may change as a result of vegetation growing via the accumulation of dead organic matter. Sometimes the thickness of the levels changes as a result of modernization work conducted on the shelters. Sometimes, as a result, various artefacts may appear in the soil. Some of these items were revealed during the research, e.g., crushed brick and slag, which are examples of frequently used mineral admixtures in soil substrates of artificial soil formations [54].

In turn, rainwater retention is one of the main functions currently fulfilled by green roofs in urban areas. The annual retention rate of green roofs can range between 5% and 85% of the precipitation [55–57]. The designs of the green roof are not the only factor influencing the formation of hydrological conditions [58–60], where time distribution and precipitation intensity [61–63], climatic conditions and seasonality [64], and pre-precipitation conditions [58], as well as the roof slope [65] also play an important role, [66]. What can also affect the amount of retention on green roofs are the initial conditions before precipitation, i.e., an antecedent dry period as well as the humidity of the substrate before precipitation [67–69]. Seasonal differences in green roof water retention are caused by varied evapotranspiration [70–72]. Higher evapotranspiration in warmer seasons causes faster drying and increased retention than in colder seasons [64,67,72]. In addition, local environmental conditions, construction techniques, and proper operation significantly determine the functionality and effects of green roofs that may influence, inter alia, the water cycle in an urbanized drainage basin [73–75].

Many of the currently conducted experiments concern small models with substrates with a thickness of several centimeters (extensive roofs), whose disadvantage is lower retention capacity and faster drying compared to intensive roofs [76]. The comparison of the results of these experiments with the tested roofs in HF areas is not fully reliable. There is a lack of literature on the study of comparable areas. On the basis of the literature of the green roofs in general, one can state that the retention capacity of systems increases along with an increase in the thickness of substrate layers. A study by Liesecke [77] demonstrated that with a substrate thickness of 4 cm that a green roof retained up to 45% of the annual runoff, and when the substrate thickness was increased to 10–15 cm, this led to a relatively small increase in retention of up to 60%. Scholz-Barth [78] demonstrated that with a 6-cm substrate layer that it is possible to retain 50% of rainwater, which rises to nearly 92% with an 11-cm layer. However, Fassman-Beck and Voyde [79,80], on the basis of measurements in profiles of 50 and 70 mm, did not find any significant differences in the retention of precipitation waters. Other studies that considered not only the thickness of substrates, but also the physical properties of the substrates, showed that the latter significantly impacts an enhanced capacity for water retention [21,81]. Generally, together with increase of substrate thickness, there was an increase in the retention capacity of soils [82,83].

The general characteristics of changes in the soil moisture content of green roofs during the analyzed vegetation periods of 2017–2019 are mainly dependent on the course of the precipitation. The rainfall distribution was highly uneven during the period under study. This meant, among other phenomena, sudden, momentary increases in water resources after torrential rainfall in July 2017 and 2018. Nevertheless, despite the occurrence of these phenomena on several occasions, no harmful excess of water was discovered in the profiles of green roofs. Due to the low retention capacity of the soils, the water resources easily available to plants (pF 2.0–2-85) tend to be depleted [76]. Only in 2017, which featured high rainfall, did soil water storage remain within the level of an easily available resource for the majority of the growing season. In the remaining two years, there were long-term soil moisture deficits, especially in the dry year of 2018 when periods of water storage were observed to be approaching a level that was unavailable to plants (DPC: pF = 4.2).

The retention potential of the studied soil formations can be assessed by comparing the measured water resources with the maximum amount of water that can be temporarily absorbed by the soil without endangering vegetation with an oxygen deficit. It is assumed that the amount of soil air should not fall below 10–15% of full water capacity (FWC). It follows that, relative to the measured average water resources, the soils on shelters S-1 and S-2 can temporarily cope with atmospheric precipitation in the region of 150 mm, while shelter profile S-3 can take about 100 mm. Therefore, the tested soils are capable of absorbing precipitation and, thanks to the construction of the entire structure, excess water is effectively drained.

Common knowledge dictates that roof conditions pose a challenge to the survival and growth of plants [59], particularly when it comes to moisture stress and severe drought. It is worth underlining that the green roofs located on historical shelters are only a few meters above ground level, which is why they are not as exposed to harsh conditions as in the case of tall buildings located, for example, in city centers. Similarly, the trees growing in the vicinity shade the ground and limit evaporation, while the root mass of the combined vegetation prevents physical substrate damage caused by wind and severe rainfall. On the other hand, the most shaded S-1 roof has the poorest soil and species composition. Aspects related to sunlight require additional research, as they may significantly influence the water balance. Historically, the S-1 and S-2 roofs were the sunniest, while S-3 was partially shaded from the beginning, which is reflected in the intensity of evapotranspiration and may affect the soil water balance. The aspect of the proportion of sunny-shady areas and the vegetation layers requires further research.

In all three green roofs, the organic matter in the roof soil retains moisture and helps provide the vegetation with nutrients, while also buffering against pH change [84]. It seems that the best conditions for plant development prevail on green roofs S-2 and S-3 due to the occurrence of clay in the soil formations, which ensures the vegetation has a superior water supply, as reflected in the higher total number of species, especially in the herb layer, inhabiting them in relation to S-1.

Despite the water shortages that occur during the growing season in green roof soils, their vegetation is able to grow properly and create forest communities with typical layers. However, the intensification of droughts and the increase in temperature during the growing season in the future may lead to the gradual loss of species that are most sensitive to water shortages. Research carried out after these edifices had been in operation for more than 100 years revealed that the green roofs there, despite their simple construction, function well in terms of benefiting the entire building as well as the development of vegetation and water management of the area. The negligence of procedures usually applied in urban green areas (raking leaves, mowing the undergrowth causing, inter alia, the destruction of tree and shrub seedlings, forming and illuminating crowns, the use of herbicides, etc.) caused the accumulation of an organic-rich horizon layer, which promotes the spontaneous development of vegetation along the path of natural succession as well as settlement of the area by species well adapted to the conditions of the habitat.

During the course of the study, it was not possible to determine whether local material from the shelter construction period was used for the construction of green roof soils. In addition, extensive investigations only yielded incomplete archival documentation and inventory drawings of the shelters and their green roofs. Therefore, more detailed research is required in the future, including a large number of sites, more data, and frequent long-term investigations.

To summarize, HF protection transcends conservation issues and is important not only for historical and educational values, but also in terms of nature. It can provide an opportunity for a better understanding of historical buildings that can be treated as an important "reservoir" of past environmental data. Studies of historic green roofs on historic fortifications can provide a lot of valuable information about how these engineering structures have functioned over the years. Historical green roofs provide not only a material link with the past, but also a "connection with the future." For over a century, they have proven their effectiveness and endurance, being not only an example of sustainable green roofs, but also, and perhaps above all, indicating a possible direction in how to shape proven forms of green roofs in an era of progressing climate change.

In consequence, all future protection and development scenarios for HF facilities, including restoration techniques, should be designed not only on the basis of a potential increase in anthropopressure resulting, for example, from growing tourist traffic, but also updated as climate scenarios evolve. They should also be based on existing strategic documents related to the preservation of urban biodiversity or climate protection.

5. Summary and Conclusions

- Green roofs created on fortifications in operation for over 100 years currently play an important role in city peripherals by improving storm water quality, reducing the rate and volume of runoff, mitigating the urban island heat effect, and supporting wildlife and habitat biodiversity, as well as carbon sequestration, and providing urban dwellers with the educational benefits of small-scale ecosystems. As "wild green roofs" on ageing shelters, they constitute a place of refuge for native forest plant species and for species biodiversity and are a valuable element of the system of ecological panels and corridors.
- In order to preserve their natural values, one should limit anthropopression, which is understood as actions aimed at transforming these sustainable ecosystems.
- Protective measures should be taken to care for the vegetation and soil environment of green roofs, especially the surface horizon of the soil in which the accumulation of organic matter occurs alongside the accumulation of nutrients for plants and water retention.
- The protective measures should especially include:

  - Keeping the roof area off limits (to pedestrian traffic) as far as possible, as it was during the military use of the buildings, in order to prevent damage to the soil litter;

- Preservation of the herb and shrub layer containing, inter alia, juvenile tree stages, which enables natural renewal of the tree stand;
- Care must be taken to protect soil organisms responsible for the breakdown of organic matter, eliminating invasive species, especially *Padus serotina* and monitoring their settlement on green roofs;
- When repairing and maintaining flat roofs, the original soil should be re-used, avoiding fertilization and the introduction of new species, especially ornamental plants. This would favor the preservation of genius loci, combining historical, cultural, and natural values, constituting an interesting complement to the "blue-green infrastructure" of the city.

Author Contributions: Conceptualization, Ł.P. and T.K.; methodology, Ł.P., T.K., A.B., D.K.; software, Ł.P.; investigation, Ł.P., T.K., A.B., D.K.; resources, T.K., A.B.; data curation, Ł.P., A.B.; writing original draft preparation, Ł.P.; writing—review and editing, Ł.P., D.K.; visualization, Ł.P., T.K.; supervision, T.K., D.K., A.B.; project administration, Ł.P. All authors have read and agreed to the published version of the manuscript.

Funding: This research received no external funding.

Acknowledgments: We would like to thank the Provincial Office of Monument Preservation and the Municipal Conservation Office in Wrocław for support in this research. This project was prepared without any external foundlings or grant. The work uses meteorological data from the Wroclaw–Swojec Faculty of Agrometeorology and Hydrometeorology Observatory (WOAiHW-S), cartographic materials from the Staatsbibliothek zu Berlin–Preussischer Kulturbesitz, and aerial photos from the state archives in Wrocław. We would also like to thank the three anonymous reviewers for their comments and suggestions, Katarzyna Pałubska and Marcin Górski for their professional consultation, Steven Jones for translating this paper, and Max Kooij for correcting this paper.

Conflicts of Interest: The authors declare no conflict of interest.

References

1. Sukopp, H. Human-caused impact on preserved vegetation. *Landsc. Urban Plan.* **2004**, *68*, 347–355. [CrossRef]
2. Hüttl, R.E.; David, K.; Schneider, B.U. Historic gardens and climate change. In *sights, Desiderata and Recommendations. In Historische Gärten und Klimawandel. Eine Aufgabe Für Gartendenkmalpflege, Wissenschaft und Gesellschaft*; De Gruyter: Berlin, Germany, 2019; p. 423.
3. Megarry, W. *Future of Our Pasts: Engaging Cultural Heritage in Climate Action: Report of the ICOMOS Workings Group on Climate Change and Cultural Heritage*; International Council on Monuments and Sites: Paris, France, 2019.
4. United Nations. *Recommendation on the Historic Urban Landscape*; United Nations: Paris, France, 2011.
5. United Nations. *Transforming Our World: The 2030 Agenda for Sustainable Development*; United Nations: New York, NY, USA, 2015; Available online: https://sustainabledevelopment.un.org/post2015/transformingourworld (accessed on 10 March 2020).
6. Fluck, H.; Wiggins, M. Climate change, heritage policy and practice in England: Risks and opportunities. *Archaeol. Rev. Camb.* **2017**, *32*, 159–181. [CrossRef]
7. Pardela, Ł.; Wilkaniec, A.; Środulska-Wielgus, J.; Wielgus, K. Znaczenie terenów zieleni pofortecznej dla zieleni miejskiej Krakowa, Poznania i Wrocławia. *Nauka Przyr. Technol.* **2018**, *12*, 153–162. [CrossRef]
8. Pardela, Ł.; Kolouszek, S. *Twierdza Wrocław 1890–1918. Monografia*, 1st ed.; Via Nova: Wrocław, Poland, 2017; ISBN 978-83-64025-35-8.
9. Pawlas, K.; Rabczenko, D.; Krzeszofiak, J. Ocena Wpływu Pyłowych Zanieczyszczeń Powietrza Na Zapadalność Na Wybrane Schorzenia, Dla Obszaru Miasta Wrocławia, Uniwersytet Wrocławski: Wrocław, Poland, 2018. Available online: http://life-apis.meteo.uni.wroc.pl/images/Raport_meteo.pdf (accessed on 10 March 2020).
10. Hogg, I. *The History of Fortification*; St. Martin's Press Inc.: New York, NY, USA, 1981.
11. Johnston, J.; Newton, J. *Building Green. A guide to Using Plants on Roofs, Walls and Pavements*; Greater London Authority: London, UK, 2004.
12. Köhler, M.; Michael, C.A. Green Roof Infrastructures in Urban Areas. In *Sustainable Built Environments*; Springer: New York, NY, USA, 2012; pp. 4698–4705. ISBN 978-1-4614-5828-9.

13. Catalano, C.; Marcenò, C.; Laudicina, V.A.; Guarino, R. Thirty years unmanaged green roofs: Ecological research and design implications. *Landsc. Urban Plan.* **2016**, *149*, 11–19. [CrossRef]

14. Thuring, C.E.; Dunnet, N.P. Persistence, loss and gain: Characterising mature green roof vegetation by functional composition. *Landsc. Urban Plan.* **2019**, *185*, 228–236. [CrossRef]

15. Köhler, M. Long-Term Vegetation Research on Two Extensive Green Roofs in Berlin. *Urban Habitats* **2006**, *4*, 3–26.

16. Thuring, C.E.; Dunnet, N.P. Vegetation composition of old extensive green roofs (from 1980s Germany). *Ecol. Process.* **2014**, *3*. [CrossRef]

17. Köhler, M.; Kaiser, D. Evidence of the climate mitigation effect of green roofs—A 20-year weather study on an extensive green roof (EGR) in Northeast Germany. *Buildings* **2019**, *9*, 157. [CrossRef]

18. Ahrendt, J. Historische Grundacher: Ihr Entwicklungsgang bis zur Erfindung des Eisenbetons. Ph.D. Thesis, Technischen Universität Berlin, Berlin, Germany, 2007.

19. Besir, A.B.; Cuce, E. Green roofs and facades: A comprehensive review. *Renew. Sustain. Energy Rev.* **2018**, 915–939. [CrossRef]

20. Jim, C.Y. An archaeological and historical exploration of the origins of green roofs. *Urban For. Urban Green.* **2017**, *27*, 32–42. [CrossRef]

21. Speak, A.F.; Rothwell, J.J.; Sidley, S.J.; Smith, C.L. Rainwater runoff retention on an aged intensive green roof. *Sci. Total Environ.* **2013**, *461–462*, 28–38. [CrossRef]

22. Kreh, W. Beiträge zur Vegetationskunde von Württemberg. *Jahresh. Des Ver. Für Vaterländische Nat. Württemberg* **1945**, *97–101*, 199–219.

23. Bornkamm, R. Vegetation und vegetations-entwicklung auf kiesdächern. *Vegetatio* **1961**, *10*, 1–24. [CrossRef]

24. Ksiazek-Mikenas, K.; Köhler, M. Traits for stress-tolerance are associated with long-term plant survival on green roofs. *J. Urban Ecol.* **2018**, 1–10. [CrossRef]

25. Magill, J.D.; Midden, K.; Groningen, J.; Therrell, M. A History and Definition of Green Roof Technology with Recommendations for Future Research. Master's Thesis, Southern Illinois University, Carbondale, IL, USA, 2011.

26. Shafique, M.; Kim, R.; Rafiq, M. Green roof benefits, opportunities and challenges—A review. *Renew. Sustain. Energy Rev.* **2018**, *90*, 757–773. [CrossRef]

27. Rowe, B.D. Long-term Rooftop Plant Communities. In *Green Roof Ecosystems*; Ecological Studies 223; Springer: Cham, Switzerland, 2015; pp. 311–332. ISBN 978-3-319-14982-0.

28. Brenneisen, S. Space for urban wildlife: Designing green roofs as habitats in Switzerland. *Urban Habitats* **2006**, *4*, 27–36.

29. Landlot, E. Orchideen-Wiesen in Wollishofen (Zürich)-ein erstaunliches Relikt aus der Anfang des 20. Jahrhunderts. *Vierteljahrsschr. Nat. Ges. Zürich* **2001**, *146*, 41–51.

30. Jankowski, M.; Bednarek, R.; Jaworska, M. Soils constructed on the 19th century fortifications in Toruń. In *Technogenic Soils of Poland*; Polish Society of Soil Science: Toruń, Poland, 2013; pp. 345–357. ISBN 978-83-934096-1-7.

31. Pardela, Ł.; Kowalczyk, T. Estimated soil water storage within a historical bunker during the growth period of vegetation. *J. Water Land Dev.* **2018**, 125–130. [CrossRef]

32. Coutts, C.; Hahn, M. Green infrastructure, ecosystem services, and human health. *Int. J. Environ. Res. Public Health* **2015**, *12*, 9768–9798. [CrossRef]

33. Oberndorfer, E.; Lundholm, J.; Brass, B.; Coffman, R.R.; Doshi, H.; Dunnet, N.P.; Graffin, S.; Köhler, M.; Liu, K.K.Y.; Rowe, B. Green roofs as urban ecosystems: Ecological structures, functions, and services. *BioScience* **2007**, *57*, 823–833. [CrossRef]

34. Van Renterghem, T. Green roofs for acoustic insulation and noise reduction. In *Nature Based Strategies for Urban and Building Sustainability*; Elsevier: New York, NY, USA, 2018; pp. 167–178. ISBN 879-0-12-812150-4.

35. Whittinghill, L.J.; Rowe, B.D.; Schutzki, R.; Cregg, B.M. Quantifying carbon sequestration of various green roof and ornamental landscape systems. *Landsc. Urban Plan.* **2014**, 41–48. [CrossRef]

36. Grard, B.J.P.; Chenu, C.; Manouchehri, N.; Houot, S.; Frascaria-Lacoste, N.; Aubry, C. Rooftop farming on urban waste provides many ecosystem services. *Agron. Sustain. Dev.* **2018**, 38. [CrossRef]

37. Lundholm, J. Plant effects on green roof ecosystem services: A 10-year retrospective. In Proceedings of the Cities Alive: 13th Annual Green Roof and Wall Conference, New York, NY, USA, 5–8 October 2015; pp. 1–15. [CrossRef]

38. MacIvor, J.S. Building height matters: Nesting activity of bees and wasps on vegetated roofs. *Isr. J. Ecol. Evol.* **2016**, *62*, 88–96. [CrossRef]

39. Fernandez-Canero, R.; Gonzales-Redondo, P. Green roofs as a habitat for birds: A review. *J. Anim. Vet. Adv.* **2010**, *9*, 2041–2052. [CrossRef]

40. General Inspection of the Engineer and Pioneer Corps and of Fortresses. *Erdbau in Permanenten Landbefestigungen*; Reichsdruckerei: Berlin, Germany, 1889.

41. General Inspection of the Engineer and Pioneer Corps and of Fortresses. *Mauerbau*; Reichsdruckerei: Berlin, Germany, 1898.

42. General Inspection of the Engineer and Pioneer Corps and of Fortresses. *Maßnahmen Gegen die Erkennbarkeit von Befestigungsanlagen (Schutzmaßnahmen)*; Reichsdruckerei: Berlin, Germany, 1905.

43. Rolf, R. *Die Entwicklung des Deutschen Festungssystems seit 1870—Vollständige und Bearbeitete Ausgabe des Manuskriptes*, 1st ed.; Fortress Books: Tweede Exloërmond, The Netherlands, 2000; ISBN 90-76396-08-6.

44. Rolf, R. *Armour Forts and Trench Shelters. German Imperial Fortifications 1870–1918*, 1st ed.; PRAK Publishing: Middelburg, The Netherlands, 2017; ISBN 978-90-817095-3-8.

45. United Nations. *World Reference Base for Soil Resources 2014. International Soil Classification System for Naming Soils and Creating Legends for Soil Maps*; World Soil Resources Reports 106; United Nations: Rome, Italy, 2014; ISBN 978-92-5-108369-7.

46. Topp, G.C.; Zebchuk, W. The determination of soil-water desorption curves for soil cores. *Can. J. Soil Sci.* **1979**, 19–26. [CrossRef]

47. Klute, A. Water Retention: Laboratory Methods. In *Methods of Soil Analysis: Part 1—Physical and Mineralogical Methods*; Agricultural Research Service and USDA: Fort Collins, CO, USA, 1986; pp. 635–662. ISBN 978-0-89118-864-3.

48. *PN-R-04032:1998 Standard. Soil and Mineral Materials—Sampling and Determination of Particle Size Distribution*; The Polish Committee for Standardization: Warsaw, Poland, 1998; ISBN 83-236-1077-0.

49. Polskie Towarzystwo Gleboznawcze. *Klasyfikacja Uziarnienia Gleb i Utworów Mineralnych PTG*; Polskie Towarzystwo Gleboznawcze: Warsaw, Poland, 2008.

50. *Soil Survey Staff Keys to Soil Taxonomy*, 12th ed.; USDA-Natural Resources Conservation Service: Washington, DC, USA, 2014.

51. Kabała, C. Polish Soil Classification, 6th edition-principles, classification scheme and correlations. *Soil Sci. Annu.* **2019**, *70*, 71–97. [CrossRef]

52. Mirek, Z.; Piękoś-Mirkowa, H.; Zając, M. *Flowering Plants and Pteridophytes of Poland. A Checklist*; Biodiversity of Poland; Polish Academy of Sciences: Kraków, Poland, 2002; ISBN 83-85444-83-1.

53. Zarzycki, K.; Trzcińska-Tacik, H.; Różański, W.; Szelong, Z.; Wołek, J.; Korzeniak, U. *Ecological Indocator Values of Vascular Plants of Poland*; Biodiversity of Poland; Polish Academy of Sciences: Kraków, Poland, 2002; ISBN 83-85444-95-5.

54. Baryła, A.; Karczmarczyk, A.; Bys, A. Role of Substrates Used for Green Roofs in Limiting Rainwater Runoff. *J. Ecol. Eng.* **2018**, *19*, 86–92. [CrossRef]

55. Berndtsson, J.C. Green roof performance towards management of runoff water quantity and quality: A review. *Ecol. Eng.* **2010**, *36*, 351–360. [CrossRef]

56. Stovin, V.; Vesuviano, G.; De-Ville, S. Defining green roof detention performance. *Urban Water J.* **2017**, *14*, 574–588. [CrossRef]

57. Pęczkowski, G.; Kowalczyk, T.; Szawernoga, K.; Orzepowski, W.; Żmuda, R.; Pokładek, R. Hydrological Performance and Runoff Water Quality of Experimental Green Roofs. *Water* **2018**, *10*, 1185. [CrossRef]

58. Bengtsson, L.; Grahn, L.; Olsson, J. Hydrological function of a thin extensive green roof in southern Sweden. *Hydrol. Res.* **2005**, *36*, 259–268. [CrossRef]

59. Dunnet, N.P.; Kingsbury, N. *Planting Green Roofs and Living Walls*, 1st ed.; Timber Press: Portland, ME, USA, 2004; ISBN 978-0-88192-640-8.

60. Liu, W.; Feng, Q.; Chen, W.; Wei, W.; Deo, R.C. The influence of structural factors on stormwater runoff retention of extensive green roofs: New evidence from scale-based models and real experiments. *J. Hydrol.* **2019**, 230–238. [CrossRef]

61. Carter, T.; Rasmussen, T.C. Hydrologic behavior of vegetated roofs. *J. Am. Water Resour. Assoc.* **2006**, *42*, 1261–1274. [CrossRef]

62. Simmons, M.T.; Gardiner, B.; Windhager, S.; Tinsley, J. Green Roofs are Not Created Equal: The Hydrologic and Thermal Performance of Six Different Extensive Green Roofs and Reflective and Non-reflective Roofs in a Sub-tropical Climate. *Urban Ecosyst.* **2008**, 339–348. [CrossRef]

63. Zubelzu, S.; Rodriquez-Sinobas, L.; Andrés-Domenech, I.; Castillo-Rodriguez, J.T.; Perales-Momparler, S. Design of water reuse storage facilities in Sustainable Urban Drainage Systems from a volumetric water balance perspective. *Sci. Total Environ.* **2019**, 133–143. [CrossRef]

64. Mentens, J.; Raes, D.; Hermy, M. Green roofs as a tool for solving the rainwater runoff problem in the urbanized 21st century? *Landsc. Urban Plan.* **2006**, *77*, 217–226. [CrossRef]

65. Villarreal, E.L.; Bengtsson, L. Response of a Sedum green-roof to individual rain events. *Ecol. Eng.* **2005**, *25*, 1–7. [CrossRef]

66. Getter, K.G.; Rowe, B.D.; Andresen, J.A. Quantifying the effect of slope on extensive green roof stormwater retention. *Ecol. Eng.* **2007**, *31*, 225–231. [CrossRef]

67. Zhang, S.; Guo, Y. Analytical Probabilistic Model for Evaluating the Hydrologic Performance of Green Roofs. *J. Hydrol. Eng.* **2013**, *18*, 19–28. [CrossRef]

68. Fassman-Beck, E.; Voyde, E.; Simcock, R.; Sing Hong, Y. 4 Living roofs in 3 locations: Does configuration affect runoff mitigation? *J. Hydrol.* **2013**, 11–20. [CrossRef]

69. Hathaway, A.M.; Hunt, W.F.; Jennings, G.D. A field study of green roof hydrologic and water quality performance. *Trans. Am. Soc. Agric. Biol. Eng.* **2008**, *51*, 37–44.

70. Van Seters, T.; Rocha, L.; MacMillan, G. Evaluation of Green Roofs for Runoff Retention, Runoff Quality, and Leachability. *Water Qual. Res. J. Can.* **2009**, *44*, 33–47. [CrossRef]

71. Palla, A.; Sansalone, J.J.; Lanza, L.G. Storm water infiltration in a monitored green roof for hydrologic restoration. *Water Sci. Technol.* **2011**, *64*, 766–773. [CrossRef]

72. Graceson, A.; Hare, M.; Monaghan, J.; Hall, N. The water retention capabilities of growing media for green roofs. *Ecol. Eng.* **2013**, 328–334. [CrossRef]

73. Wong, G.K.L.; Jim, C.Y. Identifying keystone meteorological factors of green-roof stormwater retention to inform design and planning. *Landsc. Urban Plan.* **2015**, 173–182. [CrossRef]

74. Cipolla, S.S.; Maglionico, M.; Stojkov, I. A long-term hydrological modelling of an extensive green roof by means of SWMM. *Ecol. Eng.* **2016**, 876–887. [CrossRef]

75. Szota, C.; Farrell, C.; Williams, N.G.S.; Arndt, S.K.; Fletcher, T.D. Drought-avoiding plants with low water use can achieve high rainfall retention without jeopardising survival on green roofs. *Sci. Total Environ.* **2017**, *603*, 340–351. [CrossRef]

76. Stovin, V.; Vesuviano, G.; Kasmin, H. The hydrological performance of a green roof test bed under UK climatic conditions. *J. Hydrol.* **2012**, 148–161. [CrossRef]

77. Liesecke, H.J. Extensive begrunung bei 5° dachneigung. *Stadt Grün* **1999**, *48*, 337–346.

78. Scholz-Barth, K. Green Roofs: Stormwater Management from the Top Down. *Environ. Des. Constr.* **2001**, *4*, 63–69.

79. Voyde, E.; Fassman, E.; Simcock, R. Hydrology of an extensive living roof under sub-tropical climate conditions in Auckland, New Zealand. *J. Hydrol.* **2010**, *394*, 384–395. [CrossRef]

80. Fassman, E.; Simcock, R.; Voyde, E. *Extensive Green (Living) Roofs for Stormwater Mitigation. Part 1: Design and Construction*; Auckland Regional Council: Auckland, New Zeland, 2010; p. 184.

81. Hilten, R.N.; Lawrence, T.M.; Tollner, E.W. Modeling stormwater runoff from green roofs with HYDRUS-1D. *J. Hydrol.* **2008**, *358*, 288–293. [CrossRef]

82. Nardini, A.; Andri, S.; Crasso, M. Influence of substrate depth and vegetation type on temperature and water runoff mitigation by extensive green roofs: Shrubs versus herbaceous plants. *Urban Ecosyst.* **2012**, 697–708. [CrossRef]

83. Metselaar, K. Water retention and evapotranspiration of green roofs and possible natural vegetation types. *Resour. Conserv. Recycl.* **2012**, 49–55. [CrossRef]

84. Best, B.B.; Swadek, R.K.; Burgess, T.I. Soil-Based Green Roofs. In *Green Roof Ecosystems*; Ecological Studies 223; Springer: Cham, Switzerland, 2015; pp. 139–174. ISBN 978-3-319-14983-7.

 *sustainability*

Article

Runoff and Water Quality in the Aspect of Environmental Impact Assessment of Experimental Area of Green Roofs in Lower Silesia

Grzegorz Pęczkowski *, Katarzyna Szawernoga, Tomasz Kowalczyk, Wojciech Orzepowski and Ryszard Pokładek

Institute of Environmental Protection and Management, Wroclaw University of Environmental and Life Sciences, Pl. Grunwaldzki 24, 50-363 Wroclaw, Poland; katarzyna.szawernoga@upwr.edu.pl (K.S.);
tomasz.kowalczyk@upwr.edu.pl (T.K.); wojciech.orzepowski@upwr.edu.pl (W.O.);
ryszard.pokladek@upwr.edu.pl (R.P.)
* Correspondence: grzegorz.peczkowski@upwr.edu.pl

Received: 28 April 2020; Accepted: 8 June 2020; Published: 11 June 2020

Abstract: Green architecture, including green roofs, can limit the effects of urbanization. Green roofs soften the thermal effect in urban conditions, especially considering the significant increase in the European and global population and that a significant share of the age group, mainly the elderly is exposed to diseases caused by high temperatures. We studied runoff and the quality of water from green roof systems in Lower Silesia, within the area of the Agro and Hydrometeorology Station Wrocław-Swojec, in the years 2012–2016. In the study, two systems with a vegetation layer based on light expanded clay aggregate and perlite were analyzed. The studies were based on the assessment of peak flow reduction, rainwater volume preservation and peak wave reduction. The calculated maximum retention performance indicator, relative to rainfall, for perlite surfaces was up to 65%, and in relation to the control surface up to 49%. In addition, the quality of water from runoff was estimated in the conditions of annual atmospheric deposition, taking into account such indicators as electrolytic conductivity; the content of N, NO_3, NO_2, NH_4, P, PO_4; and the content of metals, Cu, Zn, Pb and Cd. The load of total nitrogen exceeded the values of concentration in rainwater and amounted to 7.17 and 13.01 mg·L^{-1} for leca and perlite, respectively. In the case of the metal content, significantly higher concentrations of copper and zinc from green surfaces were observed in relation to precipitation. For surfaces with perlite, these were 320 mg·L^{-1} and 241 mg·L^{-1}, respectively, with rainwater concentrations of 50 and 31 mg·L^{-1}.

Keywords: green roofs; integrated environmental assessment; quality of runoff water; performance system

---

1. Introduction

The increasing population and overall consumption, urban development, and the simultaneous change in the climate and extreme weather events due to the intensive planning of city development comprise the integrated society and environment rating. The growing amount of people aged 65 and over in Europe and especially in Poland represent a significant part of society that is vulnerable to high temperature diseases [1]. The urban heat island effect is the main reason behind a change in the quality of life for people living in cities areas [2–4]. This effect causes urban areas temperatures to be much higher than surrounding rural areas [5]. Another important aspect is the changing natural hydrologic systems and the impact on water quality, which is drained into the environment [6–9]. By 2050, almost 70% of the global population will live in cities [10]. Since the population is continuously growing, any problems currently facing cities will affect a staggeringly larger proportion of people over time.

Thus, finding solutions to problems like heat waves will be an integral part of future city living. One of the desired methods of reducing the effects of urbanization is green architecture, including green roofs.

Green roofs mitigate the effects that traditional building materials have on urban environments by decreasing air temperatures [11,12] and costly energy consumption [13,14]. Green roofs produce a cooling effect on roof surfaces and air temperatures through the process of evapotranspiration, the transfer of water from the soil and vegetation to the air. This endothermic process consumes heat energy, so the warmer the air is, the more evapotranspiration takes place [12].

Compact buildings will cause changes in water circulation compared to undeveloped areas. There are a variety of practices used in the management of rainwaters, but it appears that the unique properties of green roofs in terms of the reduction of runoff volume [15–19] and its quality [20–22] are of special importance in the approach to the growing urbanization. Similarly important is the reduction of noise and air pollution [23,24], the reduction of oxygen dioxide levels [25], the aesthetics of landscape, biodiversity [26,27], and improvements to the technical conditions and service life of structures [28].

At both the design and operation stages, such systems may require optimization in the aspect of hydrological features and water properties of the substrates used. For that purpose, one can use existing software allowing the analysis of the flow of water and dissolved substances in the partially unsaturated and saturated biological layer of the soil substrate [29]. The present authors analyzed the performance of the experimental roof systems during storm events in 2016. The analyses were carried out for two variants of models of green roof systems.

There are two main goals of our work. The first is the study of the hydrological performance of experimental models of green roofs based on the expanded clay and perlite of the extensive type. Inn Characteristic studies were based on the assessment of peak flow reduction, rainwater volume preservation and peak wave reduction. Hydrological conditions were simulated for the discussed model solutions. The prepared model was calibrated on the basis of experimental data, which allowed for the estimation of hydrological efficiency for the analyzed substrates. A literature review indicated promising results, with the assumption that the flow through the substrate has a unidimensional character [30–34]. The second goal was to study the impact of the areas in question on the quality of rainwater from runoff. The characteristics of nitrogen, phosphorus and some metals subject to runoff were assessed. The results were analyzed for rainfall and control area.

2. Research Methodology

2.1. Models of Green Roofs

The scope of the experiment comprised two models of extensive type green roofs with differing composition of the vegetation layer, including components improving the retention properties of the substrates. The functionality of the systems was estimated in the local climate conditions of the city of Warsaw (Wrocław; 51st 11′ N, 17st 14′ E), Poland, with a particular focus on the retention performance, which is reflected in the reliability of the system and in the runoff water quality. Based on the designed steel support structure, experimental models were built with dimensions of 1000 × 2000 mm at a height of 1 m above the surface. That height is also the reference level for meteorological station and disdrometer.

The experimental green roofs (Figure 1) with thickness of 8 cm, was composed of a geotextile layer and a gravel drainage layer with a particle size fraction of 1–2 cm and water proofing membrane for the hydroinsulation of the roof. In this case, RMS 300 Optigrün International AG production geotextile was used, allowing the retention of up to 2 $L·m^{-2}$ of water, and additionally a water proofing membrane. To provide thermal insulation for the models, extruded polystyrene (XPS) was applied in the horizontal and vertical part of the construction, with a 5 cm layer thickness. In both cases, the vegetation layer was prepared on the base of horticultural soil, the share of which was 60% v/v. The substrate was based on an expanded clay aggregate contained sand with fine and medium fraction (20% v/v) and an expanded clay aggregate with small and medium fraction of 4–8 mm (20% v/v). In the case of the substrate based on perlite, it contained 20% v/v of sand, as in the previous case; 5% v/v of expanded clay aggregate; and 15% v/v of perlite. It was characterized by suitable properties, in particular a stable structure preventing settlement, and a proper hydraulic capacity.

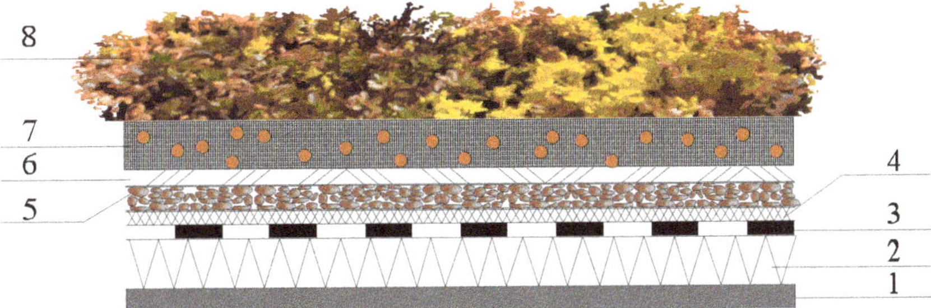

Figure 1. The construction details of green roof extensive models: substrate with expanded clay aggregate and substrate with perlite. 1—model support construction; 2—thermal insulation, extruded polystyrene (XPS); 3—insulation water proofing membrane; 4—geotextile, type RMS 300; 5—gravel layer with granulation from 1 to 2 cm; 6—filtration geotextile; 7—substrate of light expanded clay aggregate or perlite (depth substrate together with filtration geotextile and gravel layer, 80mm); 8—plants of the sedum species (*Sedum spurium, Sedum sexangulare, Sedum telephium, Sedum floriferum,* and *Sedum album*).

The substrate used as the vegetation layer was prepared using horticultural soil with a pH close to neutral (pH 6.7), sand fraction, and the specified admixtures improving the retention properties. The performed particle size analysis revealed that the substrate was a sandy loam, the properties are presented in Table 1.

Table 1. Properties of the substrates composed of leca and perlite.

Parameter	Unit	GR1 (Leca)	GR2 (Perlite)
Bulk density	$g\cdot cm^{-3}$	0.40	0.27
Specific density	$g\cdot cm^{-3}$	1.81	1.85
Water capacity *			
pF 0	mm	37.0	39.0
pF 2.0		16.5	19.8
pF 2.9		13.5	17.0
pF 4.2		3.0	4.4

* Water content was calculated for 50 mm substrate thickness.

The calculated water leachability for the vegetation layer was 22.5 mm in the case of expanded clay aggregate and 32.6 mm in the case of perlite, and the values of the leachability coefficient amounted to 40.9% and 38.4%, respectively. Air permeability, calculated as the ratio of leachability and absolute porosity, was 55.3 and 49.2%.

In the design of the vegetation layer, species of stonecrops (*Sedum*) from the family *Crassulaceae* were used: *spurium, telephium, floriferum* and *sexangulare*. The choice of *Sedum* species was dependent on extensive models, and by the plant care method, especially for the maintenance-free roof without any possibility of irrigation [35].

The measurement period began on 1 April and ended on 31 October 2016. During that time, all events with atmospheric precipitation were recorded. During the entire cycle, measurements of moisture and runoff from the analyzed surfaces were made in real time. In the case of moisture, measurements were taken using a multi-channel sensor TDR (Time Domain Reflectometry, E-TEST,

Poland, Lublin) with data recorded at 1 min intervals. The range of accuracy of the sensors was ± 0.02 cm$^3 \cdot$cm^{-3} (water). The measurements of runoff was implemented with the use of a system based on a set of tilt troughs and recoding of impulses using data loggers Hobo UX120 (model: Hobo UX120–006M, Onset Computer Corporation, Bourne, USA). One measurement impulse was equal to a runoff of 0.01 mm$\cdot$min^{-1}. The laser disdrometer (Precipitation Monitor: 5.4110.10 and LNM View software) allowed the continuous measurement of rainfall (0.005 mm$\cdot$h^{-1} minimum intensity and 0.16 mm minimum droplet size).

The retention properties of the model surfaces were designated by the retention indicator calculated in relation to rainfall or to the control surface (RPI$_{ratio}$):

$$RPI_{ratio} = 1 - \frac{\sum runoff\ GR}{\sum P}\ 100\%$$

(1)

where: runoff GR is runoff green roof area, and P is precipitation. Using Equation (1) the runoff from the green surfaces and the precipitation were determined.

2.2. Runoff Water Quality Monitoring

The quality analysis was carried out on the basis of water samples coming from runoff from model surfaces, control surfaces and precipitation. Out of all recorded events, eight enabled us to obtain a quantity allowing for determination. Rainwater for analysis was collected using adapted apparatus for measuring real-time runoff from the surface. So, the times of the beginning of the event and the times of starting the runoff and their end were known. The time between obtaining samples for analysis and their direct determination was short and met the criteria for their preparation for determination. The analyses of all indicators were carried out at the Faculty Laboratory for Environmental Research of Wrocław University of Environmental and Life Sciences. In the determination of the pollution indicators, the methods used followed the standards PN-EN 26777:1999, PN-82C-04576/08, PN-ISO 7150:2002; for phosphates, the method followed the PN-EN 1189-2000; for concentrations of metals, with the method of atomic absorption spectrometry AAS.

For the independent samples, statistical tests were conducted with the use of the t-test, and comparison was made of the quality indicators in runoff water and in rainwater. Two hypotheses were proposed: the zero hypothesis H_0: the quality indicators from the green surfaces were the same as in the precipitation and control area; and the second hypothesis H_1: the mean values of water quality indicators for the green surfaces differ from those in rainwater and in water from the control surface. The adopted level of significance was 5%. Conclusions concerning the equality of the mean values were preceded with Levene tests.

2.3. Runoff Modelling

The simulation for the experimental models was conducted with the use of the program, based on Richards' Equation (2), assuming one-dimensional flow direction, for the variant single porosity model. The standard van Genuchten hydraulic model was adopted in the solution. The hysteresis effect was not included, the variant without hysteresis gives the expected results [30].

$$\frac{\partial \theta}{\partial t} = \frac{\partial}{\partial z}\left[K(\theta) \cdot \left(\frac{\partial h}{\partial z} - 1\right)\right] - S$$

(2)

where θ is the volumetric content of water [L$^3 \cdot$L^{-3}], z is the spatial coordinate [L], t is the time [T], K is the unsaturated hydraulic conductivity [L$\cdot$T^{-1}], and S is the unit uptake of water by the plants [L$^3 \cdot$L$^{-3} \cdot$T^{-1}]. The effect of temperature gradients on the flow in a porous medium was neglected. For the description of the hydraulic properties of the soil the van Genuchten–Mualem Equation (3) was used, in the form [36–38]:

$$\theta(h) = \begin{cases} \theta_r + \dfrac{\theta_s - \theta_r}{[1 + |\alpha h|^n]^m} & h < 0 \\ \theta_s & h \geq 0 \end{cases} \tag{3}$$

$$K = K_s S_e^{1/2} \left[1 - \left(1 - S_e^{1/m} \right)^m \right]^2 \tag{4}$$

$$S_e = \frac{\theta - \theta_r}{\theta_s - \theta_r} \tag{5}$$

where θ_r is the residual moisture [$L^3 \cdot L^{-3}$]; θ_s is the saturated moisture [$L^3 \cdot L^{-3}$]; S_e is the effective saturation (5); α is the constant [L^{-1}]; n, $m = 1 - 1 \cdot n^{-1}$ is the form parameter; K_s is hydraulic conductivity [$L \cdot T^{-1}$].

The modeled vertical profiles with a drainage and vegetation layer were discretized with the use of 77 elements.

Solving Equations (3) and (4) requires the estimation of parameters θ_r, θ_s, α, n and K_s [36], the values of which on the depend type of substrate. The parameters are shown in Table 2. The values of the parameters were determined in laboratory conditions (K_s, θ_s), or adopted and calculated (θ_r, α, n) [39]. K_s was measured in several replicates in laboratory conditions. The boundary conditions at the boundary of the soil and the atmosphere can change from an unsaturated value to full soil saturation, and depend on hydrological processes (infiltration and evapotranspiration, and atmospheric precipitation), and, consequently, on the moisture and the soil substrate [30]. The initial water content before the occurrence of subsequent events was determined by means of the time domain reflectometry technique (TDR). The measurement interval was 30 seconds. The vegetation cover factor of the experimental models was included. The model's performance was evaluated statistically based on the Nash–Sutcliffe (NSE) Equation (6), and the performance and mean square error RMSE (7). The range of this statistic can be within ($-\infty$, ∞). The calculated Nash–Sutcliffe performance could be from $-\infty$ to 1. NSE = 1 corresponds to a perfect match, when NSE = 0 model the predictions were as accurate as the average observed.

$$RMSE = \sqrt{\frac{\sum\limits_{t=1}^{t=te} [q_{mea}(t) - q_{sim}(t)]^2}{\sum\limits_{t=1}^{t=te} 1}} \tag{6}$$

$$NSE = 1 - \frac{\sum\limits_{t=1}^{t=te} (q_{mea}(t) - q_{sim}(t))^2}{\sum\limits_{t=1}^{t=te} (q_{mea}(t) - q_{amea})^2} \tag{7}$$

where q_{mea} is the measured flowrate, q_{sim} is the simulated flowrate, q_{amea} is the average measured flowrate, and t_e is the end of the run off measure.

Table 2. Parameters for the green roof model.

Model	θ_r (cm³ cm⁻³)	θ_s (cm³ cm⁻³)	α (m m⁻¹)	n (–)	K_s (mm min⁻¹)
Substrate with leca GR1	0.015	0.731	0.1627	1.1298	15.95
Substrate with perliteGR2	0.001	0.787	0.0092	1.0137	0.610

Using the described and validated model, runoff simulations were conducted. The model was calibrated for measured and theoretical runoffs were analyzed for 13 June 2016.

3. Results and Discussion

The model study in the conditions of the differentiated vegetation layer in relation to the control surface was conducted in the years 2012–2016. Detailed studies conducted in 2016 showed that 27 out of 37 events caused runoff. Twelve events generated runoff with intensity below 0.1mm·min^{-1}, and two above 1mm·min^{-1}. Eight selected events from the entire observation period generated runoff, the size of which enabled the determination of indicators. In each case, runoff was initiated by rainfall larger than 10 mm. The stoppage of the runoff and its delay varied and depended not only on the type of surface, but also on the substrate moisture content and rainfall intensity. For all events (Table 3) the runoff was initiated each time, but the runoff delay and hold varied. This depended mainly on the initial humidity of the ground and the intensity of the rain. Each green roof model reduced the mean peak discharge indicator. For example, on day 5.10, rainfall with a total of 26.4 mm and an intensity of 0.11 mm·min^{-1} was retained for models based on expanded clay and perlite in the amounts of 63 and 70.1%, respectively. In another case (5.09) at a significant intensity of 1.31 mm·min^{-1}, but with a smaller sum precipitation of 13.9 mm, the holding amounts was 28.9 and 50.7%, respectively.

Table 3. Rainfall and runoff characteristics of the studied events in experimental object.

Rainfall Event	Rain Depth (mm)	Peak of Rain Intensity (mm·min^{-1})	Rain Duration (min)	Retained Volume (%)		Flow Peak (mm·min^{-1})		Peak Reduction (%)		Moisture (Initial) (% v/v)	
				Leca	Perlite	Leca	Perlite	Leca	Perlite	Leca	Perlite
13.06	22.6	1.47	367	31.4	64.9	1.41	1.22	26.1	38.9	15.7	16.6
17.06	12.2	1.1	381	311	55.1	0.92	0.87	33.6	39	20.3	23.1
31.07	10.1	0.8	1249	55.8	63.4	0.57	0.53	40.5	47.7	19.1	21.3
21.08	21.2	0.4	466	43.2	68.3	0.29	0.26	44	62.7	17	19.1
5.09	13.9	1.31	458	28.9	50.7	1.19	1.12	29.1	38.9	19.9	22.8
17.09	20.7	0.22	686	54.3	62.2	0.18	0.16	42.4	49.8	22.3	23.6
3.1	29.8	0.26	1433	62.1	74	0.21	0.19	52.2	60	23.6	24.9
5.1	26.4	0.11	1640	63	70.1	0.09	0.08	51	59.1	25.3	27.1

The analyzed surfaces delayed the start of the runoff. The calculated retention performance indicator amounted to 9 and 11 min, respectively (Figure 2).

For the event of the 13.06 used to calibrate the model, the mean square error (RMSE), (6) was 0.16 mm·min^{-1} (GR1, leca) and 0.12 (GR2, perlite), while the calculated statistics describing the relative value of residual variance, the Nash–Sutcliffe efficiency (NSE) (7)), and the assumed values at the level 0.74 (GR1) and 0.84 (GR2). A simulation can be accepted as satisfactory when NSE >0.5 (a higher model accuracy for NSE values close to 1). The calculated NSE and RMSE values for validation are presented in the Table 4. During validation, the models designed as leca- and perlite-based displayed a good representation of flow volume intensity, accurately simulating events observed in other periods. It can, therefore, be concluded that in both cases the level of matching achieved was close to good.

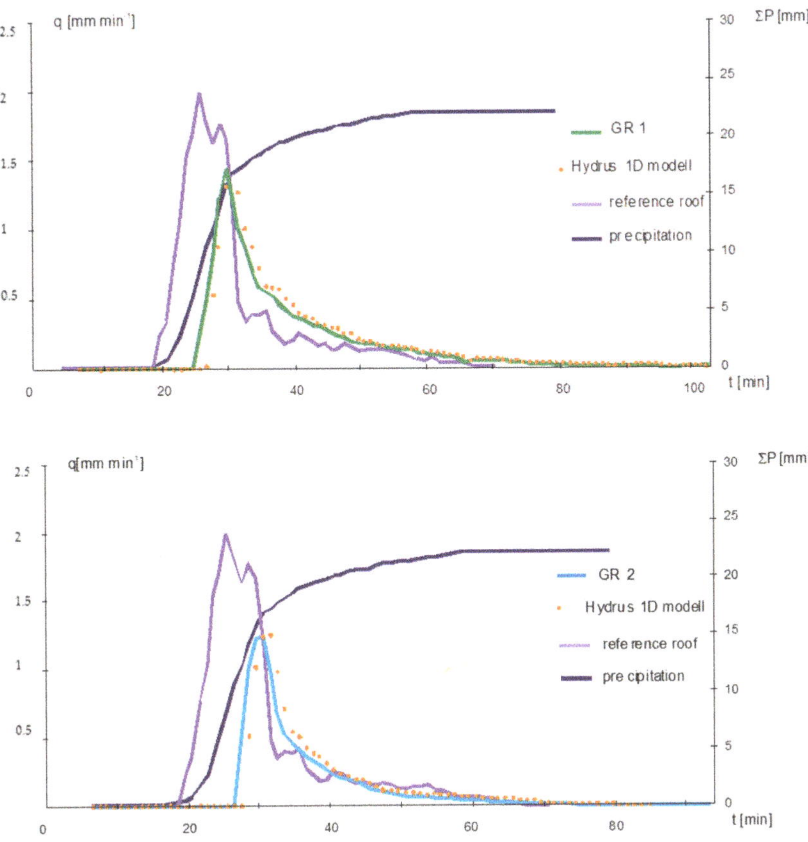

Figure 2. The intensity of runoff and accumulated precipitation for the surface GR1 and GR2 and the control surface in June 2016.

Table 4. Standard measures of model evaluation.

Date of the Event	Rain Intensity mm min⁻¹	Substrate of Leca		Substrate of Perlite	
		NSE (−)	RMSE (%)	NSE (−)	RMSE (%)
Calibration data					
13.06	1.47	0.74	0.16	0.84	0.12
Validation data					
17.06	1.1	0.75	0.15	0.83	0.12
20.06	0.2	0.65	0.028	0.69	0.023
31.07	0.8	0.79	0.07	0.82	0.05
9.08	0.05	0.57	0.009	0.63	0.008
Other events					
21.08	0.4	0.75	0.085	0.8	0.061
5.09	1.31	0.78	0.069	0.85	0.063
17.09	0.22	0.64	0.058	0.7	0.055
3.1	0.26	0.66	0.061	0.71	0.059
5.1	0.11	0.58	0.048	0.61	0.043

The water content and time and depth illustrate the conditions of infiltration in the green roof systems (Figure 3). In the conditions of high saturation of the designed profiles, one can note that in the model based on expanded clay extrudate (GR1) the state of moisture close to porosity was attained in minute 45. In the case of surface GR2 a significant level of saturation was observed already in minute 20. In this case, this may indicate an improvement of the retention capacity of the profile and of the hydrological properties of the systems as a result of the application of the admixture of perlite in the substrate.

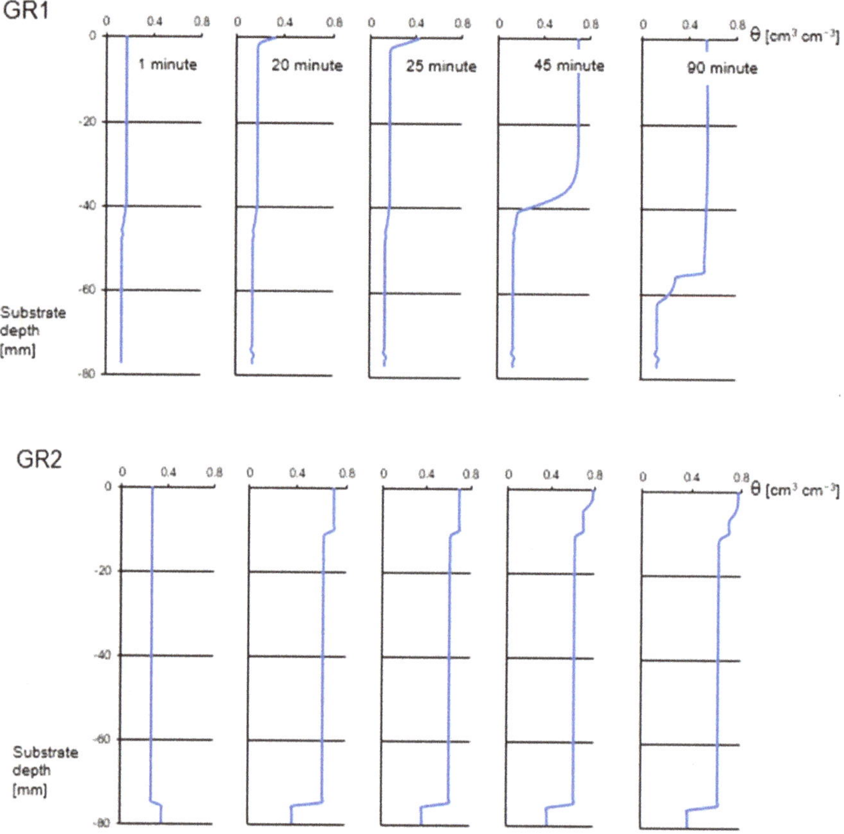

Figure 3. Water content profiles in models with leca (GR1) and perlite (GR2) at different time steps for 13 June 2016.

Studies of green roofs, reflected in their reliability in the context of retention capacity, have been conducted by other authors. Pala et al. [32] monitored the green roof of the University of Genova. They confirmed that it can significantly mitigate the generation of runoff, with the median values of retained volume and peak reduction equal to 94 and 98.7%, respectively. The authors applied a conceptual linear reservoir and a Hydrus-1D models to simulate the hydrologic behavior of the system. The simulations with both models reproduce acceptable matching capabilities the experimental measurements, as confirmed by the Nash–Sutcliffe efficiency index that was generally greater than 0.60 [32]. Wong [40] conducted a study in the region of Hong Kong, a humid and tropical climate. Thin layer (40 mm) solutions of green roofs ensured a reduction of runoff intensity at the expected level. Despite the small thickness of the layers, controlled runoff was achieved, and the retention capacity and thus good performance were obtained in a relatively short time. The methods of modelling with the use of the program Hydrus applied by Feitosa and Wilkinson [41] showed that the effectiveness of the analyzed systems, reflected in their reliability, decreased with increasing intensity and duration of rainfall. Locatelli et al., on the basis of the analyses of statistical data from three locations, concluded that in the case of a single event the peak discharge and the reduction of runoff volume decreased with the extension of the period of event recurrence [42]. Data concerning the quality of runoff water in the Swojec object were based on eight rainfall episodes. The increase in total nitrogen concentration was recorded in the samples. The data were compared with the use of t-tests, assuming a 5% level of significance. Two hypotheses were adopted: the zero hypothesis H_0: the quality indicators from the green surfaces were the same as in precipitation and control area; and the second hypothesis H_1: the mean values of the water quality indicators were different in relation to rainwater and water from the control surface.

For each of events, samples were collected from the model green roofs, from the control surface, and from atmospheric precipitation. The Analyzed indicators were total nitrogen (TN), NO_2, NO_3, NH_4, T-P, PO_4, Cu, Zn, Pb, Cd and electrolytic conductivity. Contrary to the adopted hypothesis, there was no distinct improvement of water quality in the runoff from the experimental surfaces. An increase in the level of concentration of nutrients was noted in runoff from the same surfaces. Table 5 presents the results of the water quality and mean values of the determined indicators. The load of total nitrogen in the runoff from the green roofs exceeded the concentration in rainwater and amounted to 7.17 and 13.01 $mg·L^{-1}$. Distinctly higher levels of concentration of total nitrogen for the substrate with perlite could have been a result of the composition of the substrate. It should be emphasized that a variation of the loads of that indicator was observed among all of the analyzed rainfall events, in particular for surface GR2, as seen Figure 4. In order to illustrate the observed variability of the pollutant load in relation to rainfall and control surface, frame charts were prepared. The lower and upper boundary of each box indicate respectively the 25th and 75th percentiles. Whiskers above and below each box indicate the 90th and 10th percentiles. For the NO_3 and NO_2 indicators, no significant differences were observed compared to rainwater. The recorded values of NO_2 were at trace.

Table 5. Summary of average indicators for green roof runoff, control site runoff and precipitation from the green roof experimental models in Wroclaw-Swojec.

Pollution Indicators	Green Roof Substrate in Leca	Green Roof Substrate in Perlite	Control Site	Precipitation
N $mg·L^{-1}$	7.17	13.01	5.52	5.23
NO_3–N $mg·L^{-1}$	1.96	3.93	1.72	1.68
NO_2–N $mg·L^{-1}$	0.01	0.01	0.059	0.065
NH_4–N $mg·L^{-1}$	0.15	0.10	0.128	0.187
P (total) $mg·L^{-1}$	0.25	0.26	0.207	0.246
PO_4–P $mg·L^{-1}$	0.15	0.12	0.106	0.121
Pb (total) $mg·L^{-1}$	113.8	61.5	110.2	103.6
Zn (total) $mg·L^{-1}$	237.6	241.4	32.1	31.1
Cd (total) $mg·L^{-1}$	2.3	1.2	1.3	1.3
Cu (total) $mg·L^{-1}$	220.8	320.2	54.1	50.2

The content of ammonium nitrogen NH_4 was relatively stable, both in relation to the experimental surfaces and to the rainwater (maximum 0.53 $mg·L^{-1}$). For the analyzed cases, the type of media did not affect the value of the indicator.

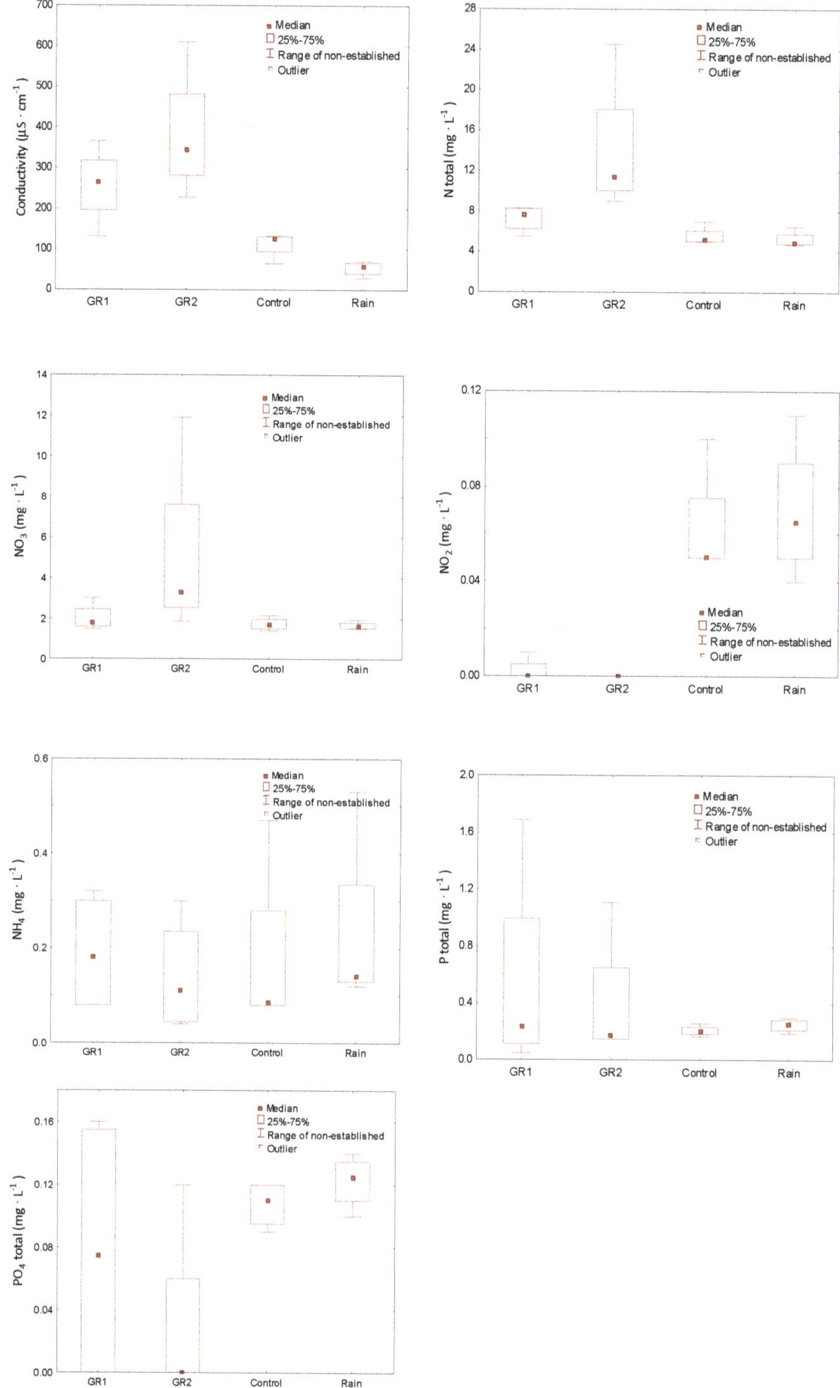

Figure 4. The range and average values for nutrients and conductivity in extensive roof experimental models in Swojec object.

The concentrations of PO_4 in the runoff from the analyzed surfaces, and in rainwater, were relatively uniform, but in the case of runoff from surface GR2 the concentrations were the lowest. Phosphorus compounds in the alkaline soil environment can be bound by calcium and manganese. This may indicate considerable concentrations discharged in the course of the vegetation season and during the winter period. The maximum concentrations of biogens can be observed after at least a dozen days from fertilization [43].

In a study concerned with the chemical composition of rainwaters, MacAvoy et al. [21] determined for 9 rainfall events higher concentrations of NO_3 and NH_4 in runoff from green roofs relative to that from a control surface. In that study it was also found that over the entire cycle of a 16-month experiment the load of nitrates was reduced by up to 32%. An experimental study conducted by Harper [44] showed that a considerable load of phosphates and nitrates was carried away in runoff in the initial period of operation of a green roof system. As a result of a nine-month operation, the load of phosphates was reduced by 5 $mg \cdot L^{-1}$ and that of nitrates by 10 $mg \cdot L^{-1}$. In studies on organic carbon by Yang and Lusk, Mitchell et al. [45], Carpenter, Kuoppamaiki, and Beecham [6,46,47] a decrease of concentrations was observed in the range from 500 to 50 $mg \cdot L^{-1}$.

In the analyzed runoff waters from the experimental surfaces, electrolytic conductivity assumed higher values in relation to runoff from the control surface and rainwater (Figure 4). Studies by other authors supported the results obtained in the experiment at Swojec. For example, Vijayaraghavan and Joshi studied the outflows from four different green roof systems and found that a better quality of runoff water was generated at a lower conductivity, relative to atmospheric precipitations [48].

The analysis of the content of metals in runoff from the green surfaces revealed several times higher concentrations of copper and zinc in comparison to those determined in rainwater. It can be assumed that in this case, the structural materials of the model (galvanized steel, zinc) or components of the substrate were a source of contamination with those elements (Table 4, Figure 5).

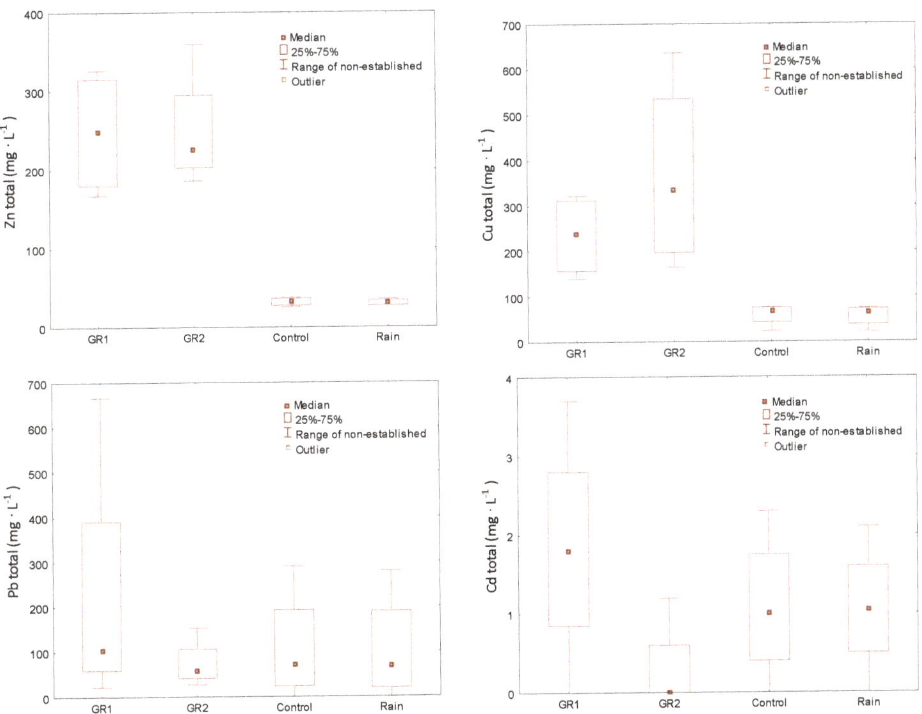

Figure 5. Range and average values for metals in extensive roof models in Swojec object.

On average, the outflow was about 30 mg·l^{-1}, Gnecco et al. [49]. A study conducted by Vijayaraghavan and Raja [50], based on artificial green roof profiles with the use of biomass demonstrated a high degree of reduction of the concentration of metals in relation to rainwater. The achieved efficiency of the removal of metals reached the level of 92%. The literature indicates that at the stage of design it is necessary to conduct an analysis of the materials used in the process of preparation of substrates. The main objective of such analysis is to identify their effects as a source of contaminants. A study conducted by Schwager et al. [51] on the leaching of certain metals and their sorption in selected substrates, and especially copper and zinc, indicated a high variation in their liberation, depending on the material. The time of equilibrium of those metals was high and amounted to 3 days

4. Conclusions

Green architecture, including green roofs, mitigate the effects of traditional building materials on urban environments. This property is directly relevant to the quality of life of society. Therefore, research into the properties of green architecture is important. A six-month study period in 2016 was analyzed. The analyses of the quality of runoff from those systems were conducted, taking into account certain pollution indicators and metals. In the study of the hydrological performance of the systems, a unidimensional model of infiltration was applied for the analysis of the hydraulic parameters. The model enabled the identification of parameters related with the infiltration that are usually determined with experimental methods. The calculated statistics for the measured and simulated values for the surfaces with expanded clay aggregates and with perlite indicated a good fit and enabled the accurate simulation of events observed in the remaining periods. The mean concentrations of nitrogen and phosphorus were higher than those determined in rainwater. The load of total nitrogen exceeded the values of concentration in rainwater and amounted to 7.17 and 13.01 mg·L^{-1} for leca and perlite, respectively. Therefore, at the level of 0.05, statistically significant differences were noted in relation to specific concentrations in rainwater and runoff from the control surface. Electrolytic conductivity assumed decidedly higher values, compared with runoff from the reference surface and atmospheric precipitation.

Excessive levels of concentration of metals, especially zinc and copper, were observed in the runoff from the green surfaces in relation both to the precipitation and to runoff from the control panel. The proper level of system performance with regard to runoff quality can be determined on the basis of longer observations. The calculated maximum retention performance indicator for the experimental green surfaces, relative to rainfall, was up to 65%, and in relation to the control surface was up to 49% (substrate with perlite). The peak discharge performance indicator was reduced by 26% in the case of leca and by 38% in the case of perlite.

Author Contributions: Developed the concept, conducted analysis of the results, final control (G.P.); analyzed and developed the results (K.S.), conducted visits at the place of the experiment (T.K.); conducted visits at the place of the experiment (W.O.); transcribed the paper, ducted literature studies (R.P.). All authors approved the final manuscript. All authors have read and agreed to the published version of the manuscript.

Funding: This research received no external funding

Conflicts of Interest: The authors declare no conflict of interest.

References

1. Conti, S.; Meli, P.; Minelli, G.; Solimini, R.; Toccaceli, V.; Vichi, M. Epidemiologic study of mortality during the summer 2003 heat wave in Italy. *Environ. Res.* **2005**, *98*, 390–399. [CrossRef]
2. Berardi, U. The outdoor microclimate benefits and energy saving resulting from green roofs retrofits. *Energy Build.* **2016**, *121*, 217–229. [CrossRef]
3. Erdem, C. Thermal regulation impact of green walls: An experimental and numerical investigation. *Appl. Energy* **2016**, *194*, 247–254. [CrossRef]
4. Kolokotsa, D.; Santamouris, M.; Zerefos, S.C. Green and cool roofs' urban heatisland mitigation potential in European climates for office buildings under freefloating conditions. *Sol. Energy* **2013**, *74*, 118–130. [CrossRef]

5. Environmental Protection Agency. Reduce Urban Heat Island Effect. 2016. Available online: https://www.epa.gov/green-infrastructure/reduce-urban-heat-island-effect (accessed on 9 June 2020).
6. Carpenter, C.M.G.; Todorov, D.; Driscoll, C.T.; Montesdeoca, M. Water quantity and quality response of a green roof to storm events: Experimental and monitoring observations. *Environ. Pollut.* **2016**, *218*, 664–672. [CrossRef] [PubMed]
7. Mohajerani, A.; Bakaric, J.; Jeffrey-Bailey, T. The urban heat island effect, its causes, and mitigation, with reference to the thermal properties of asphalt concrete. *J. Environ. Manag.* **2017**, *197*, 522–538. [CrossRef]
8. Sharma, A.; Conry, P.; Fernando, H.; Hamlet, A.; Hellmann, J.; Chen, F. Green and cool roofs to mitigate urban heat island effects in the Chicago metropolitan area: Evaluation with a regional climate model. *Environ. Res. Lett.* **2016**, *11*, 064004. [CrossRef]
9. Zhang, Q.; Miao, L.; Wang, X.; Liu, D.; Zhu, L.; Zhou, B.; Sun, J.; Liu, J. The capacity of greening roof to reduce storm water runoff and pollution. *Landsc. Urban Plan.* **2015**, *144*, 142–150. [CrossRef]
10. Population Reference Bureau. Human Population: Urbanization. Available online: http://www.prb.org/Publications/Lesson-Plans/HumanPopulation/Urbanization.aspx (accessed on 1 July 2009).
11. Solcerova, A.; Ven, F.; Wang, M.; Rijsdijk, M.; Giesen, N. Do green roof cool the air? *Build. Environ.* **2016**, *111*, 249–255. [CrossRef]
12. MacIvor, J.; Margolis, L.; Perotto, M.; Drake, J. Air temperature cooling by extensive green roofs in Toronto Canada. *Ecol. Eng.* **2016**, *95*, 36–42. [CrossRef]
13. Squier, M.; Davidson, C.I. Heat flux and seasonal thermal performance of an extensive green roof. *Build. Environ.* **2016**, *107*, 235–244. [CrossRef]
14. Tam, V.; Wang, J.; Le, K. Thermal insulation and cost effectiveness of green-roof systems: An empirical study in Hong Kong. *Build. Environ.* **2016**, *44*, 46–54. [CrossRef]
15. Fassman-Beck, E.; Voyde, E.; Simcock, R.; Hong, Y.S. 4 Living roofs in 3locations: Does configuration affect runoff mitigation? *J. Hydrol.* **2013**, *490*, 11–20. [CrossRef]
16. Deska, I.; Mrowiec, M.; Ociepa, E.; Łacisz, K. Investigation of the Influence of Hydrogel Amendment on the Retention Capacities of Green Roofs. *Ecol. Chem. Eng. S* **2018**, *25*, 373–382. [CrossRef]
17. Speak, A.F.; Rothwell, J.J.; Lindley, S.J.; Smith, C.L. Rainwater runoff retention an aged intensive green roof. *Sci. Total Environ.* **2013**, *461–462*, 28–38. [CrossRef]
18. Stovin, V.; Poë, S.; De-Ville, S.; Berretta, C. The influence of substrate and vegetation configuration on green roof hydrological performance. *Ecol. Eng.* **2015**, *85*, 159–172. [CrossRef]
19. Versini, P.A.; Ramier, D.; Berthier, E.; de Gouvello, B. Assessment of the hydrological impacts of green roof: From building scale to basin scale. *J. Hydrol.* **2015**, *524*, 562–575. [CrossRef]
20. Karczmarczyk, A.; Baryła, A.; Kożuchowski, P. Design and Development of Low P-Emission Substrate for the Protection of Urban Water Bodies Collecting Green Roof Runoff. *Sustainability* **2017**, *9*, 1795. [CrossRef]
21. MacAvoy, S.E.; Plank, K.; Mucha, S.; Williamson, G. Effectiveness of foam-based green surfaces in reducing nitrogen and suspended solids in an urban installation. *Ecol. Eng.* **2016**, *91*, 257–264. [CrossRef]
22. Razzaghmanesh, M.; Beecham, S.; Kazemi, F. Impact of green roofs onstormwater quality in a South Australian urban environment. *Sci. Total Environ.* **2014**, *470–471*, 651–659. [CrossRef]
23. Abhijith, K.V.; Kumar, P.; Gallagher, J.; Mcnabola, A.; Baldauf, R.; Pilla, F.; Broderick, B.; Di Sabatino, S.; Pulvirenti, B. Air pollution abatement performances of green infrastructure in open road and built-up street canyon environments—A review. *Atmos. Environ.* **2017**, *162*, 71–86. [CrossRef]
24. Pérez, G.; Coma, J.; Barreneche, C.; de Gracia, A.; Urrestarazu, M.; Burés, S.; Cabeza, L.F. Acoustic insulation capacity of Vertical Greenery Systems for buildings. *Appl. Acoust.* **2016**, *110*, 218–226. [CrossRef]
25. Moghbel, M.; Erfanian, S.R. Environmental benefits of green roofs on microclimate of Tehran with specific focus on air temperature, humidity and CO_2 content. *Urban Clim.* **2017**, *20*, 46–58. [CrossRef]
26. Telichenko, V.; Benuzh, A.; Mochalov, I. Landscape Architecture and green spaces in Russia. *Urban Habitats* **2017**, *117*. [CrossRef]
27. Washburn, B.; Swearingin, R.; Pullins, C.; Rice, M. Composition and Diversity of Avian Communities Using a New Urban Habitat: Green Roofs. *Environ. Manag.* **2016**, *57*, 1230–1239. [CrossRef] [PubMed]
28. Rabah, D.; Emmanuel, B.; Rafik, B. Modeling green wall interactions with street canyons for building energy simulation in urban context. *Urban Clim.* **2016**, *16*, 75–85. [CrossRef]

29. Soulis, K.X.; Valiantzas, J.D.; Ntoulas, N.; Kargas, G.; Nektarios, P.A. Simulation of green roof runoff under different substrate depths and vegetation covers by coupling a simple conceptual and a physically based hydrological model. *J. Environ. Manag.* **2017**, *200*, 434–445. [CrossRef]
30. Simůnek, J.; van Genuchten, M.T. Modeling nonequilibrium flow and transport with HYDRUS. *Vadose Zone J.* **2008**, *7*, 782–797. [CrossRef]
31. Breitmeyer, R.J.; Stewart, M.K.; Huntington, J.L. Evaluation of gridded meteorological data for calculating water balance cover storage requirements. *Vadose Zone J.* **2018**, *17*, 180009. [CrossRef]
32. Palla, A.; Gnecco, I.; Lanza, L.G. Compared performance of a conceptual and a mechanistic hydrologic models of a green roof. *Hydrol. Process.* **2012**, *26*, 73–84. [CrossRef]
33. Sandoval, V.; Bonilla, C.A.; Gironás, J.; Vera, S.; Victorero, F.; Bustamante, W.; Rojas, V.; Leiva, E.; Pastén, P.; Suárez, F. Porous media characterization to simulate water and heat transport through green roof substrates. *Vadose Zone J.* **2017**, *16*, 14. [CrossRef]
34. Qin, H.P.; Peng, Y.N.; Tang, Q.L.; Yu, S.L. A Hydrus model for irrigation management of green roofs with a water storage layer. *Ecol. Eng.* **2016**, *95*, 399–408. [CrossRef]
35. Richtlinien für Planung, Bau und Instandhaltung von begrünbaren Flächenbefestigungen. In *Forschungsgesellschaft Landschaftsentwicklung Landschaftsbau*; FLL: Bonn, Germany, 2008.
36. Soto, M.A.; Chang, H.K.; van Genuchten, M.T. Fractal-based models for the unsaturated soil hydraulic functions. *Geoderma* **2017**, *306*, 144–151. [CrossRef]
37. Sayah, B.; Rodríguez, M.G.; Juana, L. Development of one-dimensional solutions for water infiltration. Analysis and parameters estimation. *J. Hydrol.* **2016**, *535*, 226–234. [CrossRef]
38. Seboong, O.; Yun, K.K.; Jun-Woo, K. Model of Hydraulic Conductivity in Korean Residual Soils. *Water* **2015**, *7*, 5487–5502. [CrossRef]
39. Wosten, J.H.M.; Van Genuchten, M.T. Using texture and other soil properties to predict the unsaturated soil hydraulic conductivity. *Soil Sci. Soc. Am. J.* **1988**, *52*, 1762–1770. [CrossRef]
40. Wong, G.; Jim, C.Y. Identifying keystone meteorological factors of green-roof stormwater retention to inform design and planning. *Landsc. Urban Plan.* **2015**, *143*, 173–182. [CrossRef]
41. Feitosa, R.C.; Wilkinson, S. Modelling green roof stormwater response for different soil depths. *Landsc. Urban Plan.* **2016**, *153*, 170–179. [CrossRef]
42. Locatelli, L.; Ole, M.; Mikkelsen, P.S.; Arnbjerg-Nielsen, K.; Jensen, B.M.; Binning, P.J. Modelling of green roof hydrological performance for urban drainage applications. *J. Hydrol.* **2014**, *519 Pt D*, 3237–3248. [CrossRef]
43. Buffam, I.; Mitchell, M.E.; Durtsche, R.D. Environmental drivers of seasonal variation in green roof runoff water quality. *Ecol. Eng.* **2016**, *91*, 506–514. [CrossRef]
44. Harper, G.E.; Limmer, M.A.; Showalter, W.E.; Burken, J.G. Nine-month evaluation of runoff quality and quantity from an experiential green roof in Missouri, USA. *Ecol. Eng.* **2015**, *78*, 127–133. [CrossRef]
45. Mitchell, M.; Matter, S.; Durtsche, R.; Buffam, I. Elevated phosphorus: Dynamics during four years of green roof development. *Urban Ecosyst.* **2017**, *20*, 1121–1133. [CrossRef]
46. Kuoppamäki, K.; Hagner, M.; Lehvävirta, S.; Setälä, H. Biochar amendment in the green roof substrate affects runoff quality and quantity. *Ecol. Eng.* **2016**, *88*, 1–9. [CrossRef]
47. Beecham, S.; Razzaghmanesh, M. Water quality and quantity investigation of green roofs in a dry climate. *Water Res.* **2015**, *70*, 370–384. [CrossRef] [PubMed]
48. Vijayaraghavan, K.; Joshi, U.M. Can green roof act as a sink for contaminants. A methodological study to evaluate runoff quality from green roofs. *Environ. Pollut.* **2014**, *194*, 121–129. [CrossRef]
49. Gnecco, I.; Palla, A.; Lanza, L.G.; La Barbera, P. The Role of Green Roofs as a Source/sink of Pollutants in Storm Water Outflows. *Water Resour. Manag.* **2013**, *27*, 4715–4730. [CrossRef]
50. Vijayaraghavan, K.; Raja, F.D. Pilot-scale evaluation of green roofs with Sargassum biomass as an additive to improve runoff quality. *Ecol. Eng.* **2015**, *75*, 70–78. [CrossRef]
51. Schwager, J.; Schaal, L.; Simonnot, M.; Claverie, R.; Ruban, V.; Morel, J.L. Emission of trace elements and retention of Cu and Zn by mineral and organic materials used in green roofs. *J. Soils Sediments Prot. Risk Assess. Remediat.* **2015**, *15*, 1789–1801. [CrossRef]

 *sustainability*

Technical Note

Climate: An R Package to Access Free In-Situ Meteorological and Hydrological Datasets For Environmental Assessment

Bartosz Czernecki [1], Arkadiusz Głogowski [2,*] and Jakub Nowosad [3]

[1]   Department of Climatology, Faculty of Geographical and Geological Sciences, Adam Mickiewicz University, 61-680 Poznań, Poland; nwp@amu.edu.pl

[2]   Institute of Environmental Protection and Development, Wrocław University of Environmental and Life Sciences, 50-375 Wrocław, Poland

[3]   Institute of Geoecology and Geoinformation, Faculty of Geographical and Geological Sciences, Adam Mickiewicz University, 61-680 Poznań, Poland; nowosad@amu.edu.pl

*   Correspondence: arkadiusz.glogowski@upwr.edu.pl

Received: 2 December 2019; Accepted: 30 December 2019; Published: 3 January 2020

Abstract: Freely available and reliable meteorological datasets are highly demanded in many scientific and business applications. However, the structure of publicly available databases is often difficult to follow, especially for users who only deal with this kind of dataset on occasion. The "climate" R package aims to fill this gap with an easy-to-use interface for downloading global meteorological data in a fast and consistent way. The package provides access to different sources of in-situ meteorological data, including the Ogimet website, atmospheric vertical sounding gathered at the University of Wyoming's webpage, and hydrological and meteorological measurements collected by the Institute of Meteorology and Water Management—National Research Institute (i.e., Polish Met Office). This article also provides a quick overview of the key functionalities available within the climate R package, and gives examples of an efficient and tidy workflow of meteorological data within the R based environment. The automation procedures included in the packages allow one to download data in a user-defined time resolution (from hourly to annual), for a user-defined time span, and for a specified group of stations or countries. The package also contains metadata, including a list of available stations, their geospatial information, and measurement descriptions with their units. Finally, the obtained datasets can be processed in R or exported to external tools (e.g., spreadsheets or GIS software).

Keywords: R; open-source software; dataset; meteorology; climate; SYNOP; geospatial information

---

1. Introduction

Meteorological conditions are key factors in many areas of human activity such as agriculture, transport, power engineering, insurance and risk assessment [1], industrial and marketing planning [2], tourism, sport, mass events [3,4], national security, and many more where atmospheric conditions may have a direct or indirect impact [5–8]. Besides the financial and safety relevance of meteorological and hydrological datasets [9], this kind of information is very often crucial to reliably answer a scientific problem [10], which heavily relies on the quality of meteorological dataset used in this kind of research.

National meteorological agencies collect in-situ measurements of the highest quality according to the standards of the World Meteorological Organization (WMO). They are simultaneously responsible for maintaining and sharing their archived databases. A significant part of the meteorological data is available for free from the global exchange of the surface synoptic observations (SYNOP), meteorological information used by aircraft pilots (METAR), or upper air soundings (TEMP) reports.

Even if most of this information is limited only to the main synoptic stations and covers only basic meteorological parameters, the data itself usually provides better accuracy compared to commonly applied coarse gridded reanalysis products [11,12].

The availability of meteorological archive databases varies among countries. In most cases, access to such databases is usually not free of charge. However, the near-surface meteorological information from synoptic stations all around the world is publicly available free of charge due to the exchange of meteorological reports (e.g., FM-12 code established by the WMO) and is stored online. The ogimet.com web service is one of the most popular repositories of meteorological data that is heavily based on freely available data sources from the National Oceanic and Atmospheric Administration (NOAA) archives processed in a raw and human-readable format. Most of the archive dataset starts around the second part of 1999 and is being updated immediately after new reports are available.

The dataset representing atmospheric upper layers are also collected on the NOAA's as well as an independent data repositories. In this study, a publicly available repository of the University of Wyoming (http://weather.uwyo.edu/upperair/sounding.html) was used, as it allows for downloading atmospheric data representing vertical profiles of the atmosphere on any of the global sounding stations dated back even up to 1960s. Moreover, this repository provides a quick summary of thermodynamic atmospheric indices, which also can be a useful source of information for interested groups of end-users.

Other data sources exist that can be more suited for locally targeted problems. Such an example is a data source provided by the Polish Institute of Meteorology and Water Management—National Research Institute (IMGW-PIB) that distributes their resources through an HTTP file server (https://dane.imgw.pl/). Thanks to the actions of the Polish atmospheric-related communities against limited access to collected data, the legislative changes were possible [13,14]. It ensured free access to meteorological and hydrological data for most commonly applied non-commercial use cases since January 2017. Nowadays, the way of the distribution of this operational data is one of the most liberal among European meteorological services.

A typical workflow of downloading meteorological data from a repository (e.g., Ogimet) conventionally using a web browser is to (1) select a country or station (2) for the given time range, and (3) measurement interval (i.e., hourly/daily). As a single query is limited to a few tens of rows per one search, thus creating a proper dataset requires manual and tedious routines. However, this approach is not a standard for all repositories. For example, the Polish hydro-meteorological repository requires a user to select the type of data, interval, and station of interest. Depending on the period and interval, a single (ZIP archive) file contains one- or five-years of observations with one or two files in every archive. Once the user selects the year (or five-year period), depending on the choices made earlier, they may encounter one set of files in the case of monthly synoptic data, three sets of files for the annual hydrological data, 13 sets for daily hydrological data, or about 60 sets in the case of hourly synoptic data. Each case has a separate data structure and different documentation. Overall, 23 possible cases for the meteorological and hydrological data exist, each requiring an individual approach for downloading and processing of the files. Since the beginning of 2017, the structure of these data has undergone numerous changes, which have confused some users, thereby discouraging them from using the repository.

The created package aims to supply access to the observational datasets which were missing so far among the R atmospheric community that had used mostly tools for downloading gridded or built-in datasets (e.g., ESD [15], rNOMADS [16], knmiR [17]). Partly this gap was covered by the `rdwd` package [18], however, its functionality is restricted to the products of the German Meteorological Service only. Keeping the aforementioned in mind, the main goal of the `climate` R package is to deliver a convenient way of accessing global and regional repositories containing meteorological and hydrological data. The choice of R [19] is related to the fact that this is currently one of the most popular programming languages among environmental researchers and data scientists, and simultaneously, it is free of charge. The created package aims at processing all formats of meteorological data independently of its origin in a tidy tabular form [20] that is suitable for various visualization and processing applications. Abbreviations of the variables are specified according to the WMO standards and were added to the package documentation. Relevant dictionaries attached to the `climate` package can be read by `imgw_meteo_abbrev` or `imgw_hydro_abbrev` commands. The created package also contains a database that clarifies the variables' metadata and geographical coordinates of each stations' location. Thanks to this feature, users can directly use the output data in geospatial analysis using R [21] programming language or external GIS software.

2. Methods and Materials

The `climate` package is distributed under the MIT license. However, users are obliged to follow the regulations provided on the respective webpages, as the package only provides an interface to the official repositories. The most stable version of the climate package is available at the Comprehensive R Archive Network (CRAN), while its developer version is hosted on the GitHub platform at http://rclimate.ml (mirrored to: https://github.com/bczernecki/climate), where third-party users can contribute to its further development.

2.1. Installation and User Guide

The `climate` package can be installed and run on any modern computer with the R environment version 3.1 or higher. The package was tested on a wide span of Windows instances and several Linux and Mac OS X distributions, and has positively undergone numerous tests before being published in the CRAN repository. The authors also deliberately avoided using external libraries in order to reduce possible dependencies or installation issues. The stable version of the `climate` package hosted on the official CRAN repository can be installed with the R's `install.packages("climate")` and activated using the `library(climate)` commands respectively. The development version is hosted on the GitHub platform at (https://github.com/bczernecki/climate), where all instructions for installing and using the package are provided. Additionally, users are encouraged to contribute, leave feedback, or suggest their own ideas for further improvements that may be added in future releases.

2.2. Datasets

Archived data stored at (1) www.ogimet.com, (2) the University of Wyoming's atmospheric sounding database and (3) in the official IMGW-PIB's repository, constitute the primary sources for the data in the `climate` package (Figure 1).

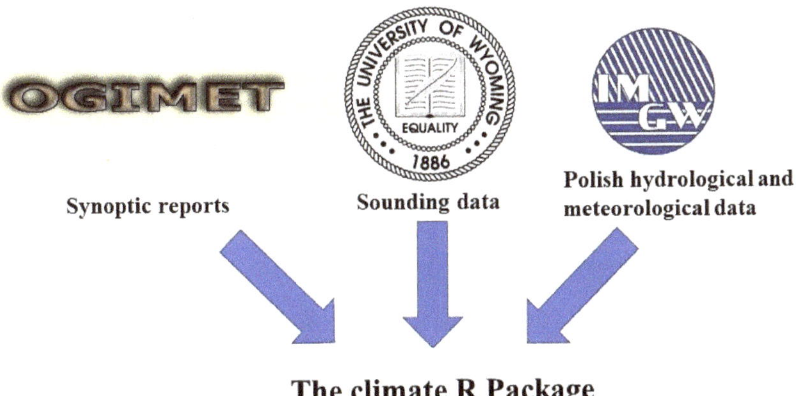

Figure 1. The data sources used in the climate R package.

The synoptic reports available in the Ogimet web service are dated back to the year 1999. This global repository shares up to 17 variables (columns) representing instantaneous measurement for an individual station in a given date and time. Data is divided into daily and hourly time intervals. It contains information for the following: 2 meters air temperature (min., max., avg.) and dew point temperature [°C], atmospheric and sea level pressure [hPa], geographical coordinates [°], altitude [m], relative humidity [%], wind speed and wind gust [km $\cdot$ h^{-1}], wind direction [direction], cloudiness [octants] and height of cloud base [km], visibility [km], sunshine duration and height of snow cover [cm].

The historical sounding (i.e., upper air from the University of Wyoming's repository) observations are not available on the Ogimet website. Therefore, this capability was added to the climate package due to the high demand for this kind of information among severe weather community, where it is commonly used for analyzing thermodynamic and kinematic atmospheric parameters [22,23]. This is also crucial information for identifying the atmospheric processes responsible for air quality problems [24]. The measurement interval is in most cases 12 hours (i.e., at 00 and 12 UTC, occasionally on some stations at 06 UTC and 18 UTC) and the data are usually available a few hours after beginning of the measurements. The sounding (also known as "rawinsonde") data has 11 columns representing the instantaneous measurement of the atmospheric vertical profile for a single station and time. It contains information for the following parameters: atmospheric pressure [hPa], altitude [m], air temperature and dew point [°C], relative humidity [%] and mixing ratio [g $\cdot$ kg^{-1}], wind speed [knots] and wind direction [°], and thermodynamic properties along with measurement metadata.

The IMGW-PIB (i.e., Polish hydro-meteorological) dataset contains measurements back to the 1950s, and the database is continually being updated, usually on a monthly basis. The meteorological data in the repository is divided, according to the hierarchy of stations, into (1) synoptic, (2) climatological, and (3) precipitation data. The synoptic and climatological stations consist of (1) hourly, (2) daily, and (3) monthly time intervals. The precipitation stations have no measurements at an hourly interval. The synoptic data are the most extensive and contain over 100 meteorological parameters. The climate data describes four essential meteorological components: air temperature [°C], wind speed [m $\cdot$ s^{-1}], relative humidity [%], and cloudiness [octants]. The precipitation data consist of the amount of precipitation with a description of the phenomena or surface precipitation type (i.e., rain, snow, snow cover height). Due to a relatively broad range of parameters obtainable for the meteorological data, the authors have thus decided to include a "vocabulary" that contains column names (i.e., meteorological parameters) in a (1) short, (2) more descriptive, or (3) original (Polish) forms. The hydrological data in the IMGW-PIB repository contains (1) daily, (2) monthly, and (3) semi-annual/annual measurements. All hydrological data uses the hydrological year, which begins on November 1st and ends on October 31st. Regardless of the temporal resolution, the hydrological data contains measurements of the maximum, mean, and minimum for the following: water flow [m^3 $\cdot$ s^{-1}], water temperature [°C], and water level [cm]. Additionally, the daily dataset includes characteristics of the ice and overgrowth phenomena observed at the station. Similar to the meteorological dataset, a user can decide whether to add an extra description to the column names.

2.3. Core Functionality of the Climate R Package

The climate package currently consists of 21 functions with ten of them visible for the end-user (Table 1). Three of them are intended for downloading meteorological data, one for hydrological data, and four are auxiliary functions to improve the legibility and improve data exploration capabilities. Despite a relatively large number of functions that might be potentially used, there are four main functions called `meteo_ogimet`, `sounding_wyoming`, `meteo_imgw` and `hydro_imgw` that are generic wrappers for other functions. They allow for simplified downloading of any requested data in a convenient way. All available functions are documented on the package website and inside the built-in R help system where the exemplary code is also provided.

Table 1. The essential user-visible functions available in the `climate` package.

Function	Description
Meteorological data—download	
`meteo_ogimet`	A generic function for downloading hourly and daily dataset from the Ogimet repository
`sounding_wyoming`	A function for downloading sounding (i.e., upper air) data for any station in the world (i.e., vertical profiles of the atmosphere) from the University of Wyoming repository
`meteo_imgw`	A generic function for downloading hourly, daily and monthly meteorological dataset from the IMGW-PIB repository
Hydrological data—download	
`hydro_imgw`	A generic function for downloading daily, monthly, and annual hydrological dataset from the IMGW-PIB repository
Auxiliary functions and datasets	
`stations_ogimet`	A function for downloading information about all stations available for the selected country in the Ogimet repository
`nearest_stations_ogimet`	A function for downloading information about nearest stations to the selected point available for the selected country in the Ogimet repository
`imgw_meteo_stations`	Built-in metadata for meteorological stations, their geographical coordinates, and ID numbers (from the IMGW-PIB repository)
`imgw_hydro_stations`	Built-in metadata for hydrological stations, their geographical coordinates, and ID numbers (from the IMGW-PIB repository)
`imgw_meteo_abbrev`	Dictionary explaining variables available for meteorological stations (from the IMGW-PIB repository)
`imgw_hydro_abbrev`	Dictionary explaining variables available for hydrological stations (from the IMGW-PIB repository)

2.4. Ogimet Meteorological Data

The generic function for downloading decoded SYNOP reports from the Ogimet repository requires defining a set of arguments according to the schema provided below for the most generic `meteo_ogimet` function.

```
meteo_ogimet(interval, date, coords, station, precip_split)
```
where:

- **interval** - temporal resolution of the data (`"hourly"`, `"daily"`) (argument not valid for: `ogimet_hourly` and `ogimet_daily` functions)
- **date** - start and finish dates (e.g., `date = c("2018-05-01", "2018-07-01")`) — character or Date class object
- **coords** — logical argument (TRUE or FALSE); if TRUE coordinates are added
- **station** — WMO ID of meteorological station(s). Character or numeric vector
- **precip_split** — whether to split precipitation fields into 6/12/24 h, numeric fields (logical value = TRUE (default) or FALSE); valid only for an hourly time step

2.5. Sounding Data

The proposed solution is based on the decoded TEMP sounding (radiosonde) reports hosted on the University of Wyoming (http://weather.uwyo.edu) server. It contains archived data for all upper air profiling stations working globally in the WMO network. The syntax for downloading the single sounding is as follows:

```
sounding_wyoming(wmo_id, yy, mm, dd, hh)
```

This function requires a few numeric arguments:

- **wmo_id** — international WMO station code
- **year** — year
- **mm** — month
- **dd** — day
- **hh** — hour (usually radiosondes are launched at 00 and 12 UTC)

The returned object contains a list of two data frames. The first consists of measurements in a tabular form for 11 meteorological elements, while the second consists of metadata and the most fundamental thermodynamic and atmospheric instability indices.

2.6. IMGW-PIB Meteorological Data

The extended range of meteorological near-surface measurements can be achieved, usually from the regional met offices' repositories. The publicly available Polish historical meteorological dataset comprises of two sections: meteorological and actinometrical data. Each of these sections is divided into subsections depending on the observational interval. The actinometric data was not implemented in the `climate` package due to ongoing changes to the data storage, and it will be added after the final format is determined.

The `climate` package contains an interface to the Polish IMGW-PIB dataset, which can be downloaded with a very similar syntax to the global dataset described previously in a simplified way. The schema shown below describes the use of the most generic `meteo_imgw` function and contains all arguments that can be used to define requested data.

```
meteo_imgw(interval, rank, year, status, coords, station, col_names)
```

where:

- **interval** — temporal resolution of the data (`"hourly"`, `"daily"`, `"monthly"`)
- **rank** — type of the stations to be downloaded (`"synop"`, `"climate"`, or `"precip"`)
- **year** — vector of years (e.g., `1966:2000`)
- **status** — logical argument (TRUE or FALSE); for removing status of the measurements
- **coords** — logical argument (TRUE or FALSE); if TRUE coordinates are added
- **station** — vector of stations; it can be an ID of a station (numeric) or a name of a stations (capital letters)
- **col_names** — three types of column names possible: "short" — default, values with shortened names, "full" — full English description, "Polish" — original names in the dataset

It is also worth noting that most of the arguments have predefined default values to support less experienced users. For example, if the `station` argument is not given, then all available datasets (here: data for all stations) are automatically downloaded. Only the `interval`, `rank` and `year` arguments are mandatory. In case any of them is not defined, the user is given a hint on the correct syntax.

2.7. IMGW-PIB Hydrological Data

The hydrological data is available in daily, monthly, and semiannual/annual temporal resolutions. The definition of the arguments in `hydro_imgw` is an analogue to the previously described for the meteorological data, with the syntax described below:

```
hydro_imgw(interval, year, coords, value, station, col_names)
```

where:

- **interval** — temporal resolution of the data (`"daily"`, `"monthly"`, `"semiannual_and_annual"`)
- **year** — vector of years (e.g., `1966:2000`)
- **coords** — logical argument TRUE or FALSE; if TRUE coordinates are added
- **value** — type of data (can be: state — `"H"`, flow — `"Q"`, or temperature — `"T"`).
- **station** — vector of stations; it can be an ID of a station (numeric) or a name of a stations (capital letters)
- **col_names** — three types of column names possible: `"short"` — default, values with shortened names, `"full"` — full English description, `"polish"` — original names in the dataset

3. Results

The purpose of this section is to show the capabilities of the created R package. The following subsections provide examples for types of analyses that can be performed using the `climate` R package together with other R packages available on CRAN.

3.1. Ogimet Meteorological Data—Use Case

The meteorological dataset use case provided below was based on hourly data from the Ogimet repository for the defined time frame, i.e., 2018/01/01 – 2018/12/31, for the location of Svalbard Lufthavn. The meteo_ogimet command allowed us to download 8761 observations for 22 variables (Listing 1). The `dplyr` and `openair` packages [25] were used to analyze and visualize part of downloaded results. After aggregating the data by the wind directions (the `"ddd"` column, Listing 2), converting directions into angles given in degrees, and reformatting dates' classes, it was possible to align it to format required by external packages and plot the seasonal wind roses (Figure 2).

Listing 1. Example of the data download using the climate package.

```
library(climate)
df <- meteo_ogimet(interval = "hourly", date = c("2018-01-01", "2018-12-31"),
station = "01008")
#> [1]  "01008"
#> |==================================================================| 100 %
head(df[, 2:11])
#>                   Date     TC    TdC TmaxC TminC ddd ffkmh Gustkmh  P0hPa PseahPa
#> 2 2018-12-31 23:00:00 -14.3 -19.2  <NA>  <NA> NNW  25.2    43.2 1000.5  1004.2
#> 3 2018-12-31 22:00:00 -13.7 -18.2  <NA>  <NA>  NW  21.6    32.4 1000.0  1003.8
#> 4 2018-12-31 21:00:00 -15.9 -18.5  <NA>  <NA> ESE  10.8    21.6  999.9  1003.7
#> 5 2018-12-31 20:00:00 -16.8 -20.1  <NA>  <NA>   E  18.0    25.2 1000.1  1003.9
#> 6 2018-12-31 19:00:00 -17.2 -21.7  <NA>  <NA> ESE  21.6    28.8 1000.5  1004.3
#> 7 2018-12-31 18:00:00 -18.3 -20.8 -15.5 -19.5 ESE  21.6    32.4 1000.7  1004.5
```

Listing 2. Example of a code used for creating a rose wind for Svalbard Lufthan in 2018.

```
library(climate)
# downloading data
df <- meteo_ogimet(interval = "hourly", date = c("2018-01-01", "2018-12-31"),
station = c("01008"))
library(openair) # external package for plotting wind roses
# converting wind direction from character into degrees
wdir <- data.frame(ddd = c("CAL", "N", "NNE", "NE", "ENE", "E", "ESE", "SE", "SSE",
"S", "SSW", "SW", "WSW", "W", "WNW", "NW", "NNW"),
dir = c(NA, 0:15 * 22.5), stringsAsFactors = FALSE)
# changing the date column to the format required by the openair package
df$date <- as.POSIXct(df$Date, tz = "UTC")
df <- merge(df, wdir, by= "ddd", all.x = TRUE) # joining two datasets
df$ws <- df$ffkmh/3.6 # converting to m/s from km/h
df$gust <- df$Gustkmh/3.6 # converting to m/s from km/h
windRose(mydata = df, ws = "ws", wd = "dir", type = "season", paddle = FALSE,
main = "Svalbard Lufthavn (2018)", ws.int = 3, dig.lab = 3, layout = c(4, 1))
```

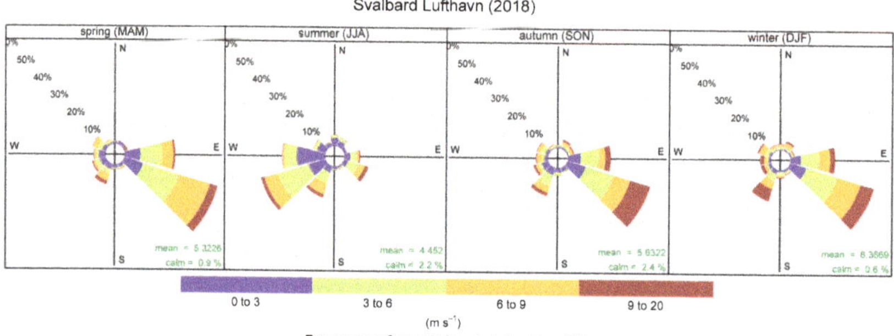

Figure 2. Seasonal wind roses for Svalbard Lufthavn in 2018 based on Ogimet dataset.

3.1.1. Searching for the Nearest Stations

The user can also use the `climate` package without knowing the station's WMO ID. The nearest synoptic stations can be found with the `nearest_ogimet_stations` function (Listing 3). It requires users to provide a pair of geographical coordinates that point to the centroid of our area of investigations. We can specify how many nearest meteorological stations an user wants to find. As a result, we get a data frame with stations metadata and distance to given coordinates. Additionally a simple map can be added with the argument `add_map = TRUE`. Exemplary results and the code is given below Figure 3.

Listing 3. Example of downloading nearest stations according to a specified location (first six nearest
stations are shown).

```
library(climate)
ns = nearest_stations_ogimet(country = "United+Kingdom", point = c(-4, 56),
no_of_stations = 50, add_map = TRUE)
head(ns)
#>     wmo_id        station_names       lon       lat alt  distance [km]
#> 29  03144           Strathallan  -3.733348  56.31667  35       46.44794
#> 32  03155             Drumalbin  -3.733348  55.61668 245       52.38975
#> 30  03148             Glen Ogle  -4.316673  56.41667 564       58.71862
#> 27  03134     Glasgow Bishopton  -4.533344  55.90002  59       60.88179
#> 35  03166   Edinburgh Gogarbank  -3.350007  55.93335  57       73.30942
#> 28  03136         Prestwick RNAS  -4.583345  55.51668  26       84.99537
```

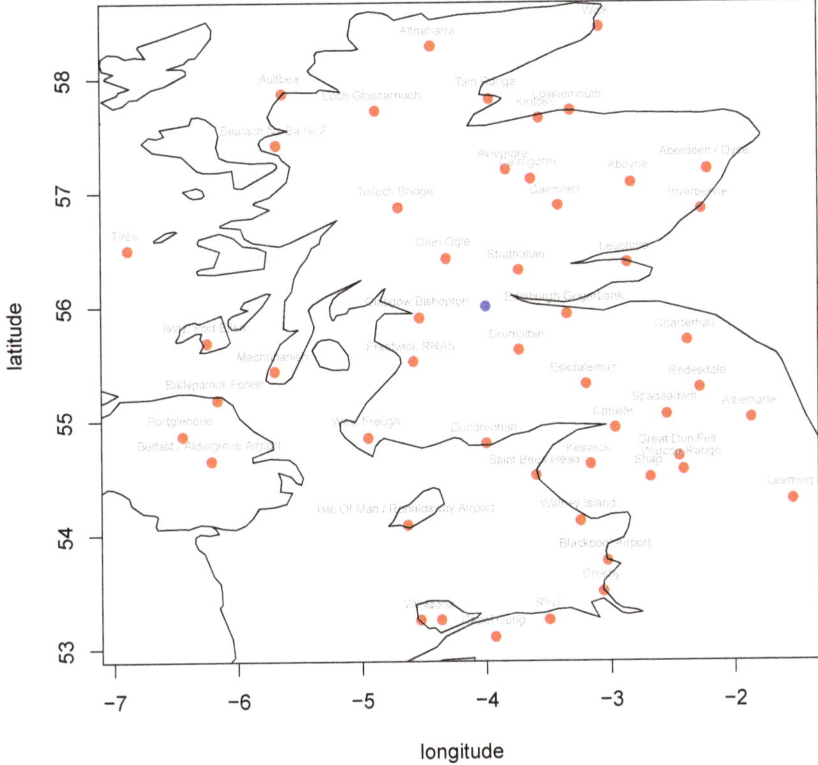

Figure 3. Example code for searching for the 50 nearest stations from the point of given coordinates (longitude 5° W, latitude 56° N) in United Kingdom. First 22 records are shown in Listing 3.

3.2. Sounding Data—Use Case

Downloading data for a single vertical profile of the atmosphere requires providing date, hour, and station's name (Listing 4). The chosen use case showed an atmospheric sounding started at 00UTC on 4th April 2019 in Łeba, Poland (Figure 4). The returned data frame from the measurements allowed users to plot temperature and humidity profiles on the Skew-T diagram generated thanks to the `RadioSonde` package [26]. It showed a strong thermal inversion up to 800–850 m a.g.l. which may strongly impact the air quality conditions in a near-surface layers [24]. The metadata and thermodynamic calculations stored in the second element of the returned list were omitted on purpose as no severe weather parameters related to atmospheric convection were detected.

Listing 4. Example of code to download sounding data, with Skew-T diagram.

```
library(climate)
library(RadioSonde) # an external package
profile <- sounding_wyoming(wmo_id = 12120,yy = 2019, mm = 4, dd = 4, hh = 0)
df <- profile[[1]]
colnames(df)[c(1, 3:4)] = c("press", "temp", "dewpt") # changing column names
RadioSonde::plotsonde(df, winds = FALSE, title = "2019-04-04 00UTC (LEBA, PL)",
col = c("red", "blue"), lwd = 3)
```

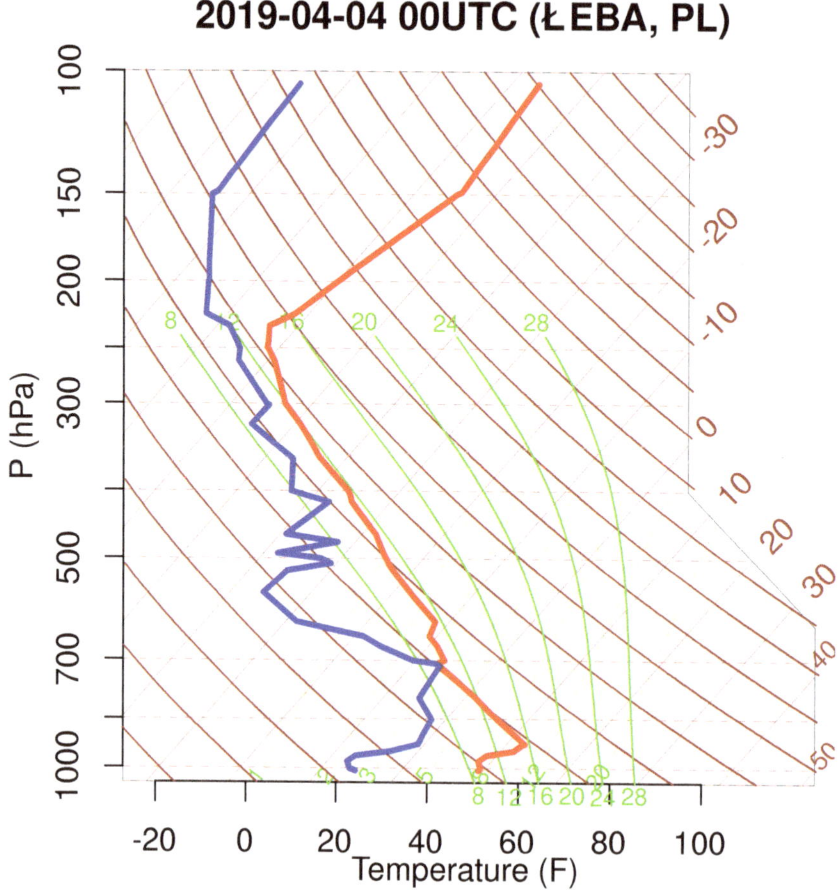

Figure 4. Example of a downloaded sounding dataset plotted on Skew-T diagram.

3.3. IMGW-PIB—Use Case

Another use case shows the possibilities of the `climate` package when coupled with the GIS and statistical capabilities of the R programming language (Figure 5). The downloaded data comprised 30 years of monthly mean air temperatures derived from the main meteorological stations in Poland. Due to the missing or suspicious diagnosed values, some data were excluded, e.g., stations' location changes during the analyzed period or having a monthly mean air temperature during the summer season of 0 °C. The next step was to create a function for calculating the slope coefficient of the linear regression model that was later applied to the whole dataset.

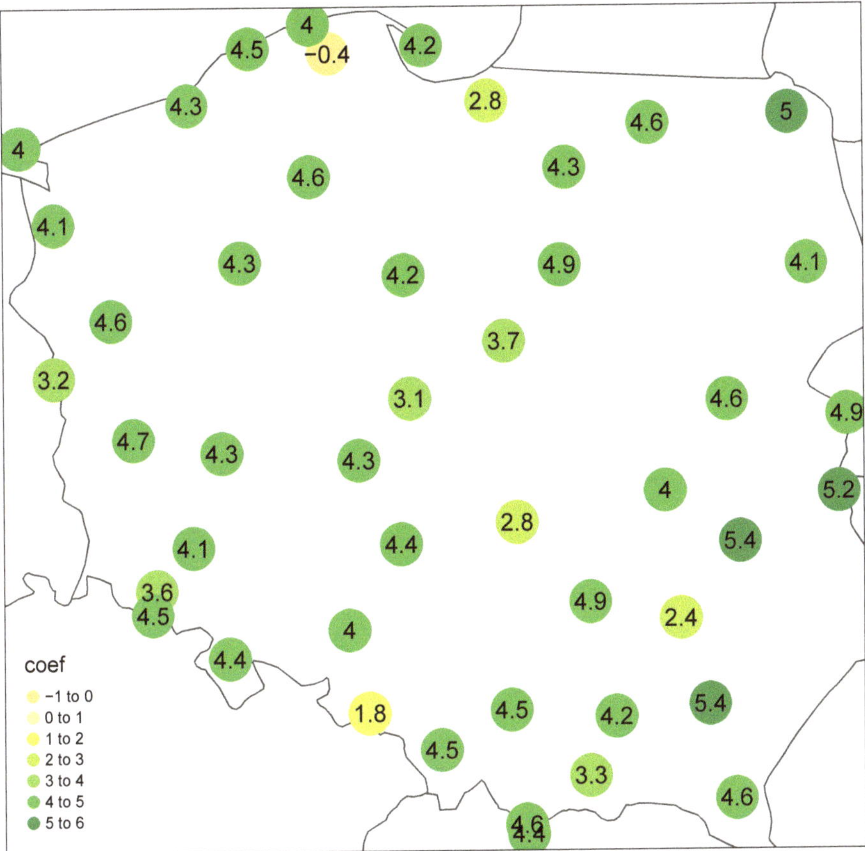

Figure 5. Air temperature trends in Poland per 100 years (°C) based on the calculated slope of linear regression for 1978-2017—IMGW-PIB dataset.

The obtained results (Listing 5) were later transformed into a spatial object using the `sf` package [27] and visualized in the form of the map using the `tmap` package [28]. The created vector layer can later be saved in any GIS format supported by the `sf` package interfacing between R and the geospatial data abstraction library drivers (GDAL). One of the major advantages of using the R programming language is being able to keep everything in one environment instead of the typical situation where three different tools are applied for (1) data preprocessing, (2) statistical analysis, and (3) spatial data visualization. Such an approach makes it possible to reduce the required time for the entire research significantly and to focus more on the obtained results. However, the user must be aware that the provided tool is only an interface for downloading the data, and that the obtained results may inherit errors from the source repositories.

Listing 5. Exemplary code for downloading, processing and visualizing data from the IMGW-PIB repository.

```r
library(ggplot2)
library(dplyr)
library(tidyr)
library(sf)
library(tmap)
library(rnaturalearth)
library(climate)
ms <- meteo_imgw("monthly", "synop", year = 1978:2017, coords = TRUE)
# calculating annual values
ms %>%
filter(!(mm > 5 && mm < 9 && t2m_mean_mon == 0)) %>%
select(station, X, Y, yy, mm, t2m_mean_mon) %>%
group_by(station, yy, X, Y) %>%
summarise(annual_mean_t2m = mean(t2m_mean_mon), n = n()) %>%
filter(n == 12) %>%
spread(yy, annual_mean_t2m) %>%
na.omit() -> trend
# extracting trends
regression <- function(x) {
df <- data.frame(yy - 1978:2017, temp = as.numeric(x))
coef(lm(temp ~ yy, data = df))[2]
}
trend$coef <- round(apply(trend[, -1:-4], 1, regression) * 100, 1)
trend <- st_as_sf(trend, coords = c("X", "Y"), crs = 4326)
# mapping the results
world <- ne_countries(scale = "medium", returnclass = "sf")
tm <- tm_shape(world) + tm_borders() +
tm_shape(trend, is.master = TRUE) + tm_dots(col = "coef", size = 4) +
tm_shape(trend) + tm_text(text = "coef")
tm
```

4. Conclusions

The `climate` R package allows users to obtain historical and most up-to-date meteorological information from both: ground and upper parts of the atmosphere. Data downloaded by `climate` gives possibilities for applying atmospheric data collected according to the WMO standards in an intuitive and fully automated way. The package is designed to be user-friendly and envisages, for the most part, environmental scientists wanting to obtain hydrological or meteorological data for research purposes in an convenient and programmable way within the R programming language. The usefulness and simplicity of the proposed solution can be especially valuable for many non-atmospheric scientists struggling with typically sophisticated and time-consuming mechanisms for accessing in-situ atmospheric data in a ready-to-use structure. The proposed solution with the `climate` package lets to save time for typical data flow in data science projects where a significant amount of time is spent on data preparation, while a core part of the computation is usually a magnitude shorter when compared to data cleaning and preprocessing [29].

Therefore for future improvements, it is planned to enlarge the `climate` R package with new local repositories so that more countries can conduct interdisciplinary research on meteorological data using a single tool, which can be targeted on a local scale in combination with global meteorological information. Also, new products (e.g., actinometric data in Poland) will be included once the IMGW-PIB repository has a mature form.

Author Contributions: Conceptualization, B.C., A.G. and J.N.; methodology, B.C., A.G. and J.N.; software, B.C., A.G. and J.N.; resources, B.C., A.G. and J.N.; writing–original draft preparation, B.C., A.G. and J.N.; visualization, B.C., A.G. and J.N. All authors have read and agreed to the published version of the manuscript.

Funding: This research was funded by the National Science Centre (NCN), Poland (Grants No.: UMO-2014/15/B/ST10/04455 & 2018/02/X/ST10/03113). The APC was funded by the Wrocław University of Environmental and Life Sciences.

Acknowledgments: The Institute of Meteorology and Water Management — National Research Institute as well as the University of Wyoming are the key sources of the data used in this paper. The authors would like to thank all volunteers and students of the Adam Mickiewicz University, Poznań who helped design the final structure of the package and have tested the developer version extensively since 2017.

Conflicts of Interest: The authors hereby declare no conflict of interest. All regulations and restrictions of data use can be found at: https://ogimet.com/, http://weather.uwyo.edu/upperair/sounding.html, and https://dane.imgw.pl/regulations, http://danepubliczne.imgw.pl.

Abbreviations

The following abbreviations are used in this manuscript:

CRAN	Comprehensive R Archive Network
GDAL	Geospatial Data Abstraction Library
IMGW-PIB	Institute of Meteorology and Water Management—National Research Institute
METAR	Meteorological information used mostly by aircraft pilots
NOAA	National Oceanic and Atmospheric Administration
SYNOP	Surface synoptic observations
TEMP	Upper air profiles
WMO	World Meteorological Organization

References

1. Schirmer, M.; Luster, J.; Linde, N.; Perona, P.; Mitchell, E.A.; Barry, D.A.; Hollender, J.; Cirpka, O.A.; Schneider, P.; Vogt, T.; et al. Morphological, hydrological, biogeochemical and ecological changes and challenges in river restoration—The Thur River case study. *Hydrol. Earth Syst. Sci.* **2014**, *18*, 2449–2462. [CrossRef]

2. Szewrański, S.; Chruściński, J.; Kazak, J.; Świąder, M.; Tokarczyk-Dorociak, K.; Żmuda, R. Pluvial flood risk assessment tool (PFRA) for rainwater management and adaptation to climate change in newly urbanised areas. *Water* **2018**, *10*, 386. [CrossRef]

3. Kendzierski, S.; Czernecki, B.; Kolendowicz, L.; Jaczewski, A. Air temperature forecasts' accuracy of selected short-term and long-term numerical weather prediction models over Poland. *Geofizika* **2018**, *35*, 19–37. [CrossRef]

4. Roshan, G.; Yousefi, R.; Błażejczyk, K. Assessment of the climatic potential for tourism in Iran through biometeorology clustering. *Int. J. Biometeorol.* **2018**, *62*, 525–542. [CrossRef] [PubMed]

5. Bryś, K.; Brys, T. The First One Hundred Years (1791–1890) of the Wrocław Air Temperature Series. In *The Polish Climate in the European Context: An Historical Overview*; Springer: Berlin/Heidelberg, Germany, 2010; pp. 485–524.

6. Głogowski, A.; Chalfen, M. Analysis of the effectiveness of the systems protecting against the impact of water damming in the river on the increase of groundwater level on the example of the Malczyce dam. In *ITM Web of Conferences*; EDP Sciences: London, UK, 2018; Volume 23, p. 00011.

7. Grinn-Gofroń, A.; Nowosad, J.; Bosiacka, B.; Camacho, I.; Pashley, C.; Belmonte, J.; De Linares, C.; Ianovici, N.; Manzano, J.M.M.; Sadyś, M.; et al. Airborne Alternaria and Cladosporium fungal spores in Europe: Forecasting possibilities and relationships with meteorological parameters. *Sci. Total Environ.* **2019**, *653*, 938–946. [CrossRef] [PubMed]

8. Czernecki, B.; Nowosad, J.; Jabłońska, K. Machine learning modeling of plant phenology based on coupling satellite and gridded meteorological dataset. *Int. J. Biometeorol.* **2018**, *62*, 1297–1309. [CrossRef] [PubMed]

9. Frei, T. Economic and social benefits of meteorology and climatology in Switzerland. *Meteorol. Appl. A J. Forecast. Pract. Appl. Train. Tech. Model.* **2010**, *17*, 39–44. [CrossRef]

10. Palmer, T.N. The economic value of ensemble forecasts as a tool for risk assessment: From days to decades. *Q. J. R. Meteorol. Soc. A J. Atmos. Sci. Appl. Meteorol. Phys. Oceanogr.* **2002**, *128*, 747–774. [CrossRef]

11. Kalnay, E.; Kanamitsu, M.; Kistler, R.; Collins, W.; Deaven, D.; Gandin, L.; Iredell, M.; Saha, S.; White, G.; Woollen, J.; et al. The NCEP/NCAR 40-year reanalysis project. *Bull. Am. Meteorol. Soc.* **1996**, *77*, 437–472. [CrossRef]

12. Miętus, M. *O przydatności rezultatów globalnych reanaliz NCEP i ERA-40 do opisu warunków termicznych w Polsce*; Instytut Meteorologii i Gospodarki Wodnej: Warsaw, Poland, 2009.

13. Council of Poland. Law of 25.02.2016. about Re-Use of Public Sector Information (Dz. U. z 2016 r., pos. 352., with Later Changes). 2016. Available online: http://prawo.sejm.gov.pl/isap.nsf/DocDetails.xsp?id=WDU20160000352 (accessed on 1 December 2019).

14. Council of Poland. Law of 20.07.2017 the Water Act (Dz. U. z 2017 r. pos. 1566). 2017. Available online: http://prawo.sejm.gov.pl/isap.nsf/DocDetails.xsp?id=WDU20170001566 (accessed on 1 December 2019).

15. Benestad, R.E.; Mezghani, A.; Parding, K.M. *'esd'-The Empirical-Statistical Downscaling tool & its visualisation capabilities*; Met Report 11/15; Norwegian Meteorological Institute: Oslo, Norway, 2015

16. Bowman, D.C.; Lees, J.M. Near real time weather and ocean model data access with rNOMADS. *Comput. Geosci.* **2015**, *78*, 88–95. [CrossRef]

17. Buishand, T.A.; De Martino, G.; Spreeuw, J.; Brandsma, T. Homogeneity of precipitation series in the Netherlands and their trends in the past century. *Int. J. Climatol.* **2013**, *33*, 815–833. [CrossRef]

18. Boessenkool, B. *rdwd: Select and Download Climate Data from 'DWD' (German Weather Service)*; R package version 1.2.0; Deutscher Wetterdienst: Offenbach, Germany, 2019

19. R Core Team. *R: A Language and Environment for Statistical Computing*; R Foundation for Statistical Computing, Vienna, Austria, 2013.

20. Wickham, H. Tidy data. *J. Stat. Softw.* **2014**, *59*, 1–23. [CrossRef]

21. Lovelace, R.; Nowosad, J.; Muenchow, J. *Geocomputation with R*; CRC Press: Boca Raton, FL, USA, 2019.

22. Blumberg, W.G.; Halbert, K.T.; Supinie, T.A.; Marsh, P.T.; Thompson, R.L.; Hart, J.A. SHARPpy: An open-source sounding analysis toolkit for the atmospheric sciences. *Bull. Am. Meteorol. Soc.* **2017**, *98*, 1625–1636. [CrossRef]

23. Taszarek, M.; Brooks, H.E.; Czernecki, B. Sounding-derived parameters associated with convective hazards in Europe. *Mon. Weather Rev.* **2017**, *145*, 1511–1528. [CrossRef]

24. Nidzgorska-Lencewicz, J.; Czarnecka, M. Winter weather conditions vs. air quality in Tricity, Poland. *Theor. Appl. Climatol.* **2015**, *119*, 611–627. [CrossRef]

25. Carslaw, D.C.; Ropkins, K. Openair—An R package for air quality data analysis. *Environ. Model. Softw.* **2012**, *27*, 52–61. [CrossRef]

26. Nychka, D.; Gilleland, E.; Zhang, L.; Hoar, T. *RadioSonde: Tools for Plotting Skew-T Diagrams and Wind Profiles*; R package version 1.4; University Corporation for Atmospheric Research: Boulder, CO, USA, 2014.

27. Pebesma, E. Simple features for R: Standardized support for spatial vector data. *R J.* **2018**, *10*, 439–446. [CrossRef]

28. Tennekes, M. tmap: Thematic Maps in R. *J. Stat. Softw.* **2018**, *84*, 1–39. [CrossRef]

29. Wickham, H.; Grolemund, G. *R for Data Science: Import, Tidy, Transform, Visualize, and Model Data*; O'Reilly Media, Inc.: Sebastopol, CA, USA, 2016.

Article

The Use of Artificial Intelligence as a Tool Supporting Sustainable Development Local Policy

Maria Mrówczyńska [1], **Małgorzata Sztubecka** [2], **Marta Skiba** [1,*], **Anna Bazan-Krzywoszańska** [1] and **Przemysław Bejga** [3]

[1] Institute of Civil Engineering, Faculty of Civil Engineering, Architecture and Environmental Engineering, University of Zielona Góra, u1 Prof. Z. Szafrana 1, 65-516 Zielona Góra, Poland
[2] Faculty of Civil and Environmental Engineering and Architecture, UTP University of Science and Technology in Bydgoszcz, Al. Prof. S. Kaliskiego 7, 85-796 Bydgoszcz, Poland
[3] Department of Pharmakology and Toxicology, Faculty of Medicine and Health Sciences, University of Zielona Góra, ul. Zyty 28, 65-046 Zielona Góra, Poland
* Correspondence: M.Skiba@aiu.uz.zgora.pl

Received: 31 May 2019; Accepted: 1 August 2019; Published: 3 August 2019

Abstract: This paper addresses the problem of noise in spa protection areas. Its aim is to determine the delimitation of the areas that exceed a permissible noise level around the sanatorium on the example of a health resort in Inowrocław. The determination of the exceedance of permissible noise levels allows us to develop directly effective local policy tools to be included in planning documents. In order to reduce noise infiltration, it is important to define environmental priorities. Taking into account their impact on the health of users in the protection area, environmental priorities enable us to introduce additional elements to street architecture. In order to properly manage space, in accordance with the idea of sustainable development, zones of environmental sensitivity—and their socio-environmental vulnerability—have been designated for assessing damage (exceeding permissible noise in health facilities) and defining methods of building resilience (proper management). This has provided the basis for a natural balance optimized for the people living in these areas. To achieve the goal above, non-linear support vector machine (SVM) networks were used. This technique allows us to classify the linearly inseparable data and to determine the optimal separation margin. The boundaries of the areas which exceeded permissible noise levels (separation margin) were estimated on the basis of noise pollution maps, created by means of the SVM technique. Thus, the study results in establishing buffer zones where it is possible to use varied land utilization in terms of form and function, as described in the planning documents. Such an activity would limit the spread of noise.

Keywords: noise; acoustic space; socio-environmental vulnerability; Support Vector Machines; spatial policy; preventive healthcare; healthcare facilities

---

1. Introduction

The effect of environmental factors on human health is inevitable. Among health-affecting factors, quite important are the physical factors of diverse nature: water, soil or air pollution [1]. Increasingly, noise is being mentioned as air pollution caused by energy. Energy propagating in the air takes the form of acoustic waves that are divided by frequency into ultrasound, audible sounds, and infrasound [2]. Noise is an undesirable, unpleasant, cumbersome, oppressive or harmful sound of excessive intensity affecting the organ of hearing, the other senses and other parts of the body or the whole body [3–7]. Proper development of the acoustic environment is based on the formation of acoustic conditions so that they are optimal from the point of view of health and human activities. Each person is characterized by an individual reception of sounds. It is the sensitivity on noise. To get to know this sensitivity it is important to properly manage and recognize social, environmental and health vulnerabilities.

"Noise sensitivity" is defined as an individual and increased probability of perceiving sounds as unpleasant [8–11]. Moreover, getting used to excessive noise levels causes indifference, which is caused by exposure to a hazardous situation [12–16].

Noise can be characterized by its source (e.g., anthropogenic or natural causes) or by specific physical values, such as power, intensity, frequency or duration. It should be noted that its relevant parameters refer not only to strictly objective values considered to be unpleasant or harmful but also, to the same extent, to the places where common sounds will cause unwanted or disturbing effects [17–19]. Organized urban spaces are treated like resources directly used by local residents. The conditions of these spaces decide also on the possibilities for creating a proper base for planners who, to a greater or lesser degree, introduce quantitative and qualitative changes in the area [20,21].

It is important then, to analyze the acoustic situation. The determination of physical parameters such as density, open spaces, their shape, the greenery and the location of the buildings has a significant effect on the noise spread in the environment [18,22]. Maps are a good tool for showing the noise distribution in the area. It is accepted in the literature that an acoustic space map can contain three different topics: a sound sources map, a map of psychoacoustic perception and a noticeable quality of sound environment [23–30]. A map of sound sources provides information on the different types of noise, including positive ones. Liu developed a map of perception of sound sources based on their spatial-temporal variation [27,28]. He qualified the sources as anthropological, biological and geographical, and showed that they are characteristic of urban landscapes and depend on their composition.

Yerli and Demir, by examining seasonal variation of noise, identified the possibility of noise reduction by greenery up to 20 dB [31]. The researchers showed the dependence of the noise level on the time of year and the time of day (daytime noise for the spring–summer is higher by 4 dB). Independently, the researchers demonstrated a relationship between the size of the paved surface (passageways) in green areas and the noise level. The impact of the characteristics of the acoustic properties of surface material used for paths on the feeling of annoyance caused by noise (perception of background factor) was also described by Vlahov [32].

Urban noise pollution is gradually increasing, mainly because of rapid industrialization and urbanization. A number of studies and reports refer to the fact that noise affects physiological and mental health [33,34] and sound space is considered to be a key factor in creating a healthy city [35–37]. According to the World Health Organization, health is defined as a state of complete physical, mental, and social well-being and not merely the absence of disease or infirmity. Particular attention should be paid to the importance of recreational and health areas due to their tasks [38]. In these areas, the noise level should be as low as possible, close to natural background levels. In Polish law spas have the lowest values of equivalent sound level. The highest permissible value is 50 dB during the day, whereas other areas have 68 dB. The Directive 2002/49/EC of the European Parliament and of the Council of 25 June 2002 relating to the assessment and management of environmental noise, indicates that it is necessary to adopt common methods to assess "environmental noise" and the definition of "limit values", in terms of harmonized indicators for the determination of noise levels. The harmful effects of noise on the human body is a complex issue and it is related to a number of its space activities. In the assessment of nuisance sounds, an indicator L_{Aeq} was used, which refers to the measured noise value at the specific time of the measurement. The obtained results show the current acoustic climate characteristic for a certain area [22,29]. The research presents on noise and its effects on human health. The results are a good output model for the extrapolation of health effects and attempts to define them in different periods of activity. The aim is the sustainable creation of such vulnerable spaces, especially in terms of noise protection. Thus, the research question is how to monitor noise in spa areas and how to counteract its spread.

Environmental noise depends on land development and usage. This paper presents, for the first time, zones, within one park, where various development and land use should be conducted. Usually, the area of the spa park is treated homogeneously. We propose to distinguish zoning and utilization zones and introduce various limitations in zones. This is necessary due to the fact that noise sources are internal and result from the intensity of use. Many studies deal with environmental noise research, most often regarded as a linear source. We use support vector machine (SVM) networks, which gives the possibility to determine zones depending on many sources of noise and the need to protect the acoustic climate of the protected space [39]. The residents and spa patients and also the inhabitants of Inowrocław, who use the attractions of the health spa, are the sources of noise in the spa area. The source of noise is, therefore, internal, anthropological, and the prevention of its spread will be supported by creating separated areas with controlled absorbency.

Mathematical multi-criteria optimization methods are usually used to plan the reduction of the impact of noise coming from various sources such as road and rail transport, construction and operations, or resulting from human life. In addition, they are based on many purposes [40–43]. In this way, an attempt is made to reduce the impact of noise along with optimizing the costs of decisions and possible implementation. The use of the support vector machine, a technique to determine the areas based on the noise measurements in spatial planning, is also new. The SVM algorithm is widely used in solving data classification problems [44,45], object recognition [46], in technical, engineering and environmental applications [47] and in medical diagnostics [48,49]. An interesting application is the use of the SVM algorithm to model ship maneuvering movements to mitigate the impact of noise problems [50]. The SVM is characterized by strong theoretical foundations based on statistics. Its modern form was developed by Vapnik and discussed in detail in [51]. The SVM technique has been gaining more and more popularity in recent years and it is still prevailing in line with the currently used trends in scientific research on the use of artificial intelligence methods.

Its application primarily solves problems of linear and non-linearly separable data classification [52–54], detects and identifies failures and errors in various systems [55,56], and assesses risk in various branches of the economy [57,58]. In this paper, the range of special zone development in the spa areas was determined by means of the SVM technique. Simultaneously, ways of developing the area in order to reduce noise spread were proposed.

2. Materials and Methods

2.1. Research Object

Inowrocław is a city in the northern-west part of Poland, located in Kuyavia-Pomerania province. It is one of 46 health resorts in Poland and it currently functions as:

- a large settlement,
- an industrial-economic and service center of a regional importance,
- an important railway junction and a significant road transport hub,
- a vocational education center,
- a health resort with a separate spa (Figure 1)

The spa function completes urban functions and generates income.

The lowland spa in Inowrocław was founded in 1875. Initially, the Health Spa and Resort (Solanki I) covered the area of 5 ha. At present, its area is approximately 85 ha. The arrangement of the Health Spa takes patients' comfort into account most of all. There, one can find numerous walking alleys, pedestrian zones and leisure areas with gazebos, brine graduation towers, mineral water drinking sites, spa houses, natural medicine institutes and other objects of public interest.

The intention of health resorts is to provide an atmosphere supporting medical treatment. Therefore, the spatial layout of health resorts should have elements unique to these types of areas. The basis is a natural therapy complex, around which the parkland is located, often of a great value due to its composition, aesthetic and cultural features. Moreover, organizing social life and entertainment plays an important role in health resorts. Most of the noise in such a type of areas is associated with human activities [59].

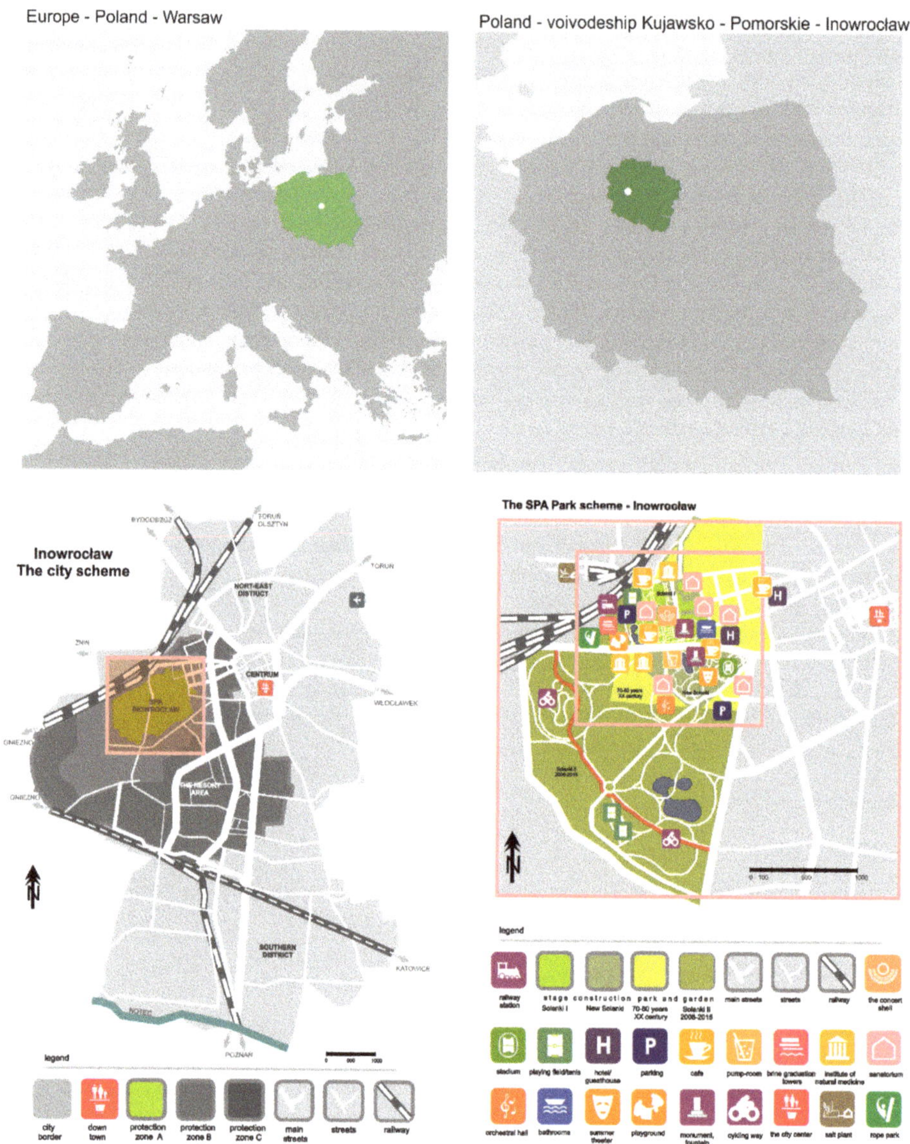

Figure 1. The map of Inowrocław, along with the health spa and resort. Source: Authors.

2.2. The Measurement of an Equivalent Sound Level

Sound level measurements were carried out in accordance with Polish legislation (PN—81/N—01306, PN-ISO 1996-1:1999) and were described also in the principles in the statement of the State Inspectorate for Environmental Protection entitled "Methods of measurement of external noise in the environment" with amendments that are adapted to the purposes of this study. Acoustic data were collected with a sampling technique approved by the standard PN-ISO 1996-2:1999 [60]. In this technique, the total duration of the measure was part of the period of reference time, which was chosen to take the variability of noise emissions into account. The measurement points were located so as to best represent the space around health resorts. It was also considered to carry out the noise measurements in conditions not exceeding the following limit values:

- temperature range from −10 °C to 40 °C,
- humidity from 25% to 98%,
- average wind speeds up to 5 m/s,
- atmospheric pressure from 940 hPa to 1060 hPa.

The designated indicator was the equivalent sound level L_{Aeq}. The measurements taken into account were the eight least favorable hours for daytime and one least favorable hour at night. The sound level was measured around the spa objects located in A zone of the health resort protection area. The buildings of five sanatoriums are located around the park in its old part. The location of the measurement points is shown in Figure 2, and the value of the measured equivalent sound level is presented in Table 1 and Figure 3.

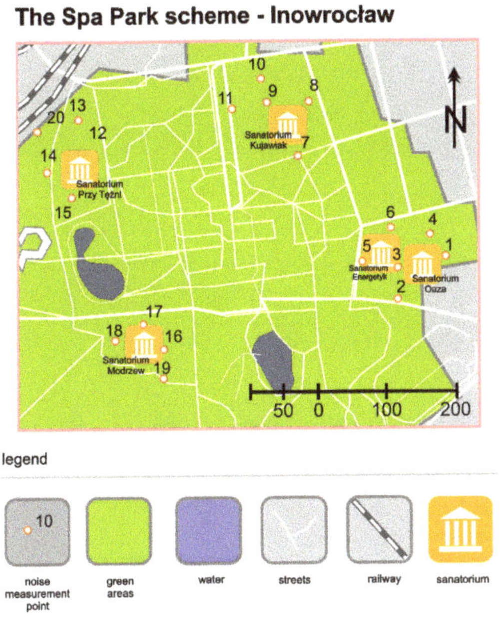

Figure 2. The maps of the health resort area with measuring points. Source: Authors.

The noise from all observable sources was measured at the designated points. The obtained results show the currently prevailing acoustic conditions of the area (Figure 3—noise contour line).

Table 1. Measured values of the equivalent sound level [dB].

Point No.	The Name of the Sanatorium	L_{Aeq} for 8 h During the Day [dB]	L_{Aeq} for 1 h During the Night [dB]	Point No.	The Name of the Sanatorium	L_{Aeq} for 8 h During the Day [dB]	L_{Aeq} for 1 h During the Night [dB]
1	Oaza	56.9	40.5	10	Kujawiak	48.5	45.1
2	Oaza	64.4	46.2	11	Kujawiak	45.7	42.4
3	Oaza	51.0	40.0	12	Przy Tężni	48.8	45.2
4	Oaza	55.9	40.1	13	Przy Tężni	54.2	46.8
2	Energetyk	64.4	46.2	14	Przy Tężni	47.9	41.6
3	Energetyk	51.0	40.0	15	Przy Tężni	54.3	45.3
5	Energetyk	58.8	47.6	20	Przy Tężni	54.6	45.8
6	Energetyk	52.0	48.1	16	Modrzew	52.8	41.5
7	Kujawiak	47.8	45.4	17	Modrzew	45.1	38.9
8	Kujawiak	62.9	50.8	18	Modrzew	47.9	40.7
9	Kujawiak	49.5	45.0	19	Modrzew	48.3	37.6

Figure 3. The maps of the health resort area with noise contour line. Source: Authors.

2.3. The SVM Neural Networks Used to Border the Land Use in Zones

Solutions that use artificial intelligence are more common [61]. The SVM classifiers, called the support vectors network technique, were developed by Vapnik [51]. It is a new approach to building and training a unidirectional network. They usually have a double-layer structure and may use different types of activation functions [62]. SVM networks lack the defects typical of multi-layer perceptron (MLP) networks, that is, the possibility of stopping the process of minimizing in one of many local minima, and network architecture arbitrarily taken at the outset, which its future capacity to generalize depends on.

Generally, in the case of linearly separable data, the SVM networks allow us to find a separation hyperplane that separates two classes [61]. In the case of linearly inseparable data, the SVM method can be used to find a hyperplane which classifies objects properly and, at the same time, has the highest possible distance from typical clusters for each class (Figure 4). In the case of linearly inseparable data, using the dimension raise, the SVM method allows us to find a curvilinear separation line with the maximum possible margin. The paper presents the problem of using a non-linear SVM network to classify data which are linearly inseparable, for example, the measurement of sound intensity in the area of Inowrocław health resort.

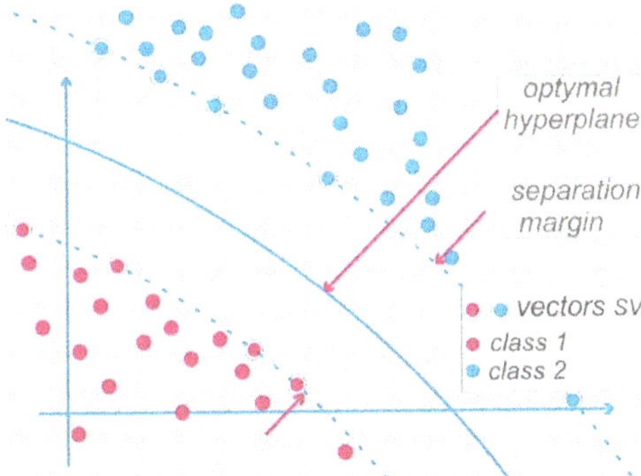

Figure 4. The optimal hyperplane with the maximum separation margin. Source: Authors.

The essence of the SVM method is the construction of an optimal hyperplane, which separates data belonging to different classes, with a maximum margin of trust (separation margin). The margin of trust shall be understood as the distance of the hyperplane from the nearest points on which support vectors will be formed (Figure 4). The points, where the support vectors are formed, are located close to the hyperplane and define its position. However, at the same time, they are the most difficult to classify. More information about the classification of linearly separable data and methods of building an optimal hyperplane can be found, inter alia, in the following works: [62–65].

A commonly used solution to classify linearly inseparable data is a projection of the original data to the function space (space of attributes), in which the data are linearly separable with a probability close to 1. Usually, the dimension of the space of attributes-K is a lot larger than the dimension of the space of the original-N, and a carried-out transformation of one space into another is a non-linear transformation [66]. The graphic presentation of the non-linear transformation of the linearly inseparable data is shown in Figure 5a,b. The linearly inseparable data in the two-dimensional space of the original-N (Figure 5a) were transformed to the space of attributes-K (Figure 5b), which is defined with the Gaussian functions (1)

$$\phi(\mathbf{x}) = \exp\left(-\frac{\|(\mathbf{x} - \mathbf{c})\|^2}{\sigma^2}\right) \tag{1}$$

where: σ—the width of the Gaussian function, $\mathbf{c}$—the Gaussian function centers, $\mathbf{x}$—the input vector.

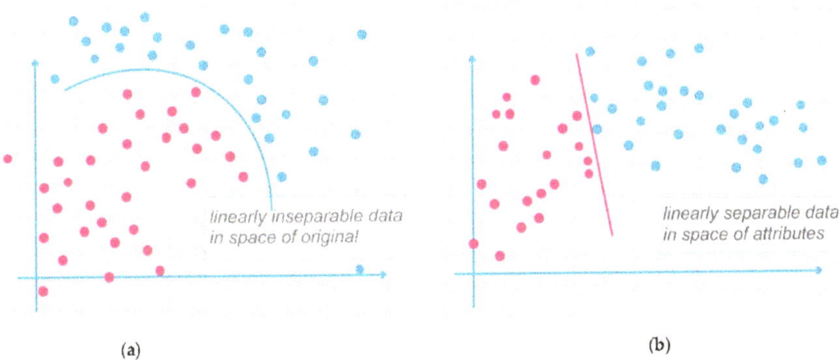

(a)

(b)

Figure 5. (a) Linearly inseparable data in space of original. Source: Authors; (b) Linearly separable data in space of attributes. Source: Authors.

After a non-linear transformation, the data become linearly separable and they can be separated by one separation plane. The location of the class separating hyperplane is determined in the space of attributes (Figure 5b), and in the space of the original, only its picture can be observed (Figure 5a).

Assuming that a set of training pairs is being classified $(\mathbf{x}_i, d_i)$, $i = 1, \ldots, N$, in which the required value d_i is equal to 1 or −1 and $\mathbf{x}_i$ the input vector is projected in the space of attributes-K, the dimension is represented by a set of attributes $\phi_j(\mathbf{x})$, $j = 1, \ldots, K$. After the transformation, Equation (2) for hyperspace separating data in the space of attributes shall be recorded as

$$g(\mathbf{x}) = \sum_{j=1}^{K} w_j \phi_j(\mathbf{x}) + b = 0 \tag{2}$$

where: w_j—the weight from a neuron in the hidden layer to the output neuron (Figure 5), b—polarization that specifies the hyperplane location in relation to the origin of a coordination system.

The signal of the output neuron for the network, with the architecture shown in Figure 5, is defined by Equation (3)

$$y(\mathbf{x}) = \mathbf{w}^T \phi(\mathbf{x}) + b \tag{3}$$

Analyzing the basic structure of the SVM neural network (Figure 6) one will notice that the structure is comparable to that of the radial basis function networks (RBF) and the difference is that the basic functions $\phi(\mathbf{x})$ may take a linear, polynomial, radial or sigmoidal form.

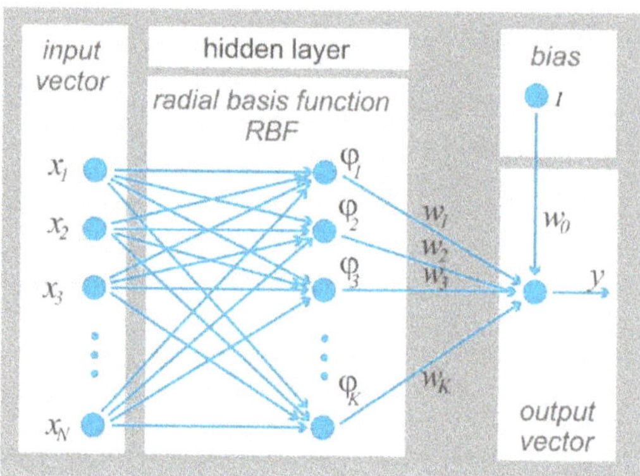

Figure 6. The basic architecture of the non-linear support vector machine (SVM) network. Source: Authors.

The aim of training the non-linear SVM network is similar to the determination of values of the vector of weights $\mathbf{w}$ so as to determine the optimal hyperplane for the linearly inseparable data, which shall minimize the probability of classification error while maintaining the condition of maximizing the separation margin. While classifying the linearly inseparable data, it is necessary to define a non-negative complementary variable λ, which is needed to reduce the current width of the separation margin. A problem defined this way is referred to as a primary problem [54,55] that shall be noted as (4)

$$\min\{\phi(\mathbf{w}, \lambda)\} = \frac{1}{2}\mathbf{w}^T\mathbf{w} + C\sum_{i=1}^{p} \lambda_i \tag{4}$$

with restrictions (5)

$$d_i\left(\mathbf{w}^T \phi(\mathbf{x})_i + b\right) \geq 1 - \lambda_i$$
$$\lambda_i \geq 0 \tag{5}$$

where: C—a parameter is taken arbitrarily by the user.

In the early stages of the non-linear SVM network training, a number of support vectors are usually equal to the number of training data. During the training process, depending on the value of the C parameter (limit value), network complexity is reduced and support vectors are formed for only some of the points. The support vectors are formed on those points for which a condition is fulfilled (6)

$$\mathbf{w}^T \phi(\mathbf{x}_i) + b = \pm 1 \tag{6}$$

It is worth noting that the higher the value of the parameter C, the narrower the separation margin and the smaller the number of support vectors. For small values of the parameter, the C network is approximated in its activity to the linear network, which extends the separation margin.

The output signal (7) of the non-linear SVM network is eventually defined as [39]

$$y(\mathbf{x}) = \mathbf{w}^T \phi(\mathbf{x}) + b = \sum_{i=1}^{P_{sv}} \alpha_i d_i K(\mathbf{x}, \mathbf{x}_i) + b \tag{7}$$

where: P_{sv}—the number of the support vectors $\mathbf{x}_i$, which is equal to the number of non-zero Lagrange multipliers, $K(\mathbf{x}, \mathbf{x}_i)$—the kernel function.

The output signal of the non-linear SVM network depends on the kernel function $K(\mathbf{x}, \mathbf{x}_i)$, not on the base function $\phi(\mathbf{x})$, as in the case of radial networks. Because of that, it becomes obvious that it is necessary to build a network with the resulting structure shown in Figure 7, whereas the most commonly used kernel functions are presented in Table 2.

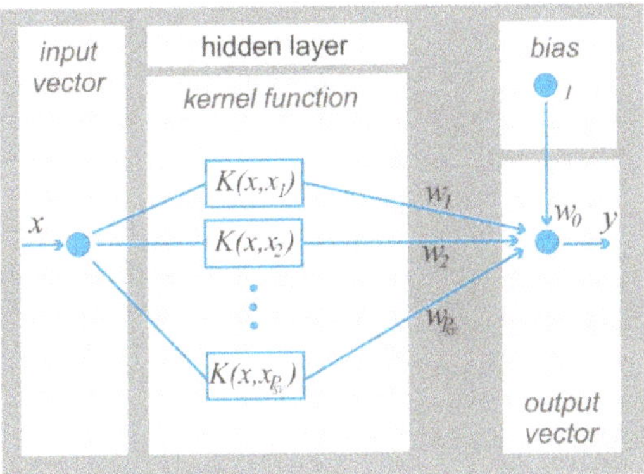

Figure 7. The architecture resulting from the non-linear SVM network. Source: Authors.

Table 2. Examples of kernel functions.

Type Kernel	Equation $K(\mathbf{x}, \mathbf{x}_i)$	Comment
linear	$K(\mathbf{x}, \mathbf{x}_i) = \mathbf{x}^T\mathbf{x}$	
polynomial	$K(\mathbf{x}, \mathbf{x}_i) = \left(\mathbf{x}^T\mathbf{x} + 1\right)^b$	b—degree polynomial
radial (Gaussian)	$K(\mathbf{x}, \mathbf{x}_i) = \left[-\left(1/2\sigma^2\right)\left(\|\mathbf{x} - \mathbf{x}_i\|^2\right)\right]$	σ—for all kernels
sigmoidal	$K(\mathbf{x}, \mathbf{x}_i) = \tanh\left(\beta_1\mathbf{x}^T\mathbf{x} + \beta_0\right)$	restrictions on the β_0 i β_1

If a linear kernel is applied, the built network is fully linear without a hidden layer. The application of a sigmoid function leads to the architecture corresponding to perceptron neural network with one hidden layer. Using the Gaussian function results in a radial basic function network, in which the number of basic functions and their centers is equated with the support vectors. Similarly, in sigmoidal networks, the number of neurons in the hidden layer is determined by the number of the support vectors.

For the tested health spa and resort, the use of SVM led to the designation of isolines of the sound level distribution in the area (Figure 7). This allowed us to distinguish the areas of "acoustic safety" (Figure 7). The task of automatic classification was based on noise intensity measurements. The results, along with the geolocation of the measurement points, were the data that were entered into the input layer. The process of classification and determination of individual zones has been carried out taking into account: the selection of a kernel function type, the selection of network parameters (architecture, C parameter) and the use of weights selection in relation to the chosen network parameters. The values of the signal in the output layer allowed us to qualify the data to appropriate classes.

Besides the distinction, the areas are characterized by an increased sound level which is adverse for users.

3. Results and Discussion

The Health Spa and Resort in Inowrocław displays many features associated with healthcare, entertainment, recreation, and relaxation. Its sound space is formed by all the events taking place there. Special attention should be given to keeping the noise level as low as possible, close to the natural background level [18,67].

In Polish legislation, the acceptable noise level in the daytime and nighttime (other sources of noise—environmental noise) for spa parks (A protection zone), is 45 dB and 40 dB. The measurement of an equivalent sound level in the area of the park for the daytime shows exceeded limit values by 7 dB, while at night the limits are exceeded on average by 4 dB. Noticeable changes in the sound level for a human is a level of 3 dB. So, the above values indicate a significant increase in sound intensity.

The information, included in the map of the acoustic climate, is primarily used to identify an existing acoustic situation in the studied area. They designate the areas where the exceedance of noise limit values occurred. The maps enable us to estimate the population size exposed to the excessive noise and to estimate possible changes to the acoustic climate in relation to the environment protection against the noise. For the greenery in the areas, such maps have not been prepared because there is a belief that those areas represent silence zones. In large parks, where the accumulation of different functions can be observed, proper zoning is essential as it improves acoustic comfort. The map showing the distribution of sounds is a helpful tool to create a protection area properly. The boundaries of the areas with the exceeded permissible noise levels (separation margin) were estimated on the basis of the noise pollution maps, created by means of the SVM technique. That is 45 dB during the night and 60 dB in the daytime and it should be applied including varied land utilization in terms of form and function, described in the planning documents. In these zones, additional elements of land utilization may be introduced to limit the noise infiltration (Figure 8).

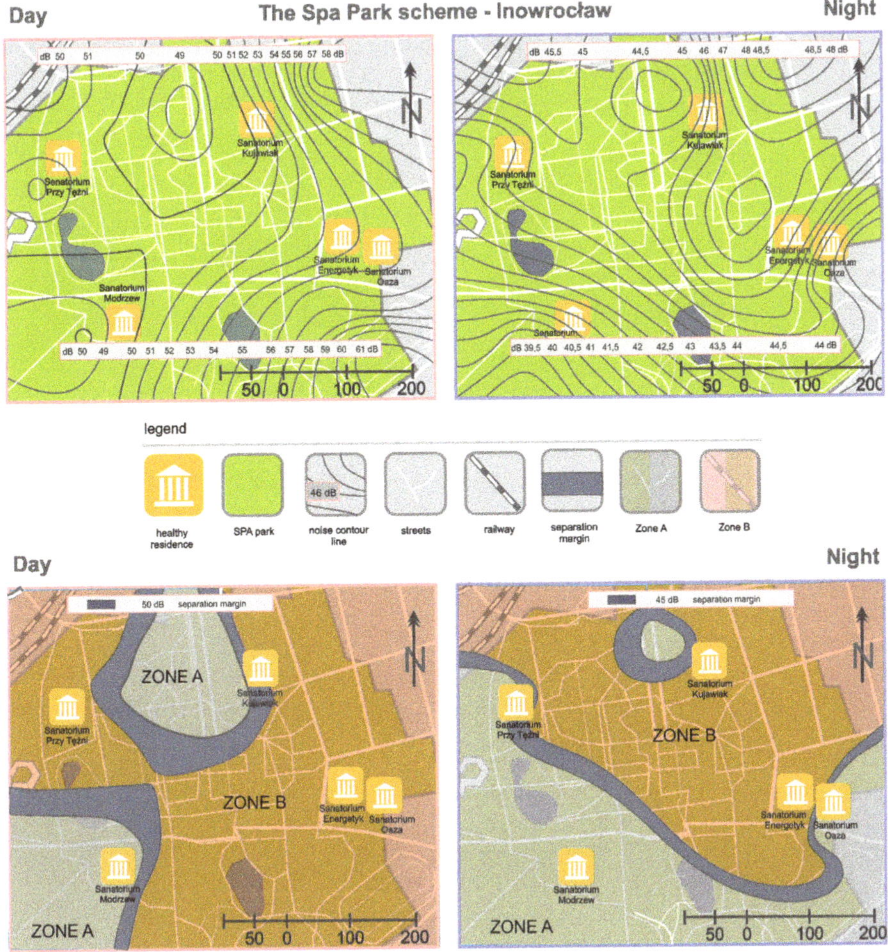

Figure 8. Acoustic zones established with the SVM neural networks. Source: Authors.

In the area around the health resort, relationships between sound, environment, and people using it, and those expressed as an acoustic comfort, were presented by Tse. In addition to the visual evaluation of landscape, the evaluation of acoustic comfort played an important role in accepting the urban park [68].

The researchers recommend including the analyses of the process of urban sprawl and its impact on the health of population, as potential determinants of health. The development of spatial policy and legal regulations should come from proper planning and integral multi-sectoral approach towards these issues, so as to protect and improve the health of population in a more and more urbanized living environment [69–72].

The existence of the problem of noise nuisance for the health spa area visitors cannot be concluded only from the measurements. Guests staying in the area may, in fact, pay attention to noise which is not taken into account in the standards. A complete assessment of these feelings can only be obtained on the basis of surveys carried out among people residing in the area. They can be the basis for determining a subjective nuisance of the received noise [73,74]. The direction of the future research should be the search for an effective policy of monitoring noise as an air pollution. In addition, it must be considered to what extent air pollution and other sources of pollution can cause short- and long-term health effects with a certain exposure to noise, and what relationships exist between health outcomes and sensitivity in the noise delivery.

The presented classification model was developed with regard to a number of spatially distributed sampling points. This method can easily be extended to the continuous systems of monitoring perception and acoustic properties of a certain location.

4. Conclusions

Maintaining the right proportions in sustainable park management requires a balance between the noise generated and the impact on the environment. Is the analyzed area vulnerable to changes, especially with regard to noise protection? Yes, but as a special area—a spa park—changes must be rational and should take into account both social and environmental aspects as well as natural aspects.

Shaping the noise climate of the greenery is influenced by many factors, such as:

- the size of the area,
- the accumulation of sound sources,
- the terrain shape,
- natural or artificial obstacles,
- the selection of sampling points,
- the number of visitors (capacity and functional program).

The tested object, the health spa and resort, is a multi-functional area comprising of objects related to healthcare, entertainment, recreation and relaxation. Increasingly, programs for the development of such areas offer diversity, adapting to the new needs of users. Health resorts change their profiles from healthcare centers to multi-functional spa, wellness and recreational places with catering and entertainment.

The space of the tested health spa area is characterized by noise infiltration between the zones. Therefore, there is the need of separating silence zones from the noisy ones by creating neutral zones. This note, which is primarily a recommendation for designing spatial systems in new areas, in the case of "Solanki" may be replaced by the introduction of acoustic partitions. However, often for aesthetic reasons, constructing engineering partitions requires masking elements such as evergreen plants. In addition, linear partitions, considering the gravity flow of air masses, can disturb natural ventilation. The composition of dense greenery and buildings is important for the noise interference issue. Noise is often generated by sports zones and playgrounds as well as by materials used for path cladding. The adjustment of a park's functional program to its absorbency may be the solution to the problem of reducing noise (anthropocentric), especially in historical parks.

Polish spatial policy, in accordance with the provisions of the Act of 27 March 2003 on spatial planning and development (Off. J. of 2016 item 778 as amended), hereinafter referred to as the SPD Act, based primarily on the findings of the planning documents, which include a municipal study of the conditions and directions of the spatial development and a local zoning plan, in accordance with article 10 paragraph 2 subparagraph 3, takes into account protected areas and policies of a protection of the environment and its resources, nature conservation, landscape conversation, including cultural landscapes and health resorts.

Poland is a country with a distinctive diversity of space functioning in Europe, but at the same time, with very severe processes of spatial order and degradation of landscape physiognomy. There are also growing problems with the management of cities and municipalities land utilization.

The paper presents an innovative approach to the classification of zones located in health spa areas, depending on the noise and done by means of the SVM technique. The proposed approach has allowed us not only to classify the areas with exceeded noise standards but also to determine the course and width of the separation margin. This margin has been identified as an area where the spread of noise to neighboring areas will be reduced by the adopted solution. Proposals of a development zone and a buffer zone for exceeded noise levels are shown in Tables 3 and 4.

Table 3. Proposals supporting spatial decision-making reinforcing the basic healthcare functions of health spa areas.

Current Status	Zone A	Separation Margin	Zone B
Park development	Park development (paths, high greenery, low greenery)	Change of park management (acoustic barriers)	Change of park management (change of track surface, change of track width depending on the boundary absorption of zones)
Sanatorium buildings	limiting the amount of acceptable space in sanatorium buildings	Slopes—earth acoustic barriers	change of building elements (windows, doors) to muted—less noisy

Table 4. Proposals of preventive actions concerning records in the documents on spatial policy and changes in the management technique.

Current Status	Reform Guidelines
residents are indifferent to the public domain management technique	strengthening the role of civic organizations to establish partnership in development policy
lack of entries in the planning documents for parks, health areas and healthcare facilities, which are based on the acoustic analysis and study of the experience of space users	• conditioning records concerning the spatial and functional management of spa areas taken from the results of the acoustic space analysis and the research relating to the experience of users; • seasonal monitoring of the area's absorbency based on the results of acoustic analyses and surveys on sound perception; • the requirement of assessing acoustic comfort in order to protect health values of the resorts
uncontrolled spatial policy system	obligatory monitoring to control the appropriateness of activities (the reduction of sound-active materials and the introduction of functional program)
inefficient ways to write the land destination and utilization in the planning documents of the areas	changes to the law in order to introduce modern and effective methods of managing the land destination and utilization in accordance with the principles of sustainable development

Greenery has a beneficial effect on the urban environment (limits the spread of dust, gases, and noise). It also provides the space necessary for recreation. The location and size of a health resort area are very important factors in determining the quality of relaxation. In larger areas, different functions are accumulated. For this reason, sports or recreational zones may appear there next to zones for silence and walking.

A special type of greenery is a health spa area. They are part of health resorts and, therefore, subject to regulations in the Act of 28 July 2005 *on health resort treatment, health resorts and Spa protection areas and health resort municipalities* (Off. J. of 2016. item. 879 as amended) and also to the Polish standard of "Health resorts". The term "health resort" is a combination of medical and recreational functions in the area. The basis of activity in these areas is a balance between various elements of the natural and cultural environment. Health spa areas are health protection zones—separated from the area of a health resort, in order to protect medicinal features and medicinal natural raw materials, which are the benefits of the environment and equipment. They are characterized by a rational shaping

Sustainability **2019**, *11*, 4199

of the landscape and their spatial arrangement should take into account proper zones. However, for such a diverse area, it is extremely difficult to separate various subareas of silence and recreation and to prevent the noise infiltration between them.

Spa objects, as sensitive areas, are not monitored in terms of the existing noise level. Sanatoriums, located mostly on the outskirts of health spa and resort areas, should offer peace and relaxation in their space. The measurements of the equivalent sound level around these buildings show, above all, that the patients are the source of the noise. Changes should include determining the limits of the health spa areas' absorbency in sensitive spaces and other ways of development such as acoustic barriers (in the zone as the separation margin) to prevent the spread of noise.

Author Contributions: All authors had an equal contribution to the creation of this article.

Funding: This research received no external funding.

Conflicts of Interest: The authors declare no conflict of interest.

References

1. Marvin Herndon, J.; Whiteside, M. Further Evidence of Coal Fly Ash Utilization in Tropospheric Geoengineering: Implications on Human and Environmental Health. *J. Geogr. Environ. Earth Sci. Int.* **2017**, *9*, 1–8. [CrossRef] [PubMed]
2. Pawlas, K.; Pawlas, N.; Boroń, M.; Szłapa, P.; Zachara, J. Infrasound and low frequency noise assessment at workplaces and environment—Review of criteria. *Environ. Med.* **2013**, *16*, 82–89. (In Polish)
3. Engel, Z. *Environmental Protection against Vibration and Noise*; Wyd. Nauk. PWN: Warszawa, Poland, 2001. (In Polish)
4. Augustyńska, D.; Engel, Z.; Kaczmarska-Kozłowska, A.; Koton, J.; Mikulski, W. Noise. Infrasound and Ultrasound. Centralny Instytut Ochrony Pracy 2009. Available online: Ergonomia.ioz.pwr.wroc.pl/download/ciop-halas-text.pdf (accessed on 23 January 2017). (In Polish).
5. Takayashi, Y.; Kanada, K.; Yonekawa, Y.; Harada, N.A. Study on the Relationship betweeen Subjective Unpleasantness and Body Surface Vibration Induced by Highlevel Low-frequency Pure Tones. *Ind. Health* **2005**, *43*, 580–587. [CrossRef] [PubMed]
6. Prasher, D. Is there evidence that environmental noise is immunotoxic? *Noise Health* **2009**, *11*, 151–155. [CrossRef] [PubMed]
7. Sørensen, M.; Andersen, Z.J.; Nordsborg, R.B.; Jensen, S.S.; Lillelund, K.G.; Beelen, R. Road Traffic Noise and Incident Myocardial Infarction: A Prospective Cohort Study. *PLoS ONE* **2012**, *7*, e39283. [CrossRef] [PubMed]
8. Van Kamp, I.; Davies, H. Noise and health in vulnerable groups: A review. *Noise Health* **2013**, *15*, 153–159. [CrossRef] [PubMed]
9. Shepherd, D.; Welch, D.; Dirks, K.N.; Mathews, R. Exploring the relationship between noise sensitivity, annoyance and health-related quality of life in a sample of adults exposed to environmental noise. *Int. J. Res. Public Health* **2010**, *7*, 3579–3594. [CrossRef]
10. Seltenrich, N. Wind Turbines: A Different Breed of Noise? *Environ. Health Perspect* **2014**, *122*, A20–A25. [CrossRef]
11. Crichton, F.; Dodd, G.; Schmid, G.; Petrie, K.J. Framing sound: Using expectations to reduce environmental noise annoyance. *Environ. Res.* **2015**, *142*, 609–614. [CrossRef]
12. Holzman, D. Vehicle Motion Alarms: Necessity, Noise Pollution or Both? *Environ. Health Perspect* **2011**, *119*, A30–A33. [CrossRef]
13. Stansfeld, S.A.; Matheson, M.P. Noise pollution: Non-auditory effects on health. *Br. Med. Bull.* **2003**, *68*, 243–257. [CrossRef] [PubMed]
14. Halonen, J.I.; Vahtera, J.; Stansfeld, S.; Yli-Tuomi, T.; Salo, P.; Pentti, J.; Kivimäki, M.; Lanki, T. Associations between nighttime traffic noise and sleep: The Finnish public sector study. *Environ. Health Perspect* **2012**, *120*, 1391–1396, Erratum in *Environ. Health Perspect* **2013**, *121*, A147. [CrossRef] [PubMed]

15. Roswall, N.; Bidstrup, P.E.; Raaschou-Nielsen, O.; Jensen, S.S.; Olsen, A.; Sørensen, M. Residential road traffic noise exposure and survival after breast cancer—A cohort study. *Environ. Res.* **2016**, *151*, 814–820. [CrossRef] [PubMed]

16. Badran, M.; Yassin, B.A.; Fox, N.; Laher, I.; Ayas, N. Epidemiology of sleep disturbances and cardiovascular consequences. *Can. J. Cardiol.* **2015**, *31*, 873–879. [CrossRef] [PubMed]

17. Bush-Vishniac, I.J.; West, J.E.; Barnhill, C.; Hunter, T.; Orellana, D.; Chivukula, R. Noise levels in Johns Hopkins Hospital. *J. Acoust. Soc. Am.* **2005**, *118*, 3629–3645. [CrossRef] [PubMed]

18. Begrlund, B.; Lindvall, T.; Schwela, D.H. Table 4.1: Guideline Values for Community Noise in Specific Environments. In *Guidelines for Community Noise*; World Health Organization: Geneva, Belgium, 1999.

19. Polish Committee for Standardization. *PN-B-02151-2:2018-01 Building Acoustics—Protection against Noise in Buildings—Part 2: Requirements regarding the Permissible Sound Level in Rooms. 08.12.2018. ICS 91.120.20*; Polish Committee for Standardization: Warsaw, Poland, 2018; p. 12. (In Polish)

20. Bazan-Krzywoszańska, A.; Mrówczyńska, M.; Skiba, M. Perceptions of Residents City Zielona Góra—Mental maps. *J. Civ. Eng. Environ. Archit.* **2015**, *32*, 19–32. Available online: http://doi.prz.edu.pl/pl/pdf/biis/435 (accessed on 23 January 17). (In Polish).

21. Osman, T.; Divigalpitiyac, P.; Arima, T. Driving factors of urban sprawl in Giza. Governorate of Greater Cairo Metropolitan Region using AHP method. *Land Use Policy* **2016**, *58*, 21–31. [CrossRef]

22. Italo, C.M.G.; Stelamaris, R.B.; Paulo, H.T.Z. Influence of urban shapes on environmental noise: A case study in Aracaju—Brazil. *Sci. Total Environ.* **2011**, *412–413*, 66–76. [CrossRef]

23. Aletta, F.; Kang, J. Soundscape approach integrating noise mapping techniques: A case study in Brighton, UK. *Noise Mapp.* **2015**, *2*. [CrossRef]

24. Genuit, K.; Fiebig, A. Psychoacoustics and its benefit for the soundscape approach. *Acta Acoust. United Acoust.* **2006**, *92*, 952–958.

25. Genuit, K.; Schulte-Fortkamp, B.; Fiebig, A. Psychoacoustic mapping within the soundscape approach. *Internoise* **2008**, *2008*, 594–606.

26. Hao, Y.; Kang, J.; Krijnders, J. Integrated effects of urban morphology on birdsong loudness and visibility of green areas. *Landsc. Urban Plan.* **2015**, *137*, 149–162. [CrossRef]

27. Liu, J.; Kang, J.; Behm, H. Birdsong as an element of the urban sound environment: A case study concerning the area of Warnemünde in Germany. *Acta Acoust. United Acoust.* **2014**, *100*, 458–466. [CrossRef]

28. Liu, J.; Kang, J.; Luo, T.; Behm, H. Landscape effects on soundscape experience in city parks. *Sci. Total Environ.* **2013**, *454*, 474–481. [CrossRef] [PubMed]

29. Liu, J.; Kang, J.; Luo, T.; Behm, H.; Coppack, T. Spatio temporal variability of soundscapes in a multiple functional urban area. *Landsc. Urban Plan.* **2013**, *115*, 1–9. [CrossRef]

30. Boerenfijn, P.; Kazak, J.; Schellen, L.; van Hooff, J. A multi-case study of innovations in energy performance of social housing for older adults in the Netherlands. *Energy Build.* **2018**, *158*, 1762–1769. [CrossRef]

31. Yerli, O.; Demir, Z. Relation between urban land uses and noise. A case study in Dulze, Turkey. *Oxid. Commun.* **2016**, *39*, 732–745.

32. Vlahov, D.; Freudenberg, N.; Proietti, F. Urban as a determinant of health. *J. Urban Health* **2007**, *84*, 16–26. [CrossRef]

33. Marquis-Favre, C.; Premat, E.; Aubree, D. Noise and its effect—A review on qualitative aspect of sound. Part II: Noise and annoyance. *Acta Acoust. United Acoust.* **2005**, *91*, 626–642.

34. Honga, J.; Jeon, J. Exploring spatial relationships among soundscape variables in urbanareas: A spatial statistical modelling approach. *Landsc. Urban Plan.* **2017**, *157*, 352–364. [CrossRef]

35. Corburn, J. *Toward the Healthy City: People, Places, and the Politics of Urban Planning*; MIT Press: Cambridge, MA, USA; London, UK, 2009; Available online: https://mitpress.mit.edu/sites/default/files/titles/content/9780262513074_sch_0001.pdf (accessed on 23 February 2017).

36. Seidman, M.; Standring, R. Noise and quality of life. *Int. J. Environ. Res. Public Health* **2010**, *7*, 3730–3738. [CrossRef] [PubMed]

37. Rosenberg, B.C. Shhh! Noisy cities, anti-noise groups and neoliberal citizenship. *J. Sociol.* **2013**, *52*, 190–203. [CrossRef]

38. Passchier-Vermer, W.; Passchier, W.F. Noise exposure and public health. *Environ. Health Perspect.* **2000**, *108*, 123–131.

39. AbhijitDebnath Prasoon KumarSingh Environmental traffic noise modelling of Dhanbad township area—A mathematical based approach. *Appl. Acoust.* **2018**, *129*, 161–172. [CrossRef]

40. Berger, R.G.; Ashtiani, P.; Ollson, C.A.; Whitfield Aslund, M.; McCallum, L.C.; Leventhall, G.; Knopper, L.D. Health-based audible noise guidelines account for infrasound and low frequency noise produced by wind turbines. *Front. Public Health* **2015**, *3*, 1–14. [CrossRef] [PubMed]

41. Ho-Huu, V.; Ganić, E.; Hartjes, S.; Babić, O.; Curran, R. Air traffic assignment based on daily population mobility to reduce aircraft noise effects and fuel consumption. *Transp. Res. Part D Transp. Environ.* **2019**, *72*, 127–147. [CrossRef]

42. Ning, X.; Qi, J.; Wu, C.; Wang, W. Reducing noise pollution by planning construction site layout via a multi-objective optimization model. *J. Clean. Prod.* **2019**, *222*, 218–230. [CrossRef]

43. Williams, R.; Veirs Val Veirs, S.; Ashe, E.; Mastick, N. Approaches to reduce noise from ships operating in important killer whale habitats. *Mar. Pollut. Bull.* **2019**, *139*, 459–469. [CrossRef]

44. Richhariya, B.; Tanveer, M. EEG signal classification using universum support vector machine. *Expert Syst. Appl.* **2018**, *106*, 169–182. [CrossRef]

45. Benítez-Peña, S.; Blanquero, R.; Carrizosa, E.; Ramírez-Cobo, P. Cost-sensitive Feature Selection for Support Vector Machines. *Comput. Oper. Res.* **2019**, *106*, 169–178. [CrossRef]

46. Phangtriastu, M.R.; Harefa, J.; Tanoto, D.F. Comparison Between Neural Network and Support Vector Machine in Optical Character Recognition. *Procedia Comput. Sci.* **2017**, *116*, 351–357. [CrossRef]

47. Leong, W.C.; Kelani, R.O.; Ahmad, Z. Prediction of air pollution index (API) using support vector machine (SVM). *J. Environ. Chem. Eng.* **2019**, 103208. [CrossRef]

48. Alirezaei, M.; Akhavan Niaki, S.T.; Akhavan Niaki, S.A. A bi-objective hybrid optimization algorithm to reduce noise and data dimension in diabetes diagnosis using support vector machines. *Expert Syst. Appl.* **2019**, *127*, 47–57. [CrossRef]

49. Battineni, G.; Chintalapudi, N.; Amenta, F. Machine learning in medicine: Performance calculation of dementia prediction by support vector machines (SVM). *Inform. Med. Unlocked* **2019**, 100200. [CrossRef]

50. Wang, Z.; Zou, Z.; Soares, C.G. Identification of ship manoeuvring motion based on nu-support vector machine. *Ocean Eng.* **2019**, *183*, 270–281. [CrossRef]

51. Vapnik, V. *Statistical Learning Theory*; Wiley: New York, NY, USA, 1998.

52. Redgwell, R.D.; Szewczak, J.M.; Jones, G.; Parsons, S. Classification of Echolocation Calls from 14 Species of Bat by Support Vector Machines and Ensembles of Neural Networks. *Algorithms* **2009**, *2*, 907–924. [CrossRef]

53. Parrado-Hernández, E.; Robles, G.; Ardila-Rey, J.A.; Martínez-Tarifa, J.M. Robust Condition Assessment of Electrical Equipment with One Class Support Vector Machines Based on the Measurement of Partial Discharges. *Energies* **2018**, *11*, 486. [CrossRef]

54. Yu, H.; Gao, L.; Li, J.; Li, S.S.; Zhang, B.; Benediktsson, J.A. Spectral-Spatial Hyperspectral Image Classification Using Subspace-Based Support Vector Machines and Adaptive Markov Random Fields. *Remote Sens.* **2016**, *8*, 355. [CrossRef]

55. Vuolo, F.; Atzberger, C. Exploiting the Classification Performance of Support Vector Machines with Multi-Temporal Moderate-Resolution Imaging Spectroradiometer (MODIS) Data in Areas of Agreement and Disagreement of Existing Land Cover Products. *Remote Sens.* **2012**, *4*, 3143–3167. [CrossRef]

56. Elangovan, K.; Krishnasamy Tamilselvam, Y.; Mohan, R.E.; Iwase, M.; Takuma, N.; Wood, K.L. Fault Diagnosis of a Reconfigurable Crawling–Rolling Robot Based on Support Vector Machines. *Appl. Sci.* **2017**, *7*, 1025. [CrossRef]

57. Shin, K.; Lee, K.J.; Kim, H. Support vector machines approach to pattern detection in bankruptcy prediction and its contingency. In Proceedings of the 11th International Conference on Neural Information Processing, ICONIP 2004, Calcutta, India, 22–25 November 2004.

58. Platt, J.C. Probabilistic outputs for support vector machines and comparisons to regularized likelihood method. *Adv. Large Margin Classif.* **1999**, *10*, 61–74.

59. Sztubecka, M.; Skiba, M. Noise level arrangement in determined zones of homogenous development of green areas on the example of the spa park in Inowrocław. *Open Eng.* **2016**, *6*, 524–531. [CrossRef]

60. Polish Committee for Standardization. *PN-ISO 1996-2:1999. Acoustics—Description and measurement of environmental noise.12.03.1999.* ICS: 17.140.01; Polish Committee for Standardization: Warsaw, Poland, 1999; p. 9. (In Polish)

61. Abbod, M.F.; Linkens, D.A.; Mahfouf, M.; Dounias, G. Survey on the use of smart and adaptive engineering systems in medicine. *Artif. Intell. Med.* **2002**, *26*, 179–209. [CrossRef]
62. Mrówczyńska, M. *Study on the Selection of Numerical Intelligence Methods for Solving Surveying Problems*; Oficyna Wydawnicza Uniwersytetu Zielonogórskiego: Zielona Góra, Poland, 2015. (In Polish)
63. Osowski, S. *Neural Networks for Processing Information*; Oficyna Wydawnicza Politechniki Warszawskiej: Warszawa, Poland, 2006. (In Polish)
64. Bishop, C.M. Pattern Recognition and Machine Learning. In *Information Science and Statistics*; Springer Science + Business Media: New York, NY, USA, 2006.
65. Jankowski, N. *Ontogemic Neural Networks. About Networks Changing Their Structure*; Akademicka Oficyna Wydawnicza EXIT: Warszawa, Poland, 2003. (In Polish)
66. Skrzypczak, I.; Kokoszka, W.; Kogut, J.; Oleniacz, G. Methods of Measuring and Mapping of Landslide Areas. In Proceedings of the Conference: 3rd World Multidisciplinary Earth Sciences Symposium (WMESS), Prague, Czech Republic, 11–15 September 2017.
67. Cover, T. Geometrical and statistical properties of system of linear inequalities with applications in pattern recognition. *IEEE Trans. Electron. Comput.* **1965**, *14*, 326–334. [CrossRef]
68. Beycioglu, A.; Gultekin, A.; Aruntas, H.Y.; Gencel, O.; Dobiszewska, M.; Brostow, W. Mechanical properties of blended cements at elevated temperatures predicted using a fuzzy logic model. *Comput. Concr.* **2017**, *20*, 247–255.
69. Tse, M.S.; Chau, C.K.; Choy, Y.S.; Tsui, W.K.; Chan, C.N.; Tang, S.K. Perception of urban park soundscape, Published by the Acoustical Society of America. *J. Acoust. Soc. Am.* **2012**, *131*. [CrossRef]
70. King, G.; Roland-Mieszkowski, M.; Jason, T.; Rainham, D.G. Noise Levels Associated with Urban Land Use. *J. Urban Health* **2012**, *89*, 1017–1030. [CrossRef]
71. Hassana, A.M.; Lee, H. Toward the sustainable development of urban areas: An overview of global trends in trials and policies. *Land Use Policy* **2015**, *48*, 199–212. [CrossRef]
72. Foraster, M.; Eze, I.; Vienneau, D.; Brink, M.; Cajochen, C.; Caviezel, S.; Héritier, H.; Schaffner, E.; Schindler, C.; Wanner, M.; et al. Long-term transportation noise annoyance is associated with subsequent lower levels of physical activity. *Environ. Int.* **2016**, *91*, 341–349. [CrossRef]
73. Świąder, M.; Szewrański, S.; Kazak, J.; van Hoof, J.; Lin, D.; Wackernagel, M.; Alves, A. Application of Ecological Footprint Accounting as a Part of an Integrated Assessment of Environmental Carrying Capacity: A Case Study of the Footprint of Food of a Large City. *Resources* **2018**, *7*, 52. [CrossRef]
74. Juszczyk, M.; Leśniak, A.; Zima, K. ANN Based Approach for Estimation of Construction Costs of Sports Fields. *Complexity* **2018**, *2018*, 7952434. [CrossRef]

MDPI

St. Alban-Anlage 66

4052 Basel

Switzerland

Tel. +41 61 683 77 34

Fax +41 61 302 89 18

www.mdpi.com

Sustainability Editorial Office

E-mail: sustainability@mdpi.com

www.mdpi.com/journal/sustainability